QCD Hard Hadronic Processes

NATO ASI Series

Advanced Science Institutes Series

A series presenting the results of activities sponsored by the NATO Science Committee, which aims at the dissemination of advanced scientific and technological knowledge, with a view to strengthening links between scientific communities.

The series is published by an international board of publishers in conjunction with the NATO Scientific Affairs Division

A	**Life Sciences**	Plenum Publishing Corporation
B	**Physics**	New York and London
C	**Mathematical and Physical Sciences**	Kluwer Academic Publishers Dordrecht, Boston, and London
D	**Behavioral and Social Sciences**	
E	**Applied Sciences**	
F	**Computer and Systems Sciences**	Springer-Verlag
G	**Ecological Sciences**	Berlin, Heidelberg, New York, London,
H	**Cell Biology**	Paris, and Tokyo

Recent Volumes in this Series

Volume 191—Surface and Interface Characterization by Electron Optical Methods
edited by A. Howie and U. Valdrè

Volume 192—Noise and Nonlinear Phenomena in Nuclear Systems
edited by J. L. Muñoz-Cobo and F. C. Difilippo

Volume 193—The Liquid State and Its Electrical Properties
edited by E. E. Kunhardt, L. G. Christophorou, and L. H. Luessen

Volume 194—Optical Switching in Low-Dimensional Systems
edited by H. Haug and L. Bányai

Volume 195—Metallization and Metal–Semiconductor Interfaces
edited by I. P. Batra

Volume 196—Collision Theory for Atoms and Molecules
edited by F. A. Gianturco

Volume 197—QCD Hard Hadronic Processes
edited by B. Cox

Volume 198—Mechanisms of Reactions of Organometallic Compounds With Surfaces
edited by D. J. Cole-Hamilton and J. O. Williams

Series B: Physics

QCD Hard Hadronic Processes

Edited by

Bradley Cox

Fermi National Accelerator Laboratory
Batavia, Illinois

Plenum Press
New York and London
Published in cooperation with NATO Scientific Affairs Division

Proceedings of a NATO Advanced Research Workshop on
QCD Hard Hadronic Processes,
held October 8–13, 1987,
in St. Croix, US Virgin Islands

Library of Congress Cataloging in Publication Data

NATO Advanced Research Workshop on QCD Hard Hadronic Processes (1987: Saint Croix, V.I.)
 QCD hard hadronic processes / edited by Bradley Cox.
  p. cm.—(NATO ASI series. Series B, Physics; vol. 197)
 "Proceedings of a NATO Advanced Research Workshop on QCD Hard Hadronic Processes, held October 8–13, 1987, in St. Croix, US Virgin Islands."
 "Published in cooperation with NATO Scientific Affairs Division."
 Bibliography: p.
 Includes index.
 ISBN 0-306-43204-8
 1. Quantum chromodynamics—Congresses. 2. Hadron interactions—Congresses. I. Cox, Bradley. II. North Atlantic Treaty Organization. Scientific Affairs Division. III. Title. IV. Series: NATO ASI series. Series B, Physics; v. 197.
QC793.3.Q35N36 1987   89-3900
539.7′548—dc20   CIP

© 1988 Plenum Press, New York
A Division of Plenum Publishing Corporation
233 Spring Street, New York, N.Y. 10013

Printed in the United States of America

ORGANIZERS

Brad Cox, Chairman	Fermilab/University of Virginia
Luigi DiLella	CERN
Keith Ellis	Fermi National Accelerator Laboratory
Tom Ferbel	University of Rochester
Chris Llewellyn-Smith	Oxford University
Karlheinz Meier	CERN
Jeff Owens	Florida State University
Klaus Pretzl	Max-Planck Institute
Leo Resvanis	University of Athens
Dominique Schiff	LAL, Orsay
William Willis	CERN

PREFACE

The Advanced Research Workshop on QCD Hard Hadronic Processes was held on 8-13 October 1987 at Hotel on the Cay, St. Croix, U.S. Virgin Islands. The underlying theme of the workshop, the first in a series, was an examination, both theoretical and experimental, of the state of understanding of Quantum Chromodynamics. Because of the pervasiveness of the strong interactions in all aspects of high energy physics, QCD is central to many problems in elementary particle physics. Therefore, this workshop was organized to provide a forum in which the theory Quantum Chromodynamics could be confronted with experiment. The workshop was organized in four sessions, each of which concentrated on a major experimental arena in which a hard QCD process can be measured experimentally. A fifth session was devoted to global issues which effect all QCD processes. Each session began with a survey of the theoretical developments in the particular area and concluded with a round table which discussed the various information presented in the course of the discussions.

A session of the workshop was devoted to the direct production of high transverse momentum photons in hadronic interactions. Data from several experiments, either completed or in progress at CERN (NA3, NA24, WA70, UA6, CCOR, R806, AFS, R110, UA1 and UA2), were discussed and the prospects for two new upcoming experiments from Fermilab (E-705, E-706) were presented. The data from the CERN experiments were examined from several different vantage points to determine the light that these experiments shed on the various questions important to the understanding of QCD. The discussion of the new Fermilab experiments revolved around their potential and what ought to be the focus of the new measurements (which were actually in progress at the time of the workshop). There were spirited debates about the state of the theory and the ability of the theory to allow the extraction of the fundamental parameters of QCD (such as Λ_{MS}). In particular, during the presentations and the round table discussion, the choice of "physical" or "optimized" scale variables for the QCD expansion was vigorously contested.

A closely related QCD process, the photoproduction of high p_t charged and neutral particles, was also discussed. Results from CERN experiments, WA69 and NA14, which measured the photoproduction of high p_t charged and neutral π's were reviewed. The data seem clearly to show the presence of point-like hard collisions of the photon with partons and are consistent with the level of hadronic production of direct photons.

A second session of the workshop dealt with the allied subject of production of leptons and weak bosons in hadronic interactions. The data on production of W and Z's in antiproton-proton collisions at CERN (UA1, UA2) and Fermilab (CDF) were presented. Data from the high mass dimuon measurements of Fermilab experiment E-615 were offered as evidence for higher twist effects (at high x_f). In addition, particularly interesting data were presented on the production of low mass lepton pairs. The data on low mass pairs appeared to agree with the direct photon data indicating that they are a manifestation of the same phenomena except with virtual rather than real photons.

The third session concerned itself with the production of jets. The data from both CERN (UA1 and UA2) and Fermilab collider experiments and the Fermilab fixed target experiments, E-557 and E-672, were discussed. Evidences for differences in quark and gluon jets from e^+e^- interactions was presented in the round table forum and compared with the results on quark-gluon jet differences measured in e^+e^- interactions. These results were in surprising agreement given the differences in experiments and the attendant experimental difficulties. The theoretical underpinnings of jet production was reviewed.

The fourth major area discussed in the workshop was the very topical issue of hadroproduction of heavy flavor. The prediction of the absolute level of the QCD cross sections for heavy flavor was debated and the sensitivity of the theoretical predictions to various parameters was discussed. Data was presented by experiment, Fermilab experiment E-691 on photoproduction of charm. Hadroproduction data was presented from fixed target experiments such as E-769 and the LEPS experiments. The hadroproduction of hidden charm was also reviewed with new data from Fermilab experiment E-537 on the observation of heavy target effects in the production of the J/ψ particle presented for the first time at this conference. The data on B production from experiments WA75 and WA78 (along with the evidence for B production from UA1) were discussed. The difficulty of making quantitative predictions of charm production because of the relatively low mass of the charm quark were contrasted with the somewhat better situation for the case of the heavier B hadrons. It is expected that the B hadroproduction will provide more quantitative tests of QCD. Data on B production from future experiments such as Fermilab experiments E-771 and E-791 should clarify the situation. The potential of these new experiments which seek to make more definitive measurements of B production were discussed.

A special session was devoted to global issues which effect all of the hard processes discussed in the sessions. Featured talks were given on higher order QCD calculations, on the present state of knowledge of hadronic structure functions, on A dependent and polarization dependent effects in QCD and on new ideas about the proper theoretical framework for calculating total cross section events.

The conference was admirably keynoted by J.D. Bjorken who gave an intriguing and somewhat different alternative intuitive pictures of several QCD processes and summarized by A. Mueller who strove mightily to give form, order and coherence to several days of intense and vigorous discussions.

Many thanks go to all of the participants who were distinguished not only by the excellence of their presentations but by the enthusiasm and vigor in their attempts to get to the heart of the problems in our present understanding of QCD. They are also to be commended for their attempts to define directions for the future.

In addition, the efforts of the organizing committee were especially critical for this workshop. Without them the relatively complete coverage of these topics, with representation of most of the involved groups in this area of high energy physics would have been impossible. The wide participation with individuals from thirty or more experimental groups together with representatives of most of the major schools of theoretical thought was due to the tireless work of the members of the organizing committee on two continents and in 15 countries.

Finally, the efficiency with which we were able to conduct this conference was due in large part to the professional coordination of day to day operations by Ms. Phyllis Hale and her assistant Ms. Joy Perington. Their tireless efforts allowed the participants to conduct their activities in the pleasant surrounds of Hotel on the Cay while still having available the facilities necessary to carry on the business of the conference.

Special thanks are due to the Scientific Affairs Division of the North Atlantic Treaty Organization for acting as the major sponsor of this Workshop and making this Workshop possible. Thanks are also due to the U.S. Department of Energy and the National Science Foundation for providing additional funding and acting as co-sponsors and to Fermilab for providing equipment and services in support of the Workshop.

As a final note, the participants were together for only a brief time, but the enthusiasm generated by the face to face discussions that started early each morning and continued on into the evening have been adequate to energize work that has continued long after the Workshop. Questions were raised and ideas were discussed that everyone took away to their widely scattered home institutions. Thus the fruit is still being born from this gathering.

Bradley Cox

CONTENTS

SPECIAL TOPICS

NEW DIRECT PHOTON EXPERIMENTS

HADRONIC JETS

HEAVY FLAVOR PRODUCTION

CONFERENCE SUMMARY

QCD: HARD COLLISIONS ARE EASY AND SOFT COLLISIONS ARE HARD

James D. Bjorken

Fermi National Accelerator Laboratory
P.O. Box 500
Batavia, Illinois 60510

INTRODUCTION

Now, are hard collisions <u>really</u> all that easy? To be sure, calculation (not to mention measurement) of hard-collision processes is far from simple, and I mean no disrespect by the title. It is meant in the physicists' sense that "what is possible to do is easy; the impossible is merely hard."

I. WHY STUDY HARD COLLISIONS?

There are probably as many answers to this question as participants in this meeting. One of them is in the title. But it may be in order in an opening talk like this to review some of the others.

1. Is the QCD Lagrangian the Correct Basis of the Strong Interactions?

There are at best only a few skeptics around who doubt the correctness of the QCD Lagrangian as basis of the strong force. I am not among them. Even though precise quantitative evidence for QCD is difficult to obtain, for me the semiquantitative and qualitative evidence that the QCD Lagrangian is correct is more persuasive than the host of highly quantitative evidence that the standard $SU(2) \times U(1)$ electroweak Lagrangian is correct. That of course is a subjective opinion, and I won't try to defend it here.

2. What does the QCD Formalism Really Mean?

Given the premise that QCD is the correct theory, there still are fundamental questions to raise. One is for theorists: given that quarks and gluons are confined, how does one <u>formulate</u> QCD as an operative field theory? No doubt the experts understand this much better than I, but I'm motivated to take some time here to express a viewpoint.

Perturbative QCD is expressed in the language of Feynman diagrams. The derivation of this perturbation expansion which I prefer is based on the Lehmann-Symanzik-Zimmerman (LSZ) formalism. In that approach[1] one assumes the physical particles are in one-to-one correspondence with the quanta appearing in the Lagrangian. This implies that the asymptotic states emergent from a collision process should be a collection of quarks and gluons, not hadrons. This is at least thinkable if the entire scattering and measurement process is restricted to distances and time scales small compared to 1 fermi.[2] This takes special measurement apparatus made out of some kind of hypothetical superheavy matter, but that to me is an "inessential complication". An S-matrix formulation of perturbative QCD to me <u>is</u> thinkable within this "small-distance" constraint.

This viewpoint suggests a criterion for applicability of perturbative QCD: if a process can be visualized as occurring in a spatial domain of dimensions small compared to a fermi and on a time scale small compared to a fermi, then it is computable in terms of perturbative QCD. The prime example is $e^+e^- \rightarrow q\bar{q} \rightarrow$ hadrons, where (at energies much greater than one inverse fermi) the minicollider and quark/gluon minidetectors can fit into such a small volume. The hadronization does not fit, but as far as σ_{tot} is concerned this doesn't matter.

How far can one generalize this idea? What is the largest space-time domain within which one can legitimately trust perturbative QCD?? And what criteria define such a domain? My own conjecture is the following:

> A space-time domain is <u>perturbative</u> (i.e. within the domain, perturbative QCD is legitimately applicable) if (and only if?) every straight-line segment lying

entirely within the domain has a proper length which is small (in absolute magnitude) compared to 1 fermi.

When this criterion (let us call it the strict perturbative criterion) is met, I cannot envisage any process within the domain which perturbative QCD could not describe. When this criterion is violated, it seems to be almost always the case that non-perturbative confinement effects enter. I say "almost" because a candidate counterexample is an isolated $Q\bar{Q}$ quarkonium state of such large mass that its dimensions are much less than 1 fermi. It can propagate over large times, violating the above criterion, while being described by a theory which is as perturbative as QED. But does perturbative QED include the hydrogen atom and positronium?

Leaving such nuances out, it is possible to characterize the nature of the strict perturbative domains somewhat further, although a general discussion may have some extra subtlety. For the classic $e^+e^- \to q\bar{q}$ case, the domain is bounded in the forward light cone by a proper-time surface $x^2 \lesssim a^2$ with a $\ll$ 1f. (Fig. 1a). The strict perturbative domains for scattering and production processes are boost-invariant regions adjacent to the future and past light cones; the boundary surfaces (Fig. 1b) can be taken to be $x^2 = a^2$ (past and future) and $x^2 = -a^2$ (spacelike).

What should one do with this? I am not sure myself. Calculations of hard processes should not depend upon mutilation of Green's functions outside the strict perturbative domains; this may provide a test of validity of perturbative calculations. Also, boundary - condition models come to mind. But I have not put much effort in pushing these notions further.

The next argument for studying QCD hard collisions is

3.   To Test The Synthesis of the Quark-Parton Model and QCD

The quark-parton model was developed outside the framework of QCD. And it is not automatic that QCD implies a parton-model. The best linkage comes in deep-inelastic lepton-hadron scattering, where the QCD moment-analysis of structure functions is elucidated in parton terms via the Altarelli-Parisi evolution equations.

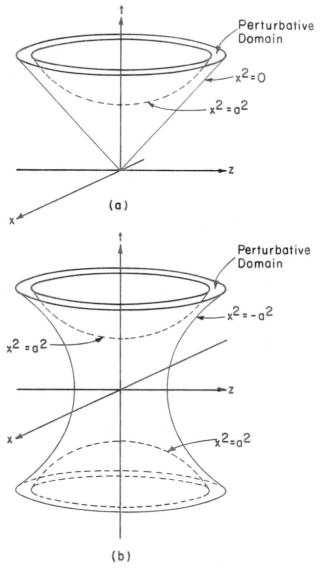

Fig. 1   A sketch of the strict perturbative domain (a) for $e^+e^- \rightarrow q\bar{q}$; and (b) for a general scattering process.

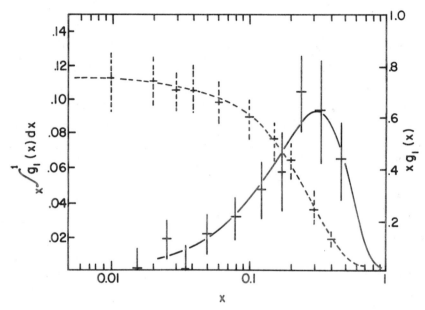

Fig. 2   Parallel-antiparallel spin asymmetry in deep inelastic muon-nucleon scattering as measured by EMC.

It seems sometimes to be forgotten that moment analyses and sum-rules based on current-algebra and/or operator-product expansions have a generality which goes beyond the parton model, and directly test QCD. Thus the recent EMC data (Fig. 2) on polarized μ-polarized proton deep-inelastic scattering,[3] which do not agree with the combination of sum-rules derived by myself[4] and by John Ellis and Bob Jaffe[5] are especially important to review critically.[6,7] There are several points to be considered: 1) Is the rate of convergence understood, both theoretically and experimentally? 2) Is Q^2 large enough, especially at small x?, and 3) Is the theoretical derivation sound, especially of the Ellis-Jaffe sum-rule, or are there loopholes? This last point is already raised by Ellis and Jaffe themselves. Everyone agrees that a measurement from polarized neutrons would be very important.

When one goes beyond $e^+e^- \rightarrow$ hadrons and deep inelastic lepton-nucleon scattering, reliance on explicit parton-model concepts increases. The next most reliable class of processes includes deep inelastic Compton scattering and two-photon production of quark pairs, both of which I regard as "safe". Less safe are processes with two quarks in the initial state and none or one in the final state. These include Drell-Yan dilepton production and direct photon production.

Why is $e^+e^- \rightarrow q\bar{q}$ safe and $q\bar{q} \rightarrow e^+e^-$ less safe? Evidently concepts such as closure, unitarity, etc. apply to an essentially complete final-state sum in the former process, allowing isolation of the primitive subprocess. But initial-state interactions are not fully summed and lead to complications. Nevertheless much has been done to classify and keep in control these effects. But this is done at the parton level, starting with, say, an Altarelli-Parisi description for the content of the initial projectiles. While this starting point is not quite from first principles, most everyone seems not to be too concerned with the lack of rigor implied.

The dominant initial-state effects are collinear "initial-state" radiation of partons and multiple-scattering. It is useful to consider what is going on in the space-time domains we introduced. Consider for definiteness the Drell-Yan process, where the strict perturbative domains extend into the past, as shown in Fig. 3. At an early time t_1, one samples the parton distribution of the initial hadron, just as would be the case in deep-inelastic scattering. (Notice the hadron size is large compared to the area of the section of the strict perturbative

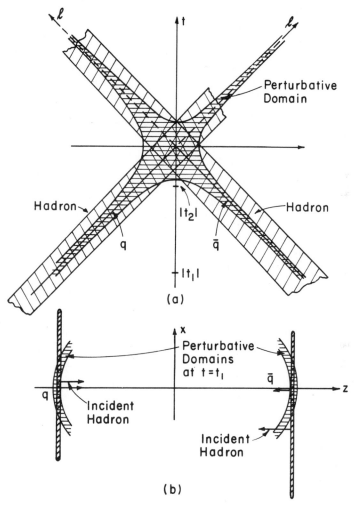

Fig. 3   Space-time picture of Drell-Yan dilepton production.

domain at time t_1). As t increases and |t| decreases, the parton distribution inside the strict perturbative domain undergoes Altarelli-Parisi evolution. The relevant Q^2 scale for that evolution depends inversely on the time scale; the scaling law is

$$|t| \sim \frac{E}{Q^2}$$

where E measures the typical energies of the relevant partons.

After some small negative time t_2, the collision process itself may be regarded to begin. The division between evolution and interaction is Lorentz-frame dependent, fuzzy, and in fact arbitrary; physics does not depend upon where the division is chosen. This is expressed by the factorization-theorems of QCD which link smoothly the physics of parton evolution with the physics of the hard collision itself.

In the region $|t| < |t_2|$, the left-moving and right-moving partons have the opportunity to interact. As I understand it, other than the hard collision itself there is nothing other than Coulomb-like interactions which are supposed to be significant in leading order. This "multiple-scattering" effect broadens the incident parton transverse-momentum distributions, and hence the p_T distribution of the dileptons. The effect evidently depends on the nature of spectator constituents in the projectiles. In the distant past these lie <u>outside</u> the strictly perturbative domain, but at times of order t_2 some move within those domains. Thus in differential Drell-Yan processes, a full specification of initial-state effects depends upon input information beyond the specification of the usual structure functions of the incident particles. However, for the process integrated over p_T, there is not expected to be corrections, provided the energy is high enough.[8]

4.   <u>What Are The Structure Functions And What Do They Mean</u>?

There are both practical and fundamental aspects to this question. On the practical side, the gluon content of hadrons needs to be pinned down. Even the nucleon needs to be understood better. The large N/P ratio, the x-dependence of the polarization asymmetry, and the possible problems with the sum rules for the spin-dependent structure function indicate that we really don't have our understanding of the nucleon structure under total control. The region of wee x (x << .05) is very important for quark and gluon constituent structure and deserves continuing attention, both theoretically and experimentally. This is clear to everyone. I will return to the subject later.

In addition, there are quite fundamental open questions having to do with the concept of "wave function" of a hadron, which underlies the parton interpretation of structure functions. The original parton-model interpretation-as do modern descendants - requires viewing a hadron at infinite momentum ("on the light cone") in order to protect this simple concept of wave-function from the essential complications of interacting relativistic quantum field theory. Even so, when one does go to infinite momentum and counts the mean number of constituents within a proton, the result comes out infinite; more specifically

$$\langle e^2 \rangle = \frac{4}{9} \langle n_{up} \rangle + \frac{1}{9} \langle n_{down} \rangle + \frac{1}{9} \langle n_{strange} \rangle = \int \frac{dx}{x} F_2(x) = \infty$$

This result is based essentially <u>on experiment</u>; no one anticipates $F_2(x) \to 0$ as $x \to 0$. Thus in the old-fashioned parton model this "wave-function" cannot be so simple.

Nevertheless, there is a lot of work in perturbative QCD based on a Fock-space description of hadrons,[9] where it is assumed that there is a nonvanishing probability of finding a finite number of constituents in an infinite-momentum hadron, even though the average number is infinite. I am too old-fashioned to accept that notion, and end up in inconclusive debate with those who do.

An argument which is used to support the Fock-space picture involves the weak interactions. Consider the decay $\tau^- \to \nu_\tau + \pi^-$. The weak Lagrangian creates only a $\bar{u}$ and d quark and this operator evidently has a finite matrix element between strong-interaction vacuum and a single pion; hence the Fock-space component of the pion appears to be manifestily present. This argument actually has negative impact to me, because it seems to apply equally well to the ancient Tamm-Dancoff Fock-space theory as applied to states of finite momenta. That theory was found to be inconsistent.[10] Nevertheless strong interactions and weak interactions peacefully coexisted for some time even before the advent of partons and QCD, thanks to current algebra. And to this day those developments remain valid.

But it is interesting to pursue the example of semileptonic τ-decay further, within the context of strictly perturbative space-time domains.

For a very high energy parent τ, the produced $\bar{q}q$ pair will lie within a perturbative domain for a long time after the decay. The time is of order E/Q^2, where Q^{-1} is some distance scale small compared to one fermi. During that time the perturbative arguments indeed should hold and few additional partons should evolve, independent of E. But then there is an additional long time required for the pion to form; the uncertainty principle alone dictates that this be[11] of order $\gamma m_\pi^{-1} = E/m_\pi^2$. This additional evolution is <u>nonperturbative</u>; it is reasonable that on average a number of wee partons of order unity attach themselves to the $\bar{u}d$ system.

Now view this in a boosted frame for which the energy scale is doubled. The dwell time of the $\bar{q}q$ in the strictly perturbative domain is doubled, as is the pion formation time. The energy of the evolved wee-parton we already considered is doubled as well; it is now semi-wee. And there is an extra 0.7 units of rapidity available in the longitudinal phase space distribution of the pionic partons. The probability of this region also being populated by an additional wee parton should again be of order unity, adding on average extra partons to this boosted state. If on average we get Δn extra partons per factor 2 increase in energy scale, the mean number in the pion configuration in a <u>boosted</u> frame with energy scale E' $\gg$ E will be

$$\bar{n} \sim (\Delta n) \frac{\ell n\ E'/E}{\ell n\ 2}$$

which diverges logarithmically as E' $\to \infty$. If this accretion of stray wee partons during the nonperturbative evolution is random in nature, as envisaged in the old parton model (which in modern language assumes that non-perturbative QCD phenomena have only short-range correlations in rapidity), the probability of having a small finite number of partons in the infinite momentum limit will diminish as in a Poisson distribution, $\sim e^{-n}$, hence as a power of momentum p as $p \to \infty$. Evidently in the infinite-momentum limit there is vanishing probability for a Fock-space configuration to exist.

This argument can be debated <u>ad</u> <u>nauseum</u> at this level. But the trouble is that I only have word pictures with little in the way of calculation to back them up. And for me Fock-space advocates have the calculations but little in the way of explanatory words to back them up. What is needed is a theoretical framework better able to penetrate beyond the strictly perturbative domains.

5.   What is the Physics of Hadronization?

There are both perturbative and non-perturbative aspects to this problem. The first is how the produced or scattered partons evolve within the strictly perturbative domain ("final-state" radiation); this is itself nontrivial, but is amenable to perturbation-theoretic analysis. But once the produced systems of partons leave the strictly perturbative space-time domain, there is much less certainty. This is the central playground for Monte-Carlo simulations, and by now, of course, the modeling is quite sophisticated.[12] But it still seems to be the case that the input physics assumptions vary a great deal model-to-model. A significant distinction occurs between "cluster" and "string" models. The "cluster" models extend the perturbative diagrammatic branching processes to the edge of (or beyond) applicability before building up the hadrons from produced partons. "String" models rely mainly on the nonperturbative mechanism of evolution and breaking of color flux-tubes, or strings which must join the outgoing quark and gluon sources.

Suppose one is conservative about the minimum, limiting distance scale Q_o^{-1} below which perturbative evolution of hadronization is reliable. One way of guessing that scale is just by looking at $e^+e^- \rightarrow$ hadrons. At or below the charm threshold region, there is no evidence for underlying jet structure; the produced hadrons are easily describable by a statistical model. Thus a minimum energy scale of $Q_o \sim$ 3-4 GeV is defensible. If one uses that, the multiplicity of radiated virtual gluons per unit rapidity (each of which must have $Q^2 \gtrsim 10\text{-}15$ GeV^2) will be too small to account for the observed multiplicity. Where does the large number of extra gluons come from? A perturbative mechanism is shown in Fig. 4; the important radiation can come from the Coulomb field itself. Such a picture is gauge-dependent, but suggestive. Certainly the non-abelian nature of the gauge theory must be relevant to the existence of hadronization, and the running of the coupling constant, the most important perturbative QCD indicator of confinement, is due (in the physical Coulomb gauge) to coupling of the transverse quanta to the instantaneous Coulomb field (Fig. 5).

From this point of view the underlying physics of "cluster" and "string" dynamics begins to show some common features. The perturbative, "cluster" approach first generates the produced entropy (the final-state quanta) via branching processes (off the Coulomb field?) while the "string" approach first generates the produced energy via growth of Coulomb flux tubes.

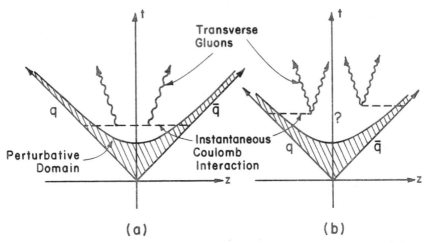

Fig. 4   Radiation of gluons from the expanding Coulomb field between receding partons.

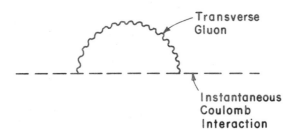

Fig. 5   The most important diagram for the running of the coupling constant in QCD.

There may be some evidence of such complementarity (rather than conflict) of the two approaches in the elegant demonstration of a depletion of particle production in three-jet e^+e^- events in the direction opposite to the emitted gluon.[13] The effect is delicate, but the comparison with the "control" process $e^+e^- \rightarrow q\bar{q}\gamma$ makes the experimental case credible, if not convincing. But it is interesting that not only the Lund "string" model predicts this effect, but also perturbative QCD calculations which take into account "color coherence". The perturbative calculations would necessarily have to treat the interparton potentials as Coulombic in nature, but it is reasonable that the Coulomb-like flux follows the quark and gluon sources in a way similar to the flux tubes.[14]

Nevertheless, these fascinating matters of principle are usually remote from practical observation. For the practicing experimentalist interested in the underlying hard processes, the word "hadronization" is often synonymous with "obscuration". However tedious the task, an understanding of hadronization processes from as close to first principles as possible is sure to pay off in the long run. We shall briefly return to the engineering aspects of hadronization later.

6.   What Are The Subprocesses and What Are Their Cross-Sections?

The first part of the question invites experimental searches for new, beyond-the-standard-model physics. It needs no motivating words or detailed discussion here. It is the latter part of the question which is the focus of attention here. How well can perturbative QCD account for conventional subprocess cross-sections? It is interesting that this problem is still as much theoretical as experimental. While impressive progress has been made in computing and codifying complicated tree-level processes, analytical[15] as well as numerical, e.g. gg → gggg, there still is a need for one-loop radiative corrections to them, e.g. in order to compare with existing data such as the beautiful UA1 analysis of 3 jet/2 jet data.

On the experimental side, there are a host of hard processes to study further. That is the business of the meeting; my superficial impression is that, from the viewpoint of an outsider like myself, the situation is in quite good shape already. But there are areas of concern, such as how to better estimate heavy-flavor production in hadron collisions.

An important feature of this topic is the need for study over a broad range of s and Q^2 in order to overdetermine, if possible, the parametrization of the hard processes, and to delineate the range of applicability of the perturbative calculations.

7. <u>What is the Effect of Nuclear Targets and Projectiles on the Hard Collision Dynamics</u>?

In the discussion of strict perturbative domains, we have already touched on the importance of the effect of nuclear matter on collision dynamics. The cleanest milieu is in deep-elastic lepton-hadron interactions; e.g. the EMC effect. This phenomenon has created much activity in the nuclear physics community. I am relatively unmoved by the phenomenon at moderate to large x; there are too many relatively mundane interpretations (e.g. an effective nucleon mass in nuclear matter a little smaller than in vacuum) to justify epic conclusions. The EMC enhancement at smaller x, x ~ 0.1, was <u>predicted</u>[16] long before the measurements, and is more interesting to me. But the most interesting regime is the shadowing region of small x which, with its subtlety and total dependence on extreme-relativistic kinematics, is too important to leave in the hands of nuclear physicists. It is good to see - at long last - the emergence of some quality data in this region.[3] A central theoretical question concerns the onset of shadowing, i.e. of an "$A^{2/3}$" dependence of the structure function on atomic number, as seen in photoproduction. If "Bethe-Heitler" photon-gluon fusion dominates (Fig. 6a), the virtual photon cross-section σ_T goes as $\alpha_s A/q^2$, and the onset of non-shadowing is for $q^2 > Q_o^2$ (with Q_o ~ 1 GeV). Furthermore the final state will typically contain two balanced high p_T jets. If the "naive" parton-model mechanism dominates (scattering from ocean quarks), the cross-section goes as $A^{2/3}/q^2$, the onset of shadowing is for $x < x_o$ ($x_o \lesssim$ .01?), and the final state typically contains a single jet along the virtual photon direction.[17] As I understand it, the EMC data tends to favor the latter interpretation, although I do not yet know how conclusive it is.

14

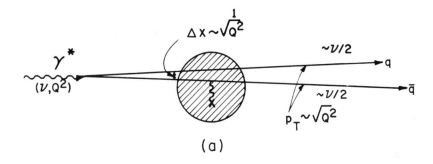

(a)

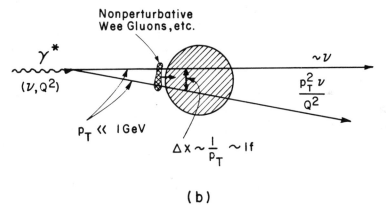

(b)

Fig. 6   Two mechanisms in deep-inelastic scattering at very small x:
(a) Bethe-Heitler photon-gluon fusion, and (b) the parton model
mechanism, each as seen in the laboratory frame.

Similar shadowing issues exist for the Drell-Yan dilepton process; again there can be very interesting effects when the outgoing dilepton has very high energy (i.e. large x_1) and moderate mass (i.e. small $x_2 = m^2/x_1 s$). The analogue of the first "$\alpha_s A$" mechanism is Bethe-Heitler bremsstrahlung of the dilepton from the fast parton due to interaction with the color Coulomb-fields in the nuclear matter. Because the fast parton efficiently penetrates even heavy nuclei, this should give a nuclear dependence going as $\alpha_s A$. The "naive" parton-model mechanism is best originally viewed in the cms frame of the dilepton. In that frame the nuclear matter is Lorentz-contracted to a thickness

$$(\Delta z)_{\text{N.M.}} \sim \frac{2 r_o A^{1/3}}{\gamma} \sim \frac{4 r_o A^{1/3}}{x_1} \frac{m}{s} M_p$$

where m is the dilepton mass.

The location of the source of the dilepton is uncertain in the longitudinal direction by an amount $(\Delta z)_{\text{source}} \sim m^{-1}$. This is large compared to the nuclear-matter distribution when

$$\frac{x_1 s}{m^2} \gg 4 M_p r_o A^{1/3} \sim 24 A^{1/3}$$

This is better expressed as

$$x_2 \ll .04 A^{-1/3}$$

Under these circumstances we find the dilepton produced from the uncontracted wee-parton cloud <u>exterior</u> to the nuclear matter (Fig. 7a); an $A^{2/3}$ yield is to be expected.

How does this latter mechanism look in the laboratory frame? It is not so different from the bremsstrahlung mechanism. Upstream of the nucleus the dilepton is radiated, leaving behind a much less energetic parton; its momentum is readily estimated to be less than that of the dilepton (or incident parton) by a factor $\sim (p_T/m)^2$, where p_T is its transverse momentum relative to the incident direction. A standard estimate of formation length puts the dilepton emission a distance $\Delta z \sim E/m^2$ upstream of the nucleus. Now comes the important feature: new gluon field is created by the transverse displacement of the secondary quark from the original direction of flight. If that displacement is

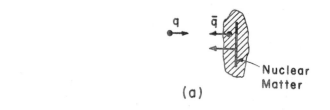

(a)

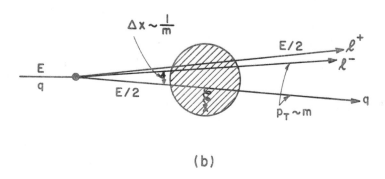

(b)

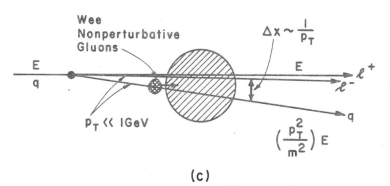

(c)

Fig. 7   A parton-model picture of shadowing in Drell-Yan dilepton
production from nuclei at very small x_2, (a) as viewed in the
dilepton rest frame, (b) in the laboratory frame for high-p_T
production ($\alpha_s A$), and (c) in the laboratory frame for low-p_T
production.

17

small compared to the confinement scale, the mechanism is perturbative. But in this case the displacement on arrival at the nuclear surface is of order 1 fermi as shown by the geometry in Fig. 7b and the simple estimate

$$\Delta x_T \sim \frac{p_T}{E} \Delta z = \frac{p_T}{E} \cdot \frac{E}{m^2} \sim \frac{1}{p_T}$$

Because of this large displacement of impact-parameters, a nonperturbative chunk of gluon field (containing wee-partons) is attached to the secondary parton. This gets absorbed on the nuclear surface, leading to the $A^{2/3}$ dependence.

I have belabored this point to again emphasize the delicacy of the argumentation: this is not mundane, grubby nuclear physics. And mundane nuclear-physics explanations for phenomena in this kinematic regime are not to be trusted. Not at all.

A-dependent effects in the final state are also of interest; outgoing partons may undergo additional multiple scattering and radiation. The simplest testing ground is again lepton-nucleon scattering, where one tests the interactions of the partially-dressed struck quark.[18]

II. SOME SPECIFIC ISSUES OF INTEREST TO ME

A. QCD Topics Not on the Agenda

Not all of hard-collision QCD is represented at this meeting. Maybe it is appropriate in these introductory remarks, however, to include some other items.

1. e^+e^- → hadrons

After the thousands of physicist-years of effort on this single reaction, I find it hard to add anything to the discussion.

2. ℓN → hadrons

Already mentioned were the questions of spin asymmetry and possible trouble with sum-rules, as well as the new evidence for nuclear shadowing at small x. Scaling-violation studies in lepto-production probably have nearly saturated until HERA gets going. But the structure of hadron final states, especially A-dependent effects, still have a long way to go, with results that would provide useful spin-off for more complex processes. The QCD gluon-radiation accompanying a hard lepton-nucleon collision will be most profitably studied at HERA, and is at best a marginal subject at fixed-target energies.

3. Exclusive QCD

Fixed-angle two body scattering processes have been rather well described by perturbative-QCD calculations based on assumed Fock-space wave functions for the colliding hadrons. There seems to be a great deal that is right about these calculations.[19] But as discussed above, I mistrust the underlying premise of Fock-space wave-functions, and strongly disagree that the predictions are rigorous consequences of QCD. The calculations may well represent the behavior of the valence components of the hadrons, but the presence of wee components may be significant, especially in modifying the overall normalizations.

An especially interesting experiment, suggested by Al Mueller,[20] is underway at Brookhaven. The perturbative, Fock-space QCD calculations imply that, on arrival at and departure from a high energy fixed-angle elastic scattering event, the valence constituents are close to each other, i.e. are small color dipoles. If this is so (and if the valence constituents are not accompanied by a wee-parton cloud), then the projectiles will not be absorbed by nuclear matter. Thus quasi-elastic fixed-angle scattering in nuclei should be much larger than a "naive" calculation assuming conventional absorption in nuclear matter of the incident and outgoing hadrons. Preliminary data[21] do show an enhancement. But not only is more data required but also a careful theoretical critique.

4. Inclusive Polarization Phenomena

Polarization data has often been the graveyard of fashionable theories. If theorists had their way they might well ban such measurements altogether out of self-protection.[22] Nowadays the exclusive QCD processes discussed above founder on polarization measurements; this remains a great challenge for the theorists.[23] The observed polarization of leading hyperons is not fully understood. In addition, it would be nice to see the systematics of heavy-flavored hadron polarization as well; this is experimentally not at all out of the realm of possibility. And even the polarization of large-x, leading nucleons in high energy nucleon-nucleon collisions is as yet not very thoroughly explored.[24]

5. Pomeron Properties

Soft collisions <u>are</u> hard. And, as explicated by the Reggeon calculus, the central element of log-s physics is the Pomeranchuk trajectory. My guess is that understanding this will be the last unsolved problem of QCD. A major experimental advance is in the offing, namely the determination of the parton structure of the Pomeron. This will be accomplished[25] via the search for jet structures in high-mass single-diffraction events; the reaction is Pomeron + proton → jet + jet + X. The amount of theoretical guidance is minimal. Those who view the Pomeron as some kind of cylinder might say it has no parton structure. Those color-blind souls who identify the Pomeron with a single gluon might give its structure function a very hard leading term. Others (myself included) might guess its structure function will look like that of a meson. But nobody to my knowledge has a decent theory.[26]

It seems to me that there is far too little attention paid to log-s physics. Indifference of theorists is in part to blame for this situation. But experimentalists at colliders don't seem to get much further than $\sigma_{e\ell}$, σ_{tot}, etc, and the negative binomial distribution. The details of event structures, over <u>all</u> rapidities, both for minimum-bias events and high-p_T events are not going to get really proper study for a long time if for no other reason that no magnetic detector exists or is contemplated[27] which has the necessary phase-space coverage.

6. Very High Multiplicity

Non-jet $\bar{p}p$ events with large $dN/d\eta$ are of interest to those who search for quark-gluon plasma. Whatever the fate of that hypothesis, this class of events - which clearly <u>exists</u> - is empirically characterized by a high initial energy density ($\sim 50 - 100$ GeV/f^3). The final hadron multiplicity is so large (say, 50 - 100 particles in 2 units of rapidity, more-or-less isotropically emitted) that the formation zone of pions and other hadrons is large, of radius $\gtrsim$ 5f. The mechanism of energy transport from the initial collision volume out to the hadronization volume may not be (and quite likely isn't) ideal hydrodynamic flow of a fluid in local thermal equilibrium. An opposite extreme, perhaps equally unlikely, might be multiple production of minijets which are experimentally unresolved. This latter mechanism might be amenable to a (perturbative?) QCD analysis, but I am not sure. I should think the fluctuation structure will in the latter case tend to be fractal and in the former case tend to be gaussian. In my opinion this subject, with its linkage to the heavy-ion program, deserves considerable attention.

B.   QCD Topics Which Are on the Agenda

As promised, there will be not too many specifics covered here; that is the business of the meeting.

1. Direct Photons

This field has matured impressively in the last few years. Much will be said and I have little to add. The relevant subprocesses are semi-clean, and a large dynamic range of s and t, along with accurate measurements, makes this a very good area for quantitative perturbative QCD studies.

Double photon production may be another interesting new frontier.

2. Dilepton Production

There is still life in this very mature subject. We have already discussed the question of shadowing, i.e. "$A^{2/3}$" behavior for production of moderate-mass leading dileptons at very high energy. It is a most interesting supplement to the electroproduction phenomena.

I also find the question of structure of the underlying event (the distribution of beam fragments), especially interesting for this process as well. It is best studied at the highest mass scale, hence in W and Z events. Existing data imply that the underlying-event topologies are similar to minimum-bias events at cms energies $\sim \sqrt{s} - m_W$. This implies a broad distribution in associated multiplicity. Events with exceptionally low associated multiplicity have special practical interest in their implications for beyond-the-standard-model physics; the large backgrounds from mundane QCD multijet phenomena might be heavily suppressed by such a cut.

What is the chance the underlying central multiplicity is in fact zero? In the Drell-Yan process, one is left with beam-jets of fractional charge and triplet color emergent from the perturbative space-time domain after the collision. Hence one must have at least the typical central multiplicity appropriate to hadronization of an elementary $q\bar{q}$ system, e.g. $W \to q\bar{q}$. There will be some probability of a large rapidity-gap occurring in that process - a probability which goes as $e^{-\Delta n}$, where $\Delta \bar{n}$ is the mean total multiplicity in the relevant $\Delta \eta$ interval. If we take $dN/d\eta \gtrsim 3$ and $\Delta \eta \gtrsim 3$, we get 10^{-3} to 10^{-4} for the probability of the rapidity gap to occur. This is too small to be practical, and the best one might hope for is an event class with dilute phase-space density. However, in gluon-gluon collisions, the situation differs; we return to that case in the next subsection.

3. <u>Hadronization of QCD Jets</u>

Clear understanding of this subject is crucial for the future. The parton-level theory deserves to be completely calculated to, say, order α_s^3 or α_s^4 (as far as is QED) provided the cost in manpower and computation is no more than 10% of that of the SSC. There is some way to go in that regard.

Again a favorite subtopic of mine has to do with structure of the underlying events in these hard collisions. Is it possible to have a large rapidity gap for the underlying event in, say, the hard process gg $\to$ H$^\circ$, where H$^\circ$ is a massive colorless particle of interest like a Higgs? If so, the event topology would imply beam fragments populating only the inner parts of the endcap detectors and the decay products of H$^\circ$ populating only the central detector elements. The argument in the previous section would imply otherwise because the right and left moving beam systems emergent from the collision are color octets.[28] However, at collider energies there is a very large probability per event for an additional semi-hard "minijet" collision ($p_T \gtrsim 5$ GeV) of gluons. Because this also involves color exchange, the combination of the two processes can occasionally lead to color-singlet systems emergent from the collision-region; hence in principle if not in practice to a rapidity gap. The more mundane language for this is that this process goes by double-Pomeron exchange as shown in Fig. 8. Were there a 2-gluon Fock-space component to the Pomeron wave function (even I can dream!), this mechanism could be justified and perhaps even computed.

An experimental test is better. The aforementioned single-diffraction experiment is one. Another, closer to the above line of thinking, is diffractive double-parton scattering. One looks for events in which there is

i) nothing in the central detector

ii) two minijets with unbalanced p_T in the left endcap detector

iii) two more in the right endcap

iv) pairwise p_T balance of each leftward jet with a rightward jet.

The Reggeon diagram is shown in Fig. 9. But it probably does not by itself express the conjectured physics very well.

Multipomeron physics is interesting!

4. Heavy Flavor Production

At the quark level, the theory for hadron production of heavy quarks appears to be rather well understood.[29] There are some caveats, however. First of all, whether the charm quark should be considered "heavy" is marginal. An effective charm-quark mass of 1.2 GeV is sometimes used to normalize the calculated cross-section to the data. This, if nothing else, will distort the kinematics of any distributions of emergent charmed hadrons.

To me, the most interesting outstanding questions have to do with hadronization. First of all, the inclusive distributions of the hadrons must differ from those of the quarks for kinematic reasons alone. This "higher twist" effect scales as m_Q^{-1}. To see this, just consider a definition of x appropriate to fixed-target experiments

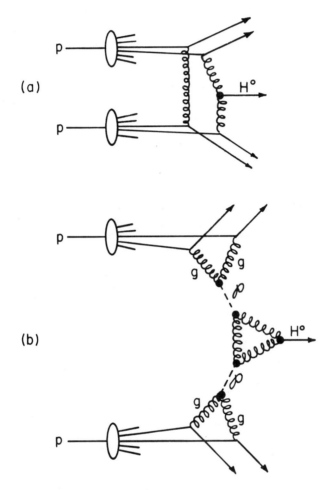

Fig. 8   "Double-pomeron" production of a massive neutral boson H^0.

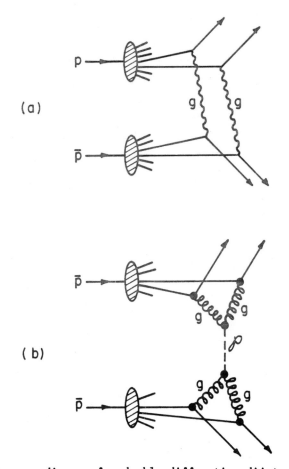

Fig. 9   Reggeon diagram for double-diffractive dijet production.

$$x_Q = \frac{(E + p_{||})_{\text{heavy quark}}}{E_{\text{beam}}}$$

$$x_H = \frac{(E + p_{||})_{\text{hadron containing }Q}}{E_{\text{beam}}}$$

x_Q and x_H scale multiplicatively under boosts; therefore boost to the Lorentz frame in which the quark Q has no longitudinal momentum. Then (neglecting transverse momenta)

$$\frac{x_H}{x_Q} \simeq \frac{m_H + p_{||}}{m_Q} = 1 + O\left(\frac{1 \text{ GeV}}{m_Q}\right)$$

On average x_H/x_Q is evidently greater than unity; thus the x-distribution of the heavy-flavored hadron is <u>broadened</u> relative to the primordial heavy-quark distribution. My own estimate of this effect is that it alone diminishes the exponent of a $(1 - x)^n$ distribution by an amount of order unity.

Another important question has to do with the relative abundance of the produced species, especially the baryon-to-meson ratio. Charmed baryon-hadroproduction appears to be quite large; the recent data[30] from Fermilab experiment E400 on central Ξ_c and $\bar{\Xi}_c$ production by incident <u>neutrons</u> is a striking example. An important feature in understanding this may be again purely kinematic in origin. For example, suppose an N or Λ is produced nearby in phase-space to a D. Simple calculation shows that the combined mass of the system remains within a few hundred MeV of the minimum value ($\gtrsim 3$ GeV) for reasonable transverse momenta ($p_T \lesssim 1$ GeV) and rapidity separations ($\Delta y \lesssim 1-2$). If for no other reason than strong final-state interactions (analogous to $\bar{K}N \rightarrow Y\pi$ near threshold)

$$N + D \rightarrow \left\{ \Lambda_c + \pi\text{'s} \right.$$

$$\Lambda + D \rightarrow \left\{ \begin{array}{l} \Xi_c + \pi\text{'s} \\ \Lambda_c + K + \pi\text{'s} \end{array} \right.$$

one can expect re-scattering into charmed baryon final states. This can also be viewed at the quark-level - a produced charmed quark together with a diquark nearby in phase-space may preferably form a charmed baryon rather than nucleon plus charmed meson because a low total mass is favored and the minimum mass of the former system is lower that the latter by more than 500 MeV. No matter which viewpoint is adopted, this feature of exothermiticity favors charmed baryon formation whenever the net baryon-number in the relevant region of phase-space is nonvanishing.

Using a simple recombination model, I have made some rough guesses for charm and bottom hadron production; similar estimates have also been done by Frankfurt and Strikman.[31] My preliminary estimates of yields are presented in Fig. 10, normalized to the primordial quark distributions. One sees factor-of-two corrections occurring for $x_F >$ 0.3 for charm and $x_F > 0.5$ for bottom.

It is important to recognize that the physics of heavy-flavor hadronization in hadron collisions is very different from corresponding hadronization phenomena in high energy e^+e^- collisions. In the latter, the final hadron has energy less than that of the parent quark and overall energy conservation is a crucial element of the dynamics. In the former case overall energy conservation is a small role, but interaction of the produced quark with its immediate environment is crucial. I think more work needs to be done on this important problem.

Another important issue is A-dependence. In general, an A^1 dependence is expected. However for production of leading charm at the highest energy, the same argumentation used for the Drell-Yan process applies, and one may anticipate an "$A^{2/3}$" dependence. This is also the case for hadroproduction of high-momentum ψ from nuclei. In both cases the mechanism again is that, as seen in the laboratory frame, a beam gluon radiates most of its momentum upstream of the nucleus into the $c\bar{c}$ system. The secondary gluon on arrival at the nucleus has limited p_T and large impact parameter relative to the original gluon trajectory. Hence there is nonperturbative evolution of wee partons which are absorbed on the nuclear surface.

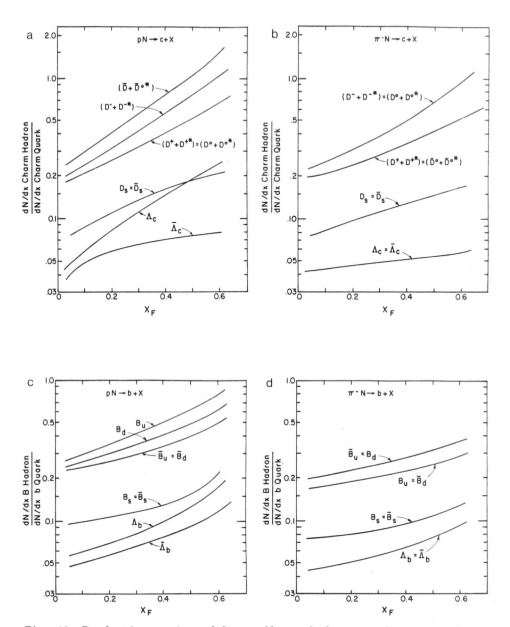

Fig. 10 Production ratios of heavy-flavor hadrons as function of x_F:
charm production (a) by protons and (b) by pions; bottom
production (c) by protons and (d) by pions. These are
normalized to the primordial heavy quark distribution.

III.   CONCLUDING REMARKS

Preparation of this talk has given me an opportunity to have a fresh look at the overall status of QCD, something I have not really done for some time. A lot is happening and the field is quite sophisticated. I therefore apologize for naivete and amateurism in this contribution. But details aside, it seems to me that the status of QCD resembles that of quantum-mechanics in, say, the 1930's. There are at least three levels at which each can be approached.

1)   Is QCD correct?

Yes, there was skepticism about the correctness of quantum theory in the 1930's. But it didn't have much impact at the practical level, and very few people asked the question. It's not so different now. With regard to this point, we can say that Giuliano Preparata is the Einstein of the 1980's.

2)   What are the experimental consequences?

This is still the Golden Age of QCD, when new subfields continue to open up and are waiting to be explored. And many basic calculations and measurements need to be done. Enough said.

3)   What do the equations mean?

It turned out to be easier to write down the Schrodinger equation and work out the consequences than figure out what it all meant. To develop the interpretation of quantum theory out to the textbook-explanation level was harder and took a long time. And, as discussed earlier, I believe it is also so with QCD: we work with Green's functions, parton distributions, and even "wave functions" and all that without fully understanding where they come from and what they mean. I think there is quite a lot left to do on foundations of QCD. Perhaps the problem is in large part in the precise formulation of the problem itself - in asking the questions the right way. I think there is something to be done; I only wish I knew how to do it.

I thank E. Berger, G. Bodwin, L. Frankfurt, and M. Strikman for vital criticisms and assistance in excision of numerous errors in an early draft of this manuscript. They are of course not responsible for any gaffes that remain.

REFERENCES

1. J. Bjorken and S. Drell, "Relativistic Quantum Fields", McGraw-Hill (New York, 1965).

2. See J. Bjorken, Proceedings of the 1979 SLAC Summer Institute on Particle Physics, ed. A. Mosher, SLAC Report SLAC-224(1980), p. 219.

3. T. Sloan, CERN preprint CERN-EP/87-188(October, 1987).

4. J. Bjorken, Phys. Rev. 148, 1467(1966).

5. J. Ellis and R. Jaffe, Phys. Rev. D9, 1444(1974).

6. F. Close, these proceedings.

7. S. Brodsky, J. Ellis, and M. Karliner, SLAC-PUB-4519(Jan. 1988), present an interesting new interpretation of these data.

8. See G. Bodwin, Phys. Rev. D31, 2616(1985) and references therein.

9. See for example S. Brodsky, Acta Physica Polonica B15, 1059(1984) and references therein.

10. I believe that in modern language this inconsistency has to do with the presence of vacuum-to-vacuum bubbles in the evolution matrix U(t), but I am not sure on this point. Infinite renormalization constants also complicate the issue.

11. Actually, a better estimate is $E/\Delta m^2$, where $\Delta m^2 = m_x^2 - m_\pi^2$ is the mass (squared) splitting between the pion and the next important massive state x. I thank M. Strikman for this comment.

12. See for example the contributions of G. Marchesini, of T. Gottschalk, of F. Paige and S. Protopopescu, and of H. Bengtsson, Proceedings of the UCLA Workshop "Observable Standard Model Physics at the SSC: Monte Carlo Simulation and Detector Capabilities," ed. H.V. Bengtsson, C. Buchanan, T. Gottschalk, and A. Soni, World Scientific(Singapore), 1986.

13. W. Hofmann, Proceedings of the XXIII International Conference on High Energy Physics, Berkeley, CA, July, 1986, ed. S. Loken, World Scientific(Singapore), 1987, Vol. II, p. 1093.

14. Y. Azimov, Y. Dokshitzer, V. Khose, and S. Troyan, Phys. Lett. 165B, 147(1985); Yad. Fiz. 43, 149(1986) and references therein.

15. S. Parke and T. Taylor, Phys. Lett. 157B, 81(1985).

16. N. Nikolaev and V. Zakharov, Phys. Lett. 55B, 197(1975).

17. J. Bjorken, in "Particles and Detectors: Festschrift for Jack Steinberger, ed. K. Kleinknecht and T.D. Lee, Springer Tracts II, 108 (Springer-Verlag, Berlin, 1986), p. 17.

18. For a nice discussion, see A. Bialas and I. Chinaj, Phys. Lett. 133B, 241(1983).

19. S. Brodsky and G. Farrar, Phys. Rev. $\underline{D11}$, 1309(1975).

20. A. Mueller, Proceedings of the 27th Rencontre de Moriond on Perturbative QCD, ed. J. Tran Thanh Van, Editions Frontieres, Gif-sur-Yvette(1987).

21. A. Carroll et. al., Penn State preprint PSU HEP/88-02.

22. G. Farrar, Phys. Rev. Letters $\underline{56}$, 1643(1986).
    The theoretical argument for banning polarization measurements is expressed in this paper: the full amplitude is A = C + D where C, the clean amplitude, is computable and D, the dirty amplitude, is not. D in magnitude is 30% of C so it only affects unpolarized data at the 10% level. Polarization data dependent on the interference term is dirty even though asymmetries up to 60% are possible.

23. See for example S. Brodsky and G. DeTeramond, SLAC preprint SLAC-PUB-4504, Dec. 1987.

24. Internal-target data from Fermilab show backward protons ($x_F \sim -0.8 \pm 0.1$) about 7% polarized at p_T = 500 MeV; this polarization disappears at p_T = 1 GeV. But what about higher p_T? And what about pp $\to$ n + x? See M. Corcoran et. al., Phys. Rev. $\underline{D22}$, 2624(1980).

25. G. Ingelman and P. Schlein, Phys. Lett. $\underline{152B}$, 256(1985).

26. An excellent analysis of the situation is presented by E. Berger, J. Collins, D. Soper, and G. Sterman, Nucl. Phys. $\underline{B286}$, 704(1987).

27. It $\underline{can}$ be done. See J. Bjorken, "Forward Spectrometers at the $\overline{SSC}$", FERMILAB-CONF-86/22.

28. For color singlet initial-state partons, the existence of a rapidity gap is very credible; an example is $W^+W^- \to H^o$ as discussed by Y. Dokshitzer, V. Khose, and S. Troyan, Proceedings of the 6th International Conference on Physics in Collision, ed. M. Derrick, World Scientific(Singapore), 1987, p. 417.

29. K. Ellis and C. Quigg, Fermilab Report, FN-445, Jan. 1987, and these proceedings.
    E. Berger, Proceedings of the 22nd Rencontre de Moriond, March 1987, and these proceedings.

30. P. Coteus et. al., Phys. Rev. Letters $\underline{59}$, 1530 (1987).

31. J. Bjorken, L. Frankfurt, and M. Strikman, "Hadronization of Heavy Quarks," in preparation.

PRELIMINARY RESULTS FROM CDF ON W, Z PRODUCTION AT THE TEVATRON COLLIDER

W.C. Carithers*

Lawrence Berkeley Laboratory
University of California
Berkeley, California 94720

INTRODUCTION

CDF is a general purpose detector featuring highly segmented electro-magnetic and hadronic calorimetry over a large region of pseudorapidity and magnetic analysis for charged tracks in the central region. The detector is shown schematically in Figure 1. The details of the detector are discussed in a series of papers to be published.(Reference2) The full detector was operational for the 1987 Collider run with the exception of the higher level triggers and half of the sampling layers in the forward hadron calorimeters.

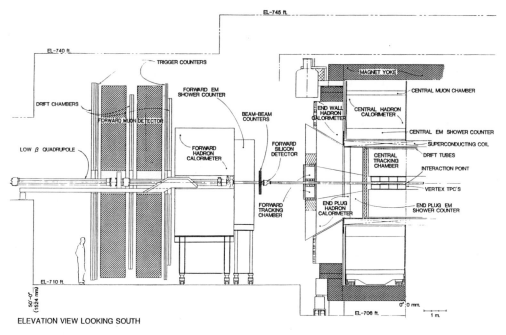

Fig. 1. Schematic drawing of CDF detector

---

* Presented on behalf of the CDF Collaboration (Reference 1)

The luminosity delivered by the Collider is shown in Figure 2. The peak luminosity increased by over an order of magnitude during the run and reliably achieved the 1987 goal of 10^{29} cm^{-2} sec^{-1} by the end of the run. The total integrated luminosity written to tape was over 30 nb^{-1}, much of it coming in the last two weeks of the run.

The trigger consisted of a logical "or" of several different requirements for different physics interests. All triggers required a in-time coincidence of beam-beam scintillation counters. In addition, any one of the following conditions would generate a trigger:

1. The total E_t (EM + hadronic) in the central region exceeded a threshold adjusted from 20–45 GeV depending on the luminosity. This trigger is primarily for jet studies.
2. The total scalar E_t in the electromagnetic section of the full calorimetry exceeded an adjustable threshold, typically 12–15 GeV. This data should yield an high P_t electron sample.
3. The coincidence of a track in the central tracking chamber of $P_t > 3$ GeV with a track in the central muon chambers as determined by a fast track processor.
4. A prescaled forward muon trigger.
5. A prescaled "minimum bias" sample requiring only the beam-beam counters.

All individual elements of the calorimeters "towers" were required to exceed a threshold of 0.1 GeV before they were allowed to participate in the E_t sum. The prescale factors and E_t threshold factors were adjusted as a function of the luminosity to keep the total data rate to approximately 1 Hz.

All 400 data tapes were processed by filter algorithms to separate good events into output streams. The output streams included: jets, electrons, muons, missing E_t, minimum bias data, and special studies. In the production pass, tracking is done locally only on a demand basis.

$W \rightarrow$ ELECTRON + NEUTRINO ANALYSIS

This section describes a very preliminary analysis covering most of the data for identifying W boson candidate events with an electron in the final state. The "electron" output stream data is defined by a requirement of an electromagnetic calorimeter cluster with $E_t > 3$ GeV and a track pointing toward that cluster. The data are then subjected to the following cuts to isolate high P_t electrons in the central region:

1. The EM cluster E_t must be at least 15 GeV.
2. The ratio of EM/hadronic energy in the cluster must exceed 0.96. In the test beam calibration of the central calorimeters this ratio typically exceeded 0.98 for real electrons.
3. An isolation cut was applied that required the ratio of the EM cluster E_t to the total E_t inside a cone typical of jet activity be greater than 0.7. The radius of the cone in eta-phi space was set to 0.7. This cut is designed to discriminate against background from jets.
4. The central EM calorimeters have strip chambers embedded near shower maximum to localize the EM shower. We require that the width of the EM shower in these chambers be less than 5 cm. and that the energy measured by the single strip chamber sample be at least 10% of the energy measured by the scintillator calorimeter.

In addition the following cuts specific to the tracking information are applied:

5. The closest distance of approach of the electron candidate track to the vertex as determined by the other tracks in the event must be less than 0.5 cm in the (drift) r-phi view and 10 cm in the (stereo) $r - z$ view.

34

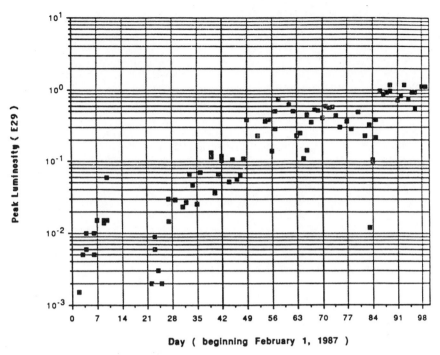

Fig. 2a. Peak luminosity for 1987 run

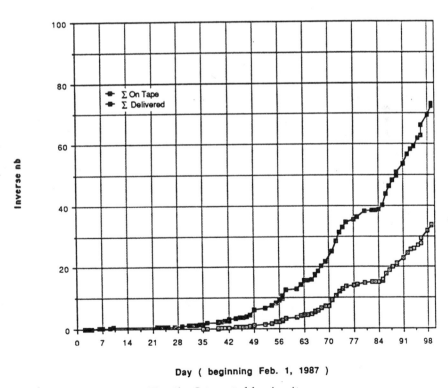

Fig. 2b. Integrated luminosity

6. The extrapolated azimuthal impact point of the electron candidate track must agree with the strip chamber to within 0.027 radians. The agreement between the impact points along the beam direction must be less than 10 cm. Both of these cuts are well outside the resolution and in fact reject very little data that survive all the other cuts.

7. Finally we require that the momentum measured by the central tracker agree with the energy measured by the calorimeters to within 5 rms resolution units. The resolution is dominated by the P_t measurement for these high P_t tracks.

The data which survive all of these cuts are shown in Figure 3 as a scatter plot of electron energy versus the missing E_t in the event. We identify W candidates as those with at least 15 GeV of missing E_t (remember that a 15 GeV cut on the electron E_t has already been imposed as part of the electron definition). We emphasize again the preliminary nature of this data since no energy corrections whatsoever have been applied. One event from the W candidate region is shown in Figure 4. The transverse mass distribution is shown in Figure 5.

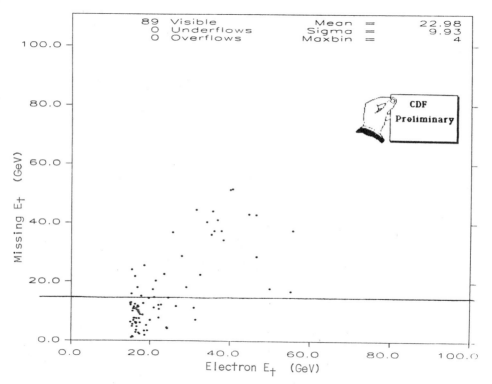

Fig. 3. Electron energy versus missing E_t as determined by the calorimeters. W candidates are considered in the region of $E_t > 15$ GeV.

$W \to$ MUON PLUS NEUTRINO

The central muon analysis suffers from lower statistics since the solid angle for muon coverage is considerably less than the calorimetric coverage for electrons. This situation is exacerbated by the loss of four sets of muon chambers due to high levels of radiation from the accelerator Main Ring.

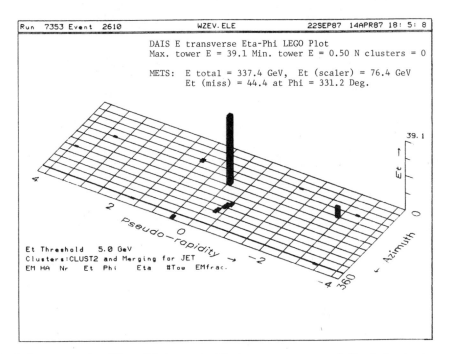

Fig. 4. Representative W candidate event. The transverse energy of all calorimeter towers in the event is shown

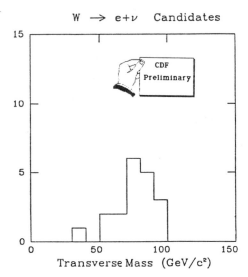

Fig. 5. Transverse mass of electron and missing E_t for W candidates.

The four layers of the muon drift chambers are equipped with charge division readout for measuring the coordinate along the beam(z) direction. Thus we can determine the impact point and local slope in the drift direction and the impact point only in the z direction.

A good central muon must then satisfy the following requirements:

1. The P_t as measured by the central tracker must be at least 5 GeV.
2. The difference in local slope measured by the muon chambers in the drift direction must agree within multiple scattering with the extrapolation of the track as measured by the central tracker.
3. The difference in impact point in both views as measured by the muon chambers and the central tracker extrapolation must agree within multiple scattering.
4. The energy measured by both the EM and hadronic calorimeters must be consistent with a minimum ionizing particle as determined by the test beam calibration of the calorimeters.

The scatter plot of muon P_t versus the missing E_t in the event is shown in Figure 6. We identify W candidates as those in the region of muon P_t greater than 15 GeV and missing E_t greater than 20 GeV. A few events are consistent with W decaying into muon final states.

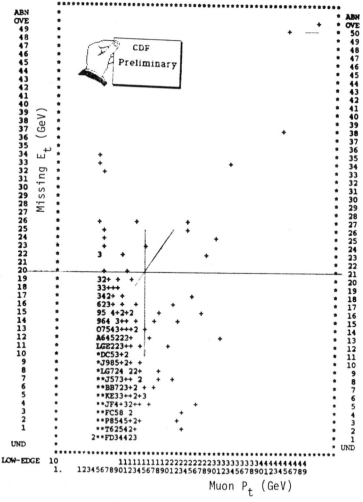

Fig. 6. Muon momentum versus missing E_t as determined by the calorimeters. W candidates are considered in the region of $E_t > 20$ GeV.

MISSING E_t

An attempt to identify W candidates has been made using the missing E_t output stream. This analysis is particularly difficult since any defect or problem with the detector tends to generate a "false" missing E_t. Using a set of hard cuts to eliminate known detector problems, a data set with missing E_t greater than 30 GeV was produced. A hand scan of this data set revealed more than 20 W candidates previously identified by other analyses.

$Z \rightarrow$ ELECTRON PAIRS

The Z analysis is still very preliminary. A hand scan of events identified by the electron-finding algorithm discussed above identified six Z candidates. One of these candidates is shown in Figure 7. Notice that one of the electron candidates is in the end plug region for this event. The mass distribution for the six candidate events is shown in Figure 8. We emphasize again that no energy corrections have been made.

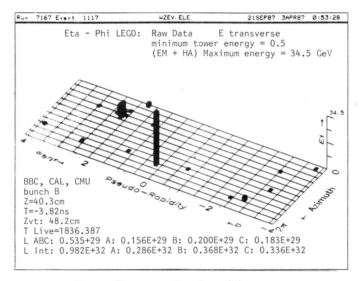

Fig. 7. Representative Z candidate event.

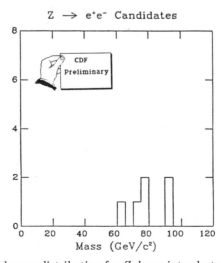

Fig. 8. Uncorrected mass distribution for Z decay into electron pair candidates.

39

CONCLUSION

The 1987 Collider run was extremely successful. Both the Tevatron and the CDF detector performed very well. A first pass through the 30 nb^{-1} of data has yielded W decay candidates through three independent analysis paths. In addition, six Z decay candidates have been identified.

REFERENCES

1. CDF Collaboration: F. Abe[p], D. Amidei[c], G. Apollinari[k], G. Ascoli[g], M. Atac[d], P. Auchincloss[n], A.R. Baden[f], A. Barbaro-Galtieri[i], V.E. Barnes[l], E. Barsotti[d], F. Bedesci[k], S. Behrends[l], S. Belforte[m], G. Bellettini[k], J. Bellinger[g], J. Bensinger[b], A. Beretvas[n], P. Berge[d], S. Bertolucci[e], S. Bhadra[g], M. Binklev[d], R. Blair[a], C. Blocker[b], J. Bofill[d], A.W. Booth[d], G. Brandenburg[f], A. Brenner[d], D. Brown[f], A. Byon[l], K.L. Byrum[q], M. Campbell[c], R. Carey[f], W. Carithers[i], D. Carlsmith[q], J.T. Carroll[d], R. Cashmore[1], F. Cervelli[k], K. Chadwick[l], T. Chapin[m], G. Chiarelli[m], W. Chinowsky[i], S. Cihangir[o], A.G. Clark[d], D. Cline[q], T. Collins[d], D. Connor[j], D. Connor[j], M. Contreras[b], J. Cooper[d], M. Cordelli[e], M. Curatolo[e], C. Day[d], R. DelFabbro[k], M. Dell'Orso[k], L. DeMortier[b], T. Devlin[n], D. DiBitonto[a], R. Diebold[a], F. Dittus[d], A. DiVirgilio[l], R. Downing[g], G. Drake[d], T. Droege[d], M. Eaton[f], J.E. Elias[d], R. Ely[i], S. Errede[g], B. Esposito[e], A. Feldman[f], B. Flaugher[n], E. Focadi[k], G.W. Foster[d], M. Franklin[g], J. Freeman[d], H. Frisch[c], Y. Fukui[h], I. Gaines[d], A.F. Garfinkel[l], P. Giannetti[2], N. Giokaris[m], P. Giromini[e], L. Gladney[j], M. Gold[i], K. Goulianos[m], J. Grimson[d], C. Grosso-Pilcher[c], J. Grunhaus[7], C. Haber[i], S.R. Hahn[j], R. Handler[q], R.M. Harris[i], J. Hauser[c], Y. Hayashide[p], T. Hessing[o], R. Hollebeek[j], L. Holloway[g], P. Hu[n], B. Hubbard[i], P. Hurst[g], J. Huth[d], M. Ito[p], J. Jaske[q], H. Jensen[d], R.P. Johnson[d], U. Joshi[n], R.W. Kadel[d], T. Kamon[o], S. Kanda[p], I. Karliner[g], H. Kautzky[d], K. Kazlauskis[n], E. Kearns[f], R. Kephart[d], P. Kesten[b], H. Keutelian[g], Y. Kikuchi[p], S. Kim[p], L. Kirsch[b], S. Kobayashi[3], K. Kondo[p], U. Kruse[g], S.E. Kuhlmann[l], A.T. Laasanen[l], W. Li[a], T. Liss[c], N. Lockyer[j], F. Marchetto[o], R. Markeloff[g], L.A. Markosky[q], M. Masuzawa[p], P. McIntyre[o], A. Menzione[k], T. Meyer[o], S. Mikamo[h], M. Miller[j], T. Mimashi[p], S. Miscetti[e], M. Mishina[h], S. Miyashita[p], H. Miyata[p], N. Mondal[q], S. Mori[p], Y. Morita[p], A. Mukherjee[d], A. Murakami[3], Y. Muraki[4], C. Nelson[d], C. Newman-Holmes[d], J.S.T. Ng[f], L. Nodulman[a], J. O'Meara[d], G. Ott[q], T. Ozaki[p], S. Palanque[6], R. Paoletti[k], A. Para[d], J. Patrick[d], R. Perchonok[d], T.J. Phillips[f], H. Piekarz[b], R. Plunkett[m], L. Pondrom[q], J. Proudfoot[a], G. Punzi[k], D. Quarrie[d], K. Ragan[j], G. Redlinger[c], R. Rezmer[a], J. Rhoades[q], L. Ristori[k], T. Rohaly[j], A. Roodman[c], H. Sanders[c], A. Sansoni[e], R. Sard[a], A. Savoy-Navarro[6], V. Scarpine[g], P. Schlabach[g], E.E. Schmidt[d], P. Schoessow[a], M.H. Schub[l], R. Schwitters[f], A. Scribano[k], S. Segler[d], M. Sekiguchi[p], P. Sestini[k], M. Shapiro[f], M. Sheaff[q], J.K. Simmons[l], P. Sinervo[j], M. Sivertz[5], K. Sliwa[d], D.A. Smith[g], R. Snider[c], L. Spencer[b], R. St.Denis[f], A. Stefanini[k], Y. Takaiwa[p], K. Takikawa[p], S. Tarem[b], D. Theriot[d], J. Ting[c], P. Tipton[i], A. Tollestrup[d], G. Tonelli[k], W. Trischuk[f], Y. Tsay[c], K. Turner[d], F. Ukegawa[p], D. Underwood[a], C. vanIngen[d], R. VanBerg[j], R. Vidal[d], R.G. Wagner[a], J. Walsh[j], T. Watts[n], R. Webb[o], T. Westhusing[g], S. White[m], V. White[d], A. Wicklund[a], H.H. Williams[j], T. Winch[q], R. Yamada[d], T. Yamanouchi[d], A. Yamashita[p], K. Yasuoka[p], G.P. Yeh[d], J. Yoh[d], and F. Zetti[k].

CDF Member Institutions:
[a] Argonne National Laboratory
[b] Brandeis University
[c] University of Chicago
[d] Fermi National Accelerator Laboratory
[e] INFN, Frascati, Italy
[f] Harvard University
[g] University of Illinois

^{h}KEK, Japan
^{i}Lawrence Berkeley Laboratory
^{j}University of Pennsylvania
^{k}INFN, University of Pisa, Italy
^{l}Purdue University
^{m}Rockefeller University
^{n}Rutgers University
^{o}Texas A&M University
^{p}University of Tsukuba, Japan
^{q}University of Wisconsin

Visitors:
^{1}Oxford University, England
^{2}INFN, Trieste, Italy
^{3}Saga University, Japan
^{4}ICRR, Tokyo University, Japan
^{5}Haverford College, Haverford, PA
^{6}CEN, Saclay, France
^{7}University of Tel-Aviv, Israel

2. Nuclear Instr. and Methods, submitted for publication.

STUDY OF DIMUONS WITH NA10

Klaus Freudenreich

IHEP, Federal Institute of Technology
Zurich, Switzerland

Abstract: The NA10 experiment [1] at the CERN SPS has collected a large statistics of dimuon pairs produced by π^--tungsten and π^--deuterium interactions at three energies. Recent results on the study of scaling violations, P_T distributions, nuclear effects and angular distributions will be presented.

1. Introduction

QCD is being tested in many processes. Generally the conclusion of such tests is that QCD works well. However, one should keep in mind the simple fact that an agreement is more easily obtained when the error bars are large. Fixed-target lepton pair experiments have now reached statistics comparable to those of DIS experiments (a few 10^5 events with $Q^2 > 16$ GeV^2). They do allow, therefore, stringent quantitative tests of QCD which are complimentary to tests in other reactions.

Scaling violations have long been seen in deeply inelastic lepton scattering. Evidence for the related scaling violations in hadronic lepton pair production has, however, been obtained until now only by a strong appeal to theory [2,3]. We shall present a comparison of invariant cross-section ratios which allows to study these violations in a model independent way. In addition, the comparison of these cross-section ratios with theoretical ones can in principle serve to determine Λ in lepton pair production.

A very stringent test of QCD is provided by the study of the triply differential cross-sections:
$$1/P_T \, d^3\sigma/dP_T dM dy$$
where high statistics − which is available now − is needed.

One aspect of another subject which can be investigated in lepton pair production from nuclear targets − the question of specifically nuclear effects − has first been seen in DIS by the EMC group [4]. In addition to this study of the dependence on the nuclear structure variable x_2, lepton pair production offers the possibility to study nuclear effects also as a function of the P_T of the lepton pair (DIS cross-sections are in most cases integrated over P_T).

Finally, the study of angular distribution provides another interesting test of QCD.

2. Direct observation of scaling violation

If data at only one beam energy are available, scaling violation can solely be exhibited with reference to theoretical models. Such analyses [2] [3] have shown that models including scaling violations yield, especially at high Q^2, better fits than the "naive" Drell-Yan model. If data at more than one energy are available scaling violations can be determined in a model independent way.

To study such scaling violations, usually the data from different energies are compared by calculating the invariant cross-sections $s d\sigma/d\tau$ or $M^3 d\sigma/dM$ as a function of τ. According to QCD one expects only small scaling violations. These are much easier detected by studying ratios of invariant cross-sections. An additional advantage is that several systematic errors drop out in such ratios. For the study of scaling violations a large lever arm in s is desirable. Combining proton-induced dilepton data from fixed-target and storage-ring experiments meets this goal. Due to the large normalization errors and relatively low statistics, scaling has been verified in this way only within a factor two [5]. In a study of scaling violations within a single experiment, only the relative normalization matters. In addition, it is much easier to ensure that the comparison is done within the same kinematical cuts. The data collected by the NA10 collaboration in $\pi^- \, W \rightarrow \mu^+\mu^- + X$ at three energies [6,7] made such a study possible. Here the relatively small span in s was compensated by the very high statistics (see Table 1) available and the good control of systematic errors.

Table 1. Summary of data used for scaling studies

i	π^- momentum [GeV/c]	s_i [GeV^2]	Number of events	target lengths [cm]
1	286	538.2	84,400	5.6 , 12.0
2	194	364.8	150,000	5.6 , 12.0
3	140	262.4	45,900	12.

The NA10 data were analyzed by means of ratios of scaling cross-sections, viz.:

$$R_{ij} = s_i \, d\sigma/d\sqrt{\tau}(s_i) \, / \, s_j \, d\sigma/d\sqrt{\tau}(s_j),$$

The principal remaining correction here was for nuclear reinteractions in the targets, which affect both the shape and the absolute norm of these ratios. The three possible ratios R_{ij} are shown as a function of $\sqrt{\tau}$ in Figure 1. One specific advantage of the NA10 spectrometer is exploited here. Since by construction the acceptance was essentially the same at the three energies, the integration over variables other than $\sqrt{\tau}$ ensured automatically that the same kinematical regions were covered. Only a cut in x_F ($-0.1 \leq x_F < 0.5$) had to be made to ensure the common kinematical coverage. Since the mass windows of the Υ-region which had to be excluded correspond at each energy to different windows in $\sqrt{\tau}$, simply omitting these Υ-regions would have created large $\sqrt{\tau}$ gaps. These gaps were filled in by the following procedure:

The parameters A_i, α_i and β_i of a function $f_i(\sqrt{\tau}) = A_i(\sqrt{\tau})^{\alpha_i}(1-\sqrt{\tau})^{\beta_i}$ were derived, at each energy s_i separately, from a fit to the measured continuum cross-sections. With the help of these fitted functions f_i one computed interpolated continuum cross-sections (with their errors) in each Υ-region. The cross-section ratios were then calculated; to avoid ratios of two interpolated cross-sections, one point in the R_{23} ratio had to be omitted. The ratios containing one interpolated value are indicated by open circles in Figure 1.

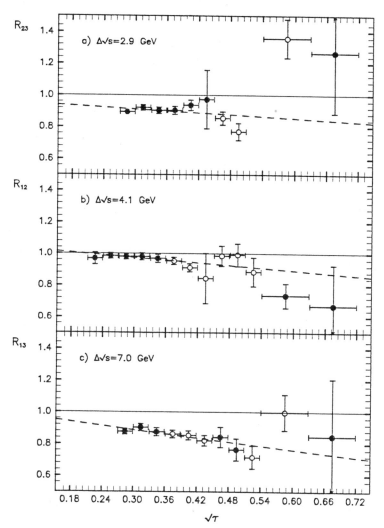

Figure 1. R_{ij} as a function of $\sqrt{\tau}$
1:$\sqrt{s} = 23.2$ GeV, 2:$\sqrt{s} = 19.1$ GeV, 3:$\sqrt{s} = 16.2$ GeV.

For perfect scaling the three ratios should be equal to one and not depend on $\sqrt{\tau}$, as indicated by the full line. The observed ratios are however clearly not constant, but fall with increasing $\sqrt{\tau}$, as is expected for QCD scaling violations. A joint fit of the form

$$R_{ij}(\sqrt{\tau}) = a_{ij} + b \times (\sqrt{s_i} - \sqrt{s_j}) \times \sqrt{\tau}$$

to the two ratios R_{12} and R_{13} constraining the third one by $R_{23} = R_{13}/R_{12}$ yielded a slope parameter $b = -0.065$. Because of correlations on the one hand between interpolated and measured cross-section values and between the ratios themselves on the other hand it was not possible to assign a standard deviation to this slope parameter. To estimate the statistical probability with which the data reject the hypothesis of perfect scaling, the following stochastic procedure (suggested by F. James) was used: The data analysis described above was repeated a very large number of times with artificially created (statistically fluctuating) cross-section values which were calculated from theoretical cross-sections which respected scaling. These "synthetic" cross-sections were calculated at all values of s and $\sqrt{\tau}$ where data existed with statistical errors according to the actual data. Interpolated cross-sections in the Υ-regions were then obtained in the same way as described above and the same type of fit to the ratios was performed each time. In this way a distribution of the slope parameter b was obtained which is shown in Figure 2.

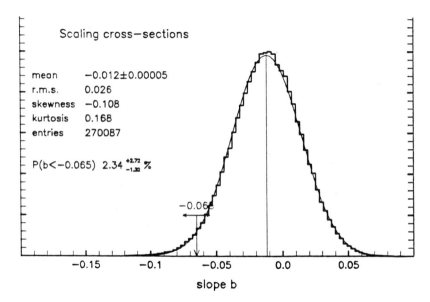

Figure 2. Distribution of the slope b.

Due to a small bias the distribution does not peak at $b = 0$ but at $b = -0.012$. The area of the distribution with $b \leq -0.065$ (the value found in the fit to the real data) represents 2.3% of the total. The systematic error on b due to reinteraction corrections and normalization uncertainties is $\Delta b = 0.010$. Taking into account this systematic error, the area with $b \leq -0.055$ represents 5.1%. The data therefore exclude scaling with a probability of 95%.

That this scaling violation is not due to a difference in the x_F-dependencies can be seen from Figure 3. It shows the ratios of the invariant cross-sections

$$R'_{ij}(x_F) = s_i \times d\sigma_i/dx_F / s_j \times d\sigma_j/dx_F$$

as a function of x_F for two regions in $\sqrt{\tau}$: one below and one above the Υ-region (except at 140 GeV/ c). Within the errors, the ratios show no x_F dependence.

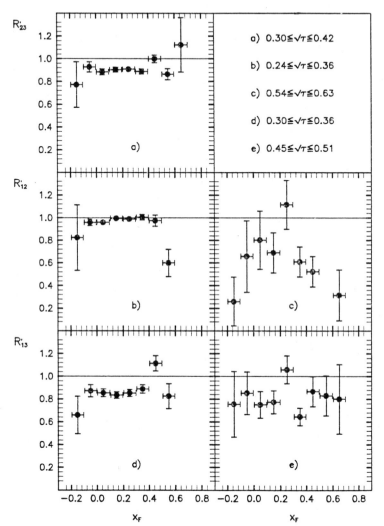

Figure 3. R'_{ij} as a function of x_F
1: $\sqrt{s} = 23.2$ GeV, 2: $\sqrt{s} = 19.1$ GeV, 3: $\sqrt{s} = 16.2$ GeV.

The above analysis is model-independent. It is however still of interest to compare the experimental cross-section ratios to theoretical predictions, which depend but weakly on the specific choice of structure functions; their main dependence coming from the QCD scaling violation parameter Λ. A comparison therefore allows a determination of the latter. The theoretical cross-sections were calculated within the SGEA approximation [9] with different values of Λ and a χ^2 value was calculated for each value of Λ. Figure 4 shows the χ^2-distribution as a function of Λ. It yields $\Lambda = (115 \pm^{145}_{92})$MeV. In order to take into account the main (statistical) error[1] on the normalization of the experimental cross-sections the above procedure was repeated by allowing for the normalization of each experimental cross-sections to vary by one standard deviation. In this way a value of $\Lambda = (148 \pm^{160}_{80})$MeV was obtained. Although the value of Λ found here is affected with a large error it clearly is well compatible with the Λ-values determined in DIS [8].

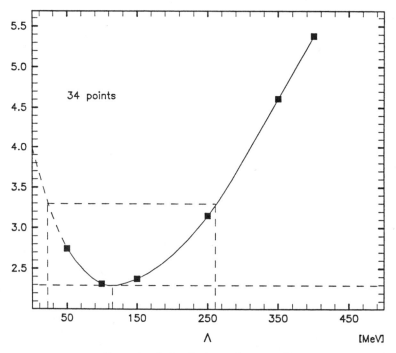

Figure 4. χ^2-distribution as function of Λ.

3. QCD tests in P_T distributions

The P_T-distributions provide the most direct test of QCD because the large P_T's are purely due to processes involving gluons. The high-statistics available at present allow not only the verification of predictions for average quantities like $<P^2_T>$, but also on the triply differential cross-sections $1/P_T$ $d^3\sigma/dP_TdMdy$. The NA10 collaboration has the largest statistics. These data — 115,000 μ-pair events with $M > 4.5$ GeV/c^2 from π^- on W at 194 GeV/c and 40,000 events (half the available statistics) with $M > 4.5$ GeV/c^2 from π^- on W at 286 GeV/c — were first been presented at the 1986 Berkeley Conference [11,12]. In the following the observed triply differential cross-section

$$1/P_T \ d^3\sigma/dP_TdMdy$$

will be compared to a QCD calculation [9] done in the SGE approximation. Apart from being differential and not integral, this calculation includes contrary to the previous one [13,14] also nuclear corrections (these will be discussed in more detail in the next section). The multiple scattering of the

---

[1] It is due to the statistics in the determination of the trigger counter efficiency.

incoming quarks inside the W-nucleus has been taken into account. This rescattering effect is important at low mass. Since the authors of Ref.[9] used the Duke-Owens parametrization of the proton and the pion [16], the EMC-effect had to be taken into account as well. Since the EMC-effect becomes noticable for $x_2 > 0.3$ and because of the correlation between x_2 and $M - M^2 = x_1 x_2 s$ — this second correction is important at high masses. The comparison of this model with the NA10 P_T distributions is shown in Figure 5, again for two pion momenta, viz. 194 and 286 GeV/c.

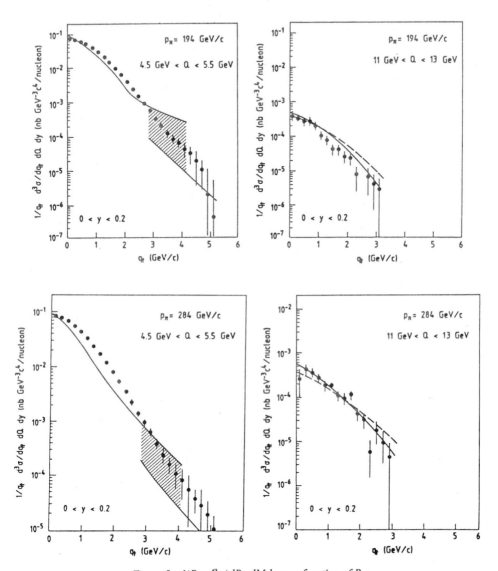

Figure 5. $1/P_T \, d^3\sigma/dP_T dM dy$ as a function of P_T

The cross-section has been calculated for $<y> = 0.1$ at $<M> = 5$ and 12 GeV/c², each time at $<y> = 0.1$. Around $P_T = 3$ GeV/c the theoretical curve splits into two for the low mass bins. This is so because the SGEA calculation becomes unreliable at large P_T. The lower theoretical curve starting at $P_T = 3$ GeV/c represents the contributions of the first-order Compton term alone. This term is clearly insufficient to describe the high P_T data, and this is maybe the strongest hint for the need of higher order contributions to the Compton term which — up to now — have not yet been calculated to second order. Although the authors of Ref. [9] conclude that their model gives a "good description of the data", the χ^2 between their model and the data is still **unsatisfactory** even for $P_T < 3$ GeV/c. It is only with respect to a first comparison between a calculation [15] — which successfully described both the P_T distributions of the bosons at the collider and those of the lepton pairs at the ISR — and the NA10 data which was presented at the 1986 Berkeley Conference that this new calculation has reduced the disagreement. These high-statistics differential p_T cross-sections at $\sqrt{s} \approx 20$ GeV remain a challenge for perturbative QCD.

4. Differential nuclear effects

The study of lepton pairs in fixed target experiments has mainly been done with heavy nuclear targets in order to collect high statistics. Except for the storage ring experiments, only one fixed target experiment [17] has collected data on hydrogen, viz. a few hundred events with $M > 4.1$ GeV/c² at 150 GeV/c and at 280 GeV/c π^- H_2. These data allowed to determine the overall A-dependence and the K-factor, but were clearly insufficient to reveal the small, differential EMC-type effects. In order to obtain the largest possible statistics with a light target, the NA10 collaboration used deuterium instead of hydrogen, gaining a factor ~ 1.6 in rate. Deuterium was also chosen because the EMC effect had originally been seen in the comparison of structure functions from deuterium with those from iron. Since the Fermi motion effects in deuterium are small and measured the interpretation of the data represents no difficulty in this respect.

The NA10 deuterium data were taken **simultaneously** with data from a W target [18,19,20]. They permitted i) a high-precision determination of the A-dependence of the cross-section, ii) the search for EMC-type effects in Drell-Yan pairs, and iii) a study of nuclear effects on P_T. The analysis was based on the statistics given in Table 2.

Table 2. Number of events studied for nuclear effects

π^- momentum [GeV/c]	W events	d events
286	49,600	7,800
140	29,300	3,200

The A-dependence of the cross-section was parametrized as:

$$\sigma = A^\alpha \, \sigma_0(A/Z),$$

where σ_0, a function of A/Z, corrects for the difference in the neutron/proton ratio for W and d. The exponent α was determined in the common $\sqrt{\tau}$ range ($0.27 \leq \sqrt{\tau} < 0.37$), yielding the following values:

at 286 GeV/c: $\alpha = 0.998 \pm 0.007 \pm 0.013$ and
at 140 GeV/c: $\alpha = 0.980 \pm 0.006 \pm 0.013$.

The values from the two energies agree with each other and are well compatible with 1. The combined value for α is:

$< \alpha > = 0.988 \pm 0.005 \pm 0.013$. The agreement between the two energies becomes less good when different regions in $\sqrt{\tau}$ are used to calculate α. Then the influence of the EMC-effect, to be discussed next, becomes noticeable.

The EMC-effect, originally discovered 1983 in the deeply inelastic μ-scattering [4], has undergone some evolution in the meantime. It is now generally accepted [8] that the structure function ratio $F_2(A)/F_2(d)$ is² about 1.05 for x < 0.30 . For x > 0.30 this ratio decreases with a slope — depending on A — of about 0.50 till the region of x ~ 0.6, where Fermi motion effects take over which cause a strong increase of this ratio. In lepton pair production one expects that the nucleon structure function should exhibit the same behavior, while no effect should be seen on the structure function of the projectile. When comparing data from a heavy target with those from deuterium one therefore expects a decrease of the cross-section ratio as a function of the variables related to the quarks in the nucleon (like x_2 and $\tau = x_1 x_2$) but no effect in the variable of the quarks in the projectile (x_1) and in x_F[3] should be seen. Figure 6 shows the ratio

$$\sigma(\pi^- W \to \mu^+ \mu^- + X) \, / \, \sigma(\pi^- d \to \mu^+ \mu^- + X)$$

as a function of x_2, $\sqrt{\tau}$, x_F and x_1; $\sigma(\pi^- W \to \mu^+ \mu^- + X)$ has been corrected for the neutron excess (7%). In this Figure the 140 GeV/c and 286 GeV/c are combined, fixing the ratio at 140 GeV/c to be the same as the one at 286 GeV/c in the common $\sqrt{\tau}$-domain.

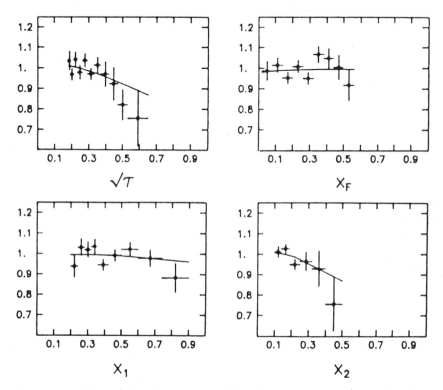

Figure 6. $\sigma(\pi^- W \to \mu^+ \mu^- + X)/\sigma(\pi^- d \to \mu^+ \mu^- + X)$ as a function of $\sqrt{\tau}$, x_F, x^1 and x_2.

As expected, the ratio is constant as a function of x_1, while it clearly drops as a function of x_2. The ratio drops also as a function of $\sqrt{\tau}$, while it is rather constant as a function of x_F. A straight line was fitted to each ratio distribution. Table 3 shows the slopes from these fits.

---

[2] There are indications that this ratio decreases for very small values of x (shadowing).

[3] Although x_F depends like τ on x_2, small and large values of x_2 can — contrary to τ, because of the acceptance — yield the same value of x_F.

Table 3. Straight line fits

variable	Slope (286 + 140 GeV/c)
x_2	$-0.55 \pm 0.21 \pm 0.12$
$\sqrt{\tau}$	$-0.34 \pm 0.15 \pm 0.08$
x_F	$0.06 \pm 0.10 \pm 0.07$
x_1	$-0.08 \pm 0.09 \pm 0.07$

As expected, the slopes in x_2 and $\sqrt{\tau}$ differ from zero, whole those in x_1 and x_F are consistent with zero. The different behaviour in x_1 and x_2 is another confirmation of the factorization hypothesis. The full lines in Figure 6 are the ratios predicted on the basis of the most recent determination [21] of the EMC-effect. To compute them, the quark distributions $q^W(x_2)$ in the W nucleus were modified according to the ratio measured in DIS:

$$q^W(x_2) = F_2^{Fe}(x_2)/F_2^d(x_2) \times q^d(x_2)$$

This prediction agrees quite well with the data except for the highest bins in the $\sqrt{\tau}$ ratios. This may be due to the disagreement already observed at large $\sqrt{\tau}$ in the structure function analysis [2].

Considering the same dilepton cross-section ratio as a function of the transverse momentum P_T allows one to study another type of nuclear effects. Figure 7 shows the corresponding plot. Both at 286 GeV/c and at 140 GeV/c the ratio is not constant but rises with increasing P_T. This rise is reflected in the difference $<P_T^2>_W - <P_T^2>_d$ equal to $0.15 \pm 0.03 \pm 0.03$ [GeV/c]^2 at 286 GeV/c and equal to $0.16 \pm 0.03 \pm 0.03$ [GeV/c]^2 at 140 GeV/c. An experimental effect due to the scattering of the incident pion in the long W target can be excluded since the data from 5.6 cm and 12 cm long W targets show no difference: $<P_T^2>_{W12} - <P_T^2>_{W5,6} = -0.01 \pm 0.03$ [GeV/c]^2. The increase of P_T must therefore originate within the nucleus; it can be explained by the rescattering of the incoming quarks within the nucleus. This behavior is in qualitative agreement with earlier predictions [22,10] which had been triggered by the observation [23] of an anomalous A-dependence of the hadronic production of high p_T particles. According to the same model, this $<P_T^2>$ difference should be twice as large for J/ψ production, because in addition to the incoming quarks the J/ψ, a strongly interacting particle, may also undergo rescattering. This is in fact observed in the NA10 data: $<P_T^2>_{J\psi W} - <P_T^2>_{J\psi d} = 0.29 \pm 0.02$ [GeV/c]^2. The dependence of the cross-section ratio as a function of P_T agrees rather well with a recent calculation [9] which takes into account these rescattering effects in the SGEA framework, the one already used to describe the P_T-distributions.

5. Angular distributions

Hadronic lepton pair production in the continuum proceeds mainly via a virtual, massive photon. The angular distribution from a vector particle decaying into two spin-1/2 particles can be written [24] as:

$$d\sigma/d\Omega \sim 1 + \lambda\cos^2\theta + \mu\sin2\theta\cos\phi + 1/2\,\nu\sin^2\theta\cos2\phi \,,$$

θ and ϕ are the polar and azimuthal angles of the spin-1/2 particles in the rest frame of the vector particle and λ,μ and ν are the parameters to be determined. In the case of the "naive" Drell-Yan model a virtual transverse photon is produced, implying $\lambda = 1$ and $\mu = \nu = 0$. Within QCD, non-perturbative [25] and perturbative [26] predictions have been made.

In perturbative QCD deviations of less than 5% from the $(1+\cos^2\theta)$ behaviour are predicted [26] in $0 \leq P_T \leq 3$ GeV/c, which means that values of ≤ 0.05 are expected for μ and ν. Figure 8 shows these predictions (dashed curve) together with the NA10 data at 286 GeV/c for W (full points) and d (open points) for the three parameters λ,μ,ν as a function of P_T. Here λ,μ,ν are defined in the Collins-Soper frame [27] which uses the bisector of beam and target momentum in the c.m. system as z-axis. While λ and μ are well compatible with 1 resp. 0, ν however is $\neq 0$ and rises with P_T. The relation $\lambda = 1 - 2\nu$, which is the analogue to the Callan-Gross relation [28] in DIS, is obviously violated. The d data (open points) demonstrate that the rise of ν cannot be explained as a nuclear effect. Figure 9, displaying the integrated values of the three parameters as a function of the three beam momenta, shows no energy dependence. This rise of the ν parameter with P_T has also been seen by other experiments [29,30]. The observed disagreement with the perturbative QCD predictions is **not** yet explained.

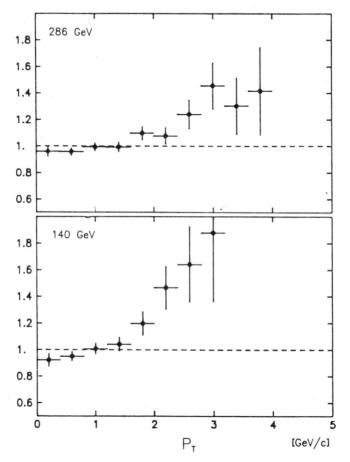

Figure 7. $\sigma(\pi^- W \to \mu^+\mu^- + X)/\sigma(\pi^- d \to \mu^+\mu^- + X)$ as a function of P_T.

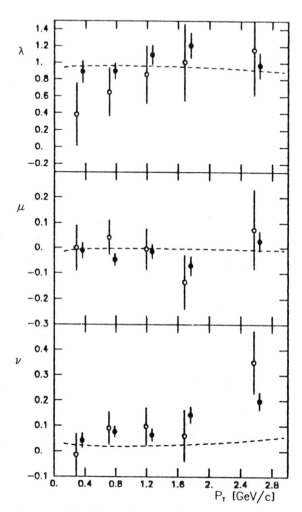

Figure 8. λ, μ and ν in the CS frame as a function of P_T at 286 GeV/c.

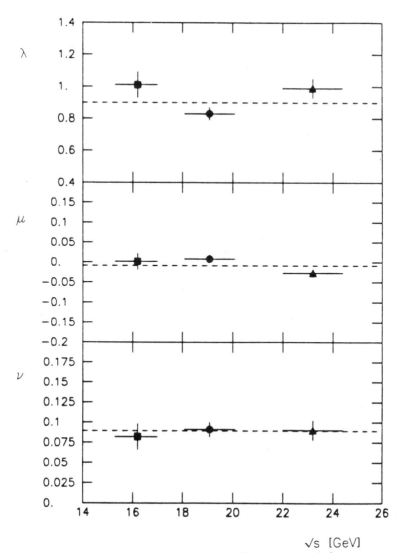

Figure 9. λ, μ and ν in the CS frame as a function of $\sqrt{s}$.

6. Conclusions on continuum production within QCD

- Scaling violations have been seen in a model independent way.
- The data are compatible with a value of Λ = 150 MeV.
- QCD fails to describe the cross-sections which are differential in P_T, in particular at high P_T.
- Two types of nuclear effects are observed: an EMC-type effect, and a rescattering effect of the incident quarks.
- The angular distributions agree with the predictions of perturbative QCD for two (λ, μ) of the three parameters, but an increase of ν with P_T − not predicted by QCD − is observed.

All that glisters is not gold!
Or QCD is not that shiny.

References

[1]   B. Betev[1a], J.J. Blaising[4b], P. Bordalo[3c], S. Borenstein[3d], Ph. Busson[3], L. Carotenuto[2], A. Degré[4b], A. Ereditato[2], S. Falciano[5e], K. Freudenreich[1f], E. Gorini[2], M. Grossmann — Handschin[5g], M. Guanziroli[5], D.A. Jensen[5h], P. Juillot[4], L. Kluberg[3], P. Lecomte[5], P. Le Coultre[5], R. Morand[4b], B. Mours[4b], A. Romana[3], R. Salmeron[3], P. Strolin[2], H. Suter[5], V.L. Telegdi[5], C. Vallée[3], J. Varela[3c], G. Viertel[5] and M. Winter[4].

[1] CERN, Geneva
[2] University of Naples, Naples
[3] Ecole Polytechnique, Palaiseau
[4] CRN and University, Strasbourg
[5] E.T.H., Zurich

a) Permanent address: INRNE, Sofia, Bulgaria
b) Now at LAPP, Annecy — le — Vieux, France
c) Now at CFMC — INIC, Lisbon, Portugal
d) Permanent address: C.U.N.Y., Jamaica, New York, USA
e) Now at INFN, Rome, Italy
f) Now at ETH, Zurich, Switzerland
g) Now at Univers. of ZH, Zurich, Switzerland
h) Permanent address: Univers. of Mass., Amherst, USA

[2]   B. Betev et al., Z. Phys. C28 (1985) 15.
[3]   H.B. Greenlee et al., Phys. Rev. Lett. 55 (1985) 1555.
[4]   J.J. Aubert et al., Phys. Lett. 123B (1983) 275.
[5]   K. Freudenreich, Proceedings of the 22nd Int. Conf. on High Energy Physics, Leipzig 1984, Volume I (1984) 276.
[6]   J.J. Blaising et al. (NA10), paper #10820 submitted to the 23rd Intern. Conference on High-Energy Physics, Berkeley (1986)
[7]   M. Winter, Ph.D. thesis, University of Strasbourg (1987), unpublished.
[8]   R. Voss, Rapporteur's talk at the Symposium on Leptons and Photons in Hamburg (1987).
[9]   P. Chiappetta and H.J. Pirner, Nucl. Phys. B291 (1987) 765.
[10]   G. T. Bodwin, S. J. Brodsky and G. P. Lepage, Phys. Rev. Lett. 47 (1981) 1799.
[11]   P. Bordalo et al. (NA10), paper #11185 submitted to the 23rd Intern. Conference on High-Energy Physics, Berkeley (1986)
[12]   K. Freudenreich, Proceedings of the 23rd Intern. Conference on High-Energy Physics, Berkeley (1986) 1113.
[13]   Y. Gabellini and J.L. Meunier, Nice preprint NTH 85/3.
[14]   P. Chiappetta et al., Nucl. Phys. B207 (1982) 251.
[15]   G. Altarelli et al., Nucl. Phys. B246 (1984) 12.
[16]   D.W. Duke and J.F. Owens, Phys. Rev. D30 (1984) 49, 943.
[17]   J. Badier et al., Phys. Lett. 104B (1981) 335.
[18]   M. Guanziroli, Ph.D. thesis, Fed. Inst. of Techn. Zürich (1986), unpublished.
[19]   P. Bordalo et al., Phys. Lett. 193B (1987) 368.
[20]   P. Bordalo et al., Phys. Lett. 193B (1987) 373.
[21]   A.C. Benvenuti et al., CERN — EP 87/13, subm. to Phys. Lett. B.
[22]   C. Michael and G. Wilk, Z. Phys. C10 (1981) 169.
[23]   Y.B. Hsiung et al., Phys. Rev. Lett. 55 (1985) 457 and references therein.
[24]   R.J. Oakes, Nuovo Cimento 44A (1966) 440.
[25]   E.L. Berger and S.J. Brodsky, Phys. Rev. Lett. 42 (1979) 940;
        E.L. Berger, Phys. Lett. 89B (1980) 241; Z. Phys. C4 (1980) 289;
        S.J. Brodsky et al., Proceedings of the Drell Yan Workshop, Fermilab (1982) 187.
[26]   P. Chiappetta and M. Le Bellac, Z. Phys. C32 (1986) 521.
[27]   J.C. Collins and D.E. Soper, Phys. Rev. D16 (1977) 2219.
[28]   C.G. Callan and D.J. Gross, Phys. Rev. Lett. 22 (1969) 156.
[29]   J. Badier et al., Z. Phys. C11 (1981) 11.
[30]   K.T. McDonald, Talk at the Advanced Research Workshop on QCD Hard Hadronic Processes, St. Croix, Virgin Islands (1987) and preprint DOE/ER/3072 — 43 (1987).

QCD IN THE LIMIT $x_F \to 1$ AS STUDIED IN THE REACTION $\pi^- N \to \mu^+ \mu^- X$

Kirk T. McDonald

Joseph Henry Laboratories
Princeton University
Princeton, NJ 08544

ABSTRACT

In Fermilab experiment E 615 we[1] have measured the production of high-mass muon pairs with 80- and 252-GeV pion beams. The relatively large momentum of quarks in pions allows us the study the QCD subprocess $q\bar{q} \to \mu^+ \mu^-$ in the kinematic limit $x_\pi^{\bar{q}} \to 1$. The data are consistent with many of the features of a higher-twist analysis, particulary the departure of the muon-pair angular distribution from the standard form $1 + \cos^2 \theta_t$. Analysis of J/ψ and ψ' production in the same data sample gives further evidence for interesting effects as $x_F \to 1$, which we attribute to the process $q\bar{q} \to c\bar{c}$.

INTRODUCTION

In a previous experiment[2] we were the first to measure the valence-quark distribution in the pion, via a Drell-Yan-model[3] analysis of the reaction $\pi N \to \mu^+ \mu^- X$. The valence-quark distribution is approximately

$$q_\pi^v(x_\pi) \sim 1 - x_\pi$$

for large x_π, the fraction of the pion's momentum carried by the quark. This is in contrast to the well-known result that in the proton

$$q_p^v(x_p) \sim (1 - x_p)^3$$

at large x_p. Thus with the pion the kinematic limit $x_\pi \to 1$ is rather accessible at presently available energies. In this limit the pion is nearly a single quark.

Theoretical impetus to pursue the limit $x_\pi \to 1$ came from a (so-called 'higher-twist') calculation by Berger and Brodsky[4] who consider that the fast quark in the pion receives its momentum via gluon exchange with other initial-state valence quarks. This leads to QCD corrections to the basic Drell-Yan subreaction $q\bar{q} \to \gamma^* \to \mu^+ \mu^-$ which manifest 2 qualitative features:

- a change in the angular distribution of the μ^+ in the muon-pair rest frame from $1 + \cos^2 \theta$ to $\sin^2 \theta$ as $x_\pi \to 1$;

- a component of the structure function $F_\pi(x_\pi)$ which is constant as $x_\pi \to 1$, with magnitude scaled by $k_T^2/M_{\mu\mu}^2$.

After suggestive evidence for the first feature was found in our previous experiment[5] we designed E 615 to study this physics in greater detail.

The apparatus is sketched in fig. 1 and described in detail elsewhere.[6] Data were collected with both 80- and 252-GeV π^- beams on a tungsten target. Results from the 80-GeV sample have been reported previously,[7] while here we give the first major results from the 252-GeV run.

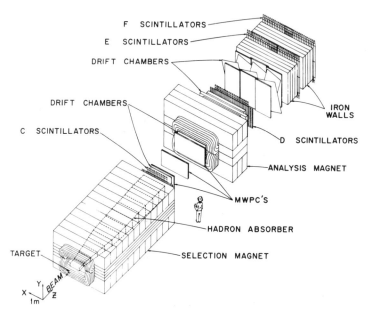

Figure 1. A view of the E 615 apparatus.

CONTINUUM ANALYSIS

First we present the pion and nucleon structure functions, for which the observed continuum muon-pair production cross section is analyzed according to

$$\frac{d\sigma}{dx_\pi dx_N} = \frac{1}{3}\sum_q \sigma_{q\bar{q}}\left[\bar{q}_\pi(x_\pi)q_N(x_N) + q_\pi(x_\pi)\bar{q}_N(x_N)\right],$$

where the annihilation cross sections for massless quarks are

$$\frac{d\sigma_{q\bar{q}}}{d\Omega} = \frac{\alpha^2 Q_q^2}{4M_{\mu\mu}^2}(1 + \cos^2\theta), \qquad \text{and} \qquad \sigma_{q\bar{q}} = \frac{4\pi\alpha^2 Q_q^2}{3M_{\mu\mu}^2},$$

and the functions $q(x)$ are the quark distributions introduced above. First-order QCD corrections in the next-to-leading-log approximation[8] modify the model calculations primarily in 2 ways:

$$\sigma_{\mu\mu} \to K\sigma_{\mu\mu} \quad \text{where } K \sim 2.5;$$

and

$$q(x) \to q(x, M_{\mu\mu}) \quad \text{with logarithmic dependence on } M.$$

In the continuum analysis the kinematic variables are related by

$$x_\pi x_N = \tau = \frac{M_{\mu\mu}^2}{s}, \qquad x_\pi - x_N = x_F = \frac{2p_L^*}{\sqrt{s}},$$

while $\sqrt{s} = 21.8$ for the 252-GeV data sample.

As our data are restricted to $x_\pi > 0.2$ the pair production is dominated by the valence quarks of the pion and we write

$$F_\pi = x_\pi \bar{u}^v(x_\pi) = x_\pi d^v(x_\pi).$$

The cross section then factorizes into

$$d\sigma_{\mu\mu} \sim F_\pi(x_\pi) G_N(x_N),$$

where

$$G_N = \frac{x_N}{9} \left[4\frac{Z}{A} u_p^v(x_N) + 4\left(1 - \frac{Z}{A}\right) d_p^v(x_N) + 5u_p^s(x_N) \right],$$

with A and Z labelling the atomic weight and atomic number of tungsten. Based on 36,000 pairs with mass above 4.05 GeV/c^2 we find the structure functions F_π and G_N shown in figs. 2 and 3.[9]

Our data for the nucleon structure function G_N, shown as solid circles in fig. 2, agree with the trend observed in the CERN NA3 experiment on muon-pair production,[10] as well as that inferred from deep-inelastic neutrino scattering.[11,12] Our result for the bin $0.04 < x_N < 0.06$ lies significantly above the trend of the higher x_N data. The pairs in this bin have mass very close to the cut at 4.05 GeV/c^2. A study is presently underway to relax this cut, which requires separation of pairs from decay of the J/ψ and ψ' resonances from those due to the Drell-Yan mechanism.

Our results for the pion structure function (solid circles in fig. 3) confirm and extend the trend seen in other experiments.[2,10,13] As our acceptance is largest at high x_π, in contrast to the other experiments, we can address the question of the behavior of F_π as $x_\pi \to 1$. For this we fit the data to the form

$$F_\pi \sim x_\pi^\alpha (1 - x_\pi)^\beta + \frac{2\gamma}{9M^2},$$

suggested by the Berger-Brodsky model.[4] The fitted parameters are given in the table:

Table 1

E_π	K	α	β	γ (GeV/c)^2	χ^2/d.o.f.
80 GeV	–	0.40 (fixed)	1.37 ± 0.07	0.50 ± 0.14	132/126
252 GeV	2.9 ± 0.14	0.37 ± 0.02	1.22 ± 0.03	0.60 ± 0.24	365/329
252 GeV†	2.8 ± 0.10	0.36 ± 0.02	1.15 ± 0.03	0.44 ± 0.16	348/331

† includes QCD evolution

The parameters for our 80-GeV sample are from ref. 6 in which $x_\pi > 0.4$ and it was not possible to measure the K factor. The parameters α, β, and γ in the third line of the table are actually functions of M. For this we follow the prescription of QCD

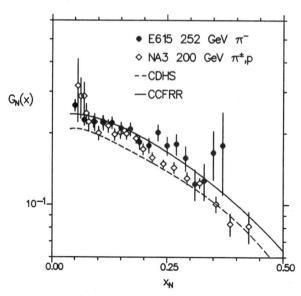

Figure 2. Measurements of the nucleon structure function, $G_N(x_N)$: • = this experiment; ◇ = NA3;[10] dashed curve = CDHS;[11] solid curve = CCFRR.[12]

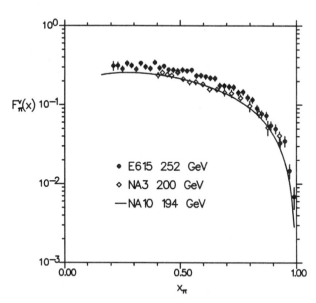

Figure 3. Measurements of the pion structure function, $F_\pi(x_\pi)$: • = this experiment; ◇ = NA3;[10] solid curve = NA10.[13]

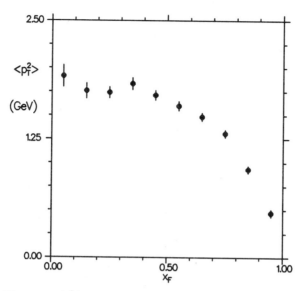

Figure 4. $\langle p_T^2 \rangle$ of the muon pairs as a function of x_F.

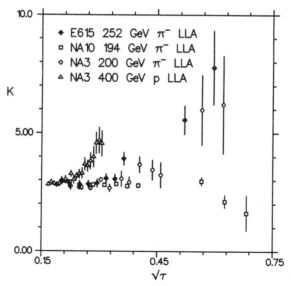

Figure 5. The K factor defined in the text as a function of $\sqrt{\tau} = M/\sqrt{s}$: $\diamond$ = this experiment; $\square$ = NA10;[13] $\circ$ = NA3[10].

evolution given by Buras and Gaemers;[14] the stated values are evaluated at $M = 4.5$ GeV/c².

Our evidence for a constant component of F_π as $x_\pi \to 1$, as measured by γ, is of 2.5 to 3 standard deviations in significance. The suggestion of Berger and Brodsky is that

$$F_\pi \sim (1 - x_\pi)^2 + \frac{2}{9} \frac{\langle k_T^2 \rangle}{M^2},$$

where $\langle k_T^2 \rangle$ is the square of the intrinsic transverse momentum of the quarks. Our values of $\beta \sim 1.2$ are not in close agreement with the model, while our nonzero values for γ lend support to it.

The square of the transverse momentum of the muon pairs is shown in fig. 4 as a function of x_F. It takes a clear drop towards the value 0.5 (GeV/c)² as $x_F \to 1$. A possible interpretation of this is that gluon emission, which leads to larger transverse momentum at moderate x_F, is no longer prominent at large x_F where the intrinsic transverse momentum dominates. If so, we find good consistency with the value of $\gamma \sim 0.5$ (GeV/c)² determined in the structure function analysis and interpreted in the Berger-Brodsky model.

In the structure function analysis we only consider pairs with $4.05 < M < 8.55$ GeV/c² to avoid contamination from resonances. Above the Υ family, we have 156 events with $M_{\mu\mu} > 11$ GeV/c². In fig. 5 we compare these high-mass events with the trend of other experiments,[10,13] plotting the K factor as a function of $\sqrt{\tau} \equiv M/\sqrt{s}$. For this we define the K factor as $d\sigma(\text{observed})/d\sqrt{\tau}$ divided by $d\sigma(\text{Drell-Yan})/d\sqrt{\tau}$, including QCD evolution in the structure functions but not in the normalization of the Drell-Yan calculation. Our results indicate a rise in the K factor with $\sqrt{\tau}$, in contrast to the report of the CERN NA10 experiment.[13] The issue is complicated by the fact that the highest mass events come from the high-energy tail of the beam spectrum, or from nucleons with Fermi motion towards the beam particle. In particular, all Drell-Yan experiments attempt to correct the results for Fermi motion in slightly varying ways before comparing with calculations (while in the deep-inelastic-scattering experiments no such attempt is made).

Figure 6 shows the results of a fit to the angular distribution of the μ^+ in the muon-pair rest frame, using the t-channel coordinate system in which the incident pion lies along the z-axis. The most general allowed form of the angular distribution is[15]

$$\frac{d\sigma}{d\Omega} \sim 1 + \lambda \cos^2 \theta + \mu \sin 2\theta \cos \phi + \frac{\nu}{2} \sin^2 \theta \cos 2\phi.$$

In the naïve Drell-Yan model with massless quarks and no transverse momentum we would have $\lambda = 1$ and $\mu = \nu = 0$. Standard QCD corrections lead to the prediction that μ is small while ν can be nonzero, but $1 - \lambda = 2\nu$.[16] The Berger-Brodsky model notes that near $x_\pi = 1$ the quark in the pion must be off-shell, leading to the prediction

$$\frac{d\sigma}{d\Omega} \sim (1 - x_\pi)^2(1 + \cos^2 \theta) + \frac{4}{9} \frac{\langle k_T^2 \rangle}{M^2} \sin^2 \theta + \frac{2}{3} \sqrt{\frac{\langle k_T^2 \rangle}{M^2}} (1 - x_\pi) \sin 2\theta \cos \phi.$$

The solid curves in fig. 6 are only to guide the eye, while the dashed curves are for the Berger-Brodsky model. At moderate x_π the sum rule $1 - \lambda = 2\nu$ does not seem

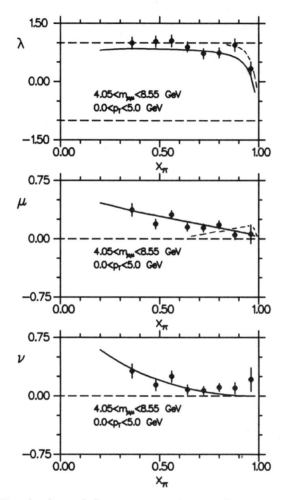

Figure 6. Fitted values of the parameters λ, μ, and ν pertaining to the angular distribution of the μ^+ about the direction of the incident pion in the muon-pair rest frame. The solid curves are only to guide the eye, while the dashed curves are from the Berger-Brodsky model.[4]

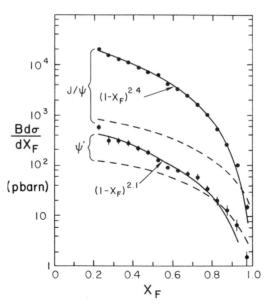

Figure 7. The cross section times branching ratio per nucleon for production of the J/ψ and ψ' resonances as a function of x_F. The solid curves are fits to the data, while the dashed curves are from the $q\bar{q} \to c\bar{c}$ annihilation model.

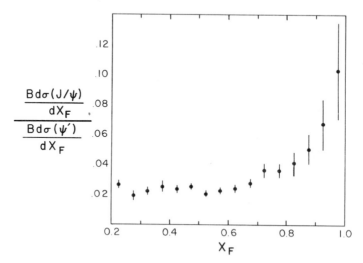

Figure 8. The ratio of the cross section (times branching ratio) for ψ' to that for J/ψ production. In the $q\bar{q}$ annihilation model this ratio has the value 0.15.

well satisfied, while at high x_τ the data are in reasonable agreement with the Berger-Brodsky model. Both μ and ν are observed to increase with transverse momentum, and are consistent with a linear dependence on this variable. Where our data overlap with those of the NA10 experiment[17] good agreement is found.

Our data on continuum muon-pair production thus broadly supports the view that a high-x_τ quark has gotten its momentum from a gluon which was emitted by another initial-state quark. This is a non-asymptotic QCD effect in that the gluon is not free, yet a well-defined calculation is possible. This contrasts with the typical application of QCD: calculations are possible only at asymptotic energies where no data exist, while the phenomena accessible at present energies cannot usually be calculated in the theory.

RESONANCE ANALYSIS

QCD effects in the reaction $\pi N \to \mu^+\mu^- X$ occur in the initial state and so are not specifically tied to the Drell-Yan mechanism. It is provocative to examine our data on resonance production at large x for additional evidence of the Berger-Brodsky effect. A first result has already been reported[18] in which we found that the decay angular distribution for $J/\psi \to \mu^+\mu^-$, while typically flat, approaches $\sin^2 \theta$ as $x_F \to 1$. If the J/ψ particles are produced via a $q\bar{q}$ annihilation of intial-state quarks this would indeed appear to be consistent with the view of Berger and Brodsky. However it is usually considered that the majority of J/ψ's are produced by gg interactions. Here we give new results that suggest the $q\bar{q}$ annihilation mechanism may dominate at large x_F.†

Figure 7 shows our measurement of the cross section times branching ratio for the J/ψ and ψ' particles as a function of x_F. The dashed curves show the calculated cross sections supposing the reaction $q\bar{q} \to c\bar{c}$ can be described by a Breit-Wigner resonance. The calculated rate includes a K factor of 2.8 as observed to hold for our continuum data. This model falls below the data at moderate x_F, but agrees well at very large x_F. Figure 8 shows the ratio of ψ' to J/ψ production as a function of x_F. The $q\bar{q}$ annihilation model predicts a value of 0.15 for this ratio, which is approached by our data at large x_F. Thus resonance production by gluon fusion appears to die out with x_F, allowing $q\bar{q}$ annihilation to appear and to manifest QCD effects like those seen in continuum production at large x_F.

Our data provide several illustrations that the accessibilty of the limit $x_F \to 1$ in πN interactions yields striking evidence for non-asymptotic, but calculable QCD effects at Fermilab energies.

REFERENCES

1. The members of the E 615 collaboration are: University of Chicago: C.E. Adolph-sen, J.P. Alexander, K.J. Anderson, J.S. Conway, J.G. Heinrich, K.W. Merritt, J.E. Pilcher, and A. Possoz; Iowa State University: E.I. Rosenberg and D.T. Simp-son; Princeton University: C. Biino, J.F. Greenhalgh, P. Kaaret, W.C. Louis, K.T. McDonald, S. Palestini, D.P. Russell, F.C. Shoemaker, and A.J.S. Smith.

† For the resonance analysis we define $x_F = p_L^\star / p_{L,\mathrm{max}}^\star$, which has a maximum value of 1. This contrasts with the definition used in the continuum analysis for which the maximum is $1 - \tau$.

2. C.B. Newman *et al.*, Determination of the Pion Structure Function from Muon-Pair Production, Phys. Rev. Lett. **42**:951 (1979).

3. S.D. Drell and T.-M. Yan, Massive Lepton-Pair Production in Hadron-Hadron Collisions at High Energies, Phys. Rev. Lett. **25**:316 (1970).

4. E.L. Berger and S.J. Brodsky, Quark Structure Functions of Mesons and the Drell-Yan Process, Phys. Rev. Lett. **42**:940 (1979); E.L. Berger, Higher-Twist Effects in QCD, Deep Inelastic Scattering, and the Drell-Yan Process, Z. Phys. C **4**:289 (1980).

5. K.J. Anderson *et al.*, Evidence for Longitudinal Photon Polarization in Muon-Pair Production by Pions, Phys. Rev. Lett. **43**:1219 (1979).

6. C. Biino *et al.*, An Apparatus to Measure the Structure of the Pion, Nucl. Instr. Meth. **A243**:323 (1986).

7. S. Palestini *et al.*, Pion Structure as Observed in the Reaction $\pi^- N \to \mu^+ \mu^- X$ at 80 GeV/c, Phys. Rev. Lett. **55**:2649 (1985).

8. G. Altarelli, R.K. Ellis, and G. Martinelli, Large Perturbative Corrections to the Drell-Yan Process in QCD, Nucl. Phys. **B157**:461 (1979); J. Kubar *et al.*, QCD Corrections to the Drell-Yan Mechanism and the Pion Structure Function, Nucl, Phys. **B175**:251 (1980).

9. For details of the continuum analysis see: J.S. Conway, Ph.D. Dissertation, U. of Chicago (1987).

10. J. Badier *et al.*, Experimental Determination of the π Meson Structure Functions by the Drell-Yan Mechanism, Z. Phys. C **18**:281 (1983).

11. H. Abramovicz *et al.*, Neutrino and Antineutrine Charged-Current Inclusive Scattering in Iron in the Energy Range $20 < E_\nu < 300$ GeV, Z. Phys. C **17**:283 (1983).

12. D.B. McFarlane *et al.*, Nucleon Structure Functions from High Energy Neutrino Interactions with Iron and QCD Results, Z. Phys. C **26**:1 (1984).

13. B. Betev *et al.*, Observation of Anomalous Scaling Violation in Muon Pair Production by 194 GeV/c π^--Tungsten Interactions, Z. Phys. C **28**:15 (1985).

14. A.J. Buras and K.J.F. Gaemers, Simple Parametrization of Parton Distributions with Q^2 Dependence Given by Asymptotic Freedom, Nucl. Phys. **B132**:249 (1978).

15. R.J. Oakes, Muon Pair Production in Strong Interactions, Nuovo Cimento **44A**:440 (1966).

16. C.S. Lam and W.-K. Tung, Parton-Model Relation without Quantum-Chromdynamic Modifications in Lepton Pair Production, Phys. Rev. D **21**:2712 (1980).

17. S. Falciano *et al.*, Angular Distribution of Muon Pairs Produced by 194 GeV/c Negative Pions, Z. Phys. C **31**:513 (1986).

18. C. Biino *et al.*, J/ψ Longitudinal Polarization from πN Interactions, Phys. Rev. Lett. **58**:2523 (1987).

SUMMARY OF DIRECT LEPTON SESSION AND ROUND TABLE

J.C. Collins

Physics Department, Illinois Institute of Technology
Chicago, IL 60616, USA

K. Freudenreich

IHEP, Federal Institute of Technology
CH-8093 Zurich, Switzerland

ABSTRACT

This session concerned the Drell-Yan process. We summarize and discuss the main issues that were covered. Most of these concern QCD questions. There are some differences between the E615 and the NA10 cross sections for Drell-Yan at large τ. Experiment and theory disagree on the angular distribution of the muons. In the round table discussion, most of the time went to discussing the question of whether experiments should or should not make a correction for Fermi motion before presenting their data.

1. INTRODUCTION

This session was concerned with the Drell-Yan process, mostly from the experimental point of view. (This process is the production in hadron-hadron collisions of any high-mass quantum that has no strong interactions. Typical cases are the production of lepton pairs via decay of a virtual photon and production of W- and Z-bosons.)

There are several reasons for interest in the Drell-Yan process. First of all, the process provides important tests of QCD. Furthermore, it allows measurements of parton distribution functions especially for cases inaccessible to lepton scattering experiments. Then, in the weak interaction sector, if we assume the validity of the standard QCD calculations of the Drell-Yan cross section, then we can use production and decay of Ws and Zs to provide measurements of parameters of the electroweak Lagrangian and ultimately to provide tests of the Weinberg-Salam theory. (For example, limits are set on the mass of the top quark.) Finally, the generalized Drell-Yan process provides a way of searching for new particles.

Data presented in this session covered all these aspects. The fixed target experiments NA10 and E615 concentrated on Drell-Yan as a probe of QCD. Their latest data improves on the quality of earlier results on π-induced muon production, and it is clear that the Drell-Yan process provides a gold mine of information and, in principle, of tests of the theory. Particularly noteworthy were data on the angular distribution of the muon, presented by both experiments; these differed substantially from the QCD predictions.

67

Although the collider experiments UA1, UA2 and CDF provided data on QCD, it was clear that their main thrust was to use Drell-Yan as a probe of the electroweak sector of the standard model and of what lies beyond. We saw measurements of parameters of the model, including limits on the mass of the top quark and on the number of neutrino species. Limits on exotic particles were presented.

The bottom line is that if the results from NA10 and E615 are taken at face value, then QCD is wrong. The results from UA1, UA2 and CDF show no deviation from the standard model either in the strong interaction part or in the weak interaction sector.

There was one theory talk scheduled, but it was not presented.

The round table discussion mostly focused on the question of whether corrections for Fermi motion should be applied to data taken on nuclear targets.

2. TERMINOLOGY

There are a couple of terminological problems that often confuse discussions on the Drell-Yan process. These concern the meanings of "structure function" and of "K-factor".

The cross section for semi-leptonic processes such as leptoproduction and Drell-Yan are expressed in terms of correlation functions of two electromagnetic (or weak) currents. The correlation functions can be expressed as linear combinations of basic tensors. Strictly speaking, structure functions are defined as the coefficients of these tensors. In both the parton model and real QCD, the structure functions are predicted in terms of parton densities ("parton distribution functions").

The parton model formulae for deep inelastic scattering give the structure functions as simple combinations of the parton densities, so that it has become common to refer to "structure function" when one really means "parton distribution". This usage is unfortunate for at least two reasons, beyond the obvious one of using one term for two concepts. First, it obscures the fact that there are higher order corrections to the formula for the structure functions – these certainly matter when one uses parton distributions defined for example in the $\overline{MS}$ scheme, as is common nowadays. Secondly, the Drell-Yan process has its own set of structure functions, of which there are four in the electromagnetic case, and one needs to refer to these in discussions of the angular distributions. In this paper we will try to use the correct terminology.

These is also confusion as to whether the parton distributions are number densities or momentum densities. We will use number densities $f_{p/H}(x)$, for which the momentum sum rule reads:

$$\sum_{\text{parton species } p} \int_0^1 dx\, x f_p(x) = 1. \qquad (2.1)$$

The momentum densities are $x f(x)$. Momentum densities are used by both the NA10 and E615 groups in their contributions to this volume.

The term "K-factor" refers to a ratio of cross sections for the same process. Originally it was the ratio of the experimental Drell-Yan cross section to the parton model prediction. At present, the "K-factor" is commonly defined as the ratio between the measured cross section and the QCD approximation used. (The approximation may or may not include higher order corrections.)

We would like to make an appeal for a more consistent terminology in the literature, so that discussions are not confused by misunderstandings.

3. QCD PROPERTIES

Perhaps the most striking result reported in the session was the apparent failure of perturbative QCD calculations to predict the angular distribution of the dimuons observed in the E615 and NA10 experiments. There was also some disagreement reported between theory and experiment as regards the transverse momentum distribution.

We should first recall that perturbative QCD calculations for the cross section for the Drell-Yan process integrated over muon angle have already provided significant and successful tests of the theory. The experimental cross sections are higher than the results of theoretical calculations that use the Born graphs for the hard scattering, often by a 'K-factor' of about two at fixed target energies. However, it was soon realized that there are large higher order corrections in the theory, which in fact today bring theory and experiment into reasonable agreement. The theoretical calculations apply both to the cross section integrated over transverse momentum of the lepton pair and to the cross section when the transverse momentum is comparable with the dimuon mass.

The standard kinds of factorization theorem can be extended to deal with the cross section when the transverse momentum is much less than the dimuon mass. This involves 'resumming large logarithms'. Again considerable success has resulted.

3.1 Angular distributions

Good data has appeared on the angular distribution of the muons. The basic parton model result is that this distribution should be the $1 + \cos^2\theta$ distribution obtained from the annihilation of massless quarks. Early in the development of QCD, it was realized that higher order graphs provided calculable deviations from this result [1]. Experiments are now accurate enough to test these predictions. According to the results presented here by the NA10 collaboration, these predictions are in serious disagreement with experiment. This confirms earlier results by the same experiment. Results presented by E615 give similar values for the angular distribution to those from NA10. (But the E615 results are not directly comparable with those of NA10, since the NA10 results are reported here in the Collins-Soper frame, whereas those from E615 are in the Gottfried-Jackson frame.)

The angular distribution can be written in its most general form as

$$\frac{dN}{d\Omega} \propto 1 + \lambda \cos^2\theta + \mu \sin 2\theta \, \cos\phi + \frac{1}{2}\nu \sin^2\theta \, \cos 2\phi. \qquad (3.1)$$

In this equation we assumed production via a virtual photon with only electromagnetic interactions. We show in Figs. 1 and 2 the latest results from NA10 and E615 for the coefficients in eq. (3.1). Clearly ν deviates substantially from its parton model value of zero.

The QCD predictions in the lowest nontrivial order $[O(\alpha_s)]$ have been known for a long time [1]. In general, calculations such as for the coefficients λ, μ and ν come from complicated formulae that must be integrated numerically over the allowed range of parton kinematic variables. But there are two dramatic simplifications in our case. First there is the relation $2\nu = 1 - \lambda$, which is satisfied up to order α_s for both of the contributing processes ($q + \bar{q} \to$ gluon $+ \gamma^*$ and $q +$ gluon $\to q + \gamma^*$). A second relation holds for the $q\bar{q}$ process, which should dominate in the case of a pion beam. This relation is

$$\lambda = \frac{2\rho^2}{\left(1 + \frac{3}{2}\rho^2\right)}, \quad \nu = \frac{\rho^2}{\left(1 + \frac{3}{2}\rho^2\right)}, \qquad (3.2)$$

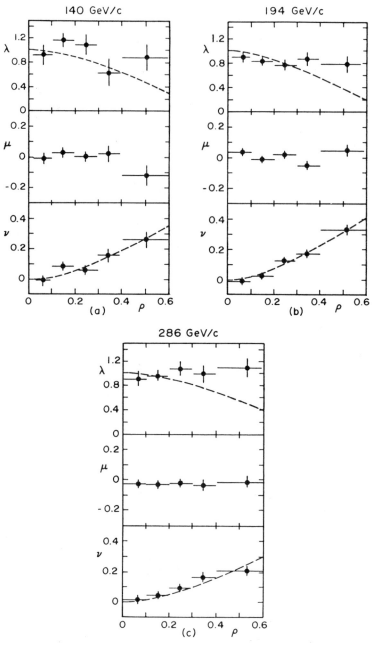

Fig. 1. Data from NA10 for angular distribution, as a function of $\rho \equiv q_T/Q$. The dashed line is a fit of the form $\nu(\rho) = \epsilon\rho^2/(1 + \epsilon\rho^2)$, $\lambda = 1 - 2\nu$. The values of ϵ are 1.46 ± 0.27, 1.84 ± 0.17 and 1.17 ± 0.17 at 140, 194 and 286 GeV.

provided that we use the CS frame.

It is not known whether either of these relations survive to higher order.

The relations (3.2) are moderately well satisfied by the data for ν, but not very well for λ, which stays too close to 1. But the determination of λ involves (for NA10 at least) much larger acceptance corrections than for ν, so one might choose to put more trust in the data for ν.

Of course, even higher order corrections can change the theoretical predictions. Unfortunately, when $q_T \to 0$, these corrections are infinite, order by order, so they cannot be ignored. For the cross section integrated over angle, there is a systematic formalism for reorganizing the perturbation series [2,3] – "soft gluon resummation" – so the divergences as $q_T \to 0$ are not by themselves a serious problem. Chiappetta and Le Bellac [4] have made calculations appropriate to the experiments we are discussing.

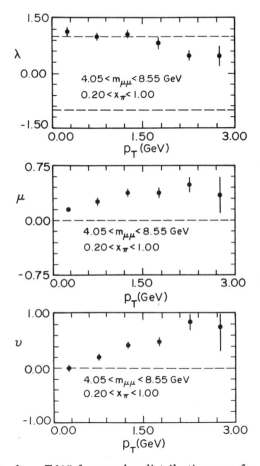

Fig. 2. Data from E615 for angular distribution as a function of q_T.

Typical of their results is that if we set $q_T = 3 \, \text{GeV}$ and $Q = 5 \, \text{GeV}$ then the cross section is higher by more than an order of magnitude than the parton model result. These values are in the region where the data shown in figs. 1 and 2 allows a useful test of eq. (3.2).

But for the angular distribution, the present state of the formalism is that we do not know how to resum the soft gluon corrections to the $1 + \cos^2 \theta$ distribution in the way that we do for the cross section integrated over angle. So what Chiappetta and Le Bellac have done is to observe that at low enough transverse momentum there is only the $1 + \cos^2 \theta$ term. They express the angular distribution in terms of the four independent structure functions (in the strict sense) for the Drell-Yan process. Let us use the set W_L, W_T, W_Δ and $W_{\Delta\Delta}$ defined in [5]. Then we have, for example:

$$\nu = \frac{2W_{\Delta\Delta}}{W_T + W_L}. \tag{3.3}$$

Now the leading order calculation is used for $W_{\Delta\Delta}$, but the full soft gluon formula is used for the denominator (which is essentially the cross section integrated over angle). In principle this is sensible: when q_T is large, the soft gluon formula is well approximated by low order graphs, and when q_T is small, coefficients like ν are small, so errors are irrelevant. 'Small' and 'large' are to be interpreted as compared with Q.

In our example earlier, $q_T = 3\,\mathrm{GeV}$ is not small compared with $Q = 6\,\mathrm{GeV}$, but the denominator of (3.3) has been increased by an order of magnitude by the soft gluon resummation, according to Chiappetta and Le Bellac. The size of ν as compared to the naïve prediction (3.2) has been reduced by this same order of magnitude. This results in the discrepancy between theory and experiment for ν.

If the Chiappetta and Le Bellac method is indeed the correct one, then QCD has failed an important test: We should be prepared to discard the theory. However, as the foregoing discussion should make clear, their application of the theory is on dangerous ground. One must improve the theoretical calculations. It may not be absolutely necessary to construct the full soft-gluon resummation procedure, for the large terms in the perturbation series that are resummed form an exponentially convergent series. This means that it would be sufficient to calculate a few higher orders. Even so this is a task that is almost infinitely easier said than done. One immediate possibility would be to examine more closely the existing order α_s^2 calculations [6] to see what clues they give for the angular distribution.

Better yet would be to analyze the general properties of the perturbation series to see what kind of generalization of the soft gluon techniques should be used for λ, μ and ν.

3.2 Scaling violations

NA10 nicely checked the predicted scaling violations. Because they ran at different energies, they could plot cross sections against $\sqrt{\tau} \equiv Q/\sqrt{s}$. This is the correct way to test scaling independently of any knowledge of parton distributions and of higher order corrections.

They used these results to compute a value of Λ_{QCD} around 150 MeV, in agreement with other determinations.

3.3 Parton Distribution in Pion; Absolute Values of Cross Section

Both NA10 and E615 used their tungsten data (from targets of different length) to extract the distribution functions of the valence quarks in the pion. The NA10 analysis was based on a continuum sample of 45900 events collected at 140 GeV/c, 150000 events at 194 GeV/c and 84400 events at 286 GeV/c. The E615 sample consisted of 4000 events from 80 GeV/c and 36000 events from 252 GeV/c. While NA10 fit their data from the three energies simultaneously, E615 fit their data at 80 and 252 GeV separately. The experimental scaling cross-section (i.e. $sd\sigma/d\tau$) from NA10 at 194 GeV/c agrees with the corresponding one from E615 at 252 GeV in the region below the Upsilon. Above the Upsilon, at large τ, they disagree with each other.

This disagreement led, after the conference, to a thorough study of the origin of this disagreement which, at the writing of this summary, is not yet finished. The determination of the parton distribution is, however, not affected by this disagreement because NA10 cut away the high τ region in their fit and E615 has too few events in that region (156 events with $M > 11\,\text{GeV}$) to influence the determination of the parton distribution.

The starting point is the standard formula for the cross section:

$$\frac{d\sigma}{dx_\pi dx_N} = K\{\frac{1}{3}\sum_q \sigma_{q\bar{q}}[f_{q/\pi}f_{\bar{q}/N} + f_{\bar{q}/\pi}f_{q/N}] + \text{possible higher order corrections}\}.$$

$$(3.4)$$

Here K, as usual, is the correction factor between the full cross section and its approximation by a truncated perturbation expansion.

In their analysis NA10 include the order α_s corrections and the soft gluon exponentiation in the form given by Chiappetta and Le Bellac [4], while E615 use the Born cross-section including structure functions evolving with Q^2. Both groups fit the valence quark distributions as well in the pion as in the nucleon. They use external inputs for the sea and gluon distributions.

Because NA10 fit data from three energies simultaneously, three separate K-factors are used, mainly to avoid the influence of fluctuations in the experimental normalization from one energy to another on the χ^2 of the global fit.

NA10 and E615 use for the pion sea and gluon momentum fraction the parameters from an earlier global fit to NA3 data obtained with proton, π^- and π^+ beams. (Today these 1750 π^+ induced events still constitute the largest π^+ sample!) The shape parameter of the gluon distribution in the pion was determined by NA10 from their Upsilon data. For the sea and gluon distributions Buras-Gaemers parameterizations both in the pion and nucleon are used. For the valence distributions NA10 assumes [7]:

$$xf_V^{NA10}(x, Q_0) = [N_V(x^{0.5} - x) + A_\pi \alpha_s^{-d_0}x](1 - x)^{\beta_0} \qquad (3.5)$$

at the starting point $Q_0 = 5\,\text{GeV}$. This parameterization is used for both the valence distribution in the pion and for the up-quark distribution in the proton. The ratio of up to down in the proton is assumed to be $0.55(1 - x)$.

The power-law at $x \to 0$ is a common assumption but may not be correct (the region of small x is even less well explored in Drell-Yan than in deeply inelastic scattering, because of the presence of the J/ψ and of the $c - \bar{c}$ background at low masses). Since the data are far from $x = 0$, this probably does not matter. However, the normalization factor N_V is fixed by the quark number sum rule. Thus if the assumption about the $x \to 0$ behavior is wrong, then the normalization and therefore the determination of the K-factor will not be correct.

The theory of J/ψ or upsilon production is yet not in good shape; it is certainly not correct to suppose that it yields reliable predictions. Therefore the use of J/ψ or upsilon production to obtain the gluon distribution is a source of unknown systematic error.

The E615 group assumes a valence distribution in the pion proportional to

$$xf_{V/\pi}^{E615} \propto x^\alpha(1 - x)^\beta. \qquad (3.6)$$

They add on a term on to allow for the Berger-Brodsky [8] mechanism. Although they do not say so, they presumably adjust the normalization to get the quark number sum rule. They do not present a parameterization for the valence distribution in the proton, but only a graph. Without fixing the normalization, it is obviously impossible to define the K factor.

NA10 fit the distributions as parameterized above, with the proper evolution with Q^2. As already mentioned, they fit an independent K factor at each energy. The E615 data at 225 GeV are used to fit the pion distribution and a K factor.

The NA10 data are fitted well for $\tau < 0.5$ with all three K factors being close to one. At large τ, however, the data fall substantially below the fitted cross-section. This region is eliminated in a second fit by a cut on $x_N < 0.40$. To minimize the influence of the pion sea parameters on the systematic error in K, NA10 cut their data also at $x_\pi > 0.35$ in this second fit. The parameters of the pion distribution obtained after these cuts are not different from those of the first global fit. The K factors also remain close to one. The overall systematic uncertainty in K is now, however, greatly reduced. It is of the order of 20% which is the same order of magnitude as the contribution to the theoretical cross-section for the soft gluon emission corrections. The main systematic uncertainty remaining is due to Λ. The E615 data obtain a good fit, but no systematic errors are given; in particular no corrections for reinteractions in their long W target had been applied at the time of the conference.

Since E615 do not include any higher order corrections in their fit, they measure a large K factor. Therefore their fit should not be regarded as a full comparison of their data to QCD, but as a convenient parameterization. They do present the K factor as a function of τ.

Since the two experiments use different approximations to QCD in their theoretical cross-sections, their results are not directly comparable. Nevertheless, both experiments find beta parameters for the pion distribution close to one and incompatible with two. Clearly, the determination of the quark distributions and the comparison of QCD to the data cannot be disentangled in such fits. Because of missing QCD predictions for the relevant fitted parameters and of the large systematic uncertainties on K the fact that K is now close to one is only a weak test of QCD.

3.4 Transverse momenta

NA10 also compared their transverse momentum distributions with those calculated by Chiappetta and Pirner [9]. The data are in disagreement with the calculations by up to a factor of two, as shown in fig. 3. The significance of this is still moot. It is known that soft gluon calculations in the relevant range of q_T are sensitive to non-perturbative properties, which are included in a properly formulated version of the soft gluon formula [3]. The non-perturbative effects include a characteristic energy dependence; there is certainly energy dependence in the difference between theory and experiment in fig. 3.

Another problem is the matching of the low and high q_T parts of the calculation. In principle this is no problem; the formalism takes care of it. But in practice, the large size of K-factor gives a problem. The low q_T calculation gives a sensible approximation to the K-factor, but the present high q_T calculation is merely a lowest-order calculation. The data confirm the need for a significant K-factor at large q_T, comparable to that for the cross section integrated over q_T. The only sensible way of comparing theory and experiment is to compute the cross-section to at least the next order. Only the non-singlet part has so far been calculated: that indeed predicts a K-factor of roughly the standard size.

Clearly more theoretical work is necessary to match the quality of the data to the quality of the calculations.

74

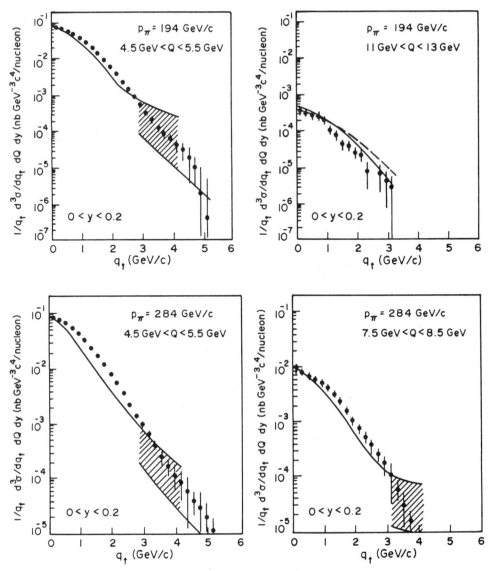

Fig. 3. Comparison of theory and (NA10) data for q_T distribution.

3.5 EMC Effect

NA10 also confirmed an EMC-like effect by comparing their data for the cross section integrated over q_T on deuterium and tungsten targets. It is a firm QCD prediction that the existence of EMC effect in lepto-production implies one for Drell-Yan (modulo the differences between valence and sea distributions, which should not be important with a pion beam).

Furthermore there is a small change in the q_T distribution between the light and heavy targets. This is certainly permitted by the theory, since the distribution of parton intrinsic transverse momentum can depend on the parent hadron. Nuclear models predict a dependence; but there is no prediction from purely perturbative methods.

The effect seen by NA10 has the appropriate sign and magnitude, as calculated by Chiappetta and Pirner [9]. (The calculation by Bodwin, Brodsky and Lepage [10] is phrased in terms of scattering of an incident quark by a nucleus, but it can be rephrased in terms of an effective intrinsic transverse momentum.)

3.6 $x_F \to 1$

Another topic was to measure the behavior of the angular distribution as $x_F \to 1$. Results were reported primarily by E615. In that region, the standard Drell-Yan cross section falls to zero, because the density of quarks in the beam goes to zero. Berger and Brodsky [8] showed that this permits a process that is normally higher twist to become the leading part of the cross section. This is where a quark out of the target scatters off the whole of the beam hadron instead of just off one of its constituents. A key prediction is that the coefficient λ in the angular distribution should go to -1 as $x_F \to 1$. The new data reconfirm this. There is disagreement with the detailed numerical predictions of Berger and Brodsky; but since the precise numbers depend on models for the transverse momentum dependence, a change in the model could improve the agreement.

The data from NA10 are not in disagreement with the data from E615, but do not go to sufficiently high x_F to test the characteristic features of the Berger-Brodsky model.

3.7 Summary

Clearly Drell-Yan experiments are providing a gold mine of QCD tests and measurements. The Q^2 is low by current standards in jet physics, but because there is no hadronization of jets to worry about, one does not need to pay the premium of an enormous energy to perform good quantitative tests.

4. ELECTROWEAK PROCESSES AND STANDARD MODEL EXTENSIONS

The UA1, UA2 and CDF collaborations measured properties of the W and Z bosons in several decay channels: $W \to e + \nu$, $\to \mu + \nu$, $\to \tau + \nu$(UA1), $Z \to e^+e^-, \mu^+\mu^-$. UA2 were also able to see the decays of W and Z to two jets.

The production characteristics of these bosons appears to be in agreement with the standard model, when one assumes that the production is the usual QCD Drell-Yan process with the standard couplings of quarks to the W and Z. Thus the cross section, the q_T distribution and the angular distribution of the decay products all appears to be about right. However, the UA2 results for the production cross sections appears high, by up to a factor of two. However, they did not regard this as significant, in view of the "large experimental and theoretical systematic uncertainties". UA1 obtained cross sections lower than the most extreme UA2 value (σB for $Z \to e^+e^-$ was measured as 116 pb $\pm$ 39 $\pm$ 11 by UA2, but as 42 pb by UA1).

Given this agreement it is possible to set limits on breakdown of the Weinberg-Salam model. Low energy weak interaction data provide a measurement of $\sin^2 \theta_W$ and G_F, and hence a prediction of M_W and M_Z. Agreement with the measurements at the colliders is satisfactory. The collider data are not accurate enough to test the one-loop corrections.

Large radiative corrections can arise if the mass of the Higgs or the mass of the top is large (or both). UA2 showed a comparison of the mass difference $M_W - M_Z$, which is measured particularly accurately, with the corrections due to loops with top quarks and Higgs bosons. There is not much sensitivity if the Higgs mass is less than a TeV; but the top mass must clearly be less than a few hundred GeV. UA1 preferred to talk about the future capability — a limit of below 150 GeV — when the ACOL is in operation.

From the widths of the W and Z could be set limits on the number of light neutrinos and on the top mass. The limit on the number of neutrinos is 3 if $m_T > 74$ GeV and 7 if one makes no restriction on the top mass.

Also presented were limits on new particles like extra Zs and supersymmetric particles that decay to leptons.

For some time, UA1 has shown a couple of events with a high p_T W in association with two high p_T jets. The invariant mass of the two jets is near the W mass. At first sight the probability for the obvious conventional background – the normal QCD process of production of jets in association with a single W – is far too low. However, as the UA1 plot of this background made clear, the measurement error on the large p_T of the W is about a factor of 2, and the data are in fact consistent with this background.

5. ROUND TABLE DISCUSSION

This discussion was chaired by Collins and Freudenreich.

5.1 Nuclear Effects

Experiments measuring the Drell-Yan reaction on heavy nuclei typically present their data after corrections for Fermi motion. The question was raised by Bjorken as to whether this is a good idea.

One motivation for the question was the substantial disagreement between NA10 and E615 on the cross section above $\sqrt{\tau}$ of about 0.5. The possibility is that the discrepancy is due to a difference in the way Fermi motion was treated. (The experimentalists discount this possibility.)

Another motivation was that besides Fermi motion there is also the EMC effect to change the cross sections on nuclei relative to nucleons; corrections for the EMC effect are not made.

Three points of view were taken during the discussion:

a. The theorists appeared unanimous that Fermi motion corrections should not be made. The concept of the parton distribution within a nucleus makes perfect sense; and the QCD prediction is that the parton distributions used are the same in any process, so there is no loss of predictive power of perturbative QCD. The relation of the nuclear distributions to the nucleonic ones is a hard problem in nuclear physics. Differences in the models for the nuclear corrections between experiments can make it hard to interpret data.

b. McDonald (E615) appeared to agree, at least in principle, but had questions as to the definitions of the kinematic variables to be used.

c. Freudenreich (NA10) disagreed strongly: He pointed out that for the calculation of a cross section, first an acceptance has to be calculated (in practice with a Monte Carlo). Since, in practice, one never can exclude, a priori, correlations between variables (either due to the dynamics of the process and/or introduced by the apparatus), a Monte Carlo acceptance integrated over some of the variables is trustworthy only if the Monte Carlo distributions agree in all variables with the measured ones. The Monte Carlo generators (containing parameters representing the assumed dynamics) are iterated until these agreements are achieved. The Fermi motion, for which one generally corrects, is part of the generators. Note that the Fermi correction plays an important role when a variable is close to its kinematical limit, e.g. for $\sqrt{s} = 19.1\,\mathrm{GeV}$ when M is greater than $11\,\mathrm{GeV}$.

Two different philosophies in analyzing data obtained with a nuclear target are possible, viz. a nuclear and a nucleonic point of view. In the first case, no Fermi motion corrections are introduced, the Monte Carlo agreement mentioned above is obtained by suitable variation of some "fudge" parameters, and all results refer (including the pion structure function!) strictly to the nucleus at hand. In the second case, the results represent, modulo EMC-type effects, nucleon properties, but only if the parameterization of the Fermi motion has been done correctly. In the first case, an outside physicist wanting to extract nucleon data would need to know the acceptances with all their correlations.

The theorists found it hard to accept Freudenreich's contention that Fermi motion corrections should be made. Collins argued as follows: Consider the normal theoretical prediction for the Drell Yan cross section

$$\frac{d\sigma}{dx_1 dx_2} = \sum_q \sigma_{q\bar{q}}[f_{\bar{q}/1}(x_1)\,f_{q/2}(x_2) + f_{q/1}(x_1)\,f_{\bar{q}/2}(x_2)] + \text{QCD corrections}, \quad (5.1)$$

where the 'QCD corrections' can be calculated in a well-defined manner from higher order graphs. This formula is correct, as it stands, for scattering off any target. In particular, it is correct for a nuclear target, with no corrections for Fermi motion or other nuclear effects, provided that one interprets f_2 as the parton distribution in a nucleus. Of course, almost all of the cross section will have x_2 between 0 and $1/A$ instead of between 0 and 1. But it is easy to agree on the convention that x_2 is to be first multiplied by A. Similarly one can agree to work with cross sections and parton distributions per nucleon.

Such an approach loses one the connection between parton distributions in a nucleus and in a nucleon. But perturbative QCD alone predicts no connection. Rather one must bring in non-perturbative QCD that is in the domain of nuclear physics. Of course, to the extent that it is a good approximation to assume that the nucleons inside a nucleus are free particles, the nucleonic and nuclear distributions are the same. Fermi motion is the most basic correction to this relation. Then one can bring in real nuclear physics in the controversial domain of the EMC effect. There is also some question as to the correctness of the form used for the Fermi motion.

Now, acceptance corrections must be used to obtain cross sections from the raw data, and since the corrections depend on the differential cross sections, it is essential to use the correct ones. Traditionally one uses (5.1) with parton distributions in a nucleon rather than a nucleus. In that case one must apply a correction for nuclear effects, in particular for Fermi motion. If one disagrees with the form of the Fermi motion correction, then one must repeat the whole experimental analysis.

The theorists cried foul to the idea that this is the only correct way to proceed. It is surely legitimate to apply (5.1) directly. Of course one must use the parton distributions in a nucleus. Then the experimentalists' job is to measure the *nuclear* parton distributions, and there is no need to apply a correction for Fermi motion in the analysis of the data. Fermi motion comes in when someone tries to predict the nuclear distributions from the nucleonic ones.

This point of view assumes that the operations

apply Fermi motion corrections

and

calculate cross section from parton distributions

commute.

Freudenreich remained unconvinced because of the acceptance correlations, pointed out above.

Here the discussion ended. Discussions later suggested the following remarks.

The resolution of the conflict may lie in the definition of the kinematic variables: We define τ as Q^2/s. For a nuclear target, we let p_N be its momentum, which is close to Ap_2, where p_2 is the onshell momentum of a single nucleon, of mass m_p. We may define variables relative to either the nucleons or the whole nucleus:

a. Relative to the nucleon, we define s to be the square of the center-of-mass energy of the pion-nucleon system: $s = (p_1 + p_2)^2 = m_\pi^2 + m_p^2 + 2Em_p$, where E is the beam energy.

b. Relative to the nucleus, we define s to be the square of the center-of-mass energy of the pion-nucleon system, divided by A: $s = (p_1 + p_N)^2/A = m_\pi^2/A + m_p^2 \times A + 2Em_p$.

Of course it is necessary to apply Fermi motion corrections to the first definition before a meaningful comparison between data on different nuclei can be made.

If masses are ignored, as is usually done by theorists, then these definitions are equivalent (aside from Fermi motion). But for a tungsten target ($A = 185$) at a beam energy of 194 GeV (typical for the experiments under discussion), the first definition gives $\sqrt{s} = 19.1$ GeV, while the second gives $\sqrt{s} = 23.0$ GeV. The mass of the nucleus is clearly not negligible, and so the theoretical discussion about treating the process simply as Drell-Yan on a whole nucleus needs modification: in general target mass effects are not under good theoretical control.

So we suggest that the resolution of the problems may be that at present energies one cannot treat Drell-Yan on a nuclear target without bringing in the nuclear physics. One must use the fact that a nucleus is a collection of nucleons, and that hard processes basically involve scattering off individual nucleons. In that case the Fermi motion correction is necessary to obtain a cross section that *at present energies* can be interpreted directly in parton model terms. We caution the reader that there is no unanimity on this issue. We also emphasize that a resolution of these problems is essential, so that theorists and experimentalists can agree on the interpretation of the data.

5.2 Disagreement on Cross Section

The data presented by NA10 and E615 on the cross section as a function of τ disagreed (at the time of the conference) by about 5 standard deviations at τ values above the upsilon (See fig. 5 of McDonald's contribution.) It was said that the cross sections as a function of $M_{\mu\mu}$ agreed within errors. Work is still being done to resolve this problem.

5.3 Angular Distribution

The discussion echoed points made earlier in this report.

5.4 Future of Drell-Yan

Freudenreich pointed out that in view of the problems in handling nuclear targets, some of which were discussed above, it would be desirable for future experiments to concentrate on measurements on hydrogen and deuterium targets. (Fermi motion is well measured for deuterium.) At present one gets high statistics only by using nuclear targets.

One would also like to have much more data with π^+ and with $\bar{p}$ beams. Current data have very poor statistics. The $\bar{p}$ case is one where the theoretical predictions are very clean. Freudenreich proposed to use the antiprotons from the CERN collider in a fixed target experiment. If provisions for the reversal of the polarity of the SPS can be made now - because one will need to transfer protons in two directions into the LHC - it will become attractive to extract antiprotons from the collider (rather than dumping them at the end of a collider run) into the beam line of a high intensity muon pair experiment. With ACOL operating at its design intensity, 6 bunches of 10^{11} $\bar{p}$'s each are expected to circulate in the collider at the beginning. After 15 hours, when the collider beams usually are dumped, the total intensity of the circulating antiprotons has dropped to half of its initial value, i.e. $3 \times 10^{11} \bar{p}$'s. Estimating an 80% extraction efficiency, $2.4 \times 10^{11} \bar{p}$'s could be delivered (in some 50 bursts) every 15 hours to a fixed target experiment like NA10. How many dimuon events could be collected with such a rate in a typical collider run (2400 hours with 70% running efficiency)? Scaling the cross section measured by E537 [11] at 125 GeV to 315 GeV gives: $\sigma(315\,\mathrm{GeV}) = 2.7 \times 10^{-34}\,\mathrm{cm^2/nucleon}$. With a 240 cm long deuterium target, the NA10 spectrometer could collect about 3700 events with $M > 4\,\mathrm{GeV}/c^2$, which is an order of magnitude more than what is available now from heavy targets.

5.5 Events with 2 Ws

UA1 had reported events where a W is produced in association with 2 high p_T jets whose mass is itself consistent with the mass of the W. A representative from UA2 reported that his experiment had looked for events in their four-jet data that looked like decay of 2 Ws to 4 jets, but had not been able to see any signal. It should be noted that the data as reported by UA1 have very large errors on the p_T measurement (a factor of $\pm 30\%$), so that their events are actually consistent with normal QCD production of lower p_T jets associated with a single W.

ACKNOWLEDGEMENTS

We would like to thank V.L. Telegdi for a critical reading of this report. This work was supported in part by the U.S. Department of Energy under contract DE-FG02-85ER-40235.

REFERENCES

1 J.C. Collins, Phys. Rev. Lett. **42**, 291 (1979); C.-S. Lam and W.-K. Tung, Phys. Rev. **D18**, 2447 (1978); K. Kajantie, J. Lindfors and R. Raitio, Nucl. Phys. **B144**, 422 (1978).

2 G. Parisi and R. Petronzio, Nucl. Phys. **B154**, 427 (1979); Yu. L. Dokshitzer, D.I. Dyakonov and S.I. Troyan, Phys. Rep. **58**, 269 (1980).

3 J.C. Collins, D.E. Soper and G. Sterman, Nucl. Phys. B **250**, 199 (1985).

4 P. Chiappetta and M. Le Bellac, Z. Phys. C, **32**, 521 (1986).

5 C.-S. Lam and W.-K. Tung, Phys. Rev. **D18**, 2447 (1978).

6 R.K. Ellis, G. Martinelli and R. Petronzio, Nucl. Phys. **B211**, 106 (1983).

7 C. Lopez and F.J. Yndurian, Nucl. Phys. **B183**, 157 (1981).

8 E.L. Berger and S.J. Brodsky, Phys. Rev. Lett. **42**, 940 (1979).

9 P. Chiappetta and H.J. Pirner, Nucl. Phys. **291**, 765 (1987)

10 G. Bodwin, S.J. Brodsky and G.P. Lepage, Phys. Rev. Lett. **47**, 1799 (1981).

11 E. Anassontzis et al., FERMILAB-Pub-87/217-E (1987).

DIRECT PHOTONS AT LARGE P_T FROM HADRONIC COLLISIONS:

A SHORT REVIEW BASED ON THE QCD ANALYSIS BEYOND LEADING ORDER

Rudolf Baier

Fakultät für Physik
Universität Bielefeld
D-4800 Bielefeld, FRG

INTRODUCTION

Almost ten years ago it was stated in a paper by Halzen and Scott[1] that "perturbative QCD predicts the observation of direct photons at large transverse momentum (p_T)". Indeed since then an impressive number of experiments[2] confirmed the production of prompt photons in hadronic collisions. Especially for proton induced production one has to note the large coverage in $\sqrt{s}$ ($20 - 630 \text{GeV}$) and p_T ($5 - 80 \text{GeV}/c$), although the experiments are rather difficult because of the low rates, e.g. prompt photon/hadronic jet $< \alpha/\alpha_S$.

The early predictions are based on the Born term (leading order) analysis in the framework of the QCD-improved parton model. Because of the pointlike coupling of the photon to quarks only two subprocesses dominate single photon production, namely the Compton: $qG \to \gamma q$ and the annihilation subprocess: $q\bar{q} \to \gamma G$. Recently the many relevant papers exploring these predictions are extensively and carefully reviewed by Owens[3].

The main concern of this contribution is the post–QCD Born term analysis, which is based on the complete QCD calculation[4] up to second order in α_S, and its detailed comparison with all available (although partly still preliminary) data[2]. Details may be found in two recent publications[5]. However the work is still in progress[6], since presently the next–to–leading order corrections to the parton (quark and gluon) densities are implemented into the prompt photon code[5] taking into account the latest high precision data of deep inelastic scattering[7] (DIS). The main part of the following presentation is based on the preliminary results of this investigation[6].

Indeed substantial efforts are required to improve on the quantitative predictive power of QCD, however "... to establish those effects (i.e. those based on the hard QCD Compton and annihilation subprocesses) would be a central test of the perturbative QCD approach ", as it was phrased by Fritzsch and Minkowski[8] in 1977.

First a short discussion of the perturbative approximation beyond the leading order is given, which intends to clarify the problem of the renormalization– and factorization scheme ambiguities[5] and which describes their "resolution". Second the theoretical results are summarized in comparison with the experimental data. The final aim of this comparison is a "precise" determination of

    i) the gluon structure function in the nucleon from $p \overset{(-)}{p} \to \gamma X$,

ii) the pion structure functions from $\pi^{\pm}p \to \gamma X$, and

iii) the QCD scale parameter $\Lambda_{\overline{MS}}$.

It is necessary to note the further obvious aim of obtaining a consistent description of all the processes, in which real (as well as virtual) photons are involved. An important example is photoproduction[9] of hadrons at large p_T.

AMBIGUITIES IN SECOND ORDER QCD PREDICTIONS

It is important to keep in mind that unphysical parameters appear in the perturbative approximation of a hard scattering cross–section beyond leading order, namely

i) the renormalization mass scale μ, which is introduced using dimensional regularization to regulate the ultra– violet divergences from the one–loop integrals (cf. Fig.1.b). The actual renormalization is performed in the $\overline{MS}$ –scheme.

ii) the factorization mass scale M, which separates hard from soft physics. At this scale the mass singularities associated with collinear emissions (cf. Fig.1.c) are separated off. By this procedure the dressed structure and fragmentation functions are introduced, which then evolve with M^2.

As an illustration the finite cross–section in the case of "non–singlet" prompt photon production, i.e. $\bar{p}p \to \gamma X - pp \to \gamma X$, is given in short–hand notation

$$Ed^3\sigma/d^3p\left(\sqrt{s}, p_T, y\right) = \sigma_{NS}^{(2)} =$$

$$a\left(\mu\right)\{\sigma_B + a\left(\mu\right)\left[b\ln\mu/\Lambda_{\overline{MS}} - 2P_{qq}\ln M/p_T\otimes\right]\sigma_B \qquad (1)$$

$$+ a\left(\mu\right)K_{q\bar{q}}\left(p_T/\Lambda_{\overline{MS}}, ...\right)\otimes\}q_V\left(M\right)\otimes q_V\left(M\right).$$

A small bremsstrahlung contribution is neglected in Eq. (1). $a\left(\mu\right)$ is given by the second order strong coupling constant α_S/π, and $b = 11/2 - N_f/3$. P_{qq} is the Altarelli–Parisi splitting function[10]. σ_B is the annihilation Born cross–section (after factorizing off the coupling $a\left(\mu\right)$). $q_V\left(M\right)$ denotes the valence quark structure function. The function $K_{q\bar{q}}$ is the genuine finite higher order term, which in this example originates from the corrections to the annihilation process. The corresponding diagrams are shown in Fig.1.

The approximation of the full function $K_{q\bar{q}}$ by constant π^2–terms in the soft–gluon approach is discussed by Contogouris and coworkers[11].

In case of the "singlet" cross–sections up to second order, i.e. for $pp-$, $p\bar{p}-$ and $\pi^{\pm}N \to \gamma + X$, additional terms[5] have to be taken into account, which arise i) from the radiative gluon corrections to Compton scattering, ii) from the gluon–gluon fusion process, $GG \to q\bar{q}\gamma$, and iii) from photon bremsstrahlung off q–q scattering.

Eq. (1) exhibits explicitly the dependence on the choice of the renormalization (RS) scheme, i.e. on μ and on $a\left(\mu\right)$, related by

$$b\ln\mu/\Lambda_{\overline{MS}} = \frac{1}{a\left(\mu\right)} + c\ln\left[\frac{ba\left(\mu\right)}{2\left(1 + a\left(\mu\right)\right)}\right], \qquad (2)$$

and c is the scheme independent second coefficient of the QCD β–function,

$$c = \left(153 - 19N_f\right)/\left(66 - 4N_f\right). \qquad (3)$$

Concerning the FS–ambiguity in addition to M one has to define the parton densities. In the following the (non–universal) physical convention[12] is used for the quark densities as they are measured via deep inelastic scattering.

$$F_2\left(x, M\right) \equiv \sum_q e_q^2 x q_V\left(x, M\right) + sea, \qquad (4)$$

including the $O\left(\alpha_S\right)$ corrections.

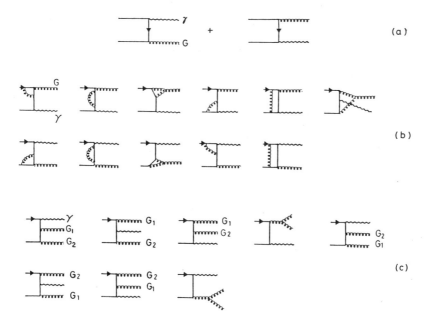

Fig.1. Feynman diagrams contributing to the annihilation subprocess. a) Born diagrams, b) virtual and c) real emission diagrams. Wavy (curly) lines denote photons (gluons).

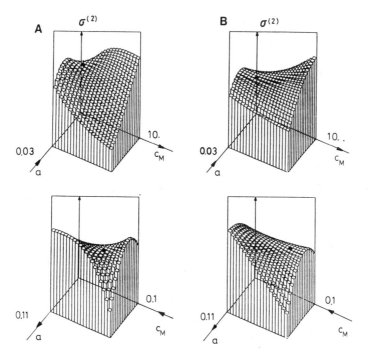

Fig.2. RS–and FS–dependence of the second order QCD invariant cross–section $\sigma^{(2)}$ as a function of $a(\mu)$ and M ($C_M = M^2/p_T^2$) for a) $pp \rightarrow \gamma X$ at $\sqrt{s} = 63\text{GeV}, p_T = 10\text{GeV}/c$ and b) $\bar{p}p \rightarrow \gamma X$ at $\sqrt{s} = 630\text{GeV}, p_T = 45\text{GeV}/c$ and $y = 0$. Two different views are shown. $\sigma^{(2)}$ is given in units of 10^{-36} (10^{-38}) cm^2/GeV^2. The saddle–point is denoted by the black square.

83

In order to get a sensible expansion for the cross–section the scales μ and M have to be of the order of the large scale present in the hard process, namely p_T, since this choice prevents the logarithms in Eq. (1) to become large. However, at finite order perturbation theory their actual values are not specified: actually, changing μ, $\mu \to \mu'$, and M, $M \to M'$, respectively, the second order cross–section has to satisfy the consistency requirement,

$$\sigma^{(2)}(\mu, M) \to \sigma^{(2)}(\mu', M') = \sigma^{(2)}(\mu, M) + O\left(a(\mu)^3\right). \tag{5}$$

Although the two sides of this Eq. (5) formally differ in terms of order a^3, which are not calculated in the considered approximation, Eq. (1), the difference matters numerically, when different RS and FS are used. Typical examples for this dependence are shown by three–dimensional plots in Fig.2, in which two different views for $\sigma^{(2)}$ versus $a(\mu)$ and M^2/p_T^2 are presented. One observes that $\sigma^{(2)}$ spans a hyperbolic surface around a saddle–point. Indeed the variations turn out to be quite large, even at large values of $p_T \simeq 50\text{GeV}/c$, as it may be deduced from the contour plots published in Ref. 5.

Therefore it is crucial to fix the scales μ and M in such a way that the approximate $\sigma^{(2)}$ becomes RS and FS invariant, as it is the case for the cross–section when calculated to all orders. Following Stevenson[13] and Politzer[14] this is achieved by applying the attractive "Principle of Minimal Sensitivity" (PMS), which is the requirement of local stability, i.e.,

$$\mu\frac{\partial \sigma^{(2)}}{\partial \mu} = 0,$$
$$M\frac{\partial \sigma^{(2)}}{\partial M} = 0, \tag{6}$$

evaluated at fixed kinematics and fixed reference $\Lambda_{\overline{MS}}$. This optimization corresponds to evaluating the QCD prompt photon cross–section σ_{OP} at the saddle–point (Fig.2).

There are a few nice features of this method to be noted:

i) in the neighborhood of the stability point the second order cross–sections are extremely flat with respect to $a(\mu)$ and M variations, e.g. varying M^2 by a few orders of magnitude the prediction stays remarkably constant when the coupling is fixed at its optimized value, $a = a_{OP}$; this stability does not hold in general in the neighborhood of the "standard" scales, $\mu = M = p_T$.

ii) the corresponding K–factor, $K = \sigma^{(2)}/\sigma^{Born}$, is controlled by the (process dependent) coupling a_{OP},

$$K_{OP} = 1 - \frac{ca_{OP}}{2(1 + ca_{OP})}$$
$$\simeq 1 - a_{OP}. \tag{7}$$

Numerically it is very close to one, namely $1 - K_{OP} < 0.10$; this is in contrast to e.g. $K \simeq 2$ for $\mu = M = p_T \simeq 10\text{GeV}/c$.

iii) at small values of $x_T = 2p_T/\sqrt{s}$ the stability conditions, Eq. (6), cannot be satisfied. As it is expected no reliable predictions can be derived for $\sigma^{(2)}$, when p_T becomes small, $p_T \le 4\text{GeV}/c$.

iv) the PMS approach leads to almost identical results as the "effective charge" scheme proposed by Grunberg[15].

OPTIMIZED RESULTS AND COMPARISON WITH DATA

In order to evaluate numerically the theoretical cross–sections for the production of prompt photons the nucleon and pion structure functions have to be specified. In the first step the two sets of parametrizations of Duke and Owens[16] for the nucleon, and of Owens[17] for the pion, respectively, are used. These sets are the result from a leading logarithm analysis of DIS, Drell–Yan pair production and production of heavy resonances. The first one (set 1) corresponds to a soft gluon distribution and a value of $\Lambda = 200$MeV, whereas the second one (set 2) corresponds to a harder gluon and a larger value of Λ, $\Lambda = 400$MeV.

Performing the comparison of the optimized predictions with all the presently available data[2] for proton as well as $\pi^\pm$ induced reactions remarkable agreement is found[5], when the predictions are based on the set 1 distributions; the set 2 gives larger rates especially in the energy and p_T range covered by the fixed target experiments, mainly because of the larger value of Λ. This comparison is summarized in Figs.3 and 4, where the ratios of data versus theory are plotted on a linear scale. Additional successfull comparisons are discussed in the prompt photon contributions[2] to this Workshop. In summary the data favour a value of $\Lambda_{\overline{MS}} \le 200$MeV, and from Figs.3 and 4 one finds $|\Delta R| \le 0.5$, where $\Delta R = \sigma_{Exp}/\sigma_{Theory} - 1$. In judging the agreement especially for the proton induced prompt photon production (Fig.3) one has to note that the compared cross–sections vary by four orders of magnitude. In general this beautiful agreement confirms the validity of the perturbative approximation.

FURTHER IMPROVEMENTS

Because of these encouraging results in applying improved QCD perturbation theory, and because of the precise prompt photon data we[6] are presently refining the analysis by implementing the second order QCD evolution[18] for the structure functions, instead of using the parametrizations of Ref. 16 and Ref. 17. This procedure then gives the consistent $O(\alpha_S^2)$ calculation. In particular we fit the recently published deep inelastic muon–proton and muon–deuterium data measured by the EMC Collaboration[19]. As it is well known the gluon distribution in the nucleon is extremely difficult to constrain from DIS data alone[20]. In contrast the prompt photon data are more selective, they disfavour the harder (set 2) gluon distribution as it is already discussed.

For our analysis we first use the following ansatz for the gluon distribution (in the universal convention[21])

$$xG\left(x, Q_0{}^2 = 2\text{GeV}^2\right) \propto (1 - x)^{\eta_G}. \tag{8}$$

The resulting χ^2 dependence from fitting the DIS data[19] is rather flat (dashed curve) in the range $4 < \eta_G < 8$ as can be seen from Fig.5 (i.e. $\chi^2/d.o.f. \simeq 1.54$). In the QCD fit only the statistical errors are included. The corresponding values for $\Lambda_{\overline{MS}}$ are indicated in order to show the Λ–gluon correlation[18]: a softer gluon distribution requires a smaller value of $\Lambda_{\overline{MS}}$. In the same Fig.5 the χ^2 versus η_G of the fit to the six (preliminary) data points for $pp \rightarrow \gamma X$ in the central region measured by the WA70 Collaboration[2] is shown. These data are used for the fit, because they are the most precise ones published up to now. The resulting solid curve has a pronounced minimum reflecting the extremely sensitive dependence of prompt photon cross–sections on the shape of the gluon distribution. The best fit is found for

$$\eta_G = 4.3 \pm 0.3, \qquad \Lambda_{\overline{MS}} = 207 \pm 30\text{MeV}. \tag{9}$$

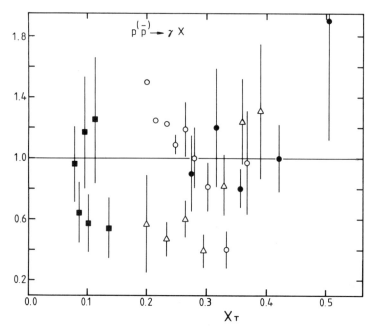

Fig.3. Ratio of the experimental to the theoretical prompt photon cross-sections versus x_T for proton induced collisions. The data[2] are from NA24, R806, CCOR and UA2 (only statistical errors are included). The optimized QCD calculation[5] is using set 1 structure functions[16].

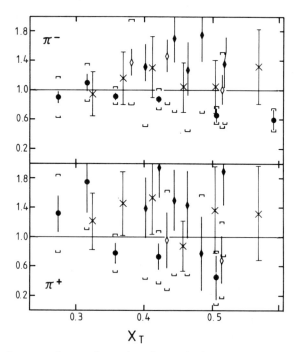

Fig.4. Ratio of the experimental to the theoretical prompt photon cross-sections versus x_T for pion induced collisions. The data[2] are from the fixed target experiments NA3, NA24 and WA70. The optimized QCD calculation[5] is using set 1 structure functions [16-17].

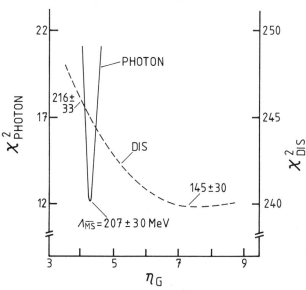

Fig.5. The χ^2 of the comparison of measured and predicted deep inelastic structure functions (dashed line) and of the cross-section (solid line) for $pp \rightarrow \gamma X$ as a function of η_G. The data for DIS are taken from EMC[19]. for prompt photons from WA70[2].

Figs.6–8 summarize the comparison of the measured prompt photon cross–sections (only the statistical errors are included) of the fixed target (WA70 and NA24) and ISR (R806 and CCOR) experiments[2] with the theoretical, always optimized (cf. Eq. (6)), expectations for different values of η_G. Instead of the χ^2 dependence here the quantity $p_T^6 E d^3\sigma/d^3p$ is plotted versus x_T on a linear scale. The data points – except a few CCOR points, which have a large systematic uncertainty – are well reproduced with the power of the gluon distribution given in Eq. (9).

In order to be able to compare the improved analysis with the results obtained before (Fig.3) we show the ratios of the prompt photon data[2] (from the WA70, NA24 and UA2 Collaborations) to the theoretical cross–sections as a function of x_T in Fig.9. The isolation cuts to select single photons applied for the UA2 data are now included (in an approximate way as described in Ref. 5). One observes that the beyond leading order corrections to the evolved structure functions do not give rise to large effects. Therefore the use of the set 1 leading order distributions of Ref. 16 as a first step for obtaining reliable predictions is justified.

With the help of the next–to–leading order structure functions one is able to further exploit the FS dependence of prompt photon cross–sections, since one can change from the non–universal (cf. Eq. (4)) to the universal convention for the quark densities: as a result the predictions differ by less than 10%, an additional bonus of the optimization method.

The corresponding analysis is also performed for the pion induced processes. Here the structure functions in the proton are fixed as described above. The gluon distribution in the pion is then determined, when the quark distributions are taken from the Drell–Yan analysis by the NA10 Collaboration[22]. With the same ansatz as given by Eq. (8) we find for the power η_G: $\eta_G = 2.13 \pm 0.11$, when the $\pi^{\pm}$ data by WA70 are fitted. The resulting χ^2 is $\chi^2 = 9.5$ for 12 data points.

We can now compare the gluon distribution (in next–to–leading order) in the nucleon, Eqs. (8)–(9), obtained from the combined fit to the DIS EMC[19] and to the prompt photon data with other recent independent determinations. There is good agreement with $G\left(x, Q^2 = 10\text{GeV}^2\right)$ as it is determined by Diemoz et al[23]. However there is disagreement with the very soft gluon – $\eta_G = 10.3 \pm 1.5$ at $Q^2 = 5\text{GeV}^2$, $\Lambda_{\overline{MS}} = 210 \pm 20\text{MeV}$, which is the result of the scaling violation analysis by the BCDMS Collaboration[24] using their preliminary high statistics DIS proton data. In this respect one has to mention that the proton structure function F_2 by EMC[19] and by BCDMS[24] differ in the region of low x, which is the sensitive region for extracting the gluon distribution. A detailed discussion of this discrepancy is given in Ref. 7. Therefore a common QCD fit to these sets of data is not possible.

In order to explore the consequences of the soft BCDMS gluon for the prompt photon spectra we[6] repeated the fit (including only the statistical errors) to these data[24] using the same evolution program[18] as for the analysis discussed before. We obtained the same result as quoted by BCDMS. Taking this soft gluon shape literally over the whole x–range, $0 < x < 1$, then the prompt photon rates for the fixed target experiments are underestimated by a factor 3–4 (as it is expected already from Fig.6, in which the dashed–dotted curve results from a very soft gluon), whereas the predictions for the $Sp\bar{p}S$ energy and p_T range remain essentially unchanged.

However it is important to note that different processes are probing the x and Q^2 dependences of the gluon distribution $G\left(x, Q^2\right)$ in different regions: presently the information obtained from DIS is essentially restricted to the region $x < 0.3$ and $Q^2 < 200\text{GeV}^2$, whereas the proton induced inclusive prompt photon yields (at 90°) are on average sensitive to the the (x, Q^2)–regions indicated by the solid curves in

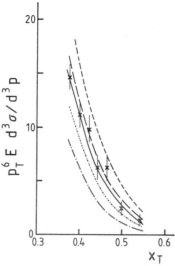

Fig.6. Comparison between theory and experiment for the invariant cross-section (multiplied by p_T^6 in units of $10^{-32} cm^2 \text{GeV}^4$) of $pp \to \gamma X$ for different values of η_G: 3.5 (dashed), 4.2 (long-dashed), 4.5 (solid), 5.0 (dotted) and 6.0 (dashed-dotted curve). The preliminary data are from the WA70 experiment[2] at $\sqrt{s} = 23 \text{GeV}$.

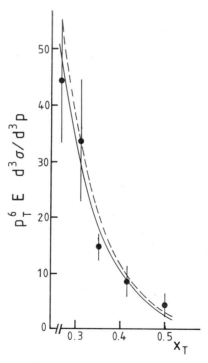

Fig.7. Comparison between theory and experiment for the invariant cross-section (multiplied by p_T^6 in units of $10^{-32} cm^2 \text{GeV}^4$) of $pp \to \gamma X$ for different values of η_G: 4.2 (dashed) and 4.5 (solid curve). The data are from the NA24 experiment[2] at $\sqrt{s} = 24 \text{GeV}$.

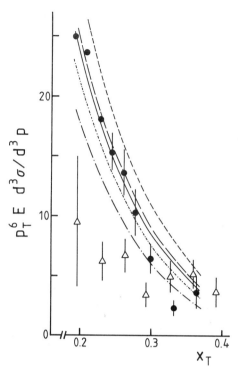

Fig.8. Comparison between theory and experiment for the invariant cross–section (multiplied by p_T^6 in units of $10^{-31} cm^2 GeV^4$) of $pp \to \gamma X$ for different values of η_G: 3.5 (dashed), 4.2 (long–dashed), 4.5 (solid), 5.0 (dotted) and 6.0 (dashed-dotted curve). The data are from the R806 (•) and CCOR (△) experiments[2] at $\sqrt{s} = 63 GeV$.

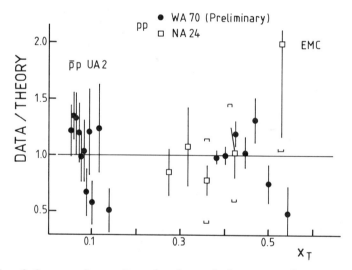

Fig.9. Ratio of the experimental to the theoretical prompt photon cross–sections versus x_T for proton induced collisions. The data[2] are from NA24, WA70 and UA2. The optimized QCD calculation[6] is using next–to–leading order fits to structure functions as described in the text with $\eta_G = 4.3$.

Fig.10. In the later case changing p_T amounts to change simultaneously the effective scale $Q^2 = M^2$ ($\simeq 0.2p_T^2$, applying optimization) as well as x, $x_{min} \leq x \leq 1.0$, but $< x > \simeq x_T$. From Fig.10 one deduces that the most interesting region to compare gluon determinations from DIS and from prompt photon spectra is the one covered by the ISR experiments: the different predictions based on the EMC-type ($\eta_G = 4.3$, cf. Eq. (9)) and on the BCDMS-type analysis are plotted in Fig.11 as solid and dashed curves, respectively. The two predictions differ by a factor of two. Although the ISR data[2] (R806 and R110) are not in favour of the very soft BCDMS gluon distribution, it may be to early to rule it out definitely, when the systematic uncertainties are taken into account.

CONCLUSIONS AND OUTLOOK

The analysis of direct photon production, mainly due to the precise data[2], reached a level, which already allows to draw quantitative conclusions, e.g. on the gluon distributions in nucleons and pions, and on the value of $\Lambda_{\overline{MS}}$. Nevertheless additional experimental data, with improved systematic errors, are required especially in the kinematical range of x_T-values overlapping with the x-range in DIS, which is sensitive to the gluon distribution, namely $0.1 < x_T < 0.3$. We are eagerly waiting for the data by the E705 and E706 Collaborations[2].

With the forthcoming data by the UA6 Collaboration[2] it may become possible to determine $\Lambda_{\overline{MS}}$ more directly from the "non-singlet" combination $\bar{p}p \to \gamma X - pp \to \gamma X$, since it is independent of the gluon.

On the other hand the continuation of measurements, which were started by the AFS Collaboration[25], of the reactions $pN \to \gamma +$ away-side jet (hadron) + X will allow to determine the gluon distribution in a more direct way than the single inclusive measurements. Concerning correlation data it is also important to mention the beautiful measurements of the two-photon final states produced in $\pi^- p$ collisions, which are reported by the NA24 and WA70 Collaborations[2].

Finally, the production of low-mass lepton pairs at large transverse momentum is of great interest. Instead of real photons, which are experimentally very difficult to resolve at high energies, these pairs may be an extremely useful signal for probing the proton structure at very small values of $x \simeq x_T$ ($x < 0.1$) with the future hadron supercolliders. Theoretically the advantage is the dominance of one subprocess, i.e. Compton scattering. The first measurements of muon-pairs are already performed by the UA1 Collaboration[26] at the CERN S$p\bar{p}$S in the mass- and p_T range, $2m_\mu - 2.5$GeV and $6 < p_T < 40$GeV/c. In addition these data are nicely described in the perturbative QCD framework by the real photon approximation[27].

In the future prompt photon production will therefore remain a significant topic in connection with accurate tests of QCD.

ACKNOWLEDMENTS

I am grateful for the hospitality of LPTHE Orsay, where this talk was prepared. In particular I would like to thank most warmly my colleagues P. Aurenche, M. Fontannaz and D. Schiff for the friendly collaboration during the last years.

It is a pleasure to thank Bradley Cox and Phyllis Hale for organizing this stimulating and enjoyable Workshop.

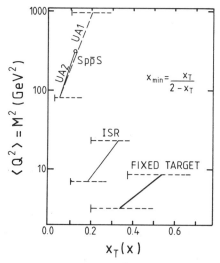

Fig.10. Illustration of the effective x and M^2 regions probed by the prompt photon experiments at the different accelerators.

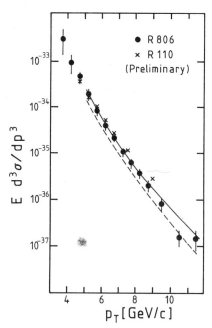

Fig.11. The invariant cross–section for $pp \rightarrow \gamma X$ versus p_T at $\sqrt{s} = 63\text{GeV}$: the optimized QCD calculations[6] are using next–to–leading order fits to structure functions as described in the text with $\eta_G = 4.3$ (solid curve) and from the fit to the BCDMS DIS–data[24] (dashed curve), respectively. The data[2] are from the ISR experiments R806 and R110.

REFERENCES

1. F. Halzen and D.M. Scott, Phys. Rev. Lett. $\underline{40}$,1117 (1978).

2. NA3 Collab., J. Badier et al., Z. Phys. $\underline{C31}$,341 (1986); F. Constantini, contribution to this workshop;
   NA24 Collab., C.De Marzo et al., Phys. Rev. $\underline{D36}$,8 (1987); P. Seyboth, contribution to this workshop;
   WA70 Collab., M. Bonesini et al., preprints CERN-EP/87-185, CERN-EP/87-222 and contributions to the Int. Europhysics Conf. on HEP, Uppsala 1987;
   M. Martin, contribution to this workshop;
   UA6 Collab., T. Cox, contribution to this workshop;
   E705 Collab., D. Wagoner, contribution to this workshop;
   E706 Collab., G. Fanourakis, contribution to this workshop;
   CCOR Collab., A.L.S. Angelis et al., Phys. Lett. $\underline{94B}$, 106 (1980); Phys. Lett. $\underline{98B}$,115 (1981);
   R806 Collab., E. Anassontzis et al., Z. Phys. $\underline{C13}$, 277 (1982);
   AFS Collab., T. Åkesson et al., Phys. Lett. $\underline{158B}$,282 (1985); J. Thompson, contribution to this workshop;
   R110 Collab., B.G. Pope, contribution to this workshop;
   UA1 Collab., B. Mours, contribution to this workshop;
   UA2 Collab., J.A. Appel et al., Phys. Lett. $\underline{176B}$, 239 (1986); K.H. Meier, contribution to this workshop;
   for a recent review, F. Richard, Rapporteur's talk at the 1987 Int. Symp. on Lepton and Photon Interactions at High Energies, Hamburg, July 1987, preprint LAL 87-48;
   for a further discussion of experiments before 1984, T. Ferbel and W.R. Molzon, Rev. Mod. Phys. $\underline{56}$, 181 (1984).

3. J.F. Owens, Rev. Mod. Phys. $\underline{59}$, 465 (1987);
   for an independent leading order analysis including bremsstrahlung contributions, E.L. Berger, E. Braaten and R.D. Field, Nucl. Phys. $\underline{B239}$, 52 (1984).

4. P. Aurenche, R. Baier, A. Douiri, M. Fontannaz and D. Schiff, Phys. Lett. $\underline{140B}$, 87 (1984).

5. P. Aurenche, R. Baier, M. Fontannaz and D. Schiff, Nucl. Phys. $\underline{B286}$, 509 (1987); P. Aurenche, R. Baier, M. Fontannaz and D. Schiff, Nucl. Phys. B, in print.

6. P. Aurenche, R. Baier, M. Fontannaz, J.F. Owens and M. Werlen, in progress.

7. For recent reviews, R. Voss, Rapporteur's talk at the 1987 Int. Symp. on Lepton and Photon Interactions at High Energies, Hamburg, July 1987, preprint CERN-EP/87-223;
   T. Sloan, Plenary Talk at the Int. Europhysics Conf. on HEP, Uppsala 1987, preprint CERN-EP/87-188.

8. H. Fritzsch and P. Minkowski, Phys. Lett. $\underline{69B}$, 316 (1977).

9. P. Aurenche, R. Baier, M. Fontannaz and D. Schiff, Nucl. Phys. $\underline{B286}$, 553 (1987); NA14 Collab., D. Treille, contribution to this workshop; WA69 Collab., E. Paul, contribution to this workshop.

10. G. Altarelli and G. Parisi, Nucl. Phys. $\underline{B126}$, 298 (1977).

11. A.P. Contogouris, N. Mebarki and H. Tanaka, Phys. Rev. D, in print, and references herein.

12. G. Altarelli, R.K. Ellis and G. Martinelli, Nucl. Phys. $\underline{B153}$, 521 (1978).

13. P.M. Stevenson, Phys. Rev. $\underline{D23}$, 2916 (1981); Nucl. Phys. $\underline{B203}$, 472 (1982).

14. H.D. Politzer, Nucl. Phys. $\underline{B194}$, 493 (1982);
    P.M. Stevenson and H.D. Politzer, Nucl. Phys. $\underline{B277}$, 758 (1986).

15. G. Grunberg, Phys. Rev. $\underline{D29}$, 2315 (1984).

16. D.W. Duke and J.F. Owens, Phys. Rev. $\underline{D30}$, 49 (1984).

17. J.F. Owens, Phys. Rev. $\underline{D30}$, 943 (1984).

18. A. Devoto, D.W. Duke, J.F. Owens and R.G. Roberts, Phys. Rev. $\underline{D27}$, 508 (1983).

19. EMC Collab., J.J. Aubert et al., Nucl. Phys. $\underline{B259}$, 189 (1985); Nucl. Phys. $\underline{B293}$, 740 (1987).

20. For a recent discussion, J.F. Owens, contribution to this workshop.

21. L. Baulieu and C. Kounnas, Nucl. Phys. $\underline{B141}$, 423 (1978);
    L. Kodaira and T. Uematsu, Nucl. Phys. $\underline{B141}$, 497 (1978).

22. NA10 Collab., K. Freudenreich, contribution to this workshop.

23. M. Diemoz, F. Ferroni, E. Longo and G. Martinelli, preprint CERN-TH.4751/87.

24. BCDMS Collab., A.C. Benvenuti et al., contribution to the Int. Europhysics Conf. on HEP, Uppsala 1987;
    M. Virchaux and A. Ouraou, private communication.

25. AFS Collab., T. Åkesson et al., Z. Phys. $\underline{C34}$, 293 (1987).

26. UA1 Collab., K. Eggert, contribution to this workshop.

27. P. Aurenche, R. Baier and M. Fontannaz, in preparation.

DIRECT PHOTON PRODUCTION FROM POSITIVE AND NEGATIVE PIONS AND PROTONS AT 200 GeV/c (NA3 COLLABORATION)

Flavio Costantini

Dipartimento di Fisica, Universita' di Pisa
and
INFN, Sezione di Pisa, Italy

ABSTRACT

Direct photon production has been studied by the NA3 Collaboration at the CERN SPS using negative and positive beams at 200 GeV/c incident on an isoscalar Carbon target. The experiment, completed in 1985, published the first fixed target measurements of the asymmetry of prompt photon production by negative and positive pions. The experiment was sensitive to photons with $2.5 \le P_T \le 6.0$ GeV/c and C.M. rapidity $-0.4 \le Y^* \le 1.2$.

NA3 final results on : a) direct photon production by $\pi^{\pm}$ and protons b) charge asymmetry measurement on direct photon production from π^- and π^+ c) direct photon pair production by $\pi^{\pm}$ and protons d) π° production measurements by $\pi^{\pm}$, K^+ and protons are recalled , commented on and compared to data recently published by other fixed target experiments .

INTRODUCTION

The discovery that high P_T prompt photons were produced in proton proton collisions [1] stimulated an intense experimental activity starting right at the beginning of the eighties.

At CERN the collaborations NA3, NA24 and WA70 presented three proposals, at the end of 1980, to measure the hadroproduction of direct photons by fixed target experiments [2]. The UA6 collaboration submitted at the same time another proposal for an experiment to be carried directly at the SPS ring with the gas jet target technique [3].

During the " Workshop for SPS future in the years 1984 -1989 " held at the end of 1982, a working group on prompt photon physics presented a picture, turned out to be impressive, of both the experimental effort undertaken at CERN and FNAL and of the status of the theory on this topic [4].

Last but not least, the UA1 and UA2 collaborations recently presented data on the prompt photon production measured at the collider, in a completely new domain of x_T [5].

The NA3 collaboration started the data taking in 1982 and presented the first preliminary data during the summer 1984 [6]. The final results of the analysis were sent for publication in the second half of 1985 [7,8,9].

As no further analysis of NA3 data has been carried since, my contribution to this workshop will be mainly focussed on two aspects :

a) I will stress the differences between one particular trigger used by NA3 (the so called "converter" trigger) and the usual calorimeter trigger, common to all the prompt photon experiments and also used by NA3 .

b) I will compare the NA3 data with the data of WA70 and NA24 that were recently published.

The complete and detailed description of the NA3 experimental set-up, of the data analysis criteria and of the physics results on the prompt photon production, the π° cross section and the photon pair production obtained with incident $\pi^{\pm}$, K^+ and protons can be found in refs [7,8,9] and references therein.

THE NA3 EXPERIMENTAL SET-UP

The main features of the experimental set-up (see Fig. 1)are here briefly recalled .

Incoming beam particles are identified by two differential Cerenkov counters (CEDARS) and by two threshold Cerenkov counters. The isoscalar targets are three identical 2 cm long graphite cylinders, spaced 15 cm apart to minimize secondary interactions and ensure optimal space determination of the event vertex.

A superconducting dipole with a field of 1.7 T and a set of 9 MWPC's for a total of 44 planes allows a spectrometer momentum resolution of $\Delta P/P = 2\times10^{-4}\,P$ (GeV/c) .

An electromagnetic calorimeter of the lead scintillator type is segmented into 40 horizontal strips (9 x 500 cm^2) symmetrically divided in two arms with respect to the beam. The calorimeter is longitudinally segmented in three stacks named $\gamma1(5X_o)$, $\gamma2(8X_o)$ and $\gamma3(12X_o)$.

The measured calorimeter energy resolution is $\Delta E/E = 0.004 + 0.22/\sqrt{E}$ (E in GeV) for single showers. The photomultiplier gains are monitored during the data taking by two independent systems of LED's.

A fine grain position measuring device (shower chamber) inserted between $\gamma1$ and $\gamma2$, in each arm of the calorimeter, provides the spatial resolution needed to separate π° decays from single photons. It consists of a single gap 4 x 2 m^2 MWPC with analog read-out of the horizontal anode wires and of both cathodes. One cathode is segmented in vertical strips 12 mm wide, the other is divided into independent pads of three sizes (5 x 5, 10 x 10, 20 x 20 cm^2), the size increasing with the distance from the beam. The pads are particularly useful in reducing the ambiguities in the geometrical reconstruction of multishower events.

The spatial resolution of the shower chamber has been measured to be $\sigma = 3$ mm for isolated showers in the energy range 8 ÷ 45 GeV. The two shower separation is about 3 cm , allowing the identification of a symmetric π° decay up to energies of 110 GeV, being 14 m the distance from the targets.

Timing information, for background rejection, is provided by both $\gamma1$ and the shower chamber.

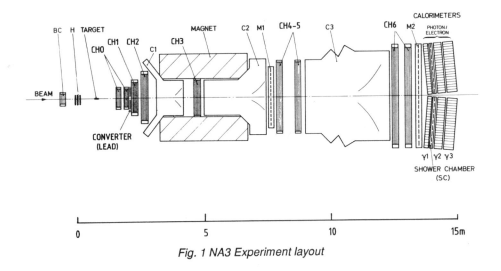

Fig. 1 NA3 Experiment layout

THE TRIGGERS

Two triggers were set for the prompt photon physics and one for the direct photon pair study.

a) The " calorimeter trigger " requires an electromagnetic energy deposition vertically localized in two adjacent strips of the calorimeter such that the corresponding vertical component of P_T , called hereafter P_{TV} , is > 3.5 GeV/c. This high threshold is set to obtain an acceptable trigger rate.

b) The " converter trigger ", peculiar to the NA3 experiment, requires the conversion of a photon in a thin lead converter (thickness 0.15 X_0 i.e. 10% conversion probability), placed just behind the target and sandwiched by MWPC's. The requirement that ensures an on-line identification of the photon conversion is based on the topology of the converted e^+e^- pair. The electron-positron pair is deflected in the horizontal plane by the vertical magnetic field of the superconducting NA3 dipole, so that the horizontal distance between the e^+ and the e^- is measured by the chambers M1 and M2 (see Fig. 2). A hardware processor reads the two homotetic chess-board triggering chambers M1, M2 recognizing in 200 ns pairs originated close to the target and having a $P_{TV} >$ 1.85 GeV/c . The electromagnetic energy deposition of the pair in the triggering strips of the calorimeter has to correspond to a $P_{TV} > 2.4$ GeV/c.

c) The "photon pair trigger" requires a $P_{TV} > 1.5$ GeV/c in one strip of the upper and in one strip of the lower part of the calorimeter. In addition no charged track has to point toward these two strips. This veto on charged particles lowers the trigger rate by more than a factor 10 ; it is expected to act more on $\pi^\circ\pi^\circ$ events than on prompt photon pair events, due to the fact that π°'s are, on average, more "accompanied" by charged particles than prompt photons [10].

It has to be stressed that the NA3 " converter trigger " selects on-line events in which one photon has actually converted. This technique is therefore basically different from the technique used by the experiments R108 and UA2 which happen to have the same name. The latter is based on the fact that the probability of observing a γ conversion is greater for the decay of a π° than for a prompt photon. Therefore the presence of prompt photons among photon pairs from π° decay is determined off line and on a statistical basis by measuring the observed non conversion probability [11].

A first advantage of the converter trigger is that, due to the parent-child relation, the P_{TV} cut on one of the two decaying photons rejects, already at the trigger level , the π° 's produced with low P_T . Given the steep P_T dependence of the π° production cross section, this cut produces the remarkable effect of rejecting on-line the much more abundantly produced low P_T π°'s. In Fig. 3 the percentage of π°'s surviving the P_{TV} cut on one of the two photons is displayed vs. the P_{TV} of the π°.

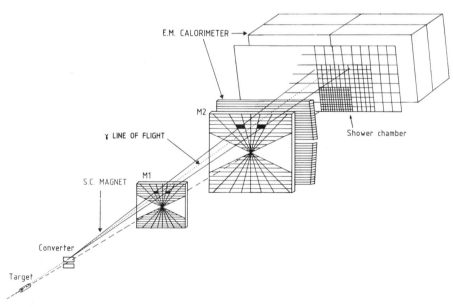

Fig. 2 NA3 " converter " trigger description. A photon produced in the target is converted in the lead converter. The resulting e+e- pair is then horizontally deflected by the S.C. dipole. The pads crossed by the pair in the two homothetic triggering chambers M1 and M2 are blackened in the drawing.

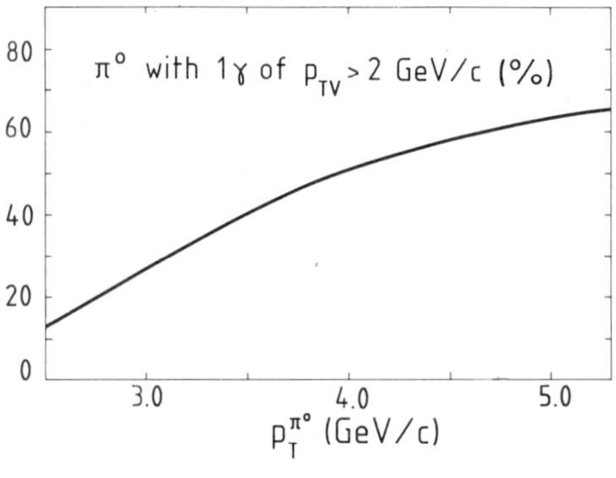

π° with 1γ of $p_{TV} > 2$ GeV/c (%)

$p_T^{\pi^\circ}$ (GeV/c)

Fig. 3 Percentage of π°'s accepted by the converter trigger vs the π° transverse momentum.

As a result the data sample collected by the converter trigger is much richer in prompt photons than the data sample collected by the calorimeter trigger, where the π° rejection cannot be made at the trigger level.

A second advantage of the converter trigger is that the photon P_T is accurately obtained off line, by measuring the momenta of the electron-positron pair in the spectrometer. The measurement of the P_T of the prompt photon is then substantially independent from the calorimeter [13], while the measurement of the P_T of π°'s relies both on the spectrometer and on the calorimeter, which measures the second photon.

The ratio γ/π° vs. P_T measured with the converter trigger, then, is not affected by the systematic error due to calorimeter non-linearity, as it happens for a pure calorimeter trigger. In fact to measure the ratio γ/π° with a simple calorimeter experiment, the energy E of a prompt photon is compared to the π° energy detected, on average, as two photons of energy E/2 each. Therefore a calorimeter non linearity folded with the steep P_T dependence of the π° cross section gives a wrong estimation of the π° contamination to the prompt photon yield [11].

The major disadvantage of the converter trigger is the low rate resulting from the fact that only 10% of all photons convert. [The thickness of the converter is a compromise between the trigger rate and the necessity of not spoiling the converted pair energy in the converter itself.]

Last, the cleanliness of the converter trigger events makes them "safer" to deal with than the calorimeter trigger events, so that the physics results obtained with the former trigger were used to cross-check the more copious results obtained by the calorimeter trigger.

CALIBRATION OF DETECTORS AND OF SIMULATION PROGRAMS

The evidence for the existence of prompt photons entirely depends on the correct simulation of the experimental apparatus and on the simulation of its triggering and detection capabilities. It is therefore of the utmost importance to have the possibility to "calibrate" not only the detectors but the Monte Carlo simulation programs as well.

The NA3 collaboration had the opportunity to regularly calibrate the whole spectrometer, i.e. the entire set of MWPC's (necessary for the converter trigger), the e.m. calorimeter and the detector to localize the e.m. showers (necessary for all triggers). The calibration was performed at the beginning of each data taking period using electron beams of known momentum ranging from 8 to 70 GeV/c. A similar shorter calibration procedure was repeated at the end of each data taking period whenever circumstances required it.

This evidently gives a far better understanding and control of the whole apparatus than is obtained when a particular detector (or part of it) is occasionally dismounted and calibrated on a test beam .

In addition the NA3 collaboration had also the opportunity, unique to my knowledge, to calibrate the detectors and the simulation programs using "in situ" tagged photons of energy ranging from 0 up to 8 GeV [12]. It has been possible in this way to actually measure the detector efficiencies and energy resolutions both for photons and electrons in an energy range where one usually relies on the extrapolations of calibrations made only with electrons at higher energies.

It should be recalled here that the knowledge of both the detection efficiency and the energy resolution for low energy photons is crucial for reconstructing π° asymmetric decays.

PHYSICS RESULTS

In order to avoid a pure reproduction of all NA3 data already published some time ago, I will present here a subset of NA3 data compared to the final ones of the other two fixed target experiments which are now available, namely the WA70 and NA24 data [14,15].

The π° invariant cross section

It is crucially important that the π° invariant cross section be accurately measured by the same experiment that measures the γ/π° ratio vs. P_T . Fig. 4 shows the " observed " γ/π° ratio vs. P_T for incident π⁻ , π⁺ and protons, for the calorimeter trigger only, and the dashed band represents the contamination due to π° 's misidentified as prompt photons. This contamination is evaluated by a Monte Carlo simulating the apparatus and its prompt photon trigger, having as input the π° cross section; it is therefore important that the π° cross section be measured by the same experiment and that the simulation programs adequately reproduce the apparatus.

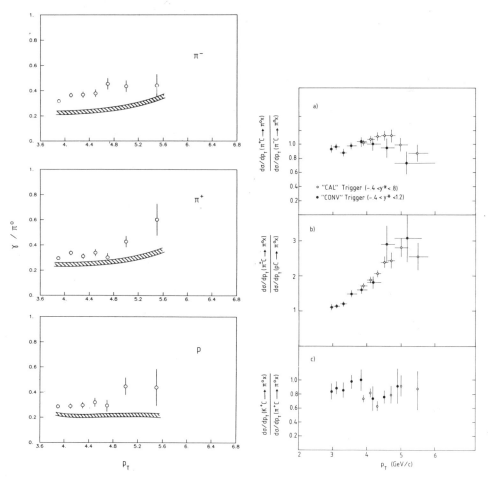

Fig. 4 " Uncorrected " γ/π° ratio vs P_T for incident negative and positive pions and protons. The dashed band represents the contamination due to π° 's misidentified as prompt photons.

Fig. 5 Ratio of differential cross sections vs P_T for π° production from various incident particles : a) π⁺/π⁻ b) π⁺/p c) K⁺/ π⁺ .

The parameters of the fit to NA3 π° invariant cross section data are shown in ref [7] ; the fit agrees remarkably well with WA70 data even in a P_T region between 5.5 and 6.5 GeV/c where no NA3 data are available [14].

The ratio of the differential cross sections for π° production from positive and negative pions vs P_T is shown in Fig. 5a integrated over the appropriate rapidity intervals. This ratio is compatible with 1, within errors, as expected from isospin invariance.

The ratio of the differential cross sections for π° production from pions and protons is shown in Fig. 5b. The increase of this ratio as a function of P_T, already known, is generally attributed to the different shape of the quark and gluon structure functions in the pion and in the proton.

The ratio of the differential cross sections for π° production from positive kaons and pions, shown in Fig. 5c, exibits a P_T independent behaviour in the range 3 ÷ 6 GeV/c . After integration over P_T and Y^* we find for such a ratio the value 0.79 ± 0.03.

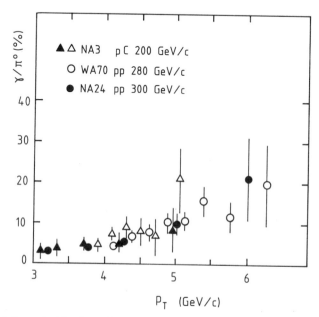

Fig. 6 " Corrected " γ/π° ratio vs P_T by incident protons for the three CERN fixed target experiments NA3, WA70, NA24 (NA3 : black triangles indicate converter trigger data and open triangles indicate calorimeter trigger data)

According to SU(3) quark compositions of mesons $\pi\pm$, π° and K+ , and to the generally accepted belief that π° 's are created by the fragmentation of initial quarks and gluons, the value of this ratio seems then to indicate that the s quark content of K mesons makes it more difficult to obtain the right combination of u and d quarks that gives a π°.

The prompt photon measurements

The ratio of prompt photon production cross section to π° production cross section $\sigma(\gamma)/\sigma(\pi^\circ)$, frequently used in publications on prompt photon studies, is shown in Fig. 6 for incident protons only and compared to the presently available measurements by other fixed target experiments. The data present a remarkable agreement among the different experiments in the √s range 20 - 24 GeV and seem to confirm that the ratio $\sigma(\gamma) / \sigma(\pi^\circ)$ increases with P_T .

The charge asymmetry in prompt photon production

The ratio of cross sections for prompt photons produced by negative and positive pions vs P_T is shown in Fig. 7 and compared to the data now available. It should be noted that this comparison is somehow arbitrary because the cross sections are measured with different targets, at different $\sqrt{s}$ and the different systematic errors are not taken into account[1].

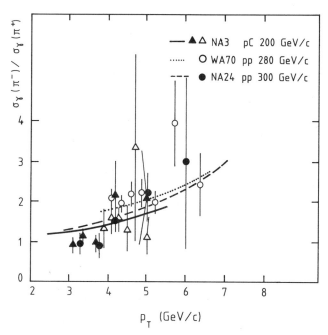

Fig. 7 Charge asymmetry by incident pions in prompt photon production vs P_T for the three CERN fixed target experiments. (NA3 : black triangles indicate converter trigger data and open triangles indicate calorimeter data. The three curves show second order QCD calculation from ref. [16] .

In addition, the compared values are taken directly from the published plots of WA70 and NA24 so that a slight error might have occurred in adapting the P_T and vertical scales. Having these words of caution in mind, the following conclusions can be safely drawn in the spirit of this workshop.

There is an excellent agreement between the data at low P_T which belong essentially to NA3 and NA24. The agreement is good also between the QCD second order predictions for these two experiments, evaluated with the same program kindly given by the authors of ref. [16].

In the higher P_T part of the plot WA70 data tend to be slightly higher than the data of the other two experiments. In fact in the P_T range 4 ÷ 7 GeV/c WA70 quotes an average value of $\sigma_\gamma(\pi^-)/\sigma_\gamma(\pi^+) = 2.$, weakly dependent on P_T.

The extrapolation at low P_T of the QCD prediction for WA70 data seems to diverge from the extrapolation of the same prediction for NA3. For WA70 the extrapolated value at low P_T seems to be between 1.5 and 2. , somehow in disagreement with the naive expectation that at low P_T the contribution of the quark annihilation diagram should vanish.

---

[1] An accurate evaluation of NA3 systematic errors is given in [8].

The prompt photon pair production

The two prompt photon differential cross section vs the transverse momentum of one photon is shown in Fig. 8 , for incident protons and negative pions. Each prompt photon has a C.M. rapidity $-0.4 \leq Y^* \leq 1.$ and a P_T greater than 1.8 GeV/c.

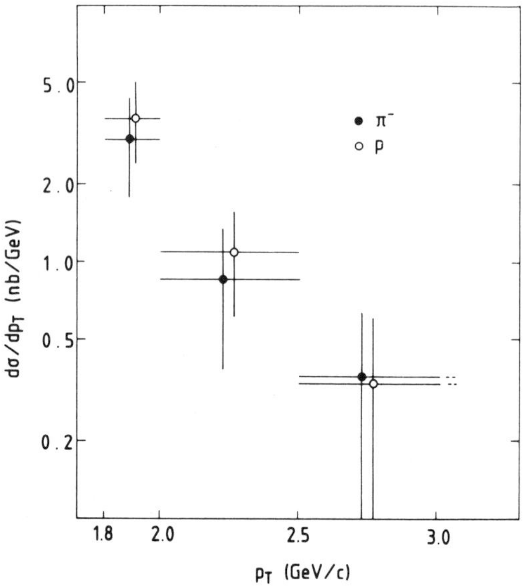

Fig. 8 Prompt photon pair differential cross section vs the P_T of one photon for incident protons and negative pions.

In table 1 the photon pair production cross sections are listed for incident negative and positive pions and protons.

TABLE 1

$\gamma\gamma$ PRODUCTION CROSS SECTION (nb)

P_T (GeV/c)	π^-	π^+	p
> 1.8	1.22 ± 0.35	0.35 ± 0.64	1.48 ± 0.38
1.8 - 2.0	0.61 ± 0.25		0.74 ± 0.25
2.0 - 2.5	0.43 ± 0.24		0.57 ± 0.23
> 2.5	0.18 ± 0.14		0.17 ± 0.13

The theoretical values of the cross sections based on a complete second order QCD calculation [17] are substantially lower; these values and a description of the effects that might cause the discrepancy can be found in ref [9].

No data comparison is possible as, to my knowledge, no other fixed target experiments have ye presented data on prompt photon pair production, and the ISR data are plotted vs the di-photon invariant mass.

CONCLUSIONS

The NA3 measurements on the prompt photon production by negative and positive pions at 200 GeV/c , published in early 1986, are compared to data now available from other two fixed target experiments.

There is a very good agreement between the three sets of experimental data and a second order QCD based calculation, indicating that the contribution of the quark annihilation process is smaller than what expected by the first order terms only.

NA3 measurements of direct photon pair production with incident π^- and protons, while showing a 3 standard deviation signal, are not yet in complete agreement with QCD calculations.

REFERENCES

1 P. Darriulat et al. Nucl. Phys. B110 (1976) 365
   For an exhaustive review see, for instance : T. Ferbel, W.R. Molzon "Direct photon production in high energy collisions ". Rev. Mod. Phys. 56 (1984) 181
2 L. Bachman et al.   CERN/SPSC/80-61 (WA70 Collaboration)
   A. Bamberger et al. CERN/SPSC/80-83 (NA24 Collaboration)
   R. Hagelberg et al. CERN/SPSC/80-106 (NA3 Collaboration)
3 J. Antille et al. CERN/SPSC/80-63 (UA6 Collaboration)
4 The conclusions of the working group on prompt photon physics, coordinated by F. Costantini, are in : " Workshop on SPS fixed target physics in the years 1984-1989 " Vol 2 CERN 83/02
5 See the Proceedings of this Workshop.
6 F. Costantini "Asymmetry measurements in direct photon production from negative and positive pions at 200 GeV/c". Proceedings of XXII-th Int. Conference on H.E. Physics Leipzig (DDR) Vol 1 (1984) 281.
7 J. Badier et al. (NA3 Collaboration) " Inclusive high P_T π° production from $\pi^\pm$ and protons at 200 GeV/c". Z. Phys. C 30 (1986) 45
8 J. Badier et al. (NA3 Collaboration) " Direct photon production from pions and protons at 200 Gev/c". Z. Phys. C 31 (1986) 341
   Y. Karyotakis Thesis Universite Paris Sud 1985 LPNHE Ecole Polytechnique.
9 J. Badier et al. (NA3 Collaboration) " Direct photon pair production from pions and protons at 200 GeV/c ". Phys. Lett. 164B (1985) 18
10 T. Akesson et al. Phys. Lett. 118B (1982) 178
11 For a detailed discussion on this subject, see the review of ref. [1] .
12 Four small beam chambers used in conjunction with a beam dipole measured the beam momentum with a precision $\Delta P/P = 0.8$ % . The momentum after the radiator was more precisely measured by the NA3 S.C. dipole. A set of three movable dipoles placed in front of the spectrometer allowed the vertical deflection of electrons so leaving "isolated" the radiated photon. For more details see also the Thesis of ref. [8] .
13 In reality a small (< 10 %) correction to the pair energy measured by the spectrometer is given by the pair bremsstrahlung in the converter itself. The energy of bremsstrahlung photons is measured in the calorimeter.
14 The physics results of the WA70 Collaboration used for the comparison are from :
   M. Bonesini et al. " High transverse momentum π° production by π^- and π^+ on protons at 280 GeV/c ". CERN EP/87-164
   M. Bonesini et al. " High transverse momentum prompt photon production by π^- and π^+ on protons at 280 GeV/c ". CERN EP/87-185
   M. Bonesini et al. " Production of high transverse momentum prompt photons and neutral pions in proton-proton interactions at 280 GeV/c ".
   Presented at the International Eurphysics Conference on H.E. Physics Uppsala 1987
15 The physics results of the NA24 Collaboration used for the comparison are from :
   C. De Marzo et al. " Measurement of direct photon production at large transverse momentum in π^-p , π^+p and pp collisions at 300 GeV/c ". Phys. Rev. D 36 (1987) 8
   C. De Marzo et al. " Measurement of π° production at large transverse momentum in π^-p , π^+p and pp collisions at 300 GeV/c ". Phys. Rev. D 36 (1987) 16
16 P. Aurenche, R. Baier, A. Douiri, M. Fontannaz, D. Schiff. Phys. Lett. 140B (1984) 8
   P. Aurenche, R. Baier, A. Douiri, M. Fontannaz, D. Schiff. Nucl. Phys. B286 (1987) 509
17 P. Aurenche et al. , preprint LPTHE Orsay 85/20

DIRECT PHOTON PRODUCTION IN pp AND p̄p COLLISIONS

P. T. Cox

The Rockefeller University
New York, NY 10021

INTRODUCTION

The simplicity of the leading-order production processes for direct (single) gamma production in pp and p̄p collisions, i.e. the QCD-Compton process $qg \rightarrow \gamma q$ and the annihilation process $\bar{q}q \rightarrow \gamma g$, is appealing, since the photon carries all the p_T of the parton collision, and there is no confusion from fragmentation. Furthermore the ratio $R = \sigma(\bar{p}p \rightarrow \gamma X)/\sigma(pp \rightarrow \gamma X)$ should be sensitive to the hardness of the gluon structure function of the nucleon. Since $\bar{q}q$ annihilation is clearly more important in p̄p than pp, due to the valence $\bar{q}$ quarks, R is expected to be larger than one, and since the gluon structure function is known to be relatively soft so that $\bar{q}q$ should become increasingly favoured over qg as p_T increases, R is expected to increase with the p_T of the emitted gamma ray.

Experiment UA6 at CERN is a fixed-target experiment using a hydrogen gas jet target in the SPS Collider p̄ (or p) beam to study high p_T π^0, η, and γ production in p̄p and pp collisions at $\sqrt{s}$ = 24.3 GeV (when the Collider beam momenta are 315 GeV/c). Electron pairs, including $J/\Psi \rightarrow e^+e^-$, can also be studied. The collaborating institutes are CERN, the University of Lausanne, the University of Michigan, and the Rockefeller University [1].

APPARATUS

The apparatus, a conventional double-arm magnetic spectrometer, is shown in fig. 1 for the case in which we are studying p̄p collisions. To study pp collisions the whole detector must be rotated by 180° with respect to the collider beams. The hydrogen jet consists of pure H_2, in clusters of ~ 10^5 molecules per cluster. It is of sufficiently low density, ~ 4×10^{14} protons per cm^3, that multiple interactions are not a problem. The small size, 3 mm transverse to the beam and 8 mm along the beam direction, gives a well-localized interaction vertex. Finally, since each bunch of beam particles traverses the jet with the SPS revolution frequency ~ 4×10^4 s^{-1}, we make efficient use of the beam particles (particularly the p̄'s). Thus the target gives us efficient parasitic operation, with high luminosity, $\mathscr{L}$: the typical $\mathscr{L}$ with p̄'s was 5.5×10^{29} cm^{-2}s^{-1}, with 4×10^{10} p̄'s stored in 3 bunches, and the typical $\mathscr{L}$ with p's was 5×10^{30} since the number of stored p's was about 10 times higher than with p̄'s. The luminosity is monitored by the rate of elastic recoil protons (or antiprotons) at 90° in the Lab, detected in a set of solid-state counters. Charged particles are detected in a set of MWPCs, placed before and after a spectrometer magnet with a field integral of 2.3 T m. Two lead/proportional-tube gas sampling calorimeters, of area 125×83 cm^2, detect electromagnetic showers, and a transition radiation detector in each arm can be used to aid electron/hadron discrimination. Each calorimeter consists of 30 lead sheets, each 4 mm thick, interleaved with alternating layers of horizontal and vertical proportional tubes of 1 cm transverse dimension and 0.5 cm thickness, filled with Ar/CO_2 at 1 atmosphere and operated at a gas gain ~ 1000. Each calorimeter is divided into 3 identical modules of 8 radiation lengths and 0.25 interaction lengths each, giving a total depth of 24 r.l.. A hodoscope of 7 horizontal strips of scintillator, each $125 \times 12 \times 1$ cm^3, located 8 r.l. deep inside each calorimeter, provides the 1st level trigger on e.m. energy deposition. The threshold on each strip is adjusted to give an approximately "equal p_T" trigger, typically with a threshold of about $p_T \geq 1$ GeV/c. The analog signals from the tubes are summed through the depth of each module, thus preserving the fine lateral segmentation (1 cm) while still allowing 3 samplings in depth. A 2nd level trigger uses the analog signals in a hardware processor, which examines the energy deposition in grids on the face of the calorimeter, and thus allows us to choose and set a higher trigger threshold for high p_T showers from γ's. The calorimeters were studied in test-beams of electrons and pions covering momenta from 10 to 100 GeV/c. The response is linear to $< 1\%$ over this range, and the energy resolution for electrons was $\sigma(E)/E = 0.34/\sqrt{E}$ at our operating voltage (1140 V). The position resolution is typically σ_x, $\sigma_y \sim$ 2 mm, improving from 3.5 mm at 10 GeV/c to 1 mm at 75 GeV/c. The 3 longitudinal samplings allow a nice separation between electrons and charged pions to be made according to the differing shower profiles.

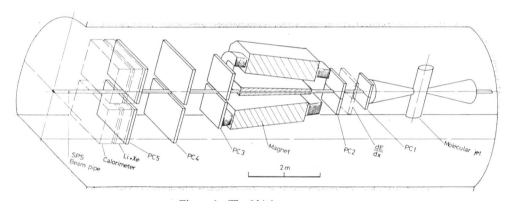

Figure 1. The UA6 apparatus.

DATA

During the 1985 collider operation we collected 1000 nb^{-1} of data in $\bar{p}p$ at 315 GeV/c, and during 1986 1600 nb^{-1} in pp at 315 GeV/c, both using the processor trigger with a threshold of $p_T \geq 2.75$ GeV/c. Just over 500 nb^{-1} of the $\bar{p}p$ data have been fully analysed, and most of the results in this talk come from these data. We have completed the most time-consuming part of the analysis for over 1000 nb^{-1} of the pp data, and we expect results within a few months. We also have 40 nb^{-1} of data in both $\bar{p}p$ and pp taken with just the hodoscope trigger with $p_T \geq 1$ GeV/c. We have already published π^0 and η results from these data [2]. Our full analysis should yield a $\bar{p}p/pp$ comparison of π^0, η, and γ production at $\sqrt{s} = 24.3$ GeV, with p_T's up to 7 GeV/c. This corresponds to an x_T ($= 2p_T/\sqrt{s}$) range from 0.25 to 0.60.

Both π^0 and η decays may contribute to sources of single γ's which must be removed from our data sample of direct γ candidates. In the two-photon decays of π^0 and η both γ's may overlap, or one γ may escape from the detector, in both cases leading to spurious direct γ events. Thus it is important to be able to measure $\pi^0 \to 2\gamma$ and $\eta \to 2\gamma$ very well.

The minimum separation between the γ's at a distance L (cm) from the decay vertex is $2mL/E$ where m is the mass (GeV/c²), and E the energy (GeV), of the π^0 or η. This gives, for E = 100 GeV (somewhat higher than our average π^0 energies), a separation of 2.7 cm at the calorimeters (10 m from the target). Thus the spatial resolution of the calorimeter is well-suited to resolve such decays.

The electronic gains of all calorimeter channels are calibrated using test pulses; the absolute energy scale is set by centering the reconstructed π^0 mass on its known value, on a run-to-run basis (i.e. at time intervals of ~ 1 hour).

Showers are reconstructed by clustering the energies deposited in the tubes, considering the horizontal and vertical views separately. Clusters are associated in depth by requiring that they project to the target position, and H and V clusters are associated by requiring approximate energy matching. Finally the invariant masses of all pairs of clusters are calculated, assuming each cluster is due to a γ ray originating at the target. A typical invariant mass distribution from pairs of clusters in the same calorimeter is shown in fig. 2 There are clear peaks due to the π^0 and η, and the mass resolutions are consistent with those expected from the energy and position resolutions of the calorimeter: $\sigma(\pi^0 \to 2\gamma) \cong 18$ MeV/c², $\sigma(\eta \to 2\gamma) \cong 35$ MeV/c². To obtain the number of η events we fit a Gaussian plus a linear background term to the η mass region. The background under the π^0 is negligible.

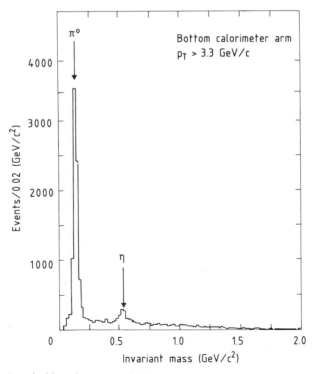

Figure 2. A typical invariant mass distribution for pairs of electromagnetic clusters in the same calorimeter.

The invariant cross sections for π^0 and η production are obtained from the measured numbers of π^0's and η's, after correcting for the detector and reconstruction efficiencies using a Monte Carlo. This generated π^0 (and η) events, decayed them and tracked the γ's through the detector. The hits in the calorimeter were replaced by real shower information, obtained from bona-fide γ showers which did reconstruct to π^0's. These events were then analysed as usual. The Monte Carlo describes the data very well, e.g. the asymmetry parameter $(E_1 - E_2)/(E_1 + E_2)$, where E_1, E_2 are the energies of the γ's from a π^0 decay, agrees well over the full range 0 to 1.

The resulting invariant cross section for π^0 production in $\bar{p}p$ collisions at $\sqrt{s} = 24.3$ GeV is shown in fig. 3, as a function of p_T of the produced π^0. Only statistical errors are shown, and there is an estimated uncertainty in the p_T scale of $\pm 1.5\%$, and an overall normalization uncertainty of $\pm 7\%$. These data correspond to 507 nb^{-1}, or about half of the $\bar{p}p$ data we have collected. Also shown are pp results [3] at a similar $\sqrt{s}$, which agree very closely. This is consistent with a small subsample of our own data [2] in which we found that the ratio of the cross sections for π^0 production in $\bar{p}p$ and pp collisions was compatible with one.

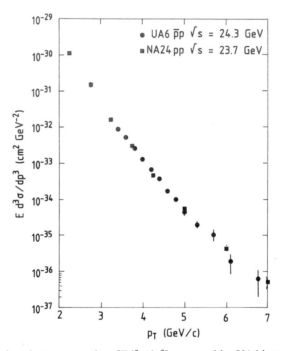

Figure 3. The invariant cross section [$Ed^3\sigma/dp^3$] measured by UA6 in $\bar{p}p \rightarrow \pi^0 X$ at $\sqrt{s} = 24.3$ GeV, and for pp $\rightarrow \pi^0 X$ at $\sqrt{s} = 23.7$ GeV from Ref. 3, as a function of p_T of the π^0.

The ratio of the invariant cross sections for η and π^0 production in $\bar{p}p$ collisions at $\sqrt{s} = 24.3$ GeV is shown in fig. 4. This ratio is calculated from the ratio of the measured numbers of events by correcting for the $\eta \rightarrow 2\gamma$ branching ratio (0.389) and using the Monte Carlo to obtain the relative π^0 to η acceptance. There appears to be no p_T dependence over the range 2 to 5 GeV/c, and our data are consistent with pp results from our own data [2] and other experiments [4].

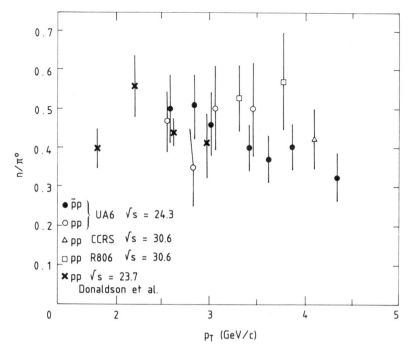

Figure 4. The ratio of the invariant cross sections for π^0 and η production in $\bar{p}p$ collisions at $\sqrt{s} = 24.3$ GeV, and in pp collisions at similar $\sqrt{s}$ values (Ref. 4).

DIRECT γ PHYSICS

The selection criteria for direct γ candidates are:

1. The cluster should not reconstruct with any other shower to a π^0 or η;

2. The cluster energy must be above 3 GeV;

3. The cluster must lie within a fiducial area of the calorimeter face; and

4. The cluster width must be narrow (which rejects high-energy π^0's and η's with both γ's merging to one cluster).

Since we do not restrict the analysis to isolated clusters, our measurement is inclusive.

The sample of candidate direct γ events, γ_{obs}, is composed of true direct γ events, γ^R, and spurious events from π^0 and η events which we have failed to reconstruct, $\gamma^F_{\pi^0}$ and γ^F_{η}:

$$\gamma_{obs} = \gamma^R + \gamma^F_{\pi^0} + \gamma^F_{\eta}.$$

With a little algebra we see that we can extract the ratio of the invariant cross sections for γ and π^0 production using the relation:

$$[\gamma/\pi^0] = [\varepsilon_{\pi^0}/\varepsilon_\gamma][(\gamma_{obs}/\pi^0) - (\gamma^F_{\pi^0}/\pi^0) - (\gamma^F_\eta/\eta) \times (\eta/\pi^0)]$$

where the RHS fractions (γ_{obs}/π^0) and (η/π^0) come directly from the data, and the other three fractions are obtained from the Monte Carlo. One can question whether the direct γ candidates are indeed photons. They certainly appear to be; for example, the longitudinal shower development of these showers is identical to that of real γ showers obtained from reconstructed π^0's. Figure 5 shows the measured ratio γ/π^0 as a function of p_T of the particle, with statistical error bars only, and also the corresponding background contribution from π^0's and η's estimated using the Monte Carlo. After subtraction of this background we obtain fig. 6. The ratio γ/π^0 clearly rises with increasing p_T, reaching a value of about 1.0 at p_T's above 6 GeV/c. Also shown are the corresponding x_T values of the data; we cover a wide range of x_T from 0.2 to 0.6.

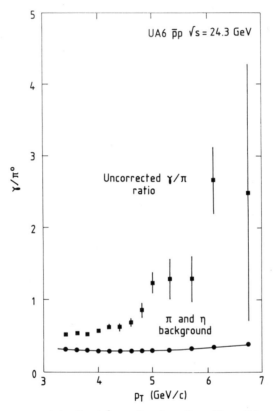

Figure 5. The uncorrected ratio γ/π^0 as a function of p_T. The solid curve is the Monte Carlo estimate of the background contribution from π^0 and η decays.

Finally we can extract the direct γ invariant cross section from this ratio using our measured π^0 invariant cross section. The result is given in fig. 7. The error bars are purely statistical; we are in the process of estimating the systematic errors, although so far we expect them to be of the same order of magnitude as the statistical ones. Also shown in fig. 7 are curves showing the predicted direct γ cross sections from the next-to-leading-order QCD calculation of Aurenche et al. [5]. This version used just three flavors of quarks. The two curves arise from using either the Set I or Set II Duke-Owens structure functions [6]. Set I includes a gluon structure function (s.f.) of the nucleon which varies as $xG(x) \sim (1 - x)^6$, whereas the Set II gluon s.f. varies as $xG(x) \sim (1 - x)^4$. The QCD predictions of references [5] and [7] for the ratio of direct γ production in $\bar{p}p$ and pp collisions still depend sensitively on the assumed gluon s.f., whereas in the measured ratio we should be relatively insensitive to systematic errors. Thus we hope that once we have completed the analysis of both our $\bar{p}p$ and pp data we can make a reasonably precise determination of the gluon structure function.

110

CONCLUSIONS

Experiment UA6 has successfully measured direct γ production in $\bar{p}p$ collisions:

- We have a clear signal of direct γ's in $\bar{p}p \rightarrow \gamma X$ at $\sqrt{s} = 24.3$ GeV, from analysis of ~ 500 nb^{-1} of data from ~ 1000 nb^{-1} collected.
- We have collected ~ 1500 nb^{-1} of pp data under the same trigger conditions (and with the same apparatus), and the data analysis is quite advanced. We expect pp results on direct γ production within a few months.
- Data-taking is to resume during the November-December SPS Collider running period. We are set up in pp mode and hope to increase our present data by a factor of 5 (to 7 pb^{-1}).
- Next year we shall rotate the apparatus in order to view $\bar{p}p$ collisions again and hope to collect 12 pb^{-1}, increasing our $\bar{p}p$ data by a factor of 12, during the 1988 Collider run.

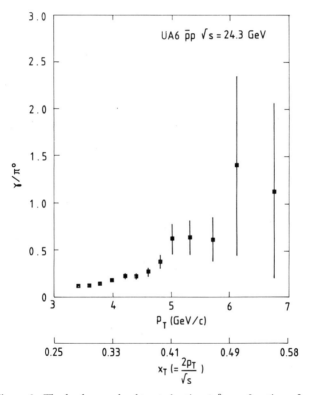

Figure 6. The background-subtracted ratio γ/π^0 as a function of p_T.

Our physics goals are of course centred on measuring the gluon structure function of the nucleon, but we also hope to study the event structure (particularly the away-side jet if possible). It has also been suggested [7] that we could make a measurement of α_s. In each case we expect small statistical errors due to the high integrated luminosities in both pp and $\bar{p}p$, and cancellation of systematic errors since we use the same apparatus for both pp and $\bar{p}p$.

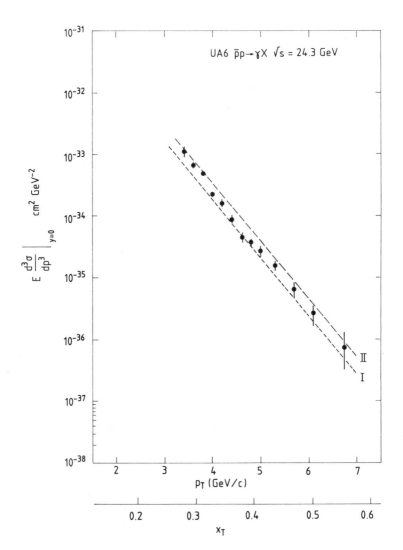

Figure 7. The invariant cross section $[Ed^3\sigma/dp^3]_{y=0}$ for $\bar{p}p \to \gamma X$ at $\sqrt{s} = 24.3$ GeV. The dashed lines are theoretical predictions (see text).

REFERENCES

[1]   The UA6 collaboration is: R. E. Breedon, L. Camilleri, R. L. Cool, P. T. Cox, G. von Dardel, L. Dick, E. C. Dukes, F. Gaille, P. Giacomelli, C. Joseph, W. Kubischta, J.-F. Loude, C. Morel, O. E. Overseth, J. L. Pages, J.-P. Perroud, P. Petersen, D. Ruegger, R. W. Rusack, G. Sozzi, G. R. Snow, L. Studer, M. T. Tran, A. Vacchi, and G. Valenti.
[2]   J. Antille et al., Phys. Lett. **B194** (1987) 568.
[3]   C. De Marzo et al., Phys. Rev. **D36** (1987) 16.
[4]   G.J. Donaldson et al., Phys. Rev. Lett. **40** (1978) 684,
        F.W. Büsser et al., Phys. Lett. **55B** (1975) 232,
        C. Kourkoumelis et al., Phys. Lett **84B** (1979) 277,
        T. Åkesson et al., Phys. Lett. **158B** (1985) 282.
[5]   P. Aurenche et al., Phys. Lett. **140B** (1984) 87, and private communication.
[6]   D.W. Duke and J.F. Owens, Phys. Rev. **D30** (1984) 49.
[7]   A.P. Contogouris and H. Tanaka, Phys. Rev. **D33** (1986) 1265, and private communication.

DIRECT PHOTON PRODUCTION BY π^-p, π^+p AND pp INTERACTION AT 280 GeV/c:

RESULTS FROM THE WA70 EXPERIMENT AT CERN

Michel Martin

University of Geneva
Geneva, Switzerland

INTRODUCTION

The W70 experiment was proposed in 1980, with the aim to study direct photon events in hadronic interactions. The experimental system consists of the CERN Omega spectrometer to which was added a fine grained electromagnetic calorimeter built for this experiment. The construction of the calorimeter, with its associated trigger electronics and calibration sytems was finished in 1983, the beam and trigger tuning as well as general conditions for data taking were also established in 1983, and data taking with full efficiency begun in 1984.

Data samples were taken in 1984,1985 and 1986 in running periods of 35 to 45 days per year. The accumulated sensitivities used for direct photon production cross sections were obtained in 1984 and 1985 and are 3.5, 1.3 and 5.2 pbarn^{-1} for π^-p, π^+p and pp interactions respectively. In addition, the 1986 data with a sensitivity of 9.2 pbarn^{-1} are now being analyzed for the determination of the two-photon production cross section in π^-p interactions.

The experimental system will be described, with details on the calorimeter, and results on prompt photon production cross sections will be given. Finally, comparison with QCD estimates based on optimised scales are shown.

EXPERIMENTAL SYSTEM

The system is shown in fig. 1.

The beam line, tuned for 280 GeV/c momentum hadrons, was run at an intensity of $2*10^7$ particles per 1.6s effective spill length, with a momentum bite of 1.8%. An aluminum and polyethylene absorber at an intermediate focus allowed an enrichment of the positive beam in pions to the level of 20%. The beam was focused on a 1m long, 25mm diameter liquid hydrogen target with a diameter of 3mm fwhm, and a divergence of 0.8 mrad. Two differential Cherenkov counters tuned on pions allowed proton and pion identification in resp 69% and 15% of the total positive beam, the remaining 16% being not unambigously identified. The contaminations of other particles in the proton and pion samples is estimated to be lower than 4% in all cases. Surrounding the beam line, sets of scintillation counters allowed the identification of triggers due to muons from the beam halo interacting in the electromagnetic calorimeter.

The spectrometer[1] is the 3m diameter, 1.5m gap Omega magnet used with a reduced field of 1.1T, with its set of multiwire proportional chambers. This consists of a total of 26 wire planes inclined of $\pm10.4^0$ relatively to the vertical direction and of 7 planes with vertical wires. In addition, two drift chambers were placed at the downstream edge of the field and a 4mx4m multiwire chamber consisting of four planes was located immediately upstream from the calorimeter.

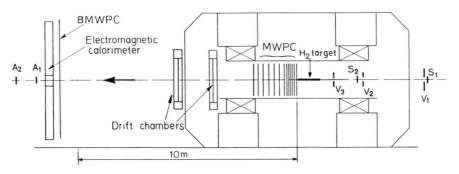

Fig. 1. The experimetal system.

The calorimeter[2] was built specifically for this experiment. It is a sampling calorimeter with a total of 24 radiation lengths of lead divided in 30 layers of 4.2mm thickness each. These are interleaved with sensitive layers consisting of teflon tubes filled with an oil-based liquid scintillator (fig 2). These tubes behave like optical fibres, with a light attenuation length of typically 1.6m, very reproducible from element to element (rms < 2%) and stable in time. The calorimeter is divided in four quadrants, each consisting of three identical segments in depth to allow hadron identification. The teflon tubes are grouped to form sensitive elements 1cm wide and 2m long. The readout of the scintillation light is done on the outside edges by a total of 3072 photomultiplier and ADC chains, providing positions and amplitudes in two orthogonal views.

In order to remove reconstruction ambiguities in shower positions for multishower events, a time of flight system[3] was installed in the upstream segment, providing a measurement of the transit time of the light along the teflon tubes. The shower position along the tubes given by this system is about 5cm for photons of energy larger than 2 GeV.

The calibration[4] was achieved by two complementary systems: an optical fiber system distributing laser light to all channels in order to follow gain variations during runs and allow a later correction, and a set of moveable ^{60}Co sources to give the relative gain equalization between all channels. The absolute gain was obtained from the data, using electrons analysed by the spectrometer as well as by a fit to the eta mass. Small corrections to the gains given by the calibrations systems were also determined from the data. This calibration technique made unnecessary the full calorimeter calibration in a test beam.

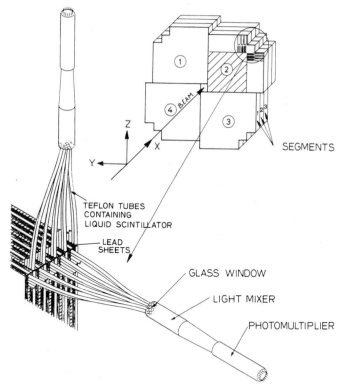

Fig. 2. The electromagnetic
         calorimeter:
         general view and construction
         detail.

The energy resolution varies with the configuration from

$$(\sigma_E/E)^2 = 0.16^2/E + 0.024^2$$

for isolated showers to

$$(\sigma_E/E)^2 = 0.28^2/E + 0.036^2$$

for all showers

The trigger logics[5] consists of a beam interaction trigger using the
signals from the scintillation counters S, V and A (fig.1.) in coincidence
with a calorimeter trigger. The calorimeter trigger selects events by a
threshold on an approximate estimate of the transverse energy. The area of
each calorimeter quadrant is divided into four horizontal and four
vertical bands, for which the signals of all photomultiplier tubes are
added with weights proportional to their distance to the beam. This gives
eight signals per quadrant, which are digitized by flash ADC's and used as
addresses to read a look-up table pre-filled to give a yes or no answer as
trigger. The time needed for this decision is 200ns total, and with an
interaction rate of 1MHz, the trigger rate was adjusted at about 50 Hz by
selecting the proper threshold, which then corresponds to about 3 GeV/c
transverse momentum. The trigger performance is shown in fig 3. Four
such look-up tables were available in parallel for each quadrant, giving
the possibility for other triggers, such as two-photon trigger for
instance.

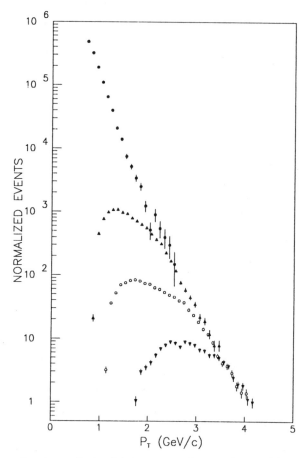

Fig. 3. Distribution of the highest p_T shower for different thresholds compared to a target interaction trigger (highest set of data points).

The data acquisition system consists of a first computer used as fast access front end and for event filtering, followed by two VAX780 for monitoring and tape writing. The rate at which events were accepted for tape writing was about 50 per 1.6s SPS effective burst.

RECONSTRUCTION AND ANALYSIS PROGRAMS

The shower reconstruction[2] is done separately for each quadrant. Different event configurations are considered, and for each one a χ^2 is calculated by using the ADC and time of flight informations. The showers of energy larger than 5 GeV are usually unambiguous, while for low energies several configurations may exist. In average, 17 showers are reconstructed per event.

The tracks reconstruction is done by the standard TRIDENT program[6]. With the configuration of multiwire chambers used here, the track reconstruction efficiency is low, varying from 35% in the beam region to 85% for low momentum tracks. An average of 12 tracks are reconstructed per event.

SELECTION OF EVENTS

Events were filtered in order to remove:

1) The accidentally overlapped events, by using the time of flight information

2) Events triggered by a high p_T hadron, by using the shower longitudinal profile as given by the three segments

3) Events with a high p_T shower corresponding to a charged track as reconstructed by the Omega spectrometer or by the 4mx4m multiwire chamber (closest distance allowed 5cm resp 3cm)

4) Halo events where a beam muon halo gave a calorimeter signal mistaken for a high p_T shower, by using the multiwire chambers informations

5) Accidental triggers and interactions not in hydrogen by requesting that the interaction vertex be reconstructed in the target fiducial volume

In this selected sample, direct photon candidates were defined as the higher p_T shower satisfying the following conditions:

1) The shower width is less than 2cm in order to remove remaining hadrons or π^0s with unresolved photons

2) The shower position is at a distance larger than 5cm from the edges of the calorimeter

3) The effective mass when combined with any other shower is outside of the π^0 or eta mass ranges (70 to 200 MeV or 450 to 550 MeV respectively) for showers of energy larger than .5 resp 2.0 GeV.

No other selection is done, in particuliar no isolation cut for the direct photon candidate is made.

ACCEPTANCE, EFFICIENCY AND BACKGROUND

The acceptance, efficiency and background were determined from two Monte Carlo simulations and also in part from the data themselves.

The Lund simulation uses the program PYTHIA[7] for parton level first order interactions, followed by JETSET[7] for fragmentation with the Lund model, together with a full simulation of the apparatus. The other simulation was a simple π^0 generation, with production cross section parametrized from the data of the present experiment[8],[9].

From these simulations, the geometrical acceptances as well as losses of good events due to the cuts to remove π^0s or etas were determined. In addition, the loss of events due to effects not simulated by Monte Carlo such as halo or overlapped events cuts were measured by the reconstructed π^0 sample rejected by these cuts and assumed to be the same for prompt photons. The result, showing the combined acceptance and efficiency as function of p_T and x_F are shown in fig 4.

117

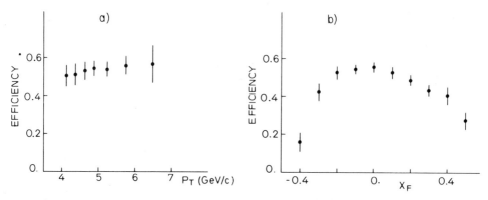

Fig. 4. Prompt photon detection efficiency: a) versus p_T and b) versus x_F

The backgroud, due mainly to unidentified π^0s or etas, has also been estimated by the Lund simulation, and is given in fig. 5. It is seen that for p_T larger than 5 GeV, the fraction of faked prompt photons in the total sample is of the order of 10% to 30% for the three data samples.

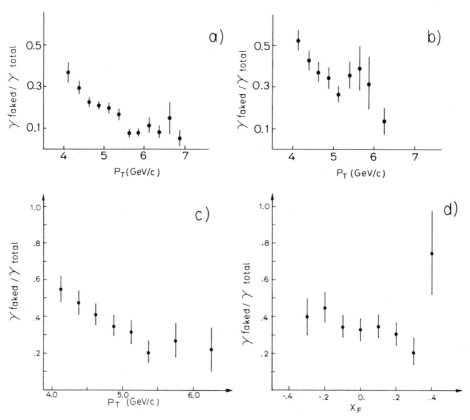

Fig. 5. Ratio of faked prompt photons / total prompt photon sample. a), b) and c): $\pi^- p$, $\pi^+ p$ and pp versus p_T. d): pp versus x_F.

RESULTS

The invariant cross section E d³σ/dp³ for prompt photon production as functions of p_T and x_F is shown in fig. 6 and 7 for production by $\pi^- p$ and $\pi^+ p$ and in fig. 8 and 9 for pp interactions. QCD estimates which include the annihilation and Compton processes are also shown in these figures: the solid and dashed curves are the predictions of second order QCD calculations using optimized scales [10] with Duke-Owens[11] structure functions sets 1 and 2. The data show that set 1 is prefered. The cross sections as two-dimensional variables are published elsewhere[8], [12].

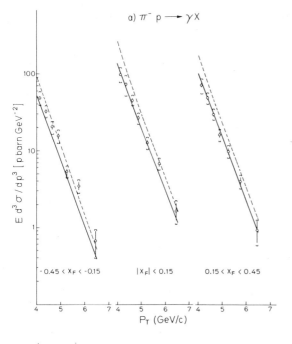

The systematic errors include the energy scale uncertainty, systematic effects in gain equalization, uncertainties in μ halo and hadron backgrounds subtraction, systematic effects in Monte Carlo estimates and uncertainties in normalization. All these effects are in the range of 5 to 15% and have been combined quadratically. The error bars in the figures are the statistical errors only, while the indicated error limits have been obtained by combining in quadrature statistical and systematic errors.

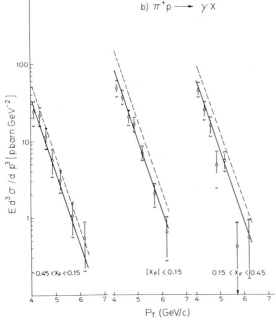

Fig. 6. Invariant cross section for prompt photon production as a function of p_T for the three ranges of x_F: $-.45 < x_F < -.15$, $-.15 < x_F < .15$ and $.15 < x_F < .45$. a) $\pi^- p$ b) $\pi^+ p$

119

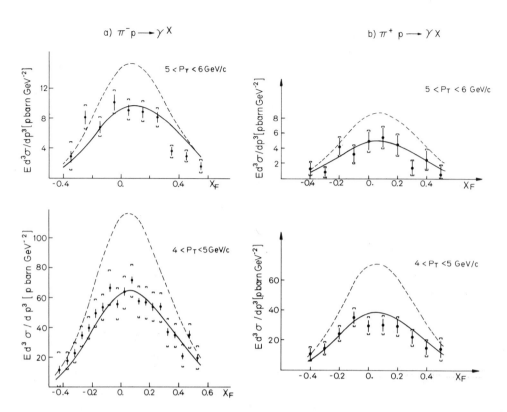

a) $\pi^- p \longrightarrow \gamma X$

b) $\pi^+ p \longrightarrow \gamma X$

Fig. 7. Invariant cross section as a function of x_F for two p_T ranges: $4.0 < p_T < 5.0$ and $5.0 < p_T < 6.0$. a) $\pi^- p$ b) $\pi^+ p$

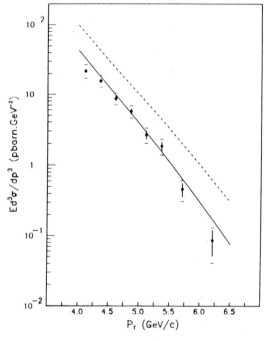

Fig. 8. Invariant cross section for prompt photon production as a function of p_T for pp interactions.

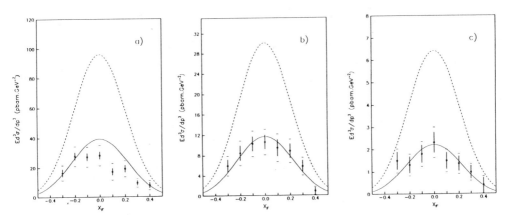

Fig. 9. Invariant cross section for prompt photon production in pp interactions as a function of x_F for three p_T ranges: a) $4.0 < p_T < 4.5$, b) $4.5 < p_T < 5.0$, c) $5.0 < p_T < 6.0$.

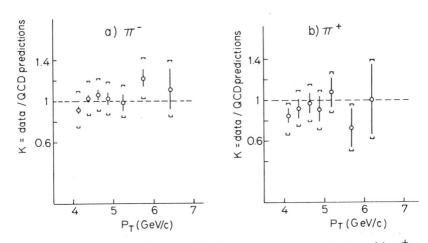

Fig. 10 Ratio data/QCD predictions versus p_T. a) $\pi^- p$ b) $\pi^+ p$

COMPARISON WITH QCD PREDICTIONS

The ratio data/QCD predictions[10] is given in fig. 10 for $\pi^- p$ and $\pi^+ p$ interactions and in fig. 11 for pp interactions. The QCD predictions are using optimized scales of ref. 10) and the structure functions set 1 of ref 11). It is seen that the agreement is good, and compatible with error estimates at the possible exception of the lowest p_T bin in pp interactions; this part of the data however, may suffer from biases due to difficulties in the trigger threshold corrections.

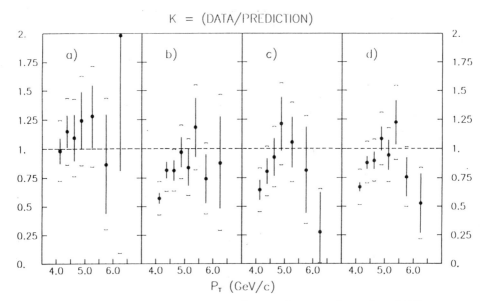

Fig. 11 Ratio data/QCD predictions for pp interactions for different x_F ranges. a) $-.35<x_F<-.15$, b) $-.15<x_F<.15$, c) $.15<x_F<.45$, d) total x_F range.

COMPARISON OF CROSS SECTIONS

The ratio of direct photon production cross sections by π^-p to π^+p and by π^+p to pp are given in fig. 12. The QCD next-to-leading order predictions for these ratios using Duke-Owens structure functions set 1 are also given, and show good agreement with the data.

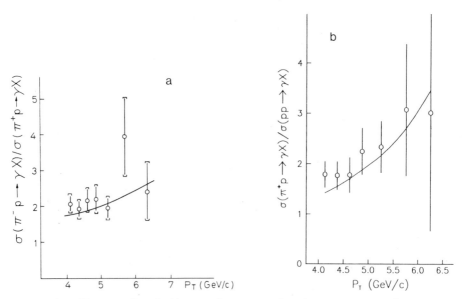

Fig. 12. ratio of direct photon production cross sections:
a) $\sigma(\pi^-p)/\sigma(\pi^+p)$ b) $\sigma(\pi^+p)/\sigma(pp)$

Finally, the difference of direct photon cross sections between π^-p and π^+p interactions is shown in fig. 13. This difference contains annihilation processes only, and is shown to be also well reproduced by QCD predictions.

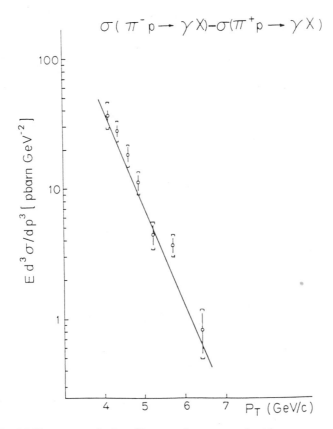

Fig. 13. Difference of the direct photon production cross sections $\sigma(\pi^-p)-\sigma(\pi^+p)$ and QCD predictions.

CONCLUSIONS

The W70 collaboration has measured the prompt photon poduction cross sections by π^-, π^+ and p on hydrogen at 280 GeV/c for transverse momenta between 4.0 and 7.0 GeV/c and Feynman x_F between -0.45 and 0.55.

A comparison of the production cross sections with QCD calculations at next-to-leading order and using structure functions set 1 of ref 11 give good agreement, inside the systematic errors.

REFERENCES

1) W. Beusch: CERN/SPSC 77-70, SPSC/T-17(1977)

2) M. Bonesini et al. "A lead-liquid scintillator electromagnetic calorimeter for direct photon physics" Nucl.Inst.Meth. A261 471 (1987)

3) M. Bonesini et al. "A 1152 channel timing system for an electromagnetic calorimeter readout" Nucl.Inst.Meth. A263 325 (1987)

4) M. Bonesini et al. "Calibration of an electromagnetic calorimeter for direct photon physics" CERN-EP/87-227 Dec 1987, submitted to Nucl.Inst.Meth.

5) M. Bonesini et al. "High p_T trigger electronics for a large orthogonal readout electromagnetic calorimeter" CERN-EP/87-200 Nov 1987, submitted to Nucl.Inst.Meth.

6) J.C. Lasalle, F. Carena, S. Pensotti Nucl. Inst.Meth. 176, 371 (1980)

7) H.U.Bengtsson, G. Ingelman, T.Sjöstrand: "The Lund Monte Carlo for high-p_T scattering, PYTHIA version 4.1, update of H.U. Bengtsson and G. Ingelman: Comp.Phys.Comm.34, 251 (1985); T.Sjöstrand: Jet fragmentation, JETSET version 6.2, update of T.Sjöstrand: Comp.Phys. Comm.27, 243,(1982)

8) M. Bonesini et al. "High transverse π^0 production by π^- and π^+ on protons at 280 GeV/c" Z. Phys. C 37, 39 (1987)

9) M. Bonesini et al. "Production of high transverse momentum prompt photons and neutral pions in proton-proton interactions at 280 GeV/c" CERN-EP/87-222, Dec 1987, to be published in Z. Phys C.

10) Aurenche, R Bayer, A. Douri, M Fontannaz and D. Schiff: Phys.Lett 140B, 87 (1984), and P. Aurenche, R. Bayer, M. Fontannaz, D. Schiff: Nucl. Phys. B286, 509 (1987)

11) D.W. Duke and J.F. Owens: Phys Rev.D30, 49 (1984); J.F. Owens: Phys.Rev. D30, 943 (1984)

12) M. Bonesini et al. "High transverse momentum prompt photon production by π^- and π^+ on protons at 280 GeV/c" CERN-EP/87-185, Oct 1987, to be published in Z. Phys. C.

RESULTS ON DIRECT PHOTON PRODUCTION FROM THE UA2 EXPERIMENT

AT THE CERN PROTON-ANTIPROTON COLLIDER

Karlheinz Meier

CERN
Geneva, Switzerland

INTRODUCTION

The UA2 experiment operating at the CERN proton-antiproton collider has accumulated an integrated luminosity of about 750 nb-1 at $\sqrt{s}$ = 630 GeV during the 1984/85 run periods. This article presents a measurement of the inclusive cross-section of high transverse momentum photons (Section 2). The associated event structure of the direct photon events is discussed in section 3. A signal for production of two direct photons in one event is observed (Section 4). The results presented here (Ref. 1) are an update of a previous publication based on less then half of the present integrated luminosity (Ref. 2).

1. PHOTON DETECTION IN UA2

The UA2 detector covers two regions of polar angle: $40° < \theta < 140°$ (central region) and $20° < \theta < 40°$, $140° < \theta < 160°$ (forward/backward region). Both regions are covered by a central tracking chamber measuring the direction of charged tracks. Also both regions are equipped with preshower detectors which in the study of photons function as active photon converters. In the central region the converter is a 1.5 radiation lengths tungsten cylinder followed by a cylindrical multiwire chamber which determines the direction of converted photons. Converted photons can be localized with a precision of 10 mr and two adjacent conversion signals are resolved if they are separated by more than 35 mr. In the forward/backward regions the preshower detector consists of a 1.4 radiation lengths lead-iron converter followed by a multi-tube proportional chamber. This chamber is able to measure the direction of a converted photon with a precision of 2 mr. An array of 480 calorimeter cells, each cell covering 15° in azimuth Φ and 0.2 units of pseudorapidity (η), measures the energy of electrons and photons. Each cell is segmented longitudinally to separate electomagnetic showers from hadron showers. In the central region the calorimeter has sufficient

thickness (4.5 absorption lengths) to contain most of the
energy in hadron showers. This is not the case in the
forward/backward calorimeter. These regions are however
equipped with magnetic spectrometers to measure the momenta
of charged tracks. A complete description of the various
detector components can be found in Ref. 3.

The copious production of high tranverse energy hadron
jets (Ref. 4) is a large source of background for direct
photon production. Some jets contain high transverse momentum
pseudoscalar mesons (π^0 and η) which may decay into two
photons. Owing to the limited space resolution photon pairs
may appear as single photons in the detector. However since
direct photons are expected to be more isolated in the event
than jet fragmentation products the fraction of direct
photons in the event sample can be increased by selecting
events containing isolated photon candidates. Due to the
different characteristics of the central and forward/backward
detectors different selection criteria are used in the two
regions.

In the central region an energy cluster with a
transverse energy in excess of 12 GeV must have a small
longitudinal and lateral size as expected for showers
originating from electrons or photons. The pattern of
photomultiplier signals must be consistent with that observed
for test-beam electrons (Ref. 3). No charged track and at
most one preshower signal may be found in a cone around the
direction defined by the event vertex and the cluster
centroid. The cone size in pseudorapidity-azimuth space is
defined as

$$\sqrt{\Delta\phi^2 + \Delta\eta^2} < 0.25$$

In the forward/backward region a single calorimeter cell
in excess of 12 GeV transverse energy is required. At most 2%
of the cell energy may be deposited in the last 6 (out of 30)
radiation lengths. Neighbour cells inside an isolation cone
must contain less than 2 GeV. The cone size defined in the
same way as for the central region is 0.53. No charged track
may point to the cell.

There are 10000 events in the central region and 6000
events in the forward/backward region with a photon candidate
satisfying the selection cuts. These events contain a certain
fraction of background photons originating from non resolved
pseudoscalar meson decays. This fraction cannot be measured
event-by-event with the UA2 detector. Instead it will be
evaluated statistically using the "conversion method". This
method determines the fraction of events in which the photon
candidate has begun showering in the preshower detector. This
fraction can be used to extract the amount of non resolved
photon pairs in the sample. Ref. 1 contains more details
about the procedure. The background fraction is found to be
about 30% at 12 GeV transverse energy, dropping to less than
5% at 30 GeV. This decrease is a consequence of the selection
criteria becoming increasingly efficient at rejecting double-
photon background with increasing transverse energy.

2. THE INVARIANT CROSS-SECTION FOR DIRECT PHOTON PRODUCTION

Experimentally the invariant inclusive cross-section for direct photons is evaluated as follows:

$$E \cdot \frac{d\sigma}{d^3p} = \frac{N \cdot (1 - b)}{L \cdot \varepsilon_g \cdot \varepsilon_c \cdot p_t \cdot \Delta p_t}$$

Here N is the number of photon candidates in a p_t bin of width Δp_t. b is the background fraction in the sample of photon candidates. L is the integrated luminosity, ε_g the geometrical acceptance and ε_c the efficiency of the selection criteria for retaining direct photon events.

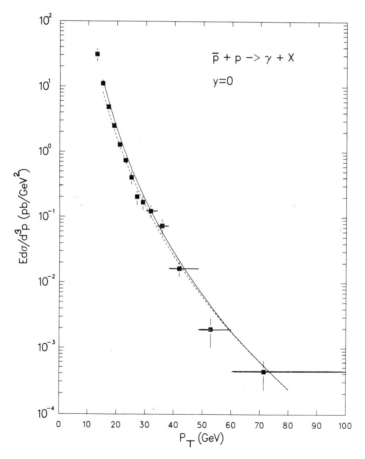

Fig. 1. The invariant cross-section for direct photon production in the central detector (see text for explanation of the curves)

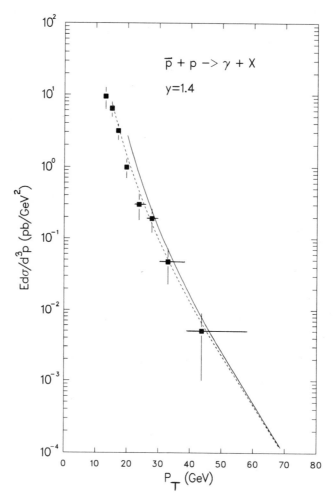

Fig. 2. The invariant cross-section for direct photon
production in the forward/backward detector (see
text for explanation of the curves)

The results are presented in figures 1 and 2. Only p_t
dependent errors are shown and they are mainly statistical.
The total systematic uncertainty on the normalization of the
cross-section is 20%. It receives the following
contributions:

- calorimeter energy scale 10% (Ref. 5)
- cut efficiency   10%
- geometrical acceptance   5%
- background calculation   12%
- integrated luminosity   8%

The results are compared with a next-to-leading order QCD
calculation (Ref. 6). The calculation uses Duke-Owens (set 1)
structure functions (Ref. 7) and the "optimized" choice of Q^2
scales (Ref. 8). The prediction includes a contribution from

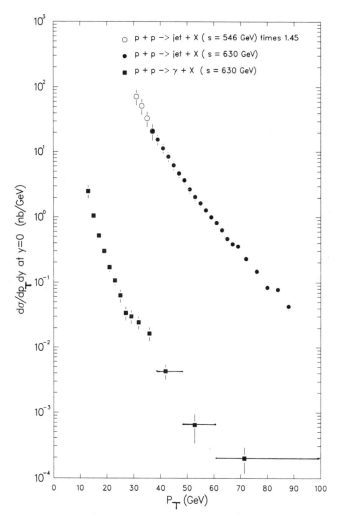

Fig. 3. The differential cross-section for direct photon
production and jet production at $\eta = 0$

bremsstrahlung off the final state quarks which is partly
suppressed in the data due to the isolation criteria. The
effect of excluding all bremsstrahlung photons with an angle
smaller than 20° to the original quark direction is shown by
the dashed curves in figures 1 and 2. At large transverse
momentum, where the inclusion of bremsstrahlung is small, an
excellent agreement between the measurement and the
calculation is observed.

It is instructive to compare the inclusive cross-section
for direct photons and jets (Ref. 4). Figure 3 shows the two
spectra both measured in the central region only. Integrated
over the intervall $p_t > 30$ GeV the ratio of the cross
sections is measured to be

$$\sigma_\gamma/\sigma_{jet} = (4.6 \pm 2.2) \cdot 10^{-4}$$

The error on this number is dominated by the uncertainty in the global energy scale for jets. By relating this number to the ratio obtained from leading order QCD calculations (Ref. 9 and Ref. 10) a value for the strong coupling constant can be determined:

$$\alpha_s \cdot K_2 / K_\gamma = 0.14 \pm 0.07$$

The K-factors express the ratio between the full (not available) QCD calculations and the used leading order calculations for direct photon and jet production.

3. THE ASSOCIATED EVENT STRUCTURE IN EVENTS CONTAINING A DIRECT PHOTON

The dominating process for direct photon production in proton-antiproton collisions at 630 GeV is described by the so called "compton diagram". In this diagram an incoming quark (antiquark) radiates a photon in a gluon field. The final state configuration is therefore expected to be a photon balanced at opposite azimuthal angle by a (quark) jet.

Here only photon candidates and jets in the central region of the detector are being considered. This corresponds to a fiducial volume $|\eta| < 0.77$. The jet detection techniques used in the central calorimeter are described in Ref. 11. The frequency of jets ($E_t > 5$ GeV) expected from a parton model calculation (Ref. 7) taking into account the geometrical coverage and the jet energy resolution in the central calorimeter is approximately 50%. The observed fraction of 46% agrees well with this number.

Figure 4 shows various ratios between the average transverse energies of the photon candidates and the jets as a function of $p_t(\gamma)$. Only events with jets of more than 5 GeV transverse energy are considered. It is seen that the photon and the highest transverse energy jet tend to balance each other. Both together carry 50% to 75% of all transverse energy detected in the rapidity interval $|\eta| < 2$ depending on $p_t(\gamma)$. The next to leading jet carries 30% to 12% of the transverse energy of the leading jet depending again on $p_t(\gamma)$. In short, the photon-jet events are dominated by two and only two objects, the photon and the leading jet.

The following comment concerns the possibility to extract the gluon structure function from the observed photon-jet cross-section. At collider energies the calculated cross-section is rather insensitive towards changes in the gluon structure function. The cross-section changes by about ±10% when varying Λ between 100 MeV and 400 MeV. The value actually used in Ref. 7 (set 1) is 200 MeV. The systematic error of the cross-section measurement is of the order of 30% (affected by uncertainties in the measurements of photon and jet energy). Therefore the present experiment does not have the necessary accuracy to perform a measurement of the the gluon structure function.

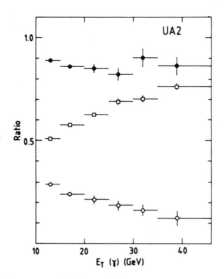

Fig. 4. full circles: ratios of transverse energies
$E_t(j_1)/E_t(\gamma)$
open squares: $E_t(\gamma) + E_t(j_1)$ normalized to the total
transverse energy detected in $|\eta| < 2$
open circles: ratios of transverse energies
$E_t(j_2)/E_t(j_1)$

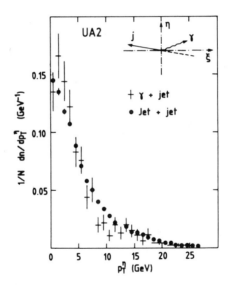

Fig. 5. The component of the photon-jet (jet-jet) transverse
momentum perpendicular to the bisector of the
photon-jet (jet-jet) axis

The observed transverse momentum of the combined system
of photon and leading jet receives contributions from two
sources. Firstly, differences in energy resolution and energy
scale for jet and photon give rise to an instrumental source
of intrinsic transverse momentum. In addition soft gluon
radiation from the initial state can introduce a physical

contribution to this quantity. The component of the total transverse momentum perpendicular to the jet direction, p_t^{η}, is less affected by instrumental biases. Figure 5 compares p_t^{η} of the photon-jet events with that of jet-jet events (Ref. 4). In both cases the sum of the transverse energies of the two involved clusters is required to exceed 40 GeV and their azimuthal separation is required to exceed 120°.

The r.m.s. value of p_t^{η} is 7.0 ± 0.2 GeV for the photon-jet events and 7.5 GeV (with a negligible statistical error) for the jet-jet events. A certain difference between the photon-jet and the jet-jet systems might be expected as a consequence of the different weights given to quarks and gluons in the structure functions. The incoming gluons are expected to radiate more than incoming quarks. This would result in a larger transverse momentum of the scattered two-body system. The experimental evidence for this is however too uncertain to confirm this effect.

4. TWO PHOTON FINAL STATES

Two possible leading order diagrams exist for two photon production in proton-antiproton collisions. The first (quark antiquark annihilation) is a pure QED diagram proportional to α_{EM}^2. The second proceeds via gluon gluon annihilation with a quark loop intermediate state. This diagram is suppressed by α_s^2 but preferred by the structure functions (gluon gluon initial state). In total both diagrams amount to about 0.4% of the direct photon production cross-section (Ref. 9).

UA2 finds 4 events with two photon candidates, each having $p_t(\gamma) > 11$ GeV and $|\eta| < 0.8$. No other high transverse momentum particles are observed in these events. Two possible sources of background have to be considered. Two jet events and photon-jet events both could be mis-identified as two photon events. Using the conversion method the background to the 4 observed events is calculated to be 0.74 ± 0.12 events. The efficiency of the selection criteria is taken as the square of the single photon efficiency. The total cross-section is then :

$$d\sigma_{\gamma\gamma}/d\eta_1/d\eta_2 = 4.1 \pm 2.1 \text{ pb } (p_t(\gamma) > 11 \text{ GeV})$$

The leading order QCD calculation of Ref. 9 yields a cross-section of 1.8 pb. Under the assumption that the K-factor for two photon production is the same as for single photon production the expected cross-section would increase to 3.7 pb.

SUMMARY

Direct photon production in proton-antiproton collisions has been studied by the UA2 experiment.

The inclusive cross-section is measured and found to be in good agreement with QCD theory. A comparison with the cross-section of inclusive jet production shows the expected ratio.

A jet balancing the transverse momentum of the photon is found with the frequency expected from the parton model. The average transverse momentum of photon-jet systems is similar to that of jet-jet systems.

A two photon signal is observed. It is of the size expected from the parton model.

Acknowledgements

I would like to thank Brad Cox for organizing this very pleasant workshop at St. Croix. The results presented are based on a thesis by Peter H. Hansen whom I wish to thank for making his material available to me. I also thank all my colleagues from UA2 for their support. Special thanks go to Annette for the careful preparation of the manuscript.

REFERENCES

1. Peter H. Hansen, Licentiatafhandling (thesis), Niels Bohr Institute, Copenhagen (1987).
2. UA2 Collaboration, J.A. Appel et al., Phys. Lett. 176B (1986) 239.
3. M. Dialinas et al., LAL Orsay preprint LAL-RT/83-14; C. Conta et al., Nucl. Instr. and Meth. 224 (1984) 65; K. Borer et al., Nucl. Instr. and Meth. 224 (1984) 29; A. Beer et al., Nucl. Instr. and Meth. 224 (1984) 360.
4. UA2 Collaboration, J.A. Appel et al., Phys. Lett. 160B (1985) 349.
5. UA2 Collaboration, R. Ansari et al., Phys. Lett. 186B (1987) 440.
6. P. Aurenche et al., Phys. Lett. 140B (1984) 87.
7. D.W. Duke and J.F. Owens, Phys. Rev. D30 (1984) 49.
8. P. Aurenche et al., Nucl. Phys. B286 (1987) 509.
9. E.L. Berger et al., Nucl. Phys. B239 (1984) 52.
10. B.L. Combridge et al., Phys. Lett. 70B (1977) 234.
11. K. Meier, contribution to these proceedings

DIRECT PHOTONS IN UA1

Benoît Mours

L.A.P.P., B.P. 909
F-74019 Annecy-le-Vieux CEDEX, France

UA1 Collaboration

Aachen - Amsterdam(NIKHEF) - Annecy(LAPP) - Birmingham - CERN - Harvard - Helsinki - Kiel - Imperial College, London - Queen Mary College, London - MIT - Padua - Paris(Coll. de France) - Riverside - Rome - Rutherford Appleton Lab. - Saclay(CEN) - Victoria - Vienna - Wisconsin Collaboration

Abstract

The production of isolated direct photons has been measured in a large transverse energy and rapidity range in proton antiproton collisions at $\sqrt{s} = 546$ GeV and 630 GeV, using the UA1 apparatus at the CERN p$\bar{\text{p}}$ Collider. The results are in good agreement with QCD predictions.

Introduction

The direct photon production in hadronic collisions is a good test of QCD since the process is well defined and complete α_s^2 QCD predictions are available. The most important experimental problem is the large background coming from π^0 (and other neutral mesons). Usually, the rejection of this background is done by a 'direct method' where the two decay photons are detected and their invariant mass is found to be consistent with the π^0 mass. A Monte Carlo program is used to compute the undetected background. This method is unusable at the Collider energy and with the UA1 detector: for this analysis, we look for photons with very high momentum (> 16 GeV) which means that the background of π^0 decaying into two photons with very small opening angle, cannot be solved with the granularity of the UA1 detector. Nevertheless, we can use a statistical method to show the direct photon signal and to reject the background. This method uses the difference between direct photon and π^0 cluster, in isolation and in longitudinal energy deposition in the calorimeter. For the isolation, which is a very powerful selection criterion, we take advantage of the fact that π^0's are produced in jets, which are the high-transverse-momentum objects produced with the highest cross-section. Since it is very improbable that jets fragment in only one particle, the π^0's are produced with some other particle and are therefore less isolated than direct photons.

We first present the data-taking conditions. Then we describe the method used to extract the direct photon signal and to compute the cross-section; we compare the results with QCD predictions and with UA2 published data. The sensitivity of the QCD predictions to various parameters is also discussed.

Data-taking

The UA1 detector has been extensively described in ref. [1] where the reader will find more details. The data-taking conditions are those used for W/Z studies. The photons are detected by lead-scintillator electromagnetic (e.m.) calorimeters, 'gondolas' for an absolute value of the rapidity $\eta < 1.5$ and 'bouchons' for $1.5 < |\eta| < 3$. The energy resolution is approximately $\sigma(E) = 0.16\sqrt{E}$, where E is in GeV.

We use data collected during the 1983 Collider run at $\sqrt{s} = 546$ GeV (integrated luminosity, $\int L \, dt = 118 \, \text{nb}^{-1}$), and during the 1984 and 1985 runs at $\sqrt{s} = 630$ GeV ($\int L \, dt = 565 \, \text{nb}^{-1}$). The first-level trigger asks for two calorimeter cells with transverse energy $E_T > 10$ GeV. The second-level trigger does a more precise energy reconstruction in the calorimeter and requests an $E_T > 10$ or 15 GeV, depending on the running conditions. Then events go through the reconstruction of the full detector information where we use that from the hadronic calorimeter to sign the e.m. characteristic of the cluster. The central detector data (a large-volume drift chamber surrounding the interaction point and operated in a homogeneous 0.7 T magnetic field) are used to compute charge isolation.

Extraction of the direct photon signal

We do the analysis [2] in three rapidity ranges, corresponding to the three parts of the detector, i.e. central region of the gondolas ($|\eta| < 0.8$), gondolas with a large incident angle ($0.8 < |\eta| < 1.4$), and bouchons ($1.6 < |\eta| < 3$). In this section, we present the method used to extract the direct photon signal for the subsample of events with $E_T > 20$ GeV, and $E_T > 16$ GeV for the bouchons.

We do a first selection which keeps all events with isolated neutral e.m. cluster. This selection includes

- technical cuts, in order to keep well-measured clusters and reject multi-π^0 clusters;
- that there be no significant missing energy, in order to suppress cosmic-rays and beam halo;
- a charge isolation requirement, in order to select neutral a cluster and to reject jets; we demand $\sum \vec{p}_T$, the vectorial sum of the transverse momentum of charged tracks detected in a cone of radius $R = \sqrt{\eta^2 + \phi^2} = 0.7$ around the cluster, to be less than the minimum of 2 GeV and 10% of the cluster E_T;
- a calorimeter isolation requirement, in order to reject jets: we demand $\sum \vec{E}_T$, the vectorial sum of the additional E_T measured in a cone of radius $R = 0.7$ around the cluster, to be less than the minimum of 2 GeV and 10% of the cluster E_T.

These selection criteria give a sample of 3664 events ($\sqrt{s} = 546$ GeV and 630 GeV), free of instrumental background, but with a non-neglibible contamination of π^0.

We evaluate this π^0 contamination by studying the event distribution as a function of four variables. The first one is E_c, defined by:

$$E_c = max\left(\sum_{R=0}^{0.7} E_T, \sum_{R=0}^{0.7} p_T \right)$$

E_c is the transverse energy in a cone of radius R = 0.7 around the γ/π^0 candidate (the γ/π^0 energy is not included in E_c). For a direct photon, E_c should correspond to the contributions from spectators and from radiated gluons in the parton collision (underlying event). For a π^0, the E_c contains, in addition, fragments of the jet and therefore the distribution should extend to higher values. To determine the E_c distribution for a pure direct photon sample, we use minimum bias events, since their energy flow is similar to that of the direct photon, if we exclude the jet regions. To compute π^0 isolation, we use two-jet ISAJET [3] Monte Carlo events with full simulation of the UA1 detector, reconstructed and selected as real data. Distributions of events taken at $\sqrt{s} = 630$ GeV are presented in fig. 1a for the three rapidity ranges. There is a peak at $E_c = 0$, corresponding to perfectly isolated events, which is what we expect for a direct photon. We fit these data distributions by a superposition of the two individual components just leaving as a free parameter F_γ, the fraction of direct photon events. The results of the fit are given in table 1. Systematic errors on pure direct photon distributions are estimated by looking at the difference between the energy flow from minimum bias events and from real direct photon candidates. For pure π^0 distributions, we change by $\pm 5\%$ the E_c value. The E_c distributions for isolated π^0 Monte Carlo events have also been checked with real isolated hadron events for which we expected a similar behaviour.

Since the size of a jet is R = 1, and E_c is taken just in a cone of radius R = 0.7, we can perform an independent measurement of the isolation by studying the variable E_r, the transverse energy measured in a ring around the γ/π^0 candidate of internal radius = 0.7 and an external radius = 1, defined by:

$$E_r = max\left(\sum_{R=0.7}^{1} E_T, \sum_{R=0.7}^{1} p_T \right).$$

As in the E_c case, E_r contains the underlying event for the direct photons and the π^0's, and jet fragments for the π^0's only. Figure 1b shows the data distribution as well as the result of the fit. Like for the E_c distribution, there is an accumulation of events at $E_r = 0$ which corresponds to direct photon signal. The values of F_γ are given in table 1. Systematic errors are computed in way similar to that used for the E_c fit.

The third variable that we study is sensitive to the longitudinal energy deposition in the e.m. calorimeter, which is segmented into four samplings. For the π^0's, we expect a shorter penetration in the calorimeter, since the two decay photons have an individual energy lower than that of a single photon. We will therefore use S_1/E, the fraction of energy deposited in the first sampling of the calorimeter. This sampling corresponds to 3.3 radiation lengths at normal incidence. The S_1/E variable is corrected for angular effect.

The pure distributions for π^0's and single photons are computed by a full simulation of the e.m. shower in the calorimeter with the GEANT Monte Carlo [4], including multi-π^0 effects. Figure 2a presents the data distributions including the result of the fit, and the pure π^0 and direct photon distributions. Values of F_γ are given in table 1. We should notice that these values of F_γ, which are in good agreement with the previous ones, are determined with completely independent criteria.

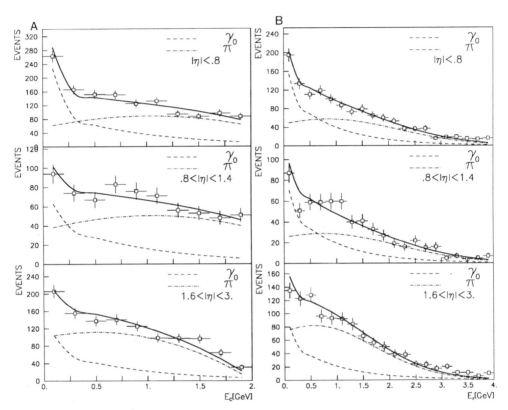

Figure 1. E_c and E_r data distributions for photon candidates with $E_T > 20$ GeV ($E_T > 16$ GeV if $|\eta| > 1.6$). The solid line is the result of the F_γ fit. It is the sum of the direct photon contribution (dashed line) and the π^0 contribution (dash-dotted line).

In order to extract the maximum information from the E_c, E_r and S_1/E variables, we define a new one as:

$$L_\gamma = \ln(P_{E_c} \times P_{E_r} \times P_{S_1/E})$$

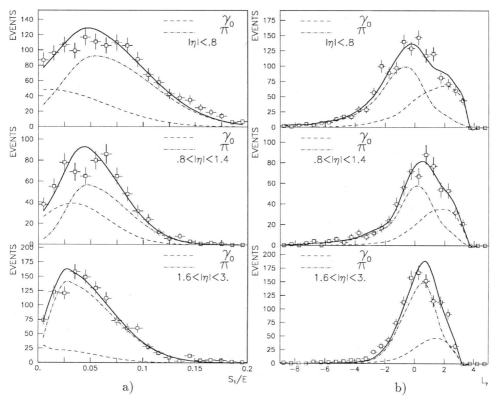

Figure 2. S_1/E and L_γ data distributions for photon candidates with $E_T > 20$ GeV ($E_T > 16$ GeV if $|\eta| > 1.6$). The solid line is the result of the F_γ fit. It is the sum of the direct photon contribution (dashed line) and the π^0 contribution (dash-dotted line).

where the P_i are the i $(E_c, E_r, S_1/E)$ probability density functions for direct photon events. Data distributions are presented in fig. 2b with, again, the result of the fit. Table 1 gives the corresponding F_γ values. For the two centre-of-mass energies and for each rapidity range, the four values of F_γ are consistent and correspond to a clear direct photon signal.

Single-photon cross-section

In order to enrich with direct photons the data sample used to compute the cross-section we add to the selection criteria described above the request for a positive value of L_γ. For both values of $\sqrt{s}$, 2350 events satisfy these selection criteria. A priori, the efficiency for the detection of the soft fragment accompanying the π^0 should increase with its energy. Then F_γ should depend on E_T. To determine this dependence, we measure F_γ for several E_T ranges by fitting the L_γ distributions in these ranges. Figure 3 presents these fits for photon candidates with $|\eta| < 0.8$. The L_γ distribution seems more compatible with the pure direct photon distribution when E_T increases.

We use a parametrization of F_γ to subtract the residual fraction of π^0 from the data distributions. This parametrization is computed by fitting the F_γ values measured in the same E_T ranges as above. Figure 4 presents these values and the parametrization used. The values of F_γ are higher than those given by table 1, since we now use the sample of events which have a positive value of L_γ. The dashed area of fig. 4 corresponds to the systematic error on F_γ.

Table 1. Fraction of direct photons (F_γ) in percent

Variable	$0 < |\eta| < 0.8$	$0.8 < |\eta| < 1.4$	$1.6 < |\eta| < 3$
F_γ for $\sqrt{s} = 546$ GeV			
E_c	$42 \pm 8 \pm 6$	$65 \pm 13 \pm 13$	$34 \pm 12 \pm 18$
E_r	$60 \pm 10 \pm 13$	$42 \pm 14 \pm 6$	$23 \pm 11 \pm 13$
S_1/E	$76 \pm 12 \pm 27$	$77 \pm 17 \pm 20$	$21 \pm 11 \pm 13$
L_γ	$47 \pm 7 \pm 9$	$48 \pm 11 \pm 12$	$33 \pm 9 \pm 12$
F_γ for $\sqrt{s} = 630$ GeV			
E_c	$42 \pm 3 \pm 6$	$32 \pm 6 \pm 7$	$28 \pm 6 \pm 15$
E_r	$45 \pm 4 \pm 10$	$44 \pm 6 \pm 7$	$27 \pm 5 \pm 15$
S_1/E	$29 \pm 4 \pm 11$	$41 \pm 7 \pm 11$	$13 \pm 5 \pm 9$
L_γ	$40 \pm 2.5 \pm 7$	$38 \pm 4 \pm 10$	$25 \pm 5 \pm 9$

After background subtraction, we correct for acceptance (83%), and selection efficiencies ($\approx 55\%$), we unsmear the energy resolution and size of the cell, and finally obtain the cross-section. Figure 5 presents the values for $\sqrt{s} = 546$ GeV and $\sqrt{s} = 630$ GeV. The systematic errors correspond only to the background correction. In addition, there is an overall uncertainty of 23%; this comes from the luminosity (15%), the calibration of the calorimeter (3% on the energy scale inducing a 16% uncertainty on the cross-section), and the efficiencies (8%). On this figure, we plot the full order-α_s^2 QCD predictions from Aurenche et al. [5]. These cross-sections are computed with set 1 structure functions from Duke and Owens [6], in the so-called 'optimized scale scheme'[7]. They are in good agreement with our data. The full line in fig. 5 is the isolated cross-section obtained by the exclusion of all quark-photon bremsstrahlung with quark-photon angle less than 57^0 (one radian); this corresponds in a first approximation to our isolation criteria at $\eta = 0$. The dashed line in fig. 5 corresponds to the total cross-section.

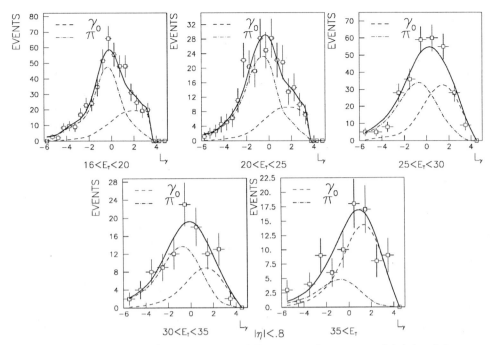

Figure 3. L_γ, distribution for several E_T ranges for events with $|\eta| < 0.8$.

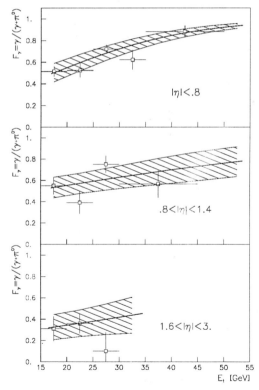

Figure 4. Fraction of direct photons as a function of E_T. The solid line is a parametrization of the points. The dashed area represents the systematic uncertainties.

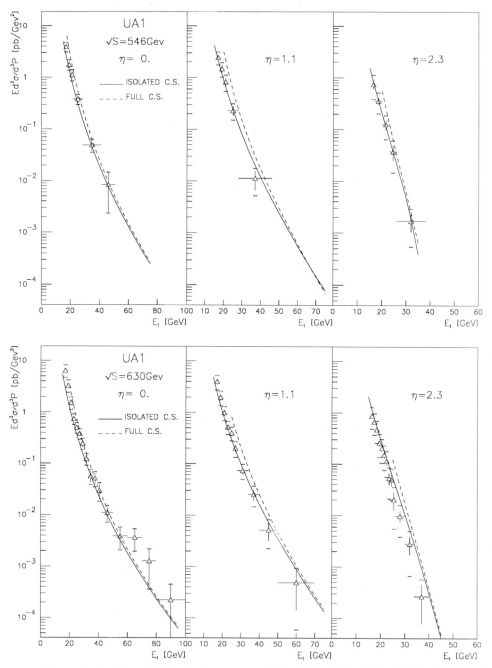

Figure 5. Isolated single photon cross-section. The dashed line is the full QCD cross-section [5]. The full line is the isolated cross-section where we exclude all quark-photon bremsstrahlung with quark-photon angle less than 57^0 (one radian).

We can compare the cross-section measured in the central region at $\sqrt{s} = 630\text{GeV}$ with the UA2 published result [8] (fig. 6). Both measurements are in good agreement within their global systematic errors (23% for UA1 and 20% for UA2). Theoretical curves are the same as before.

A more precise way of comparing the data with the theoretical predictions is to calculate their ratio. Figure 7 presents this ratio as a function of the scaling variable $x_T = 2E_T/\sqrt{s}$ for the three rapidity ranges. The background correction errors, which are E_T-dependent, are included in the error bars. In addition, there is an uncertainty of 23% on the normalization.

The reference theory used to compute the denominator is, as for the previous figure, the isolated cross-section prediction from Aurenche et al., using the Duke and Owens set 1 structure functions with $\Lambda = 200$ MeV and the optimized scale. In order to show the importance of the isolation criteria, we plot on fig. 7 the full cross-section divided by the isolated cross-section (dashed line). The isolation criteria induce a strong change in the cross-section at low values of x_T, which is also the region where the background is the most important. Therefore, the small discrepancies in this region are not significant.

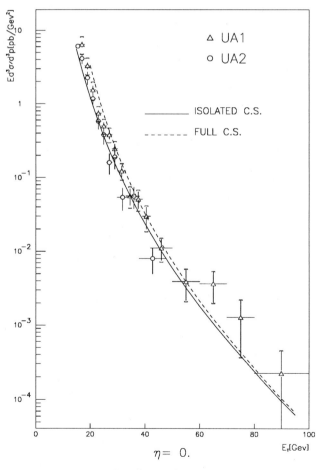

Figure 6. UA1 and UA2 isolated single photon cross-section measured at $\sqrt{s} = 630$ GeV.

As an example of the sensitivity of the model, we plot also in fig. 7 the ratios of computed cross-section with the set 2 structure functions of ref. [6] and with those of Eichten et al. [9], to the reference cross-section. The variations in the theoretical predictions are too small to allow a choice between these sets. The modification of the value of the QCD parameter Λ between 100 and 400 MeV, or the change of the optimized scale to $Q^2 = p_T^2$ or $Q^2 = 4p_T^2$ or the variation of the slope of the gluon structure function (η_G between 4 and 8 with $xG(x)=A_G(1-x)^{\eta_G}$) induce similar changes in the cross-section (with maximum variation of $\approx 30\%$). This weak sensitivity of the theory is mainly due to the high value of the photon momentum, which corresponds to a kinematical region where the variations of α_s are small. This relative insensitiveness of the theory to these various parameters makes the test of the theory a very strong one, since the predictions for the cross-section are absolute, with few degrees of freedom, and agree with the experiment.

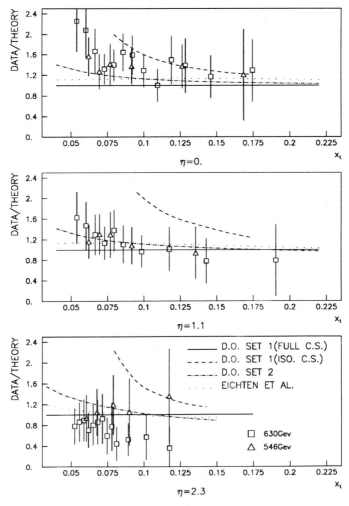

Figure 7. Data over theory. The reference theory is the isolated cross-section from Aurenche et al., using the Duke and Owens set 1 structure functions with $\Lambda = 200$ MeV and the optimized scale.

Summary

A clear signal of direct photon production in p$\bar{\text{p}}$ collision at $\sqrt{s} = 546$ GeV and $\sqrt{s} = 630$ GeV has been observed by the UA1 experiment. The isolated direct photon cross-sections are well described by QCD predictions. These predictions are in our kinematical range, weakly sensitive to QCD 'free parameters' (structure functions, Λ, choice of scale). Our result is consistent with the published direct photon data from UA2.

Acknowledgements

I would like to thank Prof. B. Cox who invited me to give this talk. I am also grateful to P. Aurenche, R. Baier, M. Fontannaz and D. Schiff for helpful discussions and for providing me with their predictions adjusted to the experimental conditions of the present study.

References

[1] UA1 Collaboration (G. Arnison et al.) Phys. Lett. **B122** (1983) 102 and (1983) 273.
[2] B. Mours, Thesis at the Université de Savoie (France) (1988).
[3] F.E. Paige and S.D. Protopopescu, ISAJET program, Brookhaven report BNL 29777 (1981).
[4] R. Brun et al., GEANT3, CERN DD/EE/84-1 (1984).
[5] P. Aurenche, R. Baier, M. Fontannaz and D. Schiff, Phys. Lett. **B176** (1984) 87 and Orsay preprint LPTHE/87-30 (1984).
[6] D.W. Duke and J.F. Owens, Phys. Rev. **D30** (1984) 49.
[7] P.M. Stevenson and H. Politzer, Nucl. Phys. **B277** (1986)758.
[8] UA2 Collaboration (J.A. Appel et al.), Phys. Lett. **B176** (1986) 239.
[9] E. Eichten et al., Rev. Mod. Phys. **56** (1984) 579.

LARGE P_T PHOTOPRODUCTION IN CERN-EXPERIMENT WA69

Ewald Paul

Physikalisches Institut
University of Bonn
Bonn, Federal Republic of Germany

INTRODUCTION

The OMEGA-photon collaboration (Bonn-CERN-Erevan-Lancaster-Manchester-RAL-Sheffield) has taken data for the processes

$\gamma p \rightarrow$ hadrons with tagged photons of $60 < E_\gamma < 175$ GeV

and

$\pi^+(K^+)p \rightarrow$ hadrons at beam momenta of 80 and 140 GeV/c.

The data taking has been completed in 1986, and the total statistics accumulated corresponds to ≈ 20 M γp events and ≈ 24M $\pi(K)p$ events, i.e. ≈ 6M per $\pi(K)$ momentum and polarity, respectively.

The data for both processes include most of the hadronic cross section: only some fraction of low multiplicity and low p_T events which, in the case of photoproduction, are dominated by e^+e^- pair production, has been partially suppressed. The choice of such an "interaction" trigger did not cost sensitivity in the range of large p_T photoproduction, i.e. the data acquisition system of the experiment could stand with such a trigger the available photon flux (of typically $7 \cdot 10^5$ energy tagged γ's per burst of 1.8 s duration).

The two parts of the experiment (i.e. photon and hadron part were carried out as far as possible with the same experimental set up (there was only an extra trigger component for vetoing e^+e^- pairs in the photon case), and essentially the same software was applied to analyze the two data sets afterwards.

Thus, experiment WA69 was particularly suited for studying photoproduction in comparison to hadroproduction; and, due to the interaction trigger, this comparison can be carried out throughout the whole p_T range from soft to hard processes.

This comparison is essential for the analysis of large p_T photoproduction in this experiment. With respect to this physics it is the aim of the experiment to quantify the pointlike-photon interactions of which are primarily interesting (Fig. 1) the Born level processes of QCD Compton and γg fusion and some Higher-Twist processes, too. However, in the single particle p_T range accessible in this experiment (typically $p_T < 5\,\text{GeV/c}$) there are other contributions from hadronlike-photon interactions which are completely dominant for the cross section at low p_T. These processes which die out with increasing p_T faster than the pointlike-photon Born level processes, are down to comparable magnitude at $p_T \leq 3\,\text{GeV/c}$ [1]. The hadronlike-photon component is mainly non-perturbative and thus not calculable reliably. (There is an admixture of a hadronic contribution due to the so-called anomalous photon structure function which can be calculated[1]). For a quantitative study of the pointlike-photon component it is necessary to estimate the non-perturbative hadronlike process using hadron-beam data assuming the validity of Vector Meson Dominance (i.e. the $q\bar{q}$ pair coupling to the photon (Fig. 2) is dominantly in a virtual ρ, ω and ϕ state) and simple quark model relations between incoming $\pi^{\pm}$ and $K^{\pm}$ on one side and the vector mesons on the other. The VDM model in combination with the quark model has been found to be adequate to a level of precision of $\approx 20\%$ for describing photoproduction at low p_T (see ref. 2). At medium p_T ($1 \leq p_T \leq 1/8\,\text{GeV/c}$) where other contributions than the hadronlike-photon component are still a small admixture[1], the validity of this model assumption is tested by experiment WA69. Finally extrapolations into the adjacent larger p_T range (i.e. $p_T > 1.8\,\text{GeV/c}$) are the basis for estimating the pointlike-photon contributions that add to the hadronlike process present there. The extrapolations are not limited to single variables (p_T, p_L etc.), but can be extended to combinations of such or even to the analysis of full event topologies.

The data analysis work of experiment WA69 is only begining. I report here first results on inclusive charged particle distributions with photon and hadron beams being analyzed typically $\approx 20\%$ and $\approx 10\%$ of the available data, respectively. In addition, preliminary results on inclusive π^0 production have been achieved which provide some impression of what might finally be obtained in this sector. Exclusive studies of events which use the capabilities of OMEGA as a multiparticle detector are in progress; results are expected in a few months.

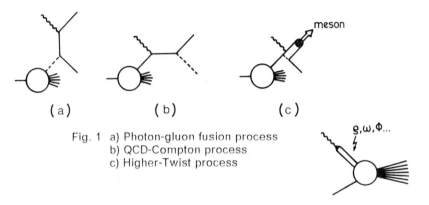

Fig. 1 a) Photon-gluon fusion process
b) QCD-Compton process
c) Higher-Twist process

Fig. 2. VDM-type process

THE EXPERIMENT

Photon and hadron beams were provided in the E1/H1 beam line in the WA of CERN.

The photon beam (Fig. 3) was produced from 450 GeV protons, extracted from the SPS, via decay photons from a first target and a 200 GeV separated e⁻ beam from a conversion in a second target which was tagged before and after producing bremsstrahlung in a third target. The resulting tagged photon beam into OMEGA covered the range from 60 to 175 GeV/c with $1/E_\gamma$ dependence (Fig. 4) and a resolution due to the tagging spectrometer performance of $DE_\gamma/E_\gamma \approx 0.1\%$. In view of this small error the energy-momentum resolution

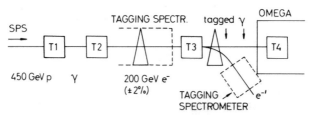

Fig. 3. Schematic photon beam line

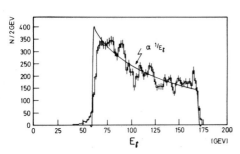

Fig. 4. Tagged photon energy distributions

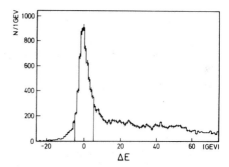

Fig. 5. Energy resolutions

of the events in OMEGA was finally dominated by the reconstruction performance of the secondary particles in the spectrometer set up. The resolution in energy which was achieved is demonstrated in Fig. 5, where we plot for the low charged multiplicity events, the difference between γp energy (i.e. the energy from the tagging system) and the energy sum of the secondaries (i.e. the energy of the charged particles measured in OMEGA): The distribution has a FWHM of ≈4 GeV averaged over the photon energy range implying that the longitudinal energy resolution is ≈2 GeV.

The hadron beams are standard unseparated beams. The facility to flag and trigger on π and K separately is provided by differential Cerenkov Counters.

The photon beam into OMEGA interacted in a 60cm H_2 target (Fig. 6). Charged particles were bent in a homogeneous magnetic dipole field and detected in a set of wire chambers downstream from the target which consisted of proportional chambers (A and B) and two large drift chambers (DC). For wide angle slow tracks another set of proportional chambers (C) was placed on both sides of the target.

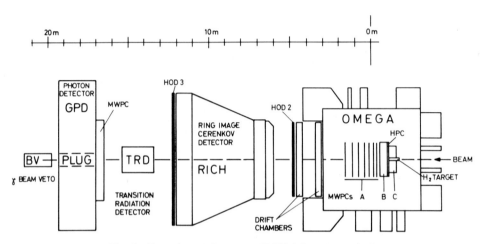

Fig. 6. Experimental set up (OMEGA Spectrometer)

Charged particles with momentum > 5 GeV/c leaving the OMEGA magnetic field pass through the Ring Imaging Cerenkov Counter (RICH) and the Transition Radiation Detector (TRD), respectively, where they are identified as pions, kaons, protons, etc. over most of the momentum range.

The photons from decays of π^0 etc. are detected in two photon detectors: the WA70 lead scintillator calorimeter (GPD), described by M. Martin in his talk at this workshop, and a lead-scintillator fibres counter (PLUG) placed in the central hole of the GPD which was built for this experiment. Both detectors together provide a rather smooth geometrical acceptance covering π^0's from small to large p_T.

In this talk emphasis will be put on large p_T physics with charged particles. So I will next discuss their measurement in the OMEGA chamber set up.

150

In the reconstruction of the charged tracks, one has to manage track finding, track fitting and the reconstruction of the event vertex in the H_2 target (not known a priori) for high multiplicity final states. The level of performance has to be so efficient that the extremely rare large p_T tracks are not overwhelmed by background of false tracks. The WA69 collaboration has made a considerable effort to improve the performance of track reconstruction in multiparticle events and to understand the level of the performance which was finally achieved. In this context extensive use was made of visual inspection of events and their details by advanced graphics tools which was developed for this purpose.

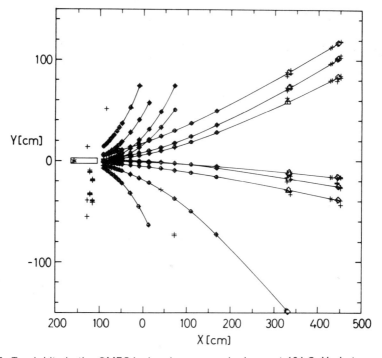

Fig. 7. Track hits in the OMEGA chambers: $\gamma p \rightarrow$ hadrons at 164 GeV photon energy. + unassociated hits; o hits associated to a track (A,B,C chambers); $\triangle/\diamondsuit$: track hits in drift chambers (DC1 and DC2); * event vertex.

As an example, Fig. 7 shows a successfully reconstructed event: the 12 forward going tracks were properly reconstructed and linked together to a common vertex. There is practically no incoherent background of hits which could confuse track finding in the forward hemisphere. A few hits have not been associated with tracks. This problem typically exists in the region close to the vertex where the distance between tracks is small. In general, the efficiency of track finding in the C chambers is lower than in the forward chambers: however, these chambers are not essential for larger momenta since such tracks have, in general, hits in the forward chambers. - The mean track reconstruction efficiency was found to be $\approx 85\%$ for momenta ≥ 0.6 GeV/c.

With respect to false large p_T tracks, the level of performance reached is summarized in the following.

There are essentially three potential background classes to be considered:

1. Spurious tracks being built up from incoherent hits due to beam halo etc.

2. Genuine tracks originating from upstream processes (scatterings in the upstream materials etc.) which then fake an event in the target (or accompany an event in the target).

3. False tracks due to a failure of pattern recognition or track reconstruction: track points could be picked up by the wrong track, a track could be broken into two parts or track elements from different tracks are erroneously linked together. (Tracks originating from secondary interactions of the particles near the primary vertex can be "absorbed" in the latter due to the imperfection of the vertex fit; however, it has been estimated that such cases do not contribute substantially to the background in the large p_T sample).

The first class of background was found to be negligible due to the low level of incoherent hits. The second class of background contributes roughly on the same level of statistics as the real large p_T tracks for $p_T \geq 2$ GeV/c. However, it could be reduced by cuts on the vertex errors.

The third class of background is more difficult to handle. Two somewhat orthogonal methods have been developed independently on the basis of careful visual inspection of events on graphics displays:

- Method A, (developed by A.T. Doyle[3]): depends on a set of geometrical cuts that should not affect the p_T distributions. Only those tracks are accepted which are detected over most of their traversal of the detector.

- Method B, (developed by H. Rotscheidt[4]): simply uses cuts on track errors.

The two methods give results which are consistent with each other. If one starts off from a sample of N_2 large p_T track candidates (after the track sample has been cleaned of second class background), about 50% of the tracks survive both methods and the overlap of the same tracks in the two final samples is $\approx 80\%$. The remaining background is estimated to be $\leq 20\%$ of the final sample.

INCLUSIVE CHARGED PARTICLE DISTRIBUTIONS

On the basis of the two methods A and B described above the two paths of parallel analysis work have been carried out on two sets of photon beam data and one set of hadron beam data. In the following, I am going to discuss the preliminary results obtained.

In one pass $\approx$3M γp and $\approx$2M πp events ($\pm$ 140 GeV/c) have been analyzed where the data samples have been cleaned using method A[3]. Inclusive charged particle distributions of $d\sigma/dp_T$ vs. p_T are shown in Fig. 8 for γp for 7 intervals of E and in Fig. 9 for π^+p and π^-p at 140 GeV/c, respectively. The full lines drawn correspond each to a fit following the parameterization of Donaldson et al.[5] (free parameter fits to the range $p_T \geq 0.75$ GeV/c, more details are given elsewhere[3]).

The difference was quantified by the ratio of the combined data from the highest 3 E ranges (so that $114 < E < 175$ GeV/c) with the combined $\pi^{\pm}$p data at 140 GeV/c. This ratio multiplied with an appropriate VDM factor[6] is shown in Fig. 10 as a function of p_T (the range of $p_T \approx 0$ was excluded because the acceptance is reduced there due to trigger biasses). The error bars are statistical only assuming that systematic errors cancel for this ratio. Whereas the γp data are entirely consistent with the πp data at the lower p_T($p_T \approx 0.75$ GeV/c) after having multiplied the latter data with the VDM factor, one observes an increasing departure from the πp data at larger p_T indicating extra contributions to the γp cross section with weaker p_T dependence than the VDM component. The ratio changes by a factor $\approx$2 for p_T rising from 1 to 3 GeV/c. This factor is similar to that observed by NA14 under slightly different kinematical conditions[7]. The most natural interpretation of this effect is that it is due to the onset of pointlike-photon processes on top of hadronlike-photon contributions. Of course, the next steps in this analysis must go towards a comprehensive and quantitative separation of the pointlike contributions from the other components and the detailed comparison to theoretical predictions.

More quantitative details have already been studied in the other (parallel) path of the data analysis. This analysis was based on 6M γp events, and the same 2M πp data. Method B was applied to achieve the final data sets[4]. For simplifying data handling at this stage, the data samples were preselected for events with at least one charged track of $p_T \geq 1$ GeV/c and $x_F > -0.2$. Apparatus and software efficiencies have been calculated using detailed Monte-Carlo. The software acceptance as function of p_T (shown in Fig. 11) is large throughout the p_T range and approximately independent of p_T.

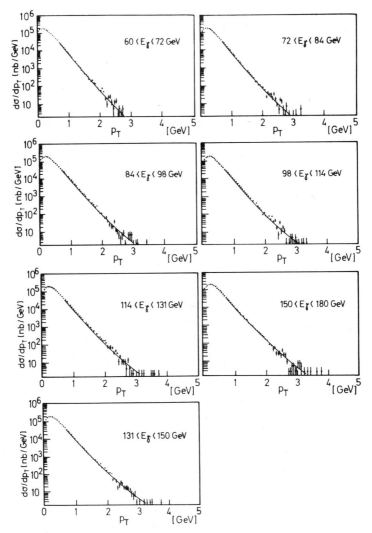

Fig. 8. $d\sigma/dp_T$ vs p_T for $\gamma p \rightarrow \pi^{\pm}X$

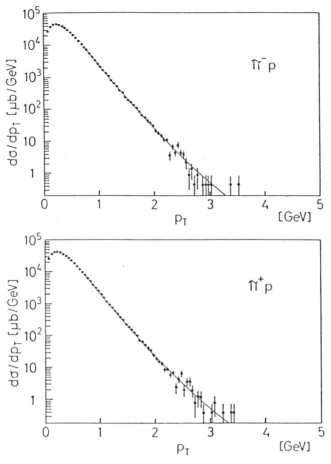

Fig. 9. $d\sigma/dp_T$ vs. p_T for $\pi^\pm p \rightarrow \pi^\pm X$

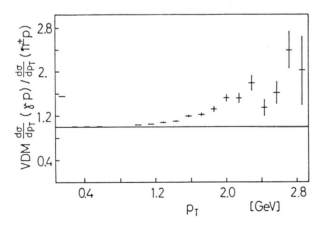

Fig. 10. Ratio of cross section as function of p_T (see text)

The preliminary inclusive charged particle invariant cross sections Ed³σ/dp³ for γp and πp data are shown in Fig. 12 as a function of p_T. For better comparison the hadron beam cross section was divided by the appropriate VDM factor[6] of 214. The errors are statistical only at this stage. For the γp data there are no previous measurements to be compared with. The overall normalization is consistent with that found by NA14 at large p_T for π^0 production on Li (no normalization of comparable charged photoproduction data is available). The $\pi p \to \pi^+ X$ data of WA69 is compared to $\pi^\pm p \to \pi^0 X$ of Donaldson et al.[5] at 100 and 200 GeV/c after having applied a factor of 2 in order to correct for the relative rate of charge to neutral π production. We note reasonable agreement with the mean values of 100 and 200 GeV/c data when allowing for systematic uncertainties of $\leq 20\%$.

First comparisons to theoretical predictions have also been carried out. As a first step, the γp data were compared to simulations using the LUND-LUCIFER program[8] which takes into account first order QCD, i.e. the Born terms in Figs. 1a,b, and calculates a full string-type hadronization. This allows us to simulate the complete multiparticle final states which we want to analyze in WA69. However, since higher order processes are not included, there is a uncertainty for the prediction of the absolute normalization of the cross sections of approximately a factor of two. This problem has been overcome recently by Aurenche et al.[1] who carried out calculations up to second-order QCD processes of the pointlike-coupling terms (and of the anomalous photon component at the LL level). The absolute normalization and the inclusive single particle distributions are finally compared to predictions at this level.

Predictions for the πp data were derived from the LUND-Twister[9] program which relies on second order QCD parameterization assuming the rigorous validity of VMD.

In Fig. 13 the p_T dependent ratio of γp and πp cross sections from this second path analysis is shown together with the ratio of the corresponding predictions. The data ratio and its rise with p_T is in agreement with the first-path data set (Fig. 10). It is also roughly consistent with the predictions on the level of the LUND-program simulations, considered here. A direct comparison of measured and predicted cross sections (not shown here) suggest moreover that the same values are approached at the largest p_T in both cases.

INCLUSIVE NEUTRAL PARTICLE DISTRIBUTIONS

The photons from the π^0 decay were detected in both photon detector components: the GPD of WA70 and the PLUG in the central hole (Fig. 6). Three cases have to be analyzed: both γ's in the PLUG or in the GPD or one γ in each detector.

At the current state of the analysis, 2M γp data have been studied with respect to inclusive π^0 production by averaging over the 3 detector configurations. A first impression of what will be achieved is shown in Fig. 14 which shows as an example the p_T distributions of π^0 with $0.2 < X_F < 0.3$ from $\approx 10\%$ of the data. The following observations can be made: The p_T values covered by WA69 are in the range from about 0 to 1.8GeV/c, i.e. for the final

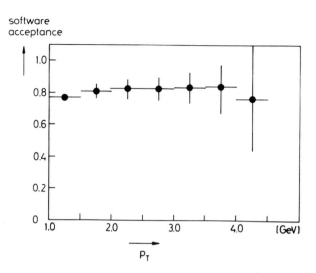

Fig. 11. Software acceptance vs. p_T

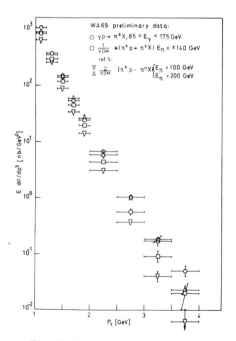

Fig. 12. Invariant cross section
$E \, d^3\sigma/dp^3$ vs. p_T ($x_F > 0$).

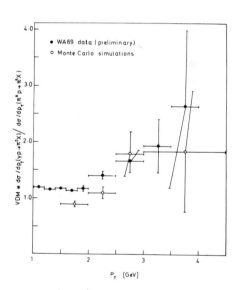

Fig. 13. Ratio of cross sections

statistics (factor 10) to ≈2.5GeV/c. The p_T slope is consistent with that of the $\pi^+p^- > \pi^0x$ data of Donaldson et al.; this is expected due to the dominance of the hadron-like photon component (assuming that the VMD model is valid).

FUTURE ANALYSIS WORK

(1)   The inclusive studies on $\gamma p \rightarrow \pi^{\pm}(K^{\pm})X$ as a function of $p_T(x_T)$ and $p_L(x_L)$ will be completed and quantified with our full data sample.

(2)   The particle identification as given by the RICH and TRD will be used to distinguish π^+ from K^+ throughout the p_T range. This will give new insight in the mechanisms of flavour production on hydrogen in the range of strangeness and charm. The final studies will also include charge asymmetries of leading particle production.

(3)   The inclusive π^0 measurements will be completed and extended to other particles like η and ρ^+.

(4)   Large p_T events will be analyzed with respect to their topological structure (exclusive event studies). In this way the photon-hadron comparison can be extended to further improve the isolation of the pointlike-photon component. In particular, Higher-Twist processes (Fig. 1c) which have a simple topological structure (the HT particle is balanced kinematically by a cluster of hadrons or current jet stemming from the fragmentation of the interacting quark) are expected to be accessible for some processes[10].

The full event reconstruction requires large acceptance throughout the kinematical space of charged and neutral secondary hadrons. The WA69 set up excludes in general the measurements of particles in the backward hemisphere of the centre of mass system which is populated by the break-up of the hydrogen target fragments. The partons involved directly in the hard interactions process are much less affected. This is demonstrated in Fig. 15 for the example of HT pion production: the x_F distribution of the directly produced pion (dashed line) which was obtained by simulation of the corresponding process (Fig. 1c)[10] is mainly in the forward hemisphere. For comparison experimental γp data of WA69 taken with an interaction trigger are shown (full line).

SUMMARY AND CONCLUSIONS

The experiment WA69 is sensitive to pointlike-photon interactions. The contributions due to this process will be separated from hadron-like-photon contributions by a comparison of the WA69 γp and π/K_p data.

As a first result, cross sections as function of p_T have measured for $\gamma p \rightarrow \pi^+(K^+)X$ and $\pi^+p \rightarrow \pi^+(K^+)X$, and first event distributions for $\gamma p \rightarrow \pi^0X$ obtained.

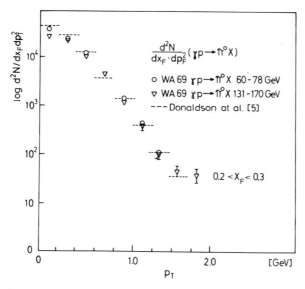

Fig. 14. Example of a p_T distribution of $\gamma p \to \pi^0 X$ for $0.2 < x_F > 0.3$ and at two energies measured in WA69 and comparison to $\pi^- p \to \pi^0 x^5$ (arbitrary normalization).

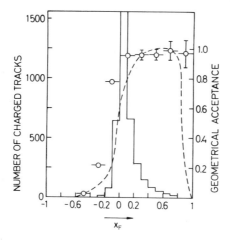

Fig. 15. x_F distributions of simulated HT pions (dashed line) and γp interaction trigger data (full line). The open circles indicate the geometrical acceptance.

More inclusive studies will be made (flavour, dependence, asymmetries, etc.) Finally, exclusive events where Higher Twist processes can be detected will be analyzed.

REFERENCES

1. P. Aurenche et al., Phys. Lett. 135 (1984) 164;
    P. Aurenche et al., LPTHE (ORSAY) 86/24, June 1986;
2. E. Paul, Proc. 1981 Int. Symp. on Lepton and Photon Interactions at High Energies, p. 301;
3. A.T. Doyle, PhD Thesis, Manchester, October 1987;
4. H. Rotscheidt, PhD Thesis, Bonn, in preparation;
5. G. Donaldson et al., Phys. Lett. 73B (1978) 375;
6. Th. Bauer et al., Rev. of Mod. Phys. 50, Nor. 2, 1978;
7. R. Barate et al., Phys. Lett. 174B (1986) 458;
8. G. Ingelman and A. Weigend, DESY 87-018, February 1987;
9. G. Ingelman, DESY 86-131, October 1986;
10. M. Benayoun et al., Nucl. Phys. B282 (1987) 433;
    A. Weigend, BN-IR86-07, Diplomarbeit, Bonn

DIRECT PHOTONS AT THE CERN ISR

Bernard G. Pope

Michigan State University
East Lansing
Michigan 48824-1116
R-110 Collaboration[*]

A hard scatter between constituent partons in hadronic interactions usually results in a system of two jets at large angles accompanied by two spectator jets. The resulting high multiplicity and the variety of contributing diagrams make the comparison with theory difficult. However, processes where one of the out-going jets is replaced by a photon often offer a more incisive comparison between experiment and theory. Photons can be produced as a result of quark-gluon collisions (Compton process) or by quark-antiquark annihilation. There are several higher order processes involving the radiation of photons from interacting quarks (bremsstrahlung process) where the photon will be close to an accompanying jet. In proton-proton collisions at ISR energies, direct photon production is predicted to be dominated by the Compton process. This implies that the recoil (away-side) jet will be a quark jet and study of the recoil jet will thus be a study of quark fragmentation. Further, it would be expected that there will be no accompanying (same-side) jet, and a measurement of the direct photon cross section will measure the gluon structure function in the interacting protons (modulo soft gluon corrections).

This report describes a study of direct photons in the transverse momentum (p_T) range 4 - 10 GeV/c produced in proton-proton collisions at the CERN ISR at a center-of-mass energy, $\sqrt{s}$, of 62.3 GeV. The observation of photons (or electron pairs) requires an open geometry which allows the measurement of charged and neutral particles associated with the photon. Thus the associated event structure can be studied.

The apparatus, shown in Fig. 1, consisted of a superconducting solenoid providing a magnetic field of 1.4 T and enclosing a system of eight cylindrical drift chambers. Four modules of lead-scintillator shower counters (14 radiation lengths thick) were also located inside the magnet. Each module was 1.5 m long, subtended 50° in azimuth and ± 1.1 units of rapidity centered at zero. In this study, direct photons were detected in either of

[*] CERN[1] - Michigan State[2] - Oxford[3] - Rockefeller[4] Collaboration.
A.L.S. Angelis[3], G. Basini[1], H-J Besch[1], R.E. Breedon[4], L. Camilleri[1], T.J. Chapin[4], R.L. Cool[4], P.T. Cox[1,4], C. von Gagern[1,4], C. Grosso-Pilcher[1], D.S. Hanna[1,4], B.M. Humphries[2], J.T. Linnemann[2,4], C. Newman-Holmes[1], R.B. Nickerson[3], B.S. Nilsen[2], N. Phinney[1,3], B.G. Pope[2], S.H. Pordes[1,4], K.J. Powell[3], R.W. Rusack[4], C.W. Salgado[2], A.M. Segar[3], S.R. Stampke[2], M.J. Tannenbaum[4], J.M. Yelton[3].

two arrays of lead-glass (SF 5) located ouside of the magnet in the angular region not covered by the shower counters.

The lead glass array used to search for direct photons was retracted to a position 2m. from the intersection region, in order to increase the discriminatory power of the apparatus against coalesced photons from π° decay. The other lead glass array was kept in its original position, 1m. from the intersection region as shown in Fig. 1. Data were taken with each array alternately retracted. The arrays were called "inside" and "outside" referring to their location with respect to the ISR ring. At our intersection, the center of mass motion (ß= 0.128) was toward the outside of the ring (or array).

Each lead glass array consisted of two walls (front and back) with a multiwire proportional strip chamber between them. The front wall of lead glass, which acted as a pre-convertor, consisted of 34 blocks each 10 cm thick (4 radiation lengths). The back wall consisted of 168 blocks each 15 cm square assembled to form an array 180 cm high, 210 cm wide and 40 cm thick (17 radiation lengths). The strip chambers had cathode strips with a 1 cm. pitch; 160 running horizontally and 192 vertically.

The r.m.s. energy resolutions, Δ E/E, of the lead glass and the shower counters were $(4.3/\sqrt{E} + 2)\%$ and $(16/\sqrt{E})\%$, respectively, with E in GeV. The r.m.s. momentum resolution of the drift chamber system was $\Delta p_T /p_T = 7\% \, p_T$ (with p_T in GeV/c). A more complete description of the apparatus and calibration method will be found in ref. 1.

The trigger required clusters of c.m. energy 4.5 GeV (with some data taken with a threshold of 3.5 GeV) in a 3 x 3 array of the back lead glass array. In addition, at least one scintillation counter of the central barrel hodoscope (A counters) was required. This ensured suppression of backgrounds from cosmic rays, upstream beam losses and beam-gas and beam-wall interactions. Finally, in order to enhance the separation of direct single photons from the large background of photons from π° decay, those B counters (in front of the electromagnetic cluster) were required to have small signals. Photons were thus required not to convert in the solenoid coil. This resulted in narrower transverse shower shapes in the strip chambers and consequent easier separation of one shower from two. The data correspond to an integrated luminosity of 8.54 x 10^{37} cm^{-2}, with a total of 450,000 events being collected.

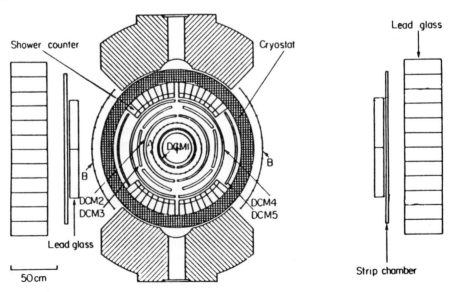

Fig. 1   The R-110 Apparatus viewed along the Beam Axis.

The analysis of these data was divided into two stages; a basic analysis of the distributions of pulse height in the lead glass arrays followed by a detailed study of the strip chambers in order to extract a single photon signal. Clusters of energy in the lead glass were examined to ensure that their shape was consistent with having been produced by the colliding proton beams (rather than, for example, upstream scattering). The ratio of pulse height deposited in the front glass array compared with the total pulse height was required to be greater than 0.14. This number, which came from test beam studies and EGS shower simulations[2], indicated that the photon(s) converted in the front glass array and might be expected to produce an adequate distribution in the strip chambers. Additional requirements for a clean sample of photon showers were the presence of at least two charged tracks in the event with an appropriate interaction vertex and the requirement that no charged track point to within ± 30 cm of the photon cluster. Additional electromagnetic clusters within ± 20 cm would also cause the event to be vetoed.

In the analysis of the strip chamber information, the variable chosen for the separation of single photon distributions from those caused by multi-photons (neutral meson decays) was $\sigma_T = \sqrt{\sigma_y^2 + \sigma_z^2}$. σ_T is the r.m.s. standard deviation of the transverse profile of the shower in the strip chamber and directly combines the horizontal and vertical cathode strip information. σ_y and σ_z are calculated from all the strip chamber information found in a 40 cm window centered at the interpolated point between the vertex and the center of the cluster in the back glass.

The signal extraction method consists of fitting the observed σ_T distribution with simulated single photon and multi-photon distributions. These latter are based upon EGS shower simulations and a detailed Monte Carlo detector simulation[2]. The EGS simulation was checked, where possible, with a sample of known single photons produced by studying the production of η mesons (where the two photons from the decay $\eta \rightarrow \gamma\gamma$ are sufficiently separated as to be distinguishable). The single photon Monte Carlo simulation program initiated electromagnetic showers in the detector (especially converting in the front glass and depositing energy in the strip chambers). A scaling algorithm, normalized using the η data, transformed the energy deposited into pulse height counts. Another Monte Carlo program produced and decayed neutral mesons in the apparatus generating multi-photon positions in the strip chambers. Table I shows the production and branching ratio assumptions that were input to the neutral meson Monte Carlo. Strip chamber responses for the appropriate energies were then taken from the EGS-generated bank of photons.

In both Monte Carlo programs (that for single photons and that for multi-photons) σ_T was calculated for each event. A frequency distribution of σ_T for each momentum bin was created for single photons and for multi-photons and compared with the data (Fig. 2 for the p_T range 4.5 - 5 GeV/c for example). A likelihood fit with one free parameter was used to obtain the fraction of single and multi-photon showers in the data sample. The results of this fit are also shown in Fig. 2. Once the fraction of single photons is obtained, acceptance and background corrections must be made in order to obtain the number of direct single photons produced. The most important background comes from asymmetric decays of neutral mesons where one of the photons missed the detector acceptance or did not produce a signal in the strip chambers. This, of course, results in a "single photon" peak from a neutral meson. After corrections, one obtains the invariant cross section for neutral meson production (including photons) and simultaneously the invariant cross section for direct single photon production. These are shown in Figs. 3 and 4 respectively, compared with previous measurements for these quantities.[3,4,5] Also shown in Fig. 4 are the theoretical predictions of Aurenche et. al.[6] with different assumptions of quark and gluon structure functions.[7]

It has been common in the literature to give the ratio of produced direct photons to $\pi°$ production or sometimes the ratio of direct photons to all neutral mesons. In Fig. 5 we present this latter ratio, γ/all, where "all" refers to all neutral mesons as defined in Table I.

Table I

NEUTRAL MESON DECAYS

particle	mass	σ/π	decay	photons	branching ratio	$\sigma.B/\pi$	%
π	.135	1.	$\gamma\gamma$	2	1.	1	61.4
η	.549	.55	$\gamma\gamma$	2	.38	.209	12.8
η	.549	.55	$\pi\pi\pi$	6	.30	.165	10.1
K_S	.498	.40	$\pi\pi$	4	.31	.124	7.6
ω	.783	.50	$\pi\gamma$	3	.09	.045	2.8
η'	.957	1.	$\eta\pi\pi$	6	.084	.084	5.2

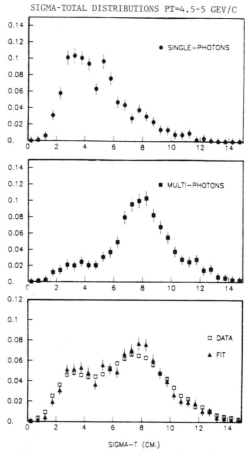

Fig. 2   Frequency Distributions of the Tranverse Profiles (σ_T) for Simulated Single Photons, Simulated Multi-Photons and the Observed Data.

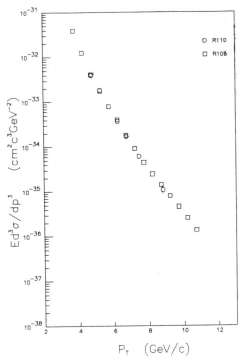

Fig. 3   Invariant Cross Section for Neutral Meson Production from this Experiment (R-110). Also shown are the results from A.L.S. Angelis et. al., Phys.Lett. 79B, 505 (1978).

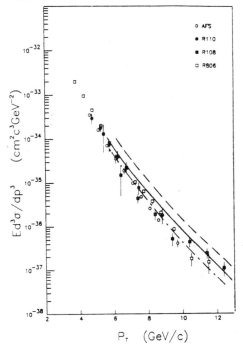

Fig. 4   Invariant Cross Section for Direct Photon Production from this Experiment (R-110). Also shown are the results from ref. 3 (R-108), ref. 4 (R-806) and ref. 5 (The Axial Field Spectrometer). Curves represent the predictions of ref. 6 with different assumptions of quark and gluon structure functions.

Shown in Fig. 5 is the ratio γ/all as measured by both "inside" and "outside" detector arrays. (Figs 3 and 4 represent the cross sections averaged over the two sides of the detector.) Due to the center of mass motion of the ISR, the photon energies, acceptances, background rates, etc. measured in the two arrays are quite different and a comparison of the cross sections gives one an estimate of possible systematic errors. This comparison and other tests lead us to assign a systematic error of about ± 0.05 in the ratio of γ/all. Also shown in Fig. 5 is the measurement of γ/all from the R-108 experiment.[3]

In order to study the event structure of direct single photons, it is necessary to obtain a separation between signal (direct photons) and background (multi-photons) on an event-by-event basis. Fig. 2 shows that such a separation is not completely achieved but that one can determine values of σ_T that might be expected to give reasonable signal-to-noise ratios. This associated-event analysis is still in progress.

We present here preliminary results for a related process, the production of massive virtual photons seen via their decay into electron pairs detected in the shower counters. The invariant mass resolution was ± 4% at m = 15 GeV/c^2. The acceptance covered the range -1.2 < y < + 1.2 where y is the rapidity of the electron pair in the c.m.

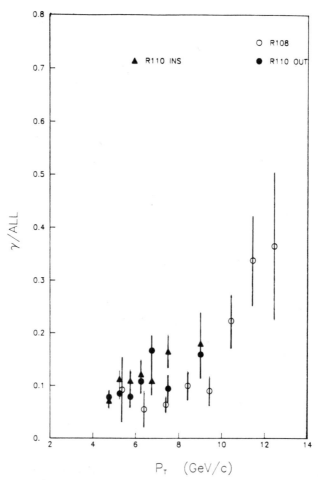

Fig. 5   The Ratio for Direct Photon Production divided by Neutral Meson Production, γ/all. Also shown are the results from ref. 3 (R-108).

The acceptance was calculated using a Monte Carlo program which included apparatus acceptance, resolution effects, and trigger requirements. Events were generated according to the quark - antiquark annihilation model of Drell and Yan[8] using the structure function distributions of Duke and Owens and with a p_T distribution used by the CFS group[9]:

$$dN/dp_T \propto p_T \left\{ 1 + (p_T/p_0)^2 \right\}^{-6}$$

Where $p_0 = 2.33 < p_T >$

The decay angular distribution was taken to be $1 + \cos^2 \theta^*$. The total acceptance varied from 10% at m = 11 GeV/c^2 to 21% at m = 20 GeV/c^2. The data correspond to an integrated luminosity of 2.47 x 10^{38} cm^{-2} (one quarter of an inverse femtobarn!) and were collected at an average instantaneous luminosity of 4 x 10^{31} cm^{-2} s^{-1} (over a four year period).

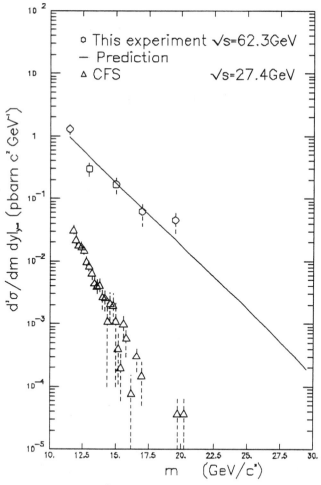

Fig. 6   The cross-section d$^2\sigma$/dmdy $_{y = 0}$ obtained in this experiment at $\sqrt{s}$ = 62.3 GeV and at $\sqrt{s}$ = 27.4 GeV by the CFS Collaboration.

167

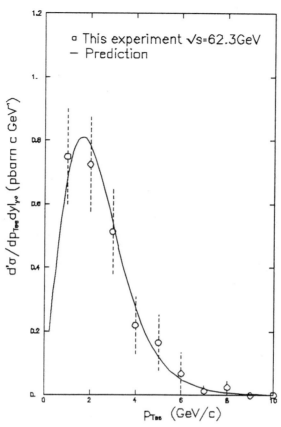

Fig. 7   The acceptance-corrected distribution of the transverse momentum of electron pairs with masses greater than 11 GeV/c^2.

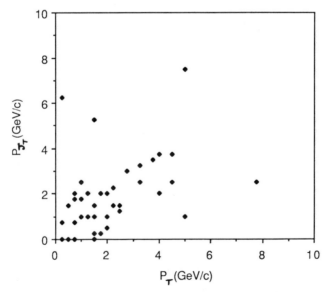

Fig. 8   The transverse momentum of the electron pair versus the net transverse momentum of the system of associated particles for events with $\delta \phi \geq 135°$.

Possible backgrounds to candidate electrons include the spatial overlap of a charged hadron and a π°, Dalitz decays, conversions, and charged hadrons that deposit most of their energy in the shower counters. Candidate electron pairs can be of opposite charge or of the same charge. The contribution of background processes to the signal of opposite charges was assumed to be given by the same-charge class. Cuts to reduce the background included the matching of track and shower (as regards to energy and position), examination of the longitudinal deposition of energy in the shower counters, and a "cleanliness" cut that required each electron to be unaccompanied by nearby tracks or showers. The combined efficiency of the cuts was estimated to be 83% per electron. After cuts, we observed 105 opposite-charge events and 10 same-charge events, yielding a net signal of 95.

The cross-section $\dfrac{d^2\sigma}{dmdy}\Big|_{y=o}$ for these events is shown in Fig. 6.

The cross section obtained at Fermilab at $\sqrt{s} = 27.4$ GeV by the CFS collaboration[9] is also shown. In order to determine the mean transverse momentum, $<p_T>$, of the electron pairs the value of $<p_T>$ used in the Monte Carlo was varied until the output p_T distribution best represented the data. In this way $<p_T>$ was found to be 2.40 ± 0.18 GeV/c. The transverse momentum distribution, corrected for acceptance is shown in Fig. 7.

As in the case of the direct real photons, studies are underway to investigate the associated event structure. The resultant momentum p_J of the associated particles was formed by adding their momentum vectorially. The difference in azimuth, $\delta\phi$, between p_J and the momentum of the electron pair was found. There appears to be some azimuthal correlation showing that the system of associated particles is produced predominantly back-to-back with the electron pair. The transverse momentum correlations between the electron pair and the associated particles were examined for events in which $\delta\phi > 135°$. A two-dimensional plot of p_T of the electron pair versus p_{J_T}, the transverse momentum of the system of associated particles is shown in Fig. 8. A correlation between these two variables is observed. The mean electron pair p_T of this sample of events is 2.7 GeV/c (no acceptance correction has been applied), while the total transverse momentum of these events (electron pair + associated particles) has a mean of 1.7 GeV/c. These features indicate a local compensation of the transverse momentum of the electron pair in the central region. These studies are continuing.

We wish to thank the ISR Division for the excellent performance of the ISR, Robert Gros for his technical help, Marie Anne Huber for expert secretarial assistance, and Barbara Taylor for typing this manuscript. This research was supported in part by the National Science Foundation under grants PHY-8214679 and PHY-8607988.

REFERENCES

1. R.B. Nickerson, D. Phil Dissertation, 1982 (Oxford University), unpublished. C.W. Salgado, Ph.D. Dissertation (Michigan State University), in preparation. See also, A.L.S. Angelis et. al., Nucl. Phys. B244 (1984), 1.

2. W.R. Nelson et. al., The EGS 4 Code System, SLAC Report -265 (1985). Low Energy Electron Transport with EGS, D.W.O. Rogers, Nuclear Instruments and Methods 227, 535 (1984). For details of the Monte Carlo simulation see the Ph.D. thesis of C.W. Salgado (Michigan State University).

3. A.L.S. Angelis et. al., Phys. Lett. 94B, 106 (1980).

4. E. Anassontzis et.al., Z. Phys. $\underline{C13}$, 277 (1982).

5. T. Akesson et. al., paper contributed to the Proceedings of the International Europhysics Conference on High Energy Physics, Bari, Italy (July 1985), p. 819.

6. P. Aurenche et. al., Phys. Lett. $\underline{140B}$, 87 (1984). See also R. Baier, "The Quark Structure of Matter", World Scientific Publ. Co., p 113 (1986). The curves of Fig. 4 are taken from the talk of R. Baier at the XVII International Symposium on Multiparticle Dynamics, Seewinkel, Austria, (June 1986).

7. D.W. Duke and J.F. Owens, Phys. Rev. $\underline{D30}$, 49 (1984).

8. S. Drell and T-M. Yan, Phys. Rev. Lett., $\underline{25}$, 316 (1970).

9. A.S. Ito et. al., Phys. Rev. $\underline{D23}$, 604 (1981).

PROMPT PHOTON PRODUCTION
IN 300 GeV/c π^-p, π^+p AND pp COLLISIONS

C. De Marzo, M. De Palma, C. Favuzzi, G. Maggi, E. Nappi,
F. Posa, A. Ranieri, G. Selvaggi and P. Spinelli
Dipartimento di Fisica dell'Universita di Bari
and Istituto Nazionale di Fisica Nucleare, Bari, Italy

A. Bamberger, M. Fuchs, W. Heck, C. Loos, R. Marx, K. Runge,
E. Skodzek, C. Weber, M. Wülker and F. Zetsche
University of Freiburg, Freiburg, Germany

V. Artemiev, Yu. Galaktionov, A. Gordeev, Yu. Gorodkov,
Yu. Kamyshkov, M. Kossov, V. Plyaskin,
V. Pojidaev, V. Shevchenko, E. Shumilov and V. Tchudakov
ITEP, Moscow, U.S.S.R.

J. Bunn*, J. Fent, P. Freund, J. Gebauer, M. Glas, P. Polakos**,
K. Pretzl, T. Schouten***, P. Seyboth,
J. Seyerlein and G. Vesztergombi****
Max-Planck-Institut für Physik und Astrophysik
München, Germany

NA24-COLLABORATION

(presented by P. Seyboth)

1. INTRODUCTION

From the measurement of prompt photon production at large transverse momentum one can obtain fairly clean information about constituent scattering dynamics, since the photon is a direct participant of the constituent scattering process. With the use of pion beams, the experiment aimed at the observation of the QCD-annihilation process. Rather precise measurements of direct photon cross sections in this experiment provide a good test of QCD, in particular since next to leading order terms are included in the QCD predictions.

Results on direct photon production have recently been published [1] and will be summarized here. Work on the analysis of two direct photon production is still in progress, therefore only preliminary results will be presented.

*PRESENT ADRESS: CERN, Geneva (Switzerland)

**PRESENT ADRESS: Bell Labs Holmdel, New Jersey (U.S.A.)

***PRESENT ADRESS: University of Nijmegen (Netherlands)

****ON LEAVE OF ABSENCE FROM CENTRAL RESEARCH INSTITUTE FOR PHYSICS, Budapest (Hungary)

2. APPARATUS AND ANALYSIS

Fig. 1 shows a schematic drawing of the apparatus. A 300 GeV/c momentum beam of negative or positive charge with up to 10^7 particles/sec impinged on a 1 m liquid H_2 target. Protons and pions in the beam were identified using two CEDAR Cerenkov counters. Protons which were misidentified as pions in the Cerenkov counter and conversely pions which were misidentified as protons were measured to be less than one per thousand. Following the H_2 target a set of proportional chambers (22 planes) was used for vertex determination. A fine grained photon position detector (PPD) [2] of 9.6 radiation lengths (X_o) thickness was located 8.12 m downstream of the target. Its sensitive area was 3x3 m^2 excluding a central hole of 0.5x0.5 m^2. It consisted of alternating layers of 1.1 X_o lead sheets and proportional tubes with 0.773 cm wire spacing, and vertical and horizontal wire orientations.

The PPD was followed by a 240 cell ring calorimeter [3] consisting of a 16 X_o lead/scintillator sandwich photon section and a 6 λ_a iron/scintillator sandwich hadron section. The acceptance of the calorimeters was between -0.8 and +0.8 in c.m.s. rapidity and 2π in azimuth. The energy flow through the 56 cm diameter central hole of the ring calorimeter was measured by a downstream calorimeter. An iron wall and a veto counter array positioned upstream of the detector were used to reduce the trigger rate due to upstream interactions and muon background.

The combined PPD/ring calorimeter system was calibrated with electrons and hadrons of 5 GeV to 170 GeV energy. The obtained energy resolution for incident electrons was $0.28/\sqrt{E/GeV}$. The final calibration was obtained by normalizing the reconstructed π^o and η masses to the expected values. The systematic uncertainty in the p_T scale was estimated to be ± 1 % and the normalisation uncertainty in the cross section determination to be ± 7 %.

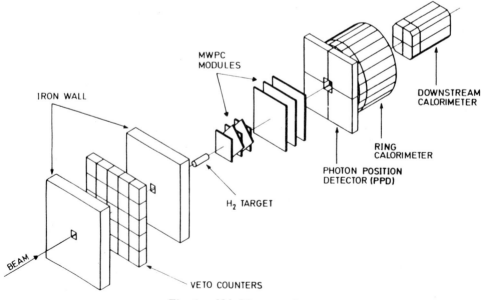

Fig. 1.   NA-24 apparatus

The data were taken at various p_T trigger thresholds to cover the full p_T range accessible. The highest trigger threshold was at $p_T > 3.75$ GeV/c. The trigger events were off-line selected by requiring: i) the reconstructed event vertex to be inside the H_2 target fiducial volume, ii) the direction of the triggering shower (as determined from the shower position in the front and the back parts of the fine grained PPD) to point at the H_2 target, iii) the total energy measured in all calorimeters to be consistent with the beam energy and iv) the shower signal to be in coincidence with the beam interaction in the hydrogen target (as determined from 30 MHz FLASH ADC's). These cuts removed most of the pile up and muon backgrounds. The sensitivity of the experiment was 1330, 190 and 450 events per nanobarn for $\pi^- p$, $\pi^+ p$ and pp collisions respectively.

In the next analysis step, photon showers were reconstructed. The two photon effective mass spectrum of those combinations containing the trigger shower are plotted in Fig. 2a, showing the π^o mass peak with a resolution of $\sigma_{m_{\pi^o}} = 16$ MeV. The asymmetry distribution of reconstructed π^o decays is plotted in Fig. 2b.

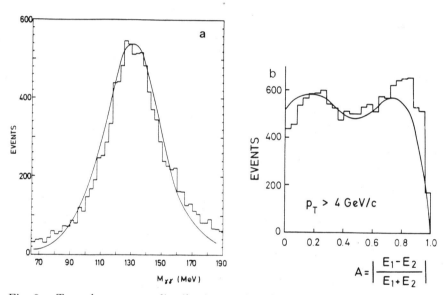

Fig. 2. Two photon mass distribution in the π^o mass region (a) and asymmetry distribution of reconstructed π^o (b). The curves show the results of the Monte Carlo simulation.

If a pair of photons was found with an invariant mass between 55 and 210 MeV (470 and 620 MeV) it was assigned to originate from a π^o (η) decay. An electromagnetic shower which could not be paired to form a π^o or η and had a width consistent with a single shower was considered a direct photon candidate.

A Monte Carlo event generator was used to estimate the detection efficiency of π^o, η and direct photons, the background to the direct photon candidates and the effects of the finite energy resolution of the calorimeters on the cross sections. This Monte Carlo program generated π^o, η and direct photon events with realistic rapidity and p_T distributions. It utilized showers obtained from tagged photon and electron beam calibration runs to simulate the detector response. The Monte Carlo events were then analysed with the same shower reconstruction program as the data.

The background to the direct photon candidates, estimated with the Monte Carlo program, is shown in Fig. 3a. The small discrepancy in reproducing the π^o decay asymmetry distribution of Fig. 2b leads to a systematic uncertainty in the background estimate. This is shown in Fig. 3b together with the measured ratio of direct photon candidates to π^o's with asymmetry <0.8. After background subtraction the direct photon yields were corrected for acceptance, detection efficiency and finite energy resolutions of the calorimeters.

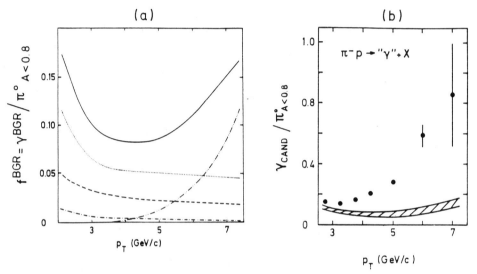

Fig. 3. (a) Monte Carlo estimated backgrounds to direct photon production: total (full), from η (dashed), from π^o with one photon not reconstructed (dotted), with one photon outside detector (dash-dotted) and with both photons coalesing (dash-double dotted).
(b) Ratio of direct photon candidates to reconstructed π^o's with asymmetry < 0.8. The hatched region shows the Monte Carlo estimated background and its systematic uncertainty.

3. RESULTS ON DIRECT PHOTON PRODUCTION

Direct photon production proceeds at the Born term level through the QCD-Compton and the QCD-annihilation processes. It was one of the aims of the experiment to provide evidence for the annihilation process. The Compton process contributes equally to direct photon production in π^-p and π^+p collisions. Due to the factor 2 larger charge of the valence anti-quark in the π^- and the factor 2 more numerous annihilation partners among the valence quarks of the proton, the annihilation process should be 8 times stronger at large x_T in π^-p collisions. Fig. 4 shows the result of the experiment for the ratio of direct photon cross sections in π^-p to π^+p collisions. This ratio rises with p_T showing the presence of the annihilation process. Recent QCD-calculations [4], [5] including next to leading order corrections agree with the measured ratio (see curves Fig. 4).

The direct photon production cross sections measured in π^-p, π^+p and pp collisions at 300 GeV are plotted in Fig. 5 as a function of p_T. The cross sections are fully inclusive since no isolation criteria on the single photon candidates were applied. The statistical errors are shown as bars and the linear sum of the statistical and the systematic errors resulting from the uncertainties in the background subtraction as brackets. In addition to these errors there is a p_T scale uncertainty of ± 1 % and a normalisation uncertainty of ± 7 % in the cross section determination which is mainly due to the applied total energy cut. The curves in Fig. 5 show the results of higher order QCD-calculations of ref. [4]. The measurements agree well with the full curve which was obtained using the QCD-scale parameter Λ=0.2 GeV, the structure function set I of Duke and Owens [6] and the optimized set of QCD scales. With Λ=0.4GeV, the structure functions of set II and optimized scales, the prediction lies above the data points (dashed curve). The optimisation results in QCD-scales of approximately $0.35 p_T^2$. A calculation with conventional scales of $4p_T^2$, Λ=0.2 GeV and the structure functions of set I is shown by the dash-dotted curve for illustration and is seen to fall below the data points. Calculations of ref. [5] are also consistent with the results of this experiment. Recent measurements by the WA-70 collaboration [7] at nearly the same energy confirm the results of this experiment.

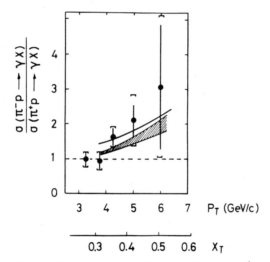

Fig. 4. Ratio of direct photon cross sections in π^-p and π^+p collisions versus p_T. QCD predictions of refs. [4 and 5] are shown by the curve and the hatched region respectively.

Fig. 6 shows the scaling behavior of the direct photon cross-sections in pp reactions and in p$\bar{\text{p}}$ reactions at the CERN Sp$\bar{\text{p}}$S collider, in which the Compton process dominates. Except for the lowest energy data from NA3 and E629, which were obtained on carbon targets, the data are well described in the form $E \cdot d^3\sigma/dp^3 = p_T^{-n} \cdot f(x_\perp)$ with n=5.15. This value of n is similar to that found for jet production and is much closer to the naive scaling value n=4 than to n$\approx$ 8 measured for π^0-production.

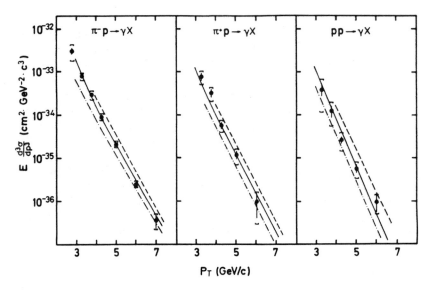

Fig. 5. Direct photon cross sections. See text for the discussion of the curves from QCD predictions.

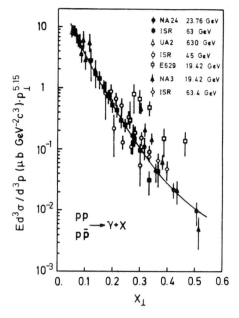

Fig. 6. Scaling plot of direct photon cross sections. Data were taken from ref. [8].

4. PRELIMINARY RESULTS ON TWO PHOTON AND TWO π^o PRODUCTION

Results shown in this section are still preliminary. Here events from both the one and the two photon triggers were used. The two photon trigger required clusters of electromagnetic energy in two different quadrants of the detector above a p_T threshold of 2 GeV/c. It ran concurrently with the one photon trigger for about half of the data taking. The analysis procedure described in section 2 was extended to all four quadrants of the detector.

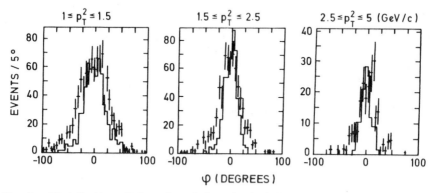

Fig. 7. Distribution of the azimuthal angle ϕ of the away side π_2^o with respect to the direction opposite the trigger π^o, with $4 \leq p_T^1 \leq 5 GeV/c$. The histogram shows predictions of the Lund model.

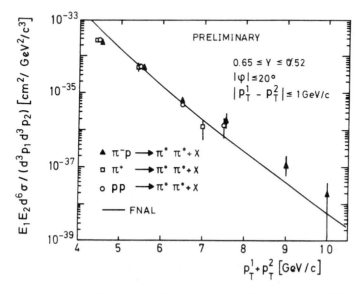

Fig. 8. Invariant cross section for back to back two π^o production with small transverse momentum difference $|p_T^1 - p_T^2| < 1 GeV/c$. For curve see text.

In order to confirm the correction and normalization procedures, two π^o production was analysed and compared to the published results. Fig. 7 shows the strong back to back azimuthal correlation in two π^o events. The distributions of the azimuthal angle ϕ of the away side π^o with respect to the direction opposite the trigger π^o are somewhat wider than those obtained from a Lund model [9] calculation using only lowest order QCD-processes (histogram in Fig. 7). The two π^o invariant cross sections for nearly back to back π^o's ($|\phi| < 20^o$) with transverse momentum differences $|p_T^1 - p_T^2| < 1 GeV/c$ are shown in Fig. 8. They are compared to pBe$\rightarrow \pi^+\pi^-+$X data from ref. [10] scaled by a factor of 1/9 for the Be target and an isospin factor of 1/2 to account for the smaller probability of producing a $\pi^o\pi^o$ pair (solid line in Fig. 8). The agreement with the pp data points is good. The π^-p data points lie higher at larger p_T, a consequence, most likely, of the harder parton spectrum in the pion.

Two photon candidate events were obtained by applying the selection procedure described in section 2 to each of the photon showers. In addition electron pairs from the Drell-Yan process (it is expected to have a cross section $\sim$ 30% of that of two photon production) were excluded by requiring that tracks should not point to both photon showers. Following ref. [11] the scaled momentum fraction z of p_T^2, the transverse momentum of the second photon, projected on the direction opposite to the transverse momentum p_T^1 of the first photon is calculated as:

$$z = -\vec{p}_T^1 \cdot \vec{p}_T^2 / |p_T^1|^2$$

The requirement $z > 0.7$ excludes low energy photon candidates which are more strongly contaminated by background. The remaining background was estimated from a Monte Carlo simulation using the observed $\pi^o\pi^o$ and $\gamma\pi^o$ rates. Table I shows the results for π^-p collisions for which the experiment has the highest sensitivity.

TABLE I. Number of $\gamma\gamma$ candidates and Monte-Carlo estimated background in π^-p reactions.

p_T^1(GeV/c)	$\gamma\gamma$ Candidates z> 0.7	Estimated background
2.5 - 3.0	18	11.5 ± 1.3
3.0 - 4.0	9	3.0 ± 0.3
4.0 - 5.0	18	0.5 ± 0.05

A signal of 13.1 ± 5.5 events (2.5 standard deviations) is observed in the range $2.5\leq p_T^1 \leq 5$ GeV/c. No candidates were found for π^+p and pp collisions corresponding to 90% confidence level upper limits of

$$\sigma(\pi^+p \rightarrow \gamma\gamma X)/\sigma(\pi^-p \rightarrow \gamma\gamma X) < 1.64$$

$$\sigma(pp \rightarrow \gamma\gamma X)/\sigma(\pi^-p \rightarrow \gamma\gamma X) < 0.65$$

These numbers are consistent with QCD calculations [11] including next to leading order corrections which predict 0.5 and 0.25 for this ratio respectively. After applying corrections for detection efficiency and resolution, the preliminary cross sections shown in Fig. 9 are obtained. These are compared to QCD-calculations using $\Lambda = 0.2 GeV/c$, set I of the Duke-Owens structure functions and the scale p_T^2. The dotted line in Fig. 9 is the result of the $q\bar{q} \to \gamma\gamma$ Born term, the full line the prediction of the next to leading order calculation of ref. [11] scaled by the rapidity acceptance factor obtained from the Born term calculation. Within the uncertainties the data agree with the predictions.

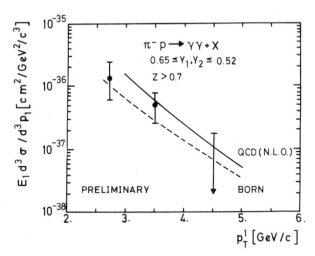

Fig. 9. Invariant cross section for two-proton production in $\pi^- p$ collisions.

5. CONCLUSION

Direct single and two proton production was studied in $\pi^- p, \pi^+ p$ and pp collisions at large transverse momentum and c.m.s. energy $\sqrt{s}=23.7\text{GeV}$. Evidence for the QCD-annihilation process was found. Within the experimental and theoretical uncertainties predictions of QCD agree well with the measured data.

REFERENCES

[1] C. De Marzo et al., Phys. Rev. **D36** (1987) 8

[2] V. Artemiev et al., NIM **224** (1984) 408

[3] C. De Marzo et al., NIM **217** (1983) 405

[4] P. Aurenche et al., Phys. Lett. **140B** (1984) 87; Z. Phys. **C29** (1985) 459,
   Nucl. Phys **B286** (1987) 509;
   preprint LPTHE Orsay 87/30, May 1987;
   see also R. Baier, these proceedings.

[5] A.P. Contogouris et al., Phys. Rev. **D35** (1987) 385

[6] D.W.Duke and J.F. Owens, Phys. Rev. **D30** (1984) 49;
   and J.F. Owens, Phys. Rev. **D30** (1984) 943.

[7] M.Bonesini et al., CERN-EP/87-185 (1987);
   see also M.Martin, these proceedings.

[8] ISR: E.Anassontzis et al., Z.Phys. **C13** (1982) 277;
A.L.S.Angelis et al.,Phys.Lett. **94B** (1980) 106;
V.Burkert, XVIIIth Rencontre de Moriond,23-29 January,1983, La Plagne.
UA2: J.A.Appel et al., Phys.Lett. **176B** (1986) 239;
E629: M.McLaughlin et al., Phys.Rev.Lett. **51** (1983) 971.
NA3: J.Badier et al., Z.Phys. **C31** (1986) 341

[9] H.U.Benggtsson and G.Ingelman, Comp.Phys.Comm. **34** (1985) 251

[10] H.Jöstlein et al., Phys.Rev. **D20** (1979) 53

[11] P.Aurenche et al., Z.Phys. **C29** (1985) 459

DIRECT PHOTON PHYSICS FROM R 806, R 807, R 808

Julia Thompson – R808 Collaboration

Physics Department
University of Pittsburgh
Pittsburgh, PA 15260

INTRODUCTION AND REVIEW

In this paper I review the contributions of the groups R806, R807, and R808 to direct photon physics, starting with the early R806 systematic studies which clearly demonstrated the existence of high P_t direct photon production, and ending with the recent gluon form factor determination from photon-jet correlations.

Table 1 summarizes the high P_t direct photon results in this paper. These are all real photon results. Results 1 - 5 are older results at 90°; results 6 and 7 are for photons at 11°; result 8 describes the most recent AFS result, the gluon distribution function determination. The apparatus pertinent to each measurement is discussed in the reference (Refs. 1-8) cited for that measurement.

The picture which emerges from the results, reinforced by theoretical calculations, is that for high P_t real photons the Compton subprocess (Fig. 1A) dominates in pp scattering for $P_t < 10$ GeV/c, with annihilation (Fig. 1B) $\leq 9\%$ (Ref. 9) and bremsstrahlung (Fig. 1C) $< 30\%$ at the 2σ level, in the range up to $\sqrt{s} = 63$GeV and $P_t \sim 10$ GeV/c. Differences in the properties of particles opposite photons and pions are consistent with the expected charge composition in gluons and quarks (gluons uncharged, quarks from protons predominantly positive) and the expected quark/gluon mixtures (100% quarks opposite γ's from Compton scattering and $\sim 40\%$ quarks opposite π's). No other differences are seen in the jets opposite γ's and π's. Thus, while we expect differences in strangeness and momentum distributions between quark and gluon jets, those differences are not striking enough to be visible in our data.

A new development which we will discuss only briefly here is the intriguing situation with the low P_t real and virtual photons (shown in Table 2). The direct electron component seems to come from a direct $e^+ e^-$ component, and both are consistent with a linear dependence on pion multiplicity. According to the general argument of Cerny, et al. (Ref. 12), such a dependence indicates that the direct electrons are probing the intermediate states of the interaction. This is of particular interest to the groups now studying high energy density states in relativistic heavy ion collisions. For the virtual photons, the low P_t excess, statistically very solid, seen by the Goldschmidt-Clermont group (Ref. 16) at $\sqrt{s} = 12$ GeV in $K^- p$ collisions is difficult to accommodate within current models (Ref. 17), since an excess at < 50 MeV/c implies a source of several Fermi's in extent, a size larger than the incident protons, or larger than seen in pion interferometry measurements of p-p interactions (Ref. 18). Perhaps these arise from rare collisions in which the intermediate state extends over a larger volume. The AFS results are consistent with, but do not require, the excess seen by the Goldschmidt-Clermont group. However, the AFS result excludes a rapid increase of such an excess with $\sqrt{s}$.

GLUON DISTRIBUTION FUNCTION

In the high P_t direct photon production region, perturbative QCD calculations are rather reliable. In conjunction with the current variety of measurements, these calculations give an impressive agreement between theory and experiment (Ref. 19). With this successful calculation of the direct photon production, one is now encouraged to use the data and theory together to put limits on QCD parameters such as the cut-off parameter Λ. One such investigation which shows promise is a determination of the distribution of gluon momenta within the proton. The full analysis by the AFS group is shown in Ref. 8. The highlights will be sketched here.

Within the picture drawn from the results described in the previous section, and summarized in Table 1, one can set (to within corrections of order 10%) the γ-jet correlated cross section at $90°$ equal to the Compton cross section:

$$\frac{d^3\sigma}{d\eta_\gamma d\eta_{jet} dP_t} = \frac{5\pi\alpha\alpha_s}{3}[\frac{G(x,Q^2)F_2^{ep}(x,Q^2)}{x^2 s^{3/2}}]$$

Second order calculations including soft gluon corrections give a k-factor of 2. (Ref. 20). Bremsstrahlung is suppressed by excluding events with charged tracks near the photon. Taking rather standard values:

$$\alpha_s = 12\pi/[25\ell n(Q^2/\Lambda^2)]$$

$$Q^2 = 4/3P_t^2$$

$$\Lambda = 0.2 GeV,$$

and F_2^{ep} from the EMC deep inelastic scattering on a hydrogen target (Ref. 21),the gluon distribution G can then be calculated, once the inclusive γ-jet cross section is measured. One possible question with the above procedure, in light of recent theoretical developments (Ref. 22) is that the full second order calculation of Aurenche, et al., using the optimized scale, prefers $Q^2 \sim 0.3P_t^2$. This will be discussed further at the conclusion of the paper.

The method of the R808 measurement was to set:

$$\frac{d^3\sigma}{d\eta_\gamma d\eta_{jet} dP_{t\gamma}} = \frac{d\sigma}{dP_{t\gamma}} f(\eta_\gamma, \eta_{jet}, P_t)$$

TABLE 1.

High P_t Direct Photon Results from R 806/R 807

Experiment	Real or Virtual γ's	$\theta\gamma$ (Approx.)	Results	Ref.
1. R806	Real	90°	Early clear direct γ signal. Agreement with QCD requires soft gluon corrections.	Ref. (1)
2. R806	Real	90°	(σ Brems. γ)/ (σ all direct γ) <0.3 (90% C.L.)	Ref. (2)
3. R806 R807 R808	Real	90°	Recoil jet: opposite γ's: $N^+/N^- = 1.27 \pm 0.06$; opposite π°'s: $N^+/N^- = 1.15 \pm 0.03$ ==> N^+/N^- quark $= 1.38 \pm 0.12$; N^+/N^- gluon $= 0.98 \pm 0.11$	Ref. (3)
4. R808	Real	90°	$(pp \to \gamma + \ldots)/pp \to \gamma + \ldots)$ ~ 1.0, $P_t < 5$ GeV/c; 2.2 ± 2, $P_t > 5$ GeV/c ==>Agrees, with poor statistics, with expectations from the annihilation and Compton subprocesses.	Ref. (4)
5. R806	Real	90°	$pp \to \gamma\gamma + \ldots$; $pp \to \gamma + $ jet $+ \ldots$ $\sim\alpha$	Ref. (5)
6. R807	Real	11°	γ/π at 11° $\approx \gamma/\pi$ at 90° ==>CIM and large forward bremsstrahlung excluded.	Ref. (6)
7. R807	Real	11°	γ, π° + recoil particle studies ==>$k_T < 750$ MeV/c; gluon spin = 1	Ref. (7)
8. R808	Real	90°	γ-jet correlations ==>soft gluon x_t distribution	Ref. (8)

a. Compton Effect

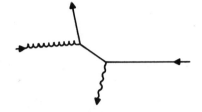

b. Quark Annihilation

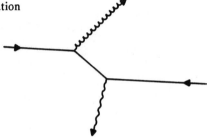

c. Bremsstrahlung

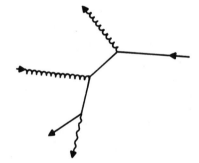

Figure 1. Direct γ Production Subprocesses

TABLE 2.

Low P_t Direct γ Results from R808

Experiment	Real or Virtual γ's	θγ (Approx.)	Results	Ref.
1. R808	Virtual	90°	Direct (e+/π) rises to ~10^{-3} at P_t 100~MeV/c	Ref. (10)
2. R808	Virtual	90°	Direct (e+/π) vs. n_π consistent with linear probes intermediate state? (Ref. (12))	Ref. (11)
3. R808	Virtual	90°	Direct (e+e⁻pairs)/π vs. n_π also linear!	Ref. (13)
4. R808	Real	90°	pp --> γ+ ... (25 MeV/c<P_t <500 MeV/c consistent with lower √s	Ref. (14)
5. R808	Real	90°	αα --> γ+ ... (100-500 MeV/c P_t) pp high E_t --> γ+ ... no excess seen	Ref. (15)

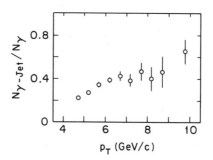

Figure 2.   Fraction of γ events with away-side jets.

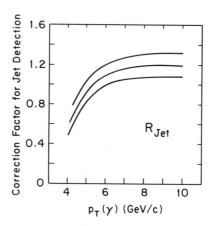

Figure 3.   Jet reconstruction efficiency from the LUND
Pythia Monte Carlo. The systematic uncertainty
is indicated by the two outer lines.

where

$$f(\eta_\gamma, \eta_{jet}, P_t) = \frac{\text{\# correlated events in } \Delta\eta_\gamma, \Delta\eta_{jet}, \Delta P_{t\gamma}}{\text{\# events with direct photon in} \Delta P_{t\gamma}}.$$

The reason for this method is that, although cross sections measured in R808 were consistent with those in R806, the normalized direct photon cross section used came from R806, which had achieved very good control of the energy scale and merging systematics. R808, however, had superior jet identification, so the fraction of photons with jets was taken from R808 (Fig. 2). Corrections were made for π° contamination of the direct γ sample. However, as we have seen in previous work, and corroborated here, the properties of jets opposite $\gamma's$ and opposite $\pi's$ are similar. The correction for π° contamination made no difference within the systematic uncertainties quoted.

TABLE 3. Systematic Contributions to Uncertainty in $\dfrac{d^3\sigma}{d\eta_\gamma d\eta_{jet} dP_t}$

Source	Systematic Error
inclusive $\frac{d\sigma_\gamma}{dP_t}$ from R806	20%
jet-finding efficiency	10%
jet-finding algorithms	15%
background in $\frac{N_{\gamma-jet}}{N_\gamma}$	5%
energy scale uncertainty	5%

A cut against bremsstrahlung was made by removing events with electromagnetic showers >180 MeV within a polar angle of 30° and an azimuthal angle of 40° of the photon. For the jets, all clusters with $|\Delta\phi| > 120°$ from the photon, $P_t > 100$ MeV and $|\eta| < 1.0$ are used in reconstructing the jet. The jet efficiency calculated from the LUND Monte Carlo is shown in Fig. 3. The reconstructed jet must be > 4 GeV, with $|\eta| < 0.4$. The $\Delta\phi$ distribution is shown in Fig. 4.

Systematic errors arising from variations of the above assumption were investigated. Contributions to systematic errors in the differential cross section are shown in Table 3. Systematic uncertainties leading to variation in $xG(x,Q^2)$ are shown in Table 4.

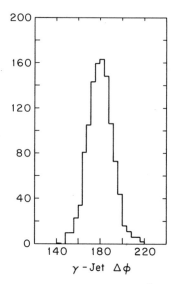

Figure 4.   Δ φ photon-jet axis (r.m.s. = 11.5⁰ and mean = 180.0⁰).

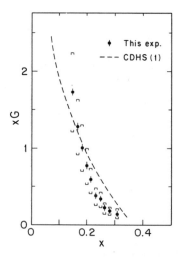

Figure 5.   The gluon distribution from the AFS data with systematic errors
indicated by the outer bars . The data are calculated
using $Q^2 = 4/3P$. The CDHS result is calculated, for
comparison, at $Q^2 = 40$ GeV^2/c^2, the middle of the Q^2
range of the data.

The results are compared with the CDHS result in Fig. 5. In this figure the Q^2 used to calculate the data points is $Q^2 = 4/3P_t^2$, but the Q^2 used for the CDHS calculation is $Q^2 = 40$ GeV^2, the midpoint of our range. The AFS result indicates a gluon softer than that from CDHS.

Changing the Q^2 scale from $Q^2 = 4/3\,P_t^2$ to $0.4(1-x)P_t^2$, the result of Aurenche's (Ref. 19) optimization structure, changes the shape of the gluon structure function results only slightly. Figure 6 shows our data, calculated for the two different Q^2 definitions, compared with the gluon structure function calculated from the Duke-Owens Set I structure functions (Ref. 25) with the matching Q^2 definition.

With either Q^2 definition our data are softer than the Duke-Owens I fit. This is seen not just in the overall normalization, but also in the slope of the data. Early analyses of our preliminary data (Ref. 24, Ref. 25) also showed the shape of our data to be somewhat steeper than other determinations of the form factor.

TABLE 4. Variation of $xG(x,Q^2)$ with Input Assumptions

Assumption	Variation
Q^2 scale $(Q^2 = P_t^2, 2P_t^2)$	$\pm 8\%$
$\Lambda(0.1 - 0.4 GeV)$	$\pm 15\%$
annihilation	$\leq 10\%$
Bremsstrahlung	$\leq 10\%$
F_2 (EMC)	$\leq 5\%$

The present situation is less clear. Preliminary results from the BCDMS group (Ref. 26) show a softer gluon distribution, more like the one we present here, but WA70 is analyzing their data and seem to be consistent with a harder distribution function than ours (Ref. 27).

The advantage of the gluon-jet correlated cross-section, in contrast to the inclusive γ cross-section which has been the standard of comparison in the past, is that the correlated cross-section comparison is sensitive to the shape as well as to the overall normalization. While it is true that some systematic errors such as energy scales and jet-finding efficiencies could change the slope of the correlated cross section, most experimental errors contribute more heavily to the normalization than to the shape. Thus, gluon-jet correlation studies should be a fertile source of comparison between experiment and theory for the coming generation of experiments.

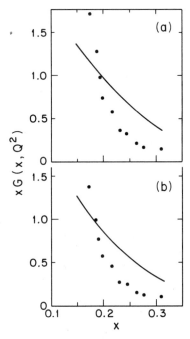

Figure 6.   Comparison of the AFS gluon distribution with the Set I of
Duke-Owens [25]. Both the AFS data and the Duke-Owens prediction
are calculated according to:

a) $Q^2 = 4/3\ P^2_t$.
b) $Q^2 = 0.4\ (1-x)\ P^2_t$, the optimized scale of Aurenche.

REFERENCES

1. E. Anasontzis, et al. Z. Phys. C13, 277 (1982).

2. T. Akesson, et al. Phys. Lett. 118B, 178 (1982).

3. T. Akesson, et al. Physica Scripta 34, 106 (1986).

4. T. Akesson, et al. Phys. Lett 158B, 282 (1985).

5. T. Akesson, et al. Z. Phys. C32, 491 (1986).

6. T. Akesson, et al. Phys. Lett 123B, 367 (1983).

7. T. Akesson, et al. Phys. Rev D31, 976 (1985).

8. T. Akesson, et al. Phys. C34, 293 (1987).

9. E. L. Berger, E. Braaten, R. D. Field, Nucl. Phys. B239, 52 (1984).

10. T. Akesson, et al. Phys. Lett. B152, 411 (1985).

11. T. Akesson et al. Phys. Lett B192, 463 (1987).

12. V. Cerny, et al., Z. Phys. C31, 163 (1986).

13. V. Hedberg, Ph.D. Thesis, Lund, 1987.

14. T. Akesson, et al., Phys Rev. D36, 2615 (1987).

15. Y. I. Choi. Ph.D. Thesis, University of Pittsburgh (1987).

16. P. V. Chliapnikov, et al., Phys. Lett. 141B, 276 (1984).

17. B. Andersson, J. D. Bjorken, private communication, 1987.

18. T. Akesson, et al., Phys. Lett. 129B, 269 (1983).
    T. Akesson, et al., Phys. Lett. 155B, 128 (1985).
    T. Akesson, et al., Phys. Lett. 187B, 420 (1987).

19. P. Aurenche, R. Baier, St. Croix Conference on QCD, October, 1987.

20. A.P. Contogouris, et al., Phys. Lett. 104B, 143 (1981).

21. J. J. Aubert, et al., Nucl. Phys. B259, 189 (1985).

22. P. Aurenche, R. Baier, M. Fontannaz, D. Schiff, LPTHE Orsay 87/30.

23. H. Abramowicz, et al., Z. Phys. C12, 289 (1982).
    D. Buchholz, private communications, 1985.

24. E. N. Argyres, A. P. Contogouris, N. Mebarki, S. D. P. Vlassopoulos, Phys. Rev. D35, 1584 (1987).

25. P. Aurenche, R. Baier, M. Fontannaz, D. Schiff, BI-TP 85/34.

26. A. Benvenuti, et al., International Lepton Photon Symposium, Hamburg, July 27-31, 1987.

27. M. Martin, St. Croix Conference on QCD, October, 1987.

CHARM PHOTOPRODUCTION AND LIFETIMES FROM THE NA14/2 EXPERIMENT

D. Treille

CERN
Geneva
Switzerland

INTRODUCTION

This experiment combines a microvertex detector made of a silicon active target and a silicon microstrip detector, for precise reconstruction of the charm vertices, and a large acceptance spectrometer where charged tracks and photons are measured.

Preliminary results are given on the observation of the decay of D^+ and D_s to $\phi\pi$, on the transverse momentum distribution of photoproduced D mesons, and on the measurement of D^0 and Λ_c lifetimes.

APPARATUS

This apparatus (Fig. 1) consists essentially of the spectrometer used in the previous NA14/1 experiment devoted to photon hard-scattering.[1]

The tagged photon beam is used in the energy range above 40 GeV ($\langle E_\gamma \rangle > \sim 80$ GeV). It interacts in a silicon target located in the first magnet (AEG). The charged particles are momentum-analysed in the silicon microstrip telescope and in 70 planes of multiwire proportional chambers (MWPCs) located before and after the second magnet (Goliath). They are identified by two Cherenkov counters (with K thresholds at 10 and 20 GeV). The photons are measured by three electromagnetic (e.m.) calorimeters.

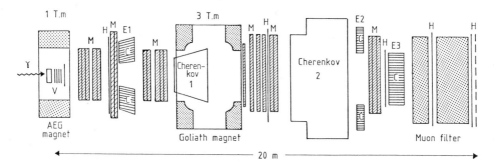

Fig. 1. The NA14/2 spectrometer

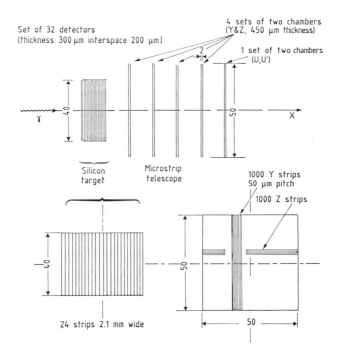

Fig. 2. The vertex detector

The items added for NA14/2 (Fig. 2) are as follows:
i) The silicon active target:[2] 32 planes, 5×4 cm^2 area, 300 μm thickness per plane, 200 μm spacing between planes, 2 mm wide strips. Each strip has an analog readout. One can thus measure the charged-particle multiplicity and its evolution along the target. Heavily ionizing fragments, which allow the main vertex or a possible reinteraction point to be located, are reconstructed as well.
ii) The silicon microstrip detectors:[3]
10 planes, 5×5 cm^2 area, 450 μm thickness per plane, 50 μm wide strips. The readout is digital (threshold at 0.35 minimum ionization). This detector worked fairly well.[4] The 1 m.i. signal was at $14 \times \sigma_{noise}$, and the transverse accuracy of one track extrapolated to the main vertex was 17 μm.

The NA14/2 experiment has registered 17 million hadronic photon interactions in 1985–86, with an open trigger on hadronic interactions. This trigger was 95% hadronic and provided an enrichment in charm of at least a factor of 2.

ANALYSIS

Filters were used to reduce the number of events to be processed through the complete pattern reconstruction program. Three independent fast selections were performed on
i) secondary vertices using the microstrips;[5]
ii) a clean silicon target pattern,[6] i.e. no heavy ionization masking the event;
iii) K^+K^- (or Kp) pairs identified in the Cherenkov counters.[7]
Cross-checks on efficiencies are possible. A few million raw data events have also been processed on the 3081E emulators at CERN.

Events of class (i) were used for general charm studies. Class (ii) provided a set of events with a double determination of vertices and enabled the coherent charm production to be studied. Class (iii) gave, for instance, an unbiased sample of $D_s \rightarrow KK\pi$ and $\Lambda_c \rightarrow pK\pi$.

SOME PRELIMINARY RESULTS

Figure 3 (partial statistics) shows our D^0 signal, allowing the measurement of the transverse momentum distribution (Fig. 4). Table 1 gives the D^0 lifetime measured using very clean subsamples.

The cleanliness of class (ii) events (Fig. 5) allows the reconstruction of many decay modes with large combinational possibilities. Figure 6, for instance, shows the $D^0 \rightarrow K\pi\pi\pi$ mode.

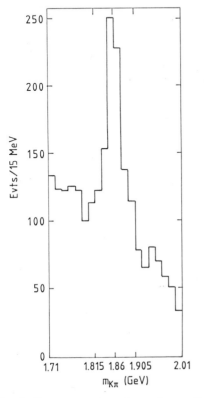

Fig. 3. $K\pi$ mass spectrum (secondary vertices' filter)

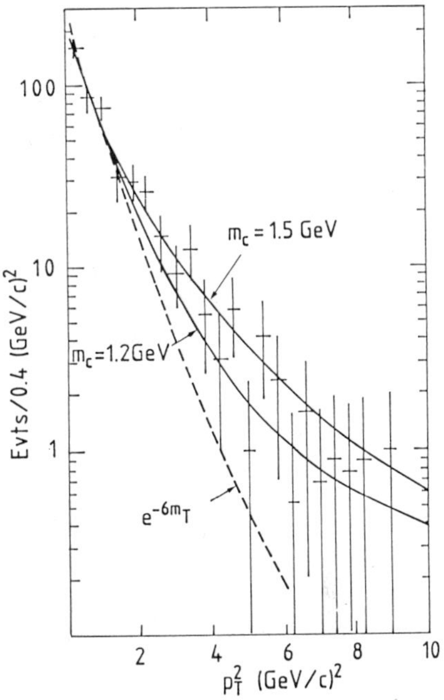

Fig. 4. D-meson transverse-momentum distribution

Table 1. The D^0 lifetime obtained from two different subsamples of our data

Channel	Flight/σ	Minimum decay time (s)	N_{signal}	$N_{backgr.}$	τ (10^{-13} s)
$D^0 \rightarrow K\pi$ (no D*) (K identified)	4	4×10^{-13}	85	40	4.42 ± 0.52
$D^0 \rightarrow K\pi$ (from D^*)	1	2×10^{-13}	88	20	4.44 ± 0.53

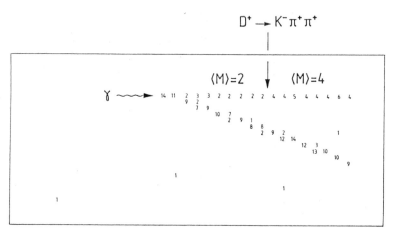

Fig. 5. Display of a D⁺D⁻ event in the active target

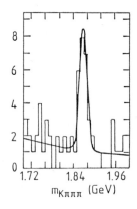

Fig. 6. $K\pi\pi\pi$ mass spectrum with $\Delta x >$ 1.5 mm (active target filter)

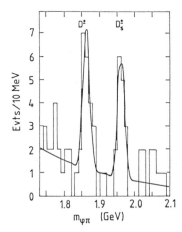

Fig. 7. $\phi\pi$ mass spectrum with $\tau > 3.5 \times 10^{-13}$ s ('KK' filter)

The $\phi\pi$ spectrum [from class (iii)] is shown in Fig. 7 with a cut of the proper time at 3.5×10^{-13} s: it displays clear signals of the D⁺ in the Cabibbo-disfavoured mode and the D$_s$ meson.

In the pKπ mode (with a non-ambiguous proton) a clean Λ_c signal is obtained (Fig. 8, on 1/3 of the statistics) despite a very small flight-distance cut ($\Delta x > 1.5$ mm). The very preliminary lifetime obtained is $(1.6^{+0.6}_{-0.4}) \times 10^{-13}$ s.

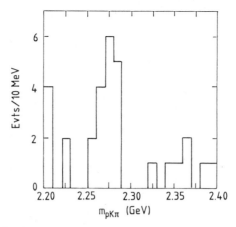

Fig. 8. pKπ mass spectrum with Δx > 1.5 mm ('pK' filter)

CONCLUSION

On the full sample we expect more than 1000 reconstructed D^0's, 50 Λ_c, etc.

Compared with the FNAL experiment E691, we cannot compete with its statistics for lifetime measurements. However, our systematics will be different (in particular because of the active target). Furthermore, our information about charm production will be obtained in a complementary photon energy range (40–120 GeV), where the various charmed particles are expected to appear in different proportions.[8]

REFERENCES

1. P. Astbury et al., Phys. Lett. **152B**:419 (1985).
2. R. Barate et al., Nucl. Instrum. Methods **A235**:235 (1985).
3. M. Cattaneo, Ph.D. thesis, Imperial College London (1987).
4. G. Barber et al., Nucl. Instrum. Methods **A253**:530 (1987).
5. C. Krafft, Thèse de doctorat, Univ. de Paris-Sud (1987).
6. M. Primout, Thèse de doctorat d'Etat, Univ. de Paris-Sud (1987).
7. C. Magneville, Thèse de doctorat d'Etat, Univ. de Paris-Sud (1987).
8. P. Roudeau, LAL 87-39 (1987), Talk given at the 2nd Topical Seminar on Heavy Flavours, San Miniato, 1987.

PHOTON HARD-SCATTERING IN THE NA14 EXPERIMENT

D. Treille

CERN
Geneva
Switzerland

In hard-scattering processes it is desirable to deal directly with the colliding partons in order to avoid complications due to the structure and fragmentation functions. One way is to consider the jets themselves. Another possibility is to observe prompt photons, either virtual (Drell–Yan) or real.

Here I will summarize the results of the hard scattering of real photons coming from the NA14/1 experiment at the CERN Super Proton Synchrotron (SPS). This experiment was completed two years ago and its results have been published;[1–5] the description of the set-up and of the analysis can be found in these publications.

The strong points of NA14 are briefly mentioned below:
- an extreme sensitivity (2–3 events per picobarn) due to the high intensity of the CERN tagged, broad electron–gamma (BEG) beam ($\langle E_\gamma \rangle \approx 80$ GeV);
- a nearly full acceptance of the spectrometer to charged and neutral particles up to $\sim 120°$ c.m. (Fig. 1);
- a good normalization ($\pm 20\%$) obtained by several independent methods (Bethe–Heitler, $\psi \rightarrow \mu^+\mu^-$, beam counting, minimum-bias trigger);
- the possibility of sending hadrons (π^-) as well as photons to the target. Hadron-induced data constitute our reference, and one of our methods for obtaining photon hard-scattering processes

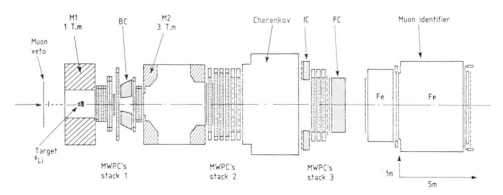

Fig. 1. Side view of the NA14 spectrometer.

was to measure departures from the hadronic behaviour obtained from our π data, which agree quite well with previous ones.[6]

Let us start with a process involving two real γ, one incident, one scattered and give our result on the inclusive distribution of photoproduced prompt photons.[1] Above the usual background sources (π^0, η, etc.) a clear prompt-photon signal is observed. This signal, plotted in Fig. 2, is in good agreement with the yield obtained from the QED Compton process[7] corrected for QCD processes.[8]

On the contrary, it is not compatible with what one would expect from gauged integrally-charged quark models.[9] Such models, which should not be confused with the already disproved non-gauged Han–Nambu versions, need hard processes involving two real photons in order to be tested. They have been criticized on purely theoretical grounds.[10] We show here that with the caveat indicated in the caption of Fig. 2, they are experimentally incorrect.

The exclusive analysis[5] fully confirms the inclusive observation.

In brief, Fig. 2 shows the first measurement of the deep-inelastic Compton process in a kinematical domain allowing a meaningful comparison with theory.

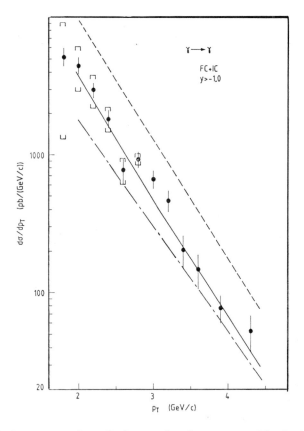

Fig. 2. The inclusive cross-section of photoproduced prompt γ. The background has been subtracted. The dash-dotted and solid curves correspond to the QED Compton Born-term and Born + correction terms, respectively. The dashed curve is 2.65 times the solid curve. Also shown are systematic errors on the subtraction of background. The statistical errors dominate for $p_T > 2.6$ GeV. In the absence of any computation of higher-order corrections in the case of integrally charged models we have simply multiplied the cross-section of Ref. 8 by the ratio of the Born terms of the integrally and fractionally charged models. This gives the upper curve.

Let us now turn to hard processes involving a fragmentation in the final state. Figures 3 show our results on photoproduction of π^0's at large p_T. In the backward region the results are only slightly larger than the vector-meson dominance (VMD) expectation deduced from incident π data. On the contrary, in the forward region it is clear that another component is absolutely necessary. When added to the VMD contribution, the QCD estimate of Aurenche et al.[11] reproduces the data rather well. As a function of rapidity, the QCD contribution shows the expected evolution: a regular rise from a small fraction (backward) to about one half of the π^0 yield in the forward region.

The photoproduction of charged pions,[3] with totally independent experimental problems, except for the normalization, gives identical results.

In our analysis we assume that the VMD contribution is given by the π yield scaled by the ratio of γ to π inelastic cross-sections. For transverse momenta lower than the range of our measurements, where no hard-scattering contribution is expected, this relationship between inclusive distributions is indeed verified.[12] As already mentioned, our own data show the right trend, but unfortunately, for technical reasons, we do not have measurements at low enough p_T to be able to check its strict validity. However, the behaviour, as a function of p_T and y, is such that the hard-scattering data at large p_T and y in our region of interest would hardly be affected by an approximation in the treatment of VMD.

In summary, Figs. 3 show the first measurements of the γ processes of the QCD Compton type.

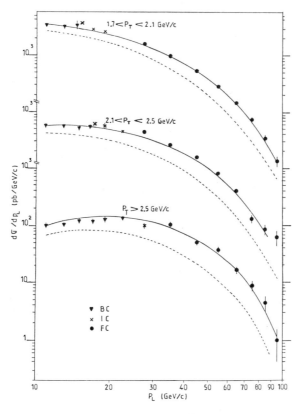

Fig. 3a. The measured cross-section for inclusive π^0 photoproduction versus longitudinal momentum for three p_T intervals: $1.7 < p_T < 2.1$ GeV/c, $2.1 < p_T < 2.5$ GeV/c, and $p_T > 2.5$ GeV/c. The dotted curves show the VMD prediction; the solid curves show the sum of VMD and the prediction of Aurenche et al. (Ref. 11).

201

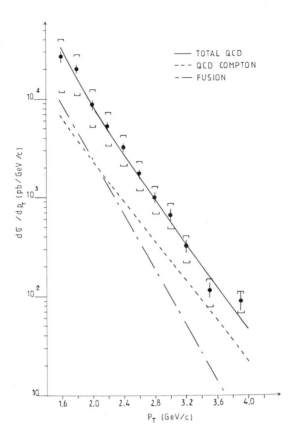

Fig. 3b. The measured cross-section for inclusive π^0 production versus p_T after subtraction of the VMD component, for $p_L > 40$ GeV/c. The three curves are derived using the computer code of Ref. 11. The dashed and dash–dotted curves correspond to Compton and fusion terms, respectively; the full curve is the total contribution. The gluon is assumed to fragment as a quark. Also shown are systematic errors due to normalization and reconstruction efficiency.

Let us now see what physical information can be obtained from these data. Ambiguity-free QCD predictions[13] are in principle a function of only two unknown parameters Λ and C; the latter determines the shape of the gluon fragmentation function

$$zD_g^{\pi^0} \approx (1 - z)^C .$$

A fit of the NA14 data[13] gives (Fig. 4) as best values

$$\Lambda_{\overline{MS}} = 0.120^{+0.105}_{-0.050} \text{ GeV},$$

$$C = 1.0^{+0.65}_{-0.60} .$$

The correlation between these quantities is obvious from Fig. 4. It is fair to say that the set of values obtained is not very compelling. For instance, $\Lambda_{\overline{MS}} = 0.2$ GeV, the preferred value from other sources,[14] would not contradict our data.

In conclusion, although our data are in good agreement with QCD, the information they bring regarding the gluon fragmentation is marginal.

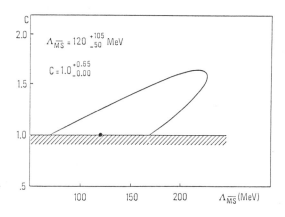

Fig. 4. Result of the best fit to the NA14 data in the forward region together with the 1 st. dev. ellipse.

We could ask ourselves, What is the nature of the limitations? The statistics are clearly limited at large p_T in spite of the impressive sensitivity; we could wish for more data at higher $\sqrt{s}$ and the FNAL Tevatron could provide them. But the poor significance of the fit described above is mostly due to the freedom to move the curves by $\sim \pm 25\%$, a number dominated by our normalization uncertainty. We should aim at a better figure for the absolute normalization. A more careful use of the Bethe–Heitler process and a more precise measurement of the γ beam using a quantameter or a beam counting device should provide it. Needless to say the other sources of systematic errors, such as the calibration of the calorimeters, also need to be treated with much care.

REFERENCES

1. P. Astbury et al., Phys. Lett. **152B**:419 (1985).
2. E. Augé et al., Phys. Lett. **168B**:163 (1986).
3. R. Barate et al., Phys. Lett. **174B**:458 (1986).
4. R. Barate et al., Z. Phys. **C33**:505 (1987).
5. E. Augé et al., Phys. Lett. **182B**:409 (1987).
6. G. Donaldson et al., Phys. Lett. **73B**:375 (1978).
7. J.D. Bjorken and E. Paschos, Phys. Rev. **185**:1975 (1969).
8. P. Aurenche et al., Z. Phys. **C24**:309 (1984).
9. For a short review, see R.M. Godbole et al., Phys. Lett. **142B**:91 (1984).
10. See, for instance, L.B. Okun' et al., Moscow report ITEP-79 (1979).
11. P. Aurenche et al., Phys. Lett. **135B**:164 (1984).
12. A.M. Breakstone et al., Univ. Calif. Santa Cruz report SCIPP 82/14 (1982).
13. P. Aurenche et al., Orsay report LPTHE 86/24 (1986).
14. F. Richard, LAL 87–48 (1987), Rapporteur talk given at the Int. Symp. on Lepton and Photon Interactions at High Energies, Hamburg, 1987.

ROUND TABLE DISCUSSION ON DIRECT PHOTONS

Chairmen :   P. Aurenche and T. Ferbel
Participants : A. Contogouris, J. Owens, L. Resvanis, D. Schiff,
                  M. Tannenbaum, and D. Treille

Beautiful experimental and theoretical work has been presented at this workshop. The experiments are very difficult, as can be inferred from the large systematic errors generally quoted. Nevertheless, six of the following methods for measuring prompt photons have been reported, and all have provided impressive results :

– Highly granular detectors with excellent spatial and energy resolutions : R806 and follow ups, NA3, NA24, WA70, UA6, E705, E706.
– Detectors relying on the statistical method for determining the probability of conversion of one and two photon clusters : R108, UA2.
– Pair spectrometer utilizing an extended-conversion technique : NA3.
– A method based on internal conversion of low mass muon pairs : UA1.
– Measurement of lateral shower structure after converter : R110.
– The statistical method, using accompanying energy in the cone and information on the shape of longitudinal energy deposition : UA1.

On the theoretical side, although predictions in the leading logarithmic approximation suffer from uncertainties at the level of a factor two or so, these uncertainties are much reduced when cross sections are calculated up to order $\alpha\alpha_s^2$, as has been demonstrated once again at this conference.

Up to now, with large systematic errors both in experiments and theory, one was tempted to conclude that comparisons of the normalization of the theory with data might not provide the most sensitive tests of QCD. Somewhat more restrictive comparisons, but stronger conclusions, were obtained when calculations on single photon spectra were compared to data over a wide kinematic range, i.e. different values of $\sqrt{s}$ and as large p_T ranges as possible. (It should be pointed out that studies of the shapes of distributions in rapidity, or angular correlations between photons and recoiling jets or hadrons, are also important and of great interest, and should be encouraged.) However, at this conference, it has become clear that enough progress has been made in the field of prompt photon physics that one should now start to worry about disagreement between theory and experiment at the 20%-30% level. This is the level of systematic error achieved in the most recent experiments and could be improved in the future.

Below we expand on these remarks, and summarize the results of the workshop.

Inclusive γ-distributions and theoretical considerations

The current theoretical situation can be summarized as follows: the first two terms of the perturbative series have been calculated for inclusive γ cross sections. The results depend on two unphysical quantities: the renormalization scale and the factorization scale. This scale ambiguity is a feature of all perturbative QCD calculations for any strong production process. At present energy and p_T-values, the numerical results are relatively sensitive to the choices used for the above scales. Consequently, a specific choice must be made in any quantitative phenomenological analysis. The optimization procedure, which can be applied to any process, leads to a well-defined specification of such scales (see R. Baier's contribution for a discussion of the advantages of the method). The results of the phenomenological analysis show that, within the optimization scheme, a consistent description of all data ($20 \leq \sqrt{s} \leq 630$ GeV, $3 \leq p_T \leq 100$ GeV/c) can be achieved with a Λ value and parton distributions in agreement with those obtained from some Deep Inelastic Scattering (DIS) experiments. Obtaining a good description of all data with a single set of parameters, consistent with DIS, can be regarded as an important achievement. Because the cross section for processes involving one photon is linear in α_s, γ physics is particularly sensitive to the value of Λ, and to the gluon distribution (for p and π^+ beams). This is in contrast with DIS, where both quantities appear only as correction terms. It should be noted that, in prompt γ reactions, there is a correlation between the value of Λ and the shape of the gluon distribution: in particular, a larger value of Λ can be accommodated, to a certain extent, by a softer gluon distribution. This is unlike the situation in DIS fits, where a larger value of Λ is correlated with a harder gluon distribution.

Concerning the shape of the gluon spectrum, we should remark that it is strongly related to the shape of the single-γ spectrum, and, clearly, the larger the p_T range covered in any experiment, the more can the gluon shape be constrained. For example, for pp interactions at $\sqrt{s} = 24$ GeV, if the cross section is normalized to 1 at $p_T = 4$ Gev/c, its value at $p_T = 7$ GeV/c differs by a factor 2, depending on whether $xG(x, 2 \text{ GeV}^2) = (1-x)^7$ or $xG(x, 2 \text{ GeV}^2) = (1-x)^5$ is used for the gluon function.

At present, it appears that the gluon distribution needed to accommodate prompt $-\gamma$ production is harder than what is derived from the recent preliminary data of BCDMS or CDHSW, but is in very good agreement with the CHARM results and recent EMC data on H_2

and D_2. The x range covered by the fixed-target direct-photon experiments is typically $x > 0.2$, whereas the DIS data probe the gluon distribution primarily for $x < 0.3$. It is then crucial to encourage precision experiments in the overlapping region of x. ISR experiment R110 or Fermilab experiment E706, with pp at 800 GeV, and $p_T > 4$ GeV/c, where intrinsic k_T-effects can be safely ignored, should be of particular interest in this respect. The published ISR results, which go down to $x \cong 0.12$, suffer from large statistical and/or systematic uncertainties, and therefore do not constrain the gluon distribution very strongly. Concerning π induced reactions, the present ratios of $(\pi^- \to \gamma)/(\pi^+ \to \gamma)$ also suffer from large uncertainties, and better statistics would be welcome at not too high $\sqrt{s}$-values (e.g. experiments E705 and E706), especially for obtaining better $\pi^+ \to \gamma$ data. Due to lack of sufficient statistics, present results do not clearly show the rise of the ratio $(\pi^- \to \gamma)/(\pi^+ \to \gamma)$ with increasing p_T, as expected from the dominance of the annihilation diagram at large p_T (WA70 and NA3 (calorimeter trigger) data are consistent with a flat ratio, whereas NA24 and NA3 (conversion trigger) show a rise within large error bars).

As a general theoretical comment, it should be mentioned that a proof that the optimization result is closest to the true result (if that is, in fact, the case) is still lacking. It has therefore been emphasized that it may be useful to also present theoretical results using the so-called standard scales.

The choice of targets and A dependence

Provided prompt-photon production on proton targets is well understood, nuclear targets can be very interesting for studying properties of gluons. In reactions where the annihilation process dominates, or is important (p or π^- beams), a gluon recoils against the large p_T photon. Since a gluon is expected to interact more strongly than a quark with nuclear matter, interesting rescattering and fragmentation effects should be observed. Energetic gluons $(E > 15$ GeV) would probably go undisturbed through nuclear matter, but less energetic ones would be expected to fragment, and their softer remnants to interact within the nucleus.

γ-jet correlations

Several strong arguments can be made in favor of studying correlations between photons and their opposite-side jets. One of these is that it allows a very direct test of the underlying partonic process. It is very interesting in this respect to study $d\sigma/dsdcos\theta$, where $\sqrt{s}$ is the energy of the γ-jet system, and θ is the angular variable (scattering angle) in the parton-parton rest-frame, which is directly related to the γ-jet rapidity difference. The cross section measurement, with both "particles" at zero rapidity, can be used to constrain the uncertainties in the structure functions and coupling constant, or to measure these. The rest of the points in the angular distribution are normalized by the value at 90°, y=0. In the leading logarithmic approximation, this normalized distribution would yield directly the constituent scattering angular distribution . In reactions dominated by QCD Compton scattering (π^+ or p beams), the $\cos\theta$ dependence should be rather flat, and substantially flatter than the same distribution for π^0-jet correlation, which is dominated by $gq \to gq$ scattering. (The pole structure of the partonic process is probed through such a study.) These comparisons could be easily performed because the γ and π^0 are measured in the same experiment. Any change in the shape of the angular distribution with s is sensitive to the role of scaling violations. It would be interesting to see evidence for parton evolution in these kinds of investigations. Such non-scaling effects are essential to explain the already existing data on π^0-π^0 and jet-jet correlations.

Once the shape of the angular distribution has been checked one can exploit the measurement of γ-jet correlations to obtain detailed information on the incoming parton distribution. This has been pioneered by the AFS collaboration in their extraction of the gluon distribution (see J. Thompson's talk). Furthermore, fixing the γ-rapidity at different values can be used to probe different x ranges. Such an approach is complementary to the information that can be extracted from single-photon inclusive measurements, and should lead to compatible results. It should be recognized that the results will depend on the choice used for scales in the phenomenological analysis (it is the quantity $\alpha_s(\mu^2)$ $F(x,M^2)$ which is measured). These affect the overall normalization, as well as the shape of the extracted x-distributions.

It is important to recall that the correlation analysis will be affected by higher-order corrections, which have not been calculated as yet. Since these play an important role in single-γ inclusive distributions, they may also affect the $\cos\theta$ distribution. (See, however, C. Maxwell's contribution, who showed that the jet-jet angular correlations obtained at the Born level at collider energies are relatively insensitive to approximate higher-order corrections; see Owens' talk for the current status of the higher order calculations.) On the experimental side, the precision of correlation measurements will be affected by the question of jet definition, which could typically introduce a smearing in x of $\delta x \cong 0.1$ in the parton distribution. Clearly, going to the highest possible energy helps in this case, since the jet becomes more collimated and better resolved.

γ-particle correlations

At low energy, the study of γ-hadron correlations may be a more appropriate approach to take, because of jet definition ambiguities. If the accompanying particle is produced at large p_T, opposite to the photon, then the z-dependence of the gluon fragmentation (for the qq annihilation-dominated processes) can be mapped out. The difference in h^+-h^-, on the other hand, will isolate the contribution from q-fragmentation.

A unique possibility of observing the fragmentation of composite colored objects occurs in reactions with prompt-γ (see A. Goshaw's contribution concerning related tests in association with heavy-flavor production). For example, in pp interactions, events with photons at large rapidities will enhance the process with a gluon coming from the backward-going proton, and scattering from a valence quark in the forward-going proton. The backward-moving system will be made up of three quarks in a color $\underline{8}$ state, whose fragmentation properties can be studied in the backward hemisphere. The forward going system will be a diquark in a color $\underline{3}$ state (as in DIS scattering). Similarly, using π beams, a study of the fragmentation of a (qq) system in a color $\underline{8}$ state can be performed (see the report from M. Martin on the WA70 experiment).

Processes related to prompt-γ production

Photoproduction reactions are related by crossing symmetry to prompt-γ reactions, and consequently a consistent picture should emerge for both classes of processes, using the same sets of parameters, before overall agreement with QCD predictions can be claimed. Thus, the deep Compton scattering process γ + hadron $\to \gamma$ + X is related to 2γ production in hadronic collisions (both processes provide tests of the magnitude of the quark charge), and γ + hadron $\to$ hadron + X is akin to single γ production. The single-hadron inclusive cross section at large p_T is very sensitive to gluon fragmentation, and various correlations can be used to successfully isolate various interesting partonic processes.

The processes quoted above have been measured by the NA14 experiment at CERN at $<s> \approx 160$ GeV^2 (see D. Treille's talk). It is essential to get statistically significant data at large p_T and large p_L values where the hadronic, non-point like, photon contribution is minimal and where perturbative predictions should give reliable results. The hadronic photon contribution can be estimated, using the Vector Meson Dominance model (VDM), directly from hadron data by scaling the inclusive cross section by the ratio of the γ to π inelastic cross sections ; it is, of course, best to get the hadron data in the same set-up, under the same running conditions, and one must therefore be able to direct hadron (π^-) beams as well as γ beams to the experiment. To check the validity of this method of estimating the VDM component, as a mere addition to the perturbative component, one has to get data in a broad region of p_T and p_L, from low values of these variables, where VDM should dominate, to the highest possible ones where it should be nearly negligible.

Finally, the sensitivity of the data to the gluon fragmentation is weakened by uncertainties in the absolute scale, due either to the normalization of the experiment or to the energy calibration of its calorimeters; to improve on this, one must achieve quite low systematic errors.

NA14 fulfilled all these requirements to a certain extent. But it should be possible, in the future, to improve on some figures. An increase in $\sqrt{s}$ and in luminosity can be provided by the Tevatron. Also, dedicated runs could allow to extend the measurements to quite low p_T. New techniques in calorimetry and,especially reliable normalization procedures of the experiment could decrease the systematic error below the 25% level, quoted by NA14, which limited the sensitivity to the gluon fragmentation function. The possibility of using hadrons and photons should certainly be kept open, as well as the exclusive nature of the measurement performed. Indeed, at high enough $\sqrt{s}$ and with sufficient luminosity, the study of topologies and correlations inside events could allow the separation of different interesting partonic processes.

CERTAIN UNCERTAINTIES IN QCD THEORY PREDICTIONS
FOR LARGE-p_T PHOTONS AND OTHER PROCESSES

A.P. Contogouris

Department of Physics
McGill University
Montreal, Quebec, Canada

INTRODUCTION

In general there are more than one scale (large variable) determining perturbative QCD quantities: E.g. the renormalization scale μ entering the running coupling $\alpha_s(\mu)$ (related with the renormalization procedure), and the factorization scale M entering the structure functions $F(x, M)$ (determining the scale violations). Until recently, for large transverse momentum (p_T) processes, simplicity suggested the choice

$$\mu^2 = M^2 = \alpha p_T^2 \tag{1}$$

where α = positive and of $0(1)$, or other choices phenomenologically equivalent; hereafter p_T will be called *physical* scale.

However, a different class of scales, called *optimal*, has also been introduced; and one of the most prominent of them is Stevenson's, according to his Principle of Minimal Sensitivity (PMS).[1,2] The procedures of determining such scales are called *optimization*.

DOES PMS OPTIMIZATION SOLVE THE PROBLEM OF SCALE(S)?

Take for simplicity $\mu = M$; then the inclusive cross section σ of some process (e.g. $\bar{p}p - pp \to \gamma X$) is a function of M/p_T. Stevenson's argument goes as follows: Suppose σ were known exactly, i.e. to all orders of perturbation; then it would be independent of M/p_T, i.e. the exact $\sigma(M/p_T)$ would be a straight line. Now, however, σ is known only up to a finite order (for $p^{\pm}p \to \gamma X$ up to $0(\alpha_s^2)$), call it $\sigma,^{(2)}$ and is not a straight line (Fig. 1). Which point of $\sigma^{(2)}$ looks more like a straight line? Clearly the point where

$$\partial\sigma^{(2)}/\partial M = 0 \qquad \text{(PMS)} \tag{2}$$

corresponding to M_{opt} (fig. 1). Then choose $\sigma^{(2)} = \sigma^{(2)}(M_{opt}/p_T)$.

However, it should be objected that the exact σ is completely unknown. If it corresponds to the dashed line of Fig. 1, then the choice M_{opt} is good. But if it corresponds to the dash-dotted line, the choice M_{opt} is poor, the best choice being M_1 or M_2, where $\sigma^{(2)}$ does not show minimal sensitivity. With the exact σ unknown, the method is without solid ground.

With two scales μ, M, $\sigma^{(2)}$ corresponds to a 2-dimensional surface. The scales μ_{opt}, M_{opt} correspond to a saddle-point.[1,3] Now the exact solution would correspond to a plane, parallel to the $\mu - M$ plane. Again the same objection is made: If the exact σ lies close to the saddle point, the choice μ_{opt}, M_{opt} is good. If, however, it lies far away, the choice is poor.

Stevenson himself states:[1] "It is not possible to prove, or even formulate, a useful theorem to this effect .... Our information does not include a bound on the error."

There is a nice feature in optimization: In many cases (but not all) with the choice μ_{opt}, M_{opt} the resulting next-to-leading order correction, say $0(\alpha_s^2)$, is significantly smaller than the corresponding Born term; i.e. the corresponding K-factor is ~ 1. This is taken to suggest rapid convergence of the perturbation series. However, nothing is known regarding the term of $0(\alpha_s^3)$. It is possible that with optimization this term and/or higher order terms grow bigger.

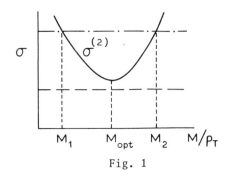

Fig. 1

There is also an unpleasant feature: On dimensional grounds μ_{opt}, M_{opt} are proportional to p_T. However, the relation with p_T is not explicit, but varies from one point of phase space to another. Otherwise stated, μ_{opt} and M_{opt} depend on all kinematic variables p_T, s and rapidity y, and exact analytic expressions do not exist. For a restricted range of p_T, s, y perhaps one can give approximate expressions, but these should vary from one range to another. To give an example, for certain direct photon reactions $A + B \rightarrow \gamma + X$ and for some range one can use the approximate expression

$$M_{opt}^2 \simeq c\, p_T^2 (1 - x_T) \qquad\qquad c = \text{const} \qquad\qquad (3)$$

Taking this at face value it is easy to see that for $x_T > 2/3$:

$$dM_{opt}/dp_T < 0 \qquad\qquad (4)$$

i.e. as p_T increase M_{opt} decreases; and since M enters the structure functions $F(x, M)$, as p_T increases the scale violations act in the wrong direction. Presumably, for $x_T > 2/3$ the expression (3) ceases to hold and should be replaced by another one. All by all, simplicity (and perhaps some intuition gained with scales like $\mu = M = p_T$) is lost.

MEANING OF OPTIMIZATION AND AMBIGUITY IN GLUON DISTRIBUTION

Take again $\mu = M$. Write the inclusive cross section calculated with the physical scale $M = p_T$ as follows:

$$\sigma(p_T) = \sigma_{Born}(p_T)(1 + C(p_T)) \qquad\qquad (5)$$

where $C(p_T)$ represents the correction. One can show that, very roughly, the inclusive cross section calculated with $M = M_{opt}$ is[2,3,4]

$$\sigma(M_{opt}) \approx \sigma_{Born}(p_T)e^{C(p_T)}; \qquad\qquad (6)$$

thus, roughly speaking, optimization amounts to exponentiation of the correction calculated with $M = p_T$.

For $A + B \rightarrow \gamma + X$, in a wide kinematic range, the correction is $C(p_T) \approx 1$; this amounts to saying that the K-factor calculated with $M = p_T$ is $K(p_T) \approx 2$. From (5) and (6), with $C(p_T) \simeq 1$ one obtains

$$\sigma(M_{opt}) \approx [e^{C(p_T)}/(1 + C(p_T))]\sigma(p_T) \simeq 1.35\sigma(p_T); \tag{7}$$

thus optimization enhances σ by $\sim 35\%$ (or even more).

However a comparable enhancement of σ (at a given scale M) is obtained by going from a soft gluon distribution to a hard one.

We illustrate this point by using recent data on $\bar{p}p \to \gamma X$, Ref. 5. Two sets of parton distributions are used: Set 1 corresponding to a soft glue (and $\Lambda = 0.2$ GeV) and Set 2 corresp. to a hard glue ($\Lambda = 0.4$). Fig 2(a) compares the data with theoretical calculations based on PMS optimization (μ_{opt}, M_{opt}). It appears that Set 2 is excluded. However, Fig. 2(b) compares the same data with calculations using the physical scale $\mu = M = p_T$. Now Set 2 rather than Set 1 is favoured.

Thus, in spite of the fact that recent data, in particular from fixed target experiments,[5,6,7] are of high quality, inherent ambiguities in the theory do not permit to resolve the ambiguity in the gluon distribution.

STABILITY, OPTIMIZATION AND APPROXIMATE CORRECTIONS

Suppose that the inclusive cross section $\sigma^{(2)}$, including complete next-to-leading order corrections, has been calculated for some process. Suppose also that an approximation to $\sigma^{(2)}$ is available which is good (i.e. to within 10–15%) at the physical scale $\mu = M = p_T$. We assert that this approximation stays good for wide variations of μ, M; this includes stability of $\sigma^{(2)}$, 1- and 2-scale PMS solutions,[3] saddle-point structure, Fastest Apparent Convergence (FAC) solutions,[8] etc.

We show this using as example the difference of inclusive cross sections for $\bar{p}p \to \gamma X$ and $pp \to \gamma X$. This is dominated by the subprocess $q\bar{q} \to \gamma g$ with $q(\bar{q})$ valence quark (antiquark) and has the form:

$$E\frac{d\sigma}{d^3p}(\bar{p}p - pp \to \gamma X) = \frac{\alpha_s(\mu)}{\pi} \int dx_a dx_b q(x_a, M)$$

$$\cdot \bar{q}(x_b, M)\{\hat{\sigma}_B \delta(1 + \frac{\hat{t} + \hat{u}}{\hat{s}}) + \frac{\alpha_s(\mu)}{\pi} f \cdot \theta(1 + \frac{\hat{t} + \hat{u}}{\hat{s}})\} \tag{8}$$

where $\hat{\sigma}_B$ stands for the Born term and f for the higher order correction. The structure of f is as follows:[3]

$$f = \ln\frac{p_T}{M} \cdot 2P_{qq} \otimes \hat{\sigma}_B - \ln\frac{p_T}{\mu} b\hat{\sigma}_B + F(x_T; x_a, x_b) \tag{9}$$

where P_{qq} is essentially[9] an Altarelli-Parisi split function, $\otimes$ denotes convolution and $b = (33 - 2N_f)/6$. The function F contains some hundred terms but no μ or M dependence.[3] Regarding the first two terms of (9) we stress that they can be written without carrying the complete calculation of f; the first term follows by integrating the Altarelli-Parisi equation and the second from the variation of the running coupling α_s.

Now we proceed with the following approximation:[4] Replace

$$F(x_T; x_a, x_b)\theta(1 + \frac{\hat{t} + \hat{u}}{\hat{s}}) \to C\hat{\sigma}_B(x_T; x_a, x_b)\delta(1 + \frac{\hat{t} + \hat{u}}{\hat{s}}) \tag{10}$$

where C is just a constant. In fact, for C we use the value obtained in the soft gluon approach[10]

$$C \simeq \frac{1}{2}C_F \pi^2 \tag{11}$$

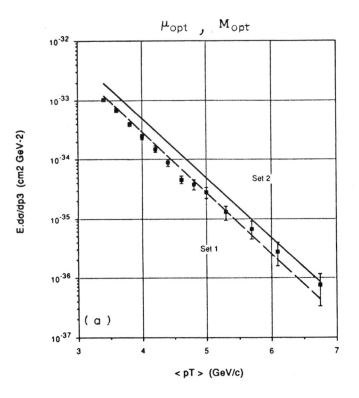

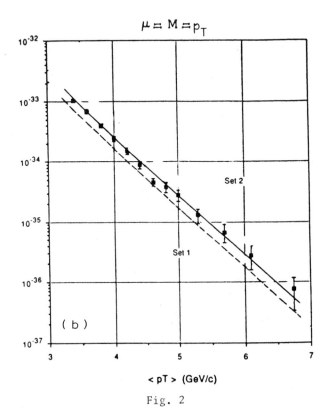

Fig. 2

The results of the calculation for $p_T = 6$ and 10 GeV are given in Fig. 3, which shows contour plots for the cross section (8). For each p_T one sets

$$M^2 = c_M p_T^2 \qquad \mu^2 = c_\mu p_T^2$$

These results are to be compared with the corresponding complete calculations of Ref. 11 (or, for $p_T = 10$ GeV with Ref. 3). In Fig. 3 the saddle-point PMS, the FAC solution, the 1-scale PMS and the physical scale solution ($\mu = M = p_T$) are denoted correspondingly by a circle, a square a triangle and an X. In all cases the similarity is striking: the results differ by $\lesssim 10\%$. One may say that the structure of the two-dimensional surfaces $\sigma^{(2)}(\mu, M)$ corresponding to the complete and to the approximate solution is almost identical for a wide range of the scales μ, M.

Why this great similarity? The reason is that the structure of $\sigma^{(2)}(\mu, M)$, including fine details, is controlled by the first two terms of Eq. (9), and, in fact, in a rather trivial manner; trivial in the sense that these terms can be introduced by hand, without recourse to the full complete calculation. More details on this are given in Ref. 4.

With the approximations (10), (11), at the physical scale $\mu = M = p_T$ one obtains the simple K-factor of the soft gluon approach[10]

$$K(p_T) \simeq 1 + \frac{\alpha_s(p_T)}{2\pi} C_F \pi^2. \tag{12}$$

The first two terms of Eq. (9) vanish identically; clearly, by working at the physical scale and neglecting them, one violates nothing.

It should be also clear that if one is interested in approximations to the complete $\sigma^{(2)}$ or if one asks the question "does the soft gluon approach work?" it is much simpler to work at the physical scale. Varying μ and M only confuses the issue.

A final point regards a systematic program for the determination of the QCD parameter Λ in a specific renormalization scheme (e.g. $\overline{MS}$)[12]. With all the terms of Eq. (9) present, there is no difficulty in using an approximation, like Eqs. (10), (11); again the results are very similar to those of complete calculations.[3]

HOW EXPERIMENT CAN GIVE AN IDEA ON THE PREFERRED SCALE

Since in many cases optimization gives an inclusive cross section $\sigma(\mu_{opt}, M_{opt})$ somewhat larger that $\sigma(p_T, p_T)$, in principle one can get some idea from experiment as to the preferred scale for a given process. Regarding large-p_T direct photons, the cross section difference for $\bar{p}p - pp \to \gamma X$ [Eq. (8)] is very appropriate, for it contains only valence distributions which are known with much precision and do not significantly differ between various sets of parton distributions. Moreover, other contributions, like photon Bremsstrahlung or partons' intrinsic transverse momentum, which involve some uncertainty, practically cancel in the difference (8), thus allowing precise theoretical predictions.

As a specific example and to show the required experimental accuracy, we have carried a calculation corresponding to the kinematic conditions of the UA6 experiment.[5] We denote by $\tilde{\sigma}$ the inclusive cross section averaged over (pseudo-) rapidity η. Fig.4 presents our results for the physical scale p_T and for two optimal scales. As usual, (see Fig. 3 and Ref. 3) the two optimal scale results are very close; and they exceed the physical scale cross section by ~ 30–40%.

Clearly, the required accuracy is quite large, in particular taking into account that Fig. 4 presents predictions for the difference $(\bar{p}p - pp) \to \gamma X$.

Also notice that if such an experiment for $(\bar{p}p - pp) \to \gamma X$ indicates preference e.g. for the optimal scales, nothing can be concluded for a different process like Drell-Yan lepton pair production or jet production at large-p_T.

215

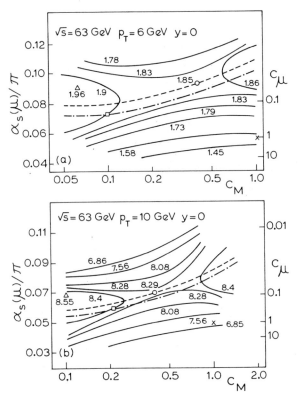

Fig. 3

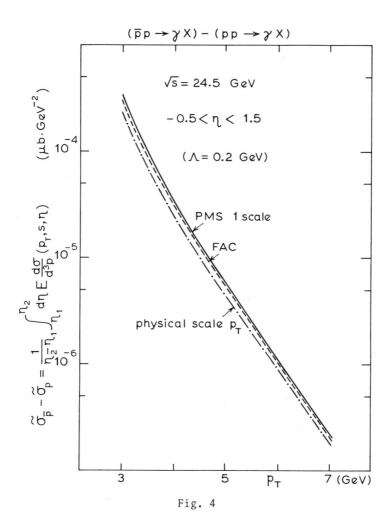

Fig. 4

PRACTICAL SUGGESTION

It is advisable to present theoretical results at the physical scale $\mu = M = p_T$. It is useful to present also optimized results, for, this gives a measure of the scale uncertainty in the theory. Together with the results at $\mu = M = p_T$ present the corresponding Born cross section, so that the K-factor is clear. This will help in many ways.

ACKNOWLEDGMENTS

We would like to thank the Organizers of the very fruitful and pleasant St. Croix Workshop, several of the participants for stimulating discussions and N. Mebarki for much help with the numerical calculations.

REFERENCES

1. P. Stevenson, Phys. Rev. D23, 2916 (1981).
2. H. Politzer, Nucl. Phys. B194, 493 (1982); P. Stevenson and H. Politzer, ibid, B277, 758 (1986).
3. P. Aurenche et al, Nucl. Phys. B286, 553 (1987).
4. A.P. Contogouris, N. Mebarki and H. Tanaka, Phys. Rev. D (in press).
5. A. Bernasconi et al (UA6 Collaboration), preprint CERN-EP/87–120.
6. C. DeMarzo et al (NA24 Collaboration), Phys. Rev. D36, 8 (1987).
7. M. Bonesini et al (WA70 Collaboration), preprint CERN-EP/87–185.
8. G. Grundberg, Phys. Lett. 95B, 70 (1980) and Phys. Rev. D29, 2315 (1984).
9. Details on the form of Eq. (9) and on the general structure of $F(x_T; x_a, x_b)$ are given in Ref. 4.
10. A.P. Contogouris and H. Tanaka, Phys. Rev. D33, 1265 (1986); A.P. Contogouris, S. Papadopoulos and J. Ralston, ibid D25, 1280 (1982) and D26, 1618 (1982); Phys. Lett. 104B, 70 (1981).
11. R. Baier, Proceedings of Meeting "The Quark Structure of Matter" (Strasbourg–Karlsruhe), World Scientific, Singapore (1986), p. 113.
12. J. Owens, Rev. Mod. Phys. 59, 465 (1987).

COMMENTS FROM DIRECT PHOTON ROUND TABLE DISCUSSION

M.J. Tannenbaum

Physics Department
Brookhaven National Laboratory*
Upton, New York 11973 USA

GENERAL COMMENTS ON THE RESULTS PRESENTED

Much beautiful experimental and theoretical work was presented, which led me to the following observations.

Systematic Errors

All experiments have them, and even the theory has them. Let's pay them more respect. Note that when a theorist says that different choices of structure functions or scales give different answers, that is a systematic error.

Very Impressive Theoretical Predictions

The predictions cover many orders of magnitude and many combinations of incident and outgoing particles. Yet, practically no data point varies from the theory beyond the quoted error (statistical and systematic). In a sense we were much more lucky 9 to 10 years ago, in 1978-1979, when the predictions disagreed with each other by factors of ≈ 100. At that time, experimentalists could try to get their data points right to the best of their ability. They didn't have the added worry of whether the points were above or below "Aurenche et al." For a flavor of the era see Ref. 1.

---

*This research has been supported in part by the U.S. Department of Energy under Contract DE-AC02-76CH00016.

The experiments are very difficult. This is indicated by the large systematic errors quoted. Nevertheless, 6 different methods for measuring direct photons have been presented here, and all give impressive results:

Granular, High Spatial and Energy Resolution Detector: R806...,
WA70, UA6, E705, E706, NA3, NA24;

Statistical Method, probability of conversion for 1 and 2 photon clusters: R108, UA2;

External Conversion, pair spectrometer: NA3;

Internal Conversion, low mass muon pairs: UA1;

Lateral Shower Structure after Converter: R110;

Statistical Method, accompanying energy in cone and shape of longitudinal energy deposition: UA1.

THE NEXT STEP

It is now quite clear that the direct photon process exists in hadron collisions, as predicted by QCD. The beauty of this process has been recognized since the very beginning: there is direct and unbiased access to one of the interacting constituents, the photon. The dominant subprocess is the QCD Compton effect.[2] The time for precision tests is upon us. Even at this meeting, people are beginning to be concerned about disagreements on the order of 5% to 10%.

It is generally accepted, by anyone who has done such a measurement, that the clearest way to test QCD is to make pair measurements, i.e. to measure both of the outgoing hard constituents.[3-8] An excellent discussion is given by Owens.[8]

Consider a high granularity, high resolution, direct photon detector, with solid angle ≈ 0.25 steradian, composed of rapidity aperture, $\Delta y = 0.25$, and azimuthal aperture, $\Delta \phi = 1$ radian, all in the p-p cm. system. This detector could be used in conjunction with a large central detector which would detect the jet from the recoiling quark. In fact, it is only necessary to detect the leading particle to determine the rapidity of the recoiling jet with adequate precision.[9] The constituent center-of-mass kinematics can be reconstructed by this method, with a precision of ≈ ±0.03 for the cosine of the constituent c.m. scattering angle, $\cos\theta^*$.

The objective of the measurement would be to map out in detail the constituent center-of-mass subprocess angular distribution for direct photon production. Similar data for neutral-pion production would be obtained simultaneously. According to QCD, the direct photons are produced by a Compton subprocess, and should exhibit that characteristic angular distribution, while the pions are produced with a t-channel pole and should show the characteristic "Rutherford Scattering" anglular distribution (see Fig. 1). The same idea is discussed by Owens in Fig. 33 of Ref. 8. A simple way to do the experiment, conceptually, is to imagine 5 different rapidity settings for the photon detector, say y = 0, 0.25, 0.5, 0.75 and 1.0. In each of these settings, configurations would be selected in which the recoiling jet is back-to-back in rapidity to the photon, within an interval of ±0.1. This procedure restricts the rapidity of the constituent subprocess c.m. system, so that it is essentially the same as the p-p c.m. system, to a precision of ≈ ±0.1.

The cross-section measurement, with both "particles" at zero rapidity, can be used to constrain the uncertainties of the structure functions and coupling constant, or to measure them.[6] This is similar to, but more precise than, the information gained from a single-particle inclusive measurement. The rest of the points in the angular distribution are normalized by the value at $\cos\theta^* = 90°$, y = 0. If it were not for higher order effects in QCD, this normalized distribution would give the constituent c.m. scattering angular distribution directly. In Fig. 1, the pure subprocess distributions are given. If the measurements were to be made, and come out this way, it would provide a simple and elegant demonstration of the validity of hard QCD, particularly for the Compton subprocess.[8]

Of course, at this meeting, we are all more sophisticated. The running of the QCD coupling constant with momentum transfer has not been included in the figure, either directly, or in the secondary effect of non-scaling in the structure functions. This touches at the heart of the higher order correction issue, namely the correct definition of Q^2 to use in the coupling constant. If $\hat{s}$ is the correct value of Q^2 for the Compton subprocess, then Q^2 will not change over the angular distribution at fixed photon energy, and the angular distribution will remain

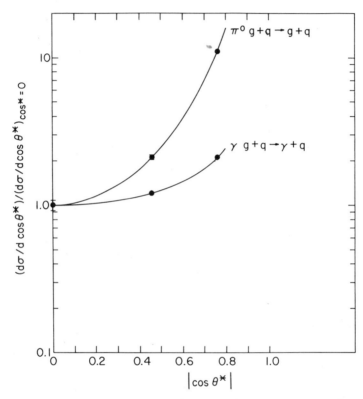

Figure 1. Constituent center-of-mass subprocess scattering angular distributions for direct photon production and neutral-pion production as indicated. The data points and error bars on the figure are to be ignored.

unchanged from that of the subprocess. It is known, of course, that the variation of the QCD coupling constant and non-scaling effects in the structure functions are essential to explain the already existing data of $\pi^0-\pi^0$ and jet-jet production.[3-5,7-8] The point of this comment is to ask the theorists how these effects will modify the photon-jet angular distributions in either of the second-order scale schemes, the optimized scale or the physical scale. Patrick, Andy go to it! The experimentalists should not rest either, let's try to get the data to 5% precision, including systematics!

REFERENCES

1. Proc. of the 1979 International Symposium on Lepton and Photon Interactions at High Energies, eds. T.B.W. Kirk and H.D.I. Abarbanel, Fermilab, Batavia, IL, pp. 589-609.
2. H. Fritzsch and P. Minkowski, Phys. Lett. 69B, 316 (1977).
3. CCOR Collaboration, R108, A.L.S. Angelis et al., Nucl. Phys. B290, 284 (1982).
4. UA1 Collaboration, G. Arnison et al., Phys. Lett. 136B, 294 (1984); 158B, 494 (1985); 177B, 244 (1986).
5. UA2 Collaboration, P. Bagnaia, et al., Phys. Lett. 144B, 283 (1984).
6. R. Baier, J. Engels and B. Petersson, Z. Phys. C2, 265 (1979); Z. Phys. C6, 309 (1980).
7. B.L. Combridge and C.J. Maxwell, Nucl. Phys. B239, 428 (1984).
8. J.F. Owens, Rev. Mod. Phys. 59, 465 (1987).
9. R. Longacre and M.J. Tannenbaum, Informal report, BNL-32888 (1983).

THE HADRONIC INTERACTION MODEL FRITIOF AND BOSE-EINSTEIN CORRELATIONS FOR STRING FRAGMENTATION

Bo Andersson

Department of Theoretical Physics
University of Lund, Sölvegatan 14A
S-223 62 Lund, Sweden

1. Introduction

In this talk I would like to discuss a set of different subjects, which we have been working with in the Lund High Energy Theory Group during the last few years.

The main dynamical tool is, as always in the Lund group, considerations of the confined force-field between colour-connected partonic excitations. We use as the model for this force-field the massless relativistic string [1]. The quark (colour 3) and antiquark (colour $\bar{3}$) degrees of freedom are localized at the endpoints of the open string state while the gluonic (colour 8) modes correspond to excitations in the interior of the force-field.

In the Lund String Fragmentation model we have studied the way such a high mass force field may decay into multiparticle final (colour singlet) hadron states. Then the high-energy momentum density obtained in a "hard" event by a set of coloured partons is on a long time scale transferred into an ordered (one-dimensional) energy-density characteristic for the string, i.e. $\kappa \sim 1$ GeV/F.

It may be worthwhile to point out that "order" also means an implicit assumption of a colour flow order (I have in another review [2] tried to give a more explicit description of the influence of this assumption) in the QCD forcefield.

The fragmentation model contains a precise space-time structure which may also be interpreted in energy-momentum space, at least as long as we keep to the longitudinal degrees of freedom. (My use of the notion of "longitudinal" - parameterized by lightcone

coordinated (t ± x) or energy-momentum components (E ± p) - as well as "transverse" -
two-dimensional vectors with subscript $_T$, e.g. $\vec{k}_T$ - will always be in accordance with intuition.
In this case we may imagine a $q\bar{q}$-pair produced by e.g. an e^+e^--annihilation reaction
spanning the string-like force-field in the "longitudinal" direction.)

It has been shown by Gottfried and Low [3] that is allowed by quantum mechanics to
discuss physics with a localization both in rapidity, basically the logarithm of the lightcone
energy-momentum, and in longitudinal space-time at the same time. The transverse
degrees of freedom which in the Lund Fragmentation Model is treated by a tunneling
process is however of an inherently quantum mechanical nature.

I would like to discuss the implications of this space-time picture on one observable
feature in hadronic physics, the Bose-Einstein correlations between equal bosons.

The success of the fragmentation model makes it tempting to extend the notion of a
string-like force field into all regions of hadronic physics. As a word of caution, let me state
that our present rather incomplete knowledge of the implications of QCD neither allows nor
prohibits such an extension.

We note that the difference between a bag of asymptotically free partons and a
well-ordered colour connected string state is not noticeable on a short-time or local scale in
the development of the final state. This is due to the fact that a string state is causal, i.e. it
takes a finite time to transfer any information on what goes on from one point to another. In
particular, if a large energy concentration is inserted in the string, then it takes some time
before the affected parts start to move away appreciably from the remainder.

As my first subject I would like to present a model for hadronic interactions, the Fritiof
model, which we have developed in the Lund group. It is based upon a surprising feature
of string dynamics, viz. that on a semi-classical level the resulting string state is essentially
the same in case one large momentum transfer Q is applied to an endpoint of a string, or if
many small momentum transfers δq_j with

$$\Sigma\ \delta q_j = Q$$

is applied over a region of the string (including the endpoint).

Supplied with the appropriate bremsstrahlung from the corresponding colour separation
and with "large p_T"-Rutherford scattering as a "natural" perturbation the Fritiof model
provides a surprisingly accurate description of hadron dynamics from PS-energies all the
way to the Sp$\bar{p}$s collider range.

I will present the Fritiof model in section 2 and in section 3 some basic notions on the
structure of the Fragmentation Model and then finally in section 4 the Bose-Einstein
correlations.

226

2. The Fritiof Model for Hadronic Interactions

2A. The Basic Dynamical Mechanism

In this section I will briefly describe a model for hadronic interactions, which has also been extended to hadron-nucleus and nucleus-nucleus collisions [14]. It has been named Fritiof after an old Nordic saga hero. Like all such guys he set himself up against the gods, was properly humiliated and in due time let into ordinary society. The model contains sufficiently much heretical stuff to resemble the beginning of the saga and it is our hope that the end will be similarly happy.

In figs. 1 and 2 a few stages are shown in the motion of a string acted upon by a large momentum transfer. In fig. 1 the momentum transfer is given to the endpoint while in fig. 2 the same momentum transfer is distributed over a region of the string (including the endpoint). In both cases the development is shown in the cms of the string. It should be noted that in fig. 1 the endpoint "by itself" drags the string all the way to the forward maximum stretching point while the remainder "together" carries the backward endpoint to maximum with a (not energy-momentum carrying) "knee" somewhere in the middle. In fig. 2 the two moving parts (note that each one carries a mass smaller than the original hadronic mass m) stretches into a very similar situation with a set of small "bends" along the string.

The transverse size of the states is in both cases of the order of m/κ, while the longitudinal size is of the order of W/κ, i.e. each one of the strings is for large masses basically flat and stretching longitudinally. While the situation in fig. 2 has been chosen in a very simple way the result is essentially the same for almost any other situation.

A dynamical motivation for a hadronic final state like the one depicted in fig. 2 can be obtained if we assume that a hadron is similar to a vortex line surrounded by a colour superconducting vacuum. In that case the main field energy will be concentrated along the vortex line core but there will also be an exponentially decreasing field stretching outwards.

If two such hadronic vortex lines would pass by each other we expect a large amount of basically elastic scatterings between the field-quanta. This will lead to a final state in which each field configuration contains a large amount of disturbances.

We note that if the elastic scatterings involve transverse momentum transfers $\delta \vec{k}_T$ then in order to preserve the mass-shell conditions there must also be longitudinal energy-momentum transfers δq_j of the order of $\sim \delta k_{Tj}^2/x_j W$ with x_j the Bjorken-fraction of the j:th parton-field quanta.

A vortex line force field hadron must behave very similar to a string. As we have learned above, the string reacts to many momentum transfers as if acted upon by a single total energy momentum transfer with

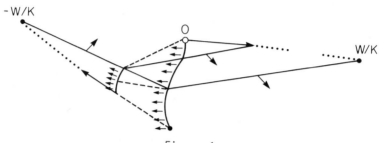

Figure 1

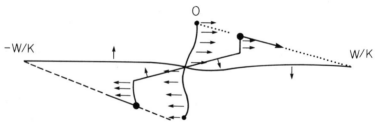

Figure 2

$$\Sigma \, \delta\vec{k}_{Tj} = \vec{Q}_T$$

$$\Sigma \, \delta q_j = Q$$

If the scatterings are incoherent we expect that $\vec{Q}_T \approx 0$ while the longitudinal momentum transfer always is a non-negative quantity.

We therefore expect that the inelastic interaction between two small-mass hadrons with large longitudinal momenta will lead to a final state with the two original small strings stretching longitudinally in a soft way to large masses.

What has happened is basically that if there is a large energy momentum inserted at an endpoint it will by itself stretch out but if there is an extended energy density around an endpoint the whole state will collectively reach the same result. It does not matter whether a finite part of this total energy is actually carried by the endpoint $\bar{q}$-particle. The result will nevertheless be the same, i.e. the whole string region will stretch together.

It may be useful to note that such a collective action-picture is not equivalent to a picture in which each parton by means of a probabilistic structure function is given once and for all a certain energy momentum fraction which is used up in the final state. Although the analogy may be of limited significance this structure function treatment always reminds me of a hadron behaving like a fragile glass container which splits up into independent pieces by the interaction.

Let me at this point remark that although we have no precise reason to expect that a valence flavour necessarily is fixed at a string endpoint we will nevertheless assume that from now on. Then one original valence flavour is carried backwards just as far as the endpoint is - but this is not because we assume that the valence flavour per se is associated with a tiny energy momentum fraction. It is due to the total collective action from the remaining parts of the vortex line string state.

Before discussing the spectrum of momentum transfers in Fritiof, let me briefly consider a simple model of a baryon affected by a momentum transfer [5]. In fig. 3 is shown a model containing three string pieces with a q-valence particle at each endpoint. There is also a junction with the property that it moves so that the tension from the three adjoining string-pieces is balanced. At rest this means that there is a 120° angle between them and the pieces move out and in along the lines exhibited.

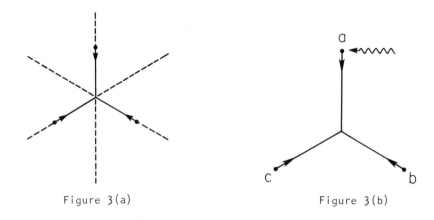

Figure 3(a)

Figure 3(b)

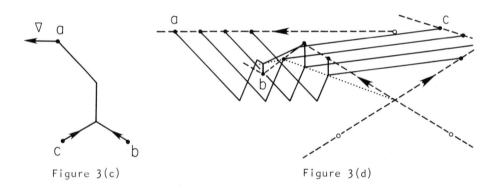

Figure 3(c)

Figure 3(d)

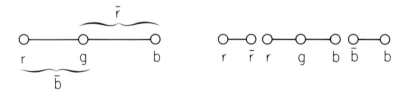

Figure 3(e)

If one of the endpoints is dragged away along the direction of Q, it takes about a quarter of a period for the other two and the junction to note what has happened. The signal is carried by a new string segment stretched from the affected q-particle to the remainder with a corner moving "downwards" to the junction with the velocity of light. It will reach the junction at the same time as the other two are about to "pass on" upwards-outwards (they now carry the energy-momentum obtained from being accelerated towards the junction).

It is obvious that for almost any angular distribution of Q, one of the two unaffected valence constituents will tend to move "after" the hit q-particle and the other "away" from it. The junction will follow in order to keep the equilibrium and the situation will almost always be a final state basically of a one-dimensional nature. There will be the hit q-particle moving along Q, one q-particle with a small adjoining string-piece moving up and down around the Q-direction and the third at the endpoint stretching backwards.

The breakup of such a system will always, as shown in fig. 3d, lead to a baryon state around the "center" q-particle together with a set of mesonic small $q\bar{q}$- strings. In practice two of the original valence flavours end up in this final baryon state rather often.

Therefore two small mass (m, m') incoming hadrons with large W_+, W'_- will lead to two final state strings with

$$(W_+, m^2/W_+, 0_T) + (m'^2/W'_-, W'_-, 0_T) \rightarrow (W_+ - Q_+, Q_-, \sim \vec{0}_T) + (Q_+, W'_- - Q_-, \sim \vec{0}_T)$$

Based upon Feynman's wee-parton spectrum and the general idea that there is no inherent scale in our considerations, we have in Fritiof assumed that the probability to obtain Q_+ and Q_- is scaling:

$$dProb \sim \frac{dQ_+}{Q_+} \frac{dQ_-}{Q_-} \qquad (1)$$

with the obvious requirements that the final state masses are larger or equal to the incoming and together smaller that the total cms energy.

The scaling spectrum means that the median final state masses are around $(W_+ W'_-)^{1/4}$ and that there are large fluctuations. Actually, remembering that a decaying string of mass M will in the main cover a rapidity region of order $2\ell nM$ with a constant number of final state particles it is obvious that the inclusive rapidity spectrum will be flattish for Fritiof over $-\ell n$ $W'_- < y < \ell n\ W_+$. The fluctuations are, however, so large that the $D/\bar{n}$ for the multiciplicity distribution is of the order of 0.5. The spectrum in eq. (1) is evidently reminiscent of "diffractive scattering". Fritiof is in this way a model of total diffraction, i.e. the final states string masses are distributed like diffractive masses but decay longitudinally.

2B. Soft Gluonic Bremsstrahlung

It is well known in QED that when a charge is accelerated, decelerated, or more generally when charges are separated (e.g. in the initial to final state) then bremsstrahlung is emitted inside a characteristic angular (or equivalently rapidity) range. Thus if e.g. an e^+e^- pair is produced and moves apart with rapidities (y_+, y_-) then photon radiation will occur inside the rapidity range $y_- < y < y_+$. Similarly when in QCD colour charges are separated there will be gluonic emission. In an e^+e^-- annihilation event to a $q\bar{q}$-pair at the cms energy W there will be gluon radiation (gluon energy-momentum $k = k_T(\exp y, \exp -y, \vec{e}_T)$ in the same light cone coordinate notation as before with the $q\bar{q}$ axis as longitudinal direction) inside the cms region:

$$-\ell n[\frac{W}{k_T}] < y < \ell n[\frac{W}{k_T}] \qquad (2)$$

Contrary to the QED situation, however, such an emitted gluon will carry a colour charge so that "afterwards" there are actually two colour separation regions, one between the (recoiling) q-particle and the gluon and another between the gluon and the (recoiling) $\bar{q}$-particle. Therefore there will be new radiation inside these regions and this will continue until finally all the coloured partons have small masses with respect to their neighbours.

Such a scheme (which, as I have explained in ref [2], is similar but nevertheless at sufficiently high energies will lead to different consequences than e.g. a Marchesini-Webber (M-W) cascade [6]) has been implemented in the Lund Colour Dipole approximation with the Monte Carlo program Ariadne [7].

It should be noted that the famous interference effects which in the M-W cascades correspond to a strong ordering in angles for the emitted gluons, in the Lund Dipole approach where each angular region is covered by a specific dipole, means: "Do not double count!" Thus each emission can only occur from a single one of the corresponding "mother-dipole" regions.

There is in connection with the possible radiation a major difference between a "hard" event and the Fritiof "soft" colour separation. This difference is related to the available energy-density.

We note that a colour antenna of the transverse size ℓ can only emit gluonic radiation of wavelengths larger than ℓ. Therefore there is an upper limit to the transverse momentum of a gluonic emission

$$\lambda = \frac{2\pi}{k_T} > \ell \qquad (3)$$

in case the available energy is spread over the region ℓ. For a hard event we expect the energy to be strongly concentrated so that there is no other upper limit than the pure kinematical one in eq. (2).

Consider fig. 4 where a Fritiof event situation is shown in the cms of one of the final state strings. A fraction α of the forward moving string section carries αP while a fraction of β of the backward moving one carries βP. Together they emit gluons inside a range corresponding to eq. (2):

$$\alpha P > k_T \exp + y$$

$$\beta P > k_T \exp - y$$

and with a k_T limited by eq. (3):

$$k_T < \frac{2\pi}{(\alpha + \beta) \cdot \ell}$$

with ℓ a transverse size of the hadron.

We immediately conclude that the most efficient value of α and β is

$$\alpha + \beta \sim \sqrt{\frac{4\pi \cosh y}{P \cdot \ell}}$$

so that k_T is limited by

$$k_T < \sqrt{\frac{P\pi}{\ell \cosh y}} \qquad (4)$$

The equation has one unknown parameter ℓ which we have found phenomenologically to be around the inverse of a hadronic mass.

The basic physics behind the limit is that if we take large values of α and β to obtain large energies, then the radiation requirement in eq. (3) is violated, and similarly if we take a small antenna, there is not enough energy to emit large k_T-gluons.

We conclude that the Fritiof soft radiation is essentially smaller in size than for a truly hard event. Nevertheless, at large hadronic interaction energies this mechanism provides a major source of both p_T and multiplicity fluctuations.

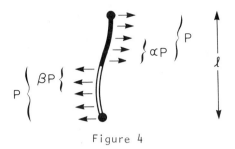

Figure 4

The emission of gluon radiation means in the Lund model that there are besides the q- and q̄-particles at the endpoint of the string a set of internal energy momentum carriers which drag out the string in different other directions. The string state becomes "bent" in such a way that the different parts of the string moves away in different directions.

Each part decays like a flat string in its own rest-frame and in that frame it will provide the same basically Poissonian amount of particles per unit of string phase space.

These different frames are now in relative motion, however, and this implies that the total multiplicity increases (this is a Lorentz contraction effect) and each particle also obtains its share of the transverse motion stemming from the gluonic energy-momenta.

There is consequently an interesting competition in the Lund fragmentation model between the production of ever more particles or to use the gluonic energy-momentum to give more transverse momentum to the particles.

2C. Large P_T Rutherford Scattering as a Perturbation in Fritiof

Up to now the Fritiof scenario contains:

1. an interaction mechanism in which a large amount of small momentum transfers results in a logarithmically distributed longitudinal energy-momentum transfer,

2. that this mechanism implies colour separation, which results in the emmision of soft gluonic radiation which tends to bend the strings and to create further fluctuations (in a correlated way) in multiplicity and transverse momentum.

According to (perturbative) QCD, colour charges from the two hadrons may also interact in accordance with Rutherford scattering, thereby producing at high energies large transverse momentum jet production. This option, which is governed by a well known cross-section $\sim k_T^{-4}$ is a natural perturbation in the Fritiof model.

It just means that one of the many small momentum transfers is large and therefore deserves special attention.

From the cross-section (we use the Pythia implementation of perturbative QCD [8]) one may compute the resulting flavour and colour of the partons involved as well as their final state energy momentum vectors k (a scattered forward moving parton) and ℓ (backward) with e.g.

$$k = (k_+, k_-, \vec{k_T})$$

Suppose for simplicity that k and ℓ correspond to a scattered (q,q) pair. Then we expect that all the many small momentum transfers will still be around and be distributed in accordance with a direct generalization of eq. (1):

$$\frac{dq_+}{q_+ + 1_+} \frac{dq_-}{q_- + k_-} \qquad (5)$$

i.e. we just pick out the parts already decided upon (ℓ_+, k_) and find the remaining energy momenta basically logarithmically distributed. Actually, as we easily see from eq. (5), q_+ and q tend to be larger than the unperturbated Q_+ and Q_- values in accordance with eq. (1).

This is physically reasonable because a large and therefore local disturbance must generally mean that more of the fields do overlap and therefore interact.

The final state is again a generalization of our earlier situation (cf fig. 5). The endpoint q-particle moves away along k, the remainder (q.) drags backward and what is left continues forward ($W_+ - \ell_+ - k_+ - q_+$).

The radiation emitted between the forward mover and the remainder backward mover will be soft as in eq. (4), but the string between the struck quark and the remainder backward mover contains large energy-density and radiates in a "hard" way, i.e. in accordance with eq. (2).

For a scattered gluon which corresponds to an interior excitation in the hadron-vortex line-string we obtain a somewhat more complex situation described in ref. [4].

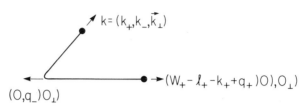

Figure 5

There are a set of problems to be faced in connection with Rutherford scattering, however. Firstly, as is well known, the cross section diverges badly for small values of k_T. We are however asking for <u>the largest</u> Rutherford scattering and this requirement leads to a cross section <u>without divergencies</u>. This is just the ordinary Zudakhov effect. Basically it corresponds to the requirement that the probability of having a scattering at $k_T > k_{T0}$ is proportional to having <u>no</u> scattering for $k_T > k_{T0}$ together with the requirement that there should be one at $k_T = k_{T0}$:

$$dP(k_T = k_{T0})\,|_{Zudakhov} = dP_0(k_T = k_{T0})\,\exp\,-\int_{k_{T0}}^{k_{Tmax}} dP_0(k'_T)$$

(dP_0 is the divergent probability.)

Therefore, as is easily understood, dP vanishes at $k_T = 0$ (in this case exponentially fast). We have nevertheless introduced a k_T-cutoff,k_{T0} in the use of the Pythia mechanism for Rutherford scattering in Fritiof, mostly in order to obtain MC-effectiveness. The parameter k_{T0} can be varied rather much without any change in the results because basically this means just a reshuffling of what to call "Rutherford scattering" and what to call "bremsstrahlung radiation".

There is merely one problem that must be faced in all perturbation theoretical situations. Any group of Feynman graphs are governed by the largest "off-shellness". Therefore in case bremsstrahlung radiation anywhere in a Rutherford scattering diagram is larger than the Rutherford momentum transfer, the radiation part should be the major part of the interaction situation. Thus "small" Rutherford scattering is anyway "drowned" by the radiation.

In order to implement this requirement we consider the radiation in accordance with the rules exhibited in the last section, in particular for the part between the remaining forward mover ("the spectator") and the backward mover. In case any gluons emitted in the process have a larger k_T than the Rutherford momentum transfer, this latter one will be allowed in at its place in the bremsstrahlung hierarchy (independently of whether it is a quark or a gluon Rutherford scattering).

In this way we use the bremsstrahlung as a "noise-level". Only those Rutherford scatterers which "stick out" above this noise level are taken seriously and allowed to "govern" the process, i.e. to be treated separately to the remaining many "small" scatterers (they are, however, always taken seriously as given momentum transfers between the two interacting hadrons).

This assumption is actually an assumption on the stability of the colour force field. In case an external disturbance is small on the level of the radiation, we assume that the field will adjust the colour ordering smoothly by suitable soft gluon exchange. Effectively the final strings always become in this way as "short" as possible (fragmenting into as few hadrons as possible).

Using these prescriptions we obtain a completely infared stable hadronic interaction model with one parameter governing the size of the radiation, "ordinary" perturbative QCD formulae for the bremsstrahlung and the Rutherford scatterings albeit with a Rutherford cutoff k_{T0}.

Finally there is at the basis the flat logarithmically distributed longitudinal energy-momentum transfers.

If the parameter k_{T0} is taken too low, however, then this spectrum becomes distorted. In particular there will be no "small" (Q_+, Q_-) values, i.e. the "diffractive parts" would vanish from the Fritiof scenario. This is another place where we are working at the moment because it is evident that the problems are related to unitarity and to the relative sizes of the different cross-sections.

The present treatment is not really satisfactory in my opinion, but anyway the value of $(k_{T0})_{min}$, where we get into difficulties is very low, around 1.5 GeV/c for $\sqrt{s}$ = 63 and 2.4 GeV/c for $\sqrt{s}$ = 540 GeV (it grows logarithmically with s).

In the next section I will show that the simple Fritiof scenario is in very good agreement with the experimental data in the whole available energy range.

2D. Some Selected Comparisons to Data and a Remark on Colour Exchange and Infrared Stability

I will start to show the multiplicity distributions from Fritiof compared to both the 250 GeV/c π^+p interactions (the EHS-collaboration) and to the UA5 collider data.

From figs. 6 and 7 we conclude that Fritiof very well describes both the main features and the large multiplicity tails. It should be noted that for these latter features it is necessary to introduce Rutherford scattering in the Fritiof model, although it is at the level of only a few GeV/c ($\sim$ 1-2 GeV/c for EHS data and 3-6 GeV/c for the collider UA5 data).

Fritiof also describes very well both the partitioning of the multiplicity distribution into different pseudorapidity bins for the UA5 data (fig. 7) and the forward-backward correlation data (fig. 8) from the same group. What is plotted in fig. 8 is the slope parameter b describing the linear increase in the multiplicity of a forward unit of pseudorapidity with the corresponding multiplicity in a backward unit and η_c is the pseudorapidity gap between the two units.

In fig. 9 I show the UA5 inclusive pseudorapidity spectrum. The height in the centre is strongly influenced by (the beginning of) Rutherford scattering while the width is directly related to the falloff in rapidity of the soft gluonic radiation.

In fig. 10 and 11, finally, I show the comparison of the Fritiof p_T spectrum both to the EHS data and the collider UA1 data.

Although space does not permit me to show it, I would like to say that Fritiof describes very well a lot of other data, e.g. the published jet profiles of UA1 as well as the "pedestal effect", i.e. the increased activity behind a large jet.

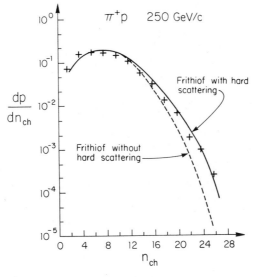

Figure 6

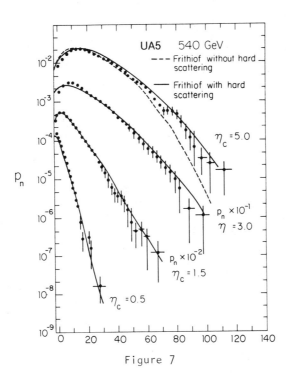

Figure 7

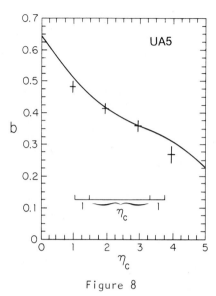

Figure 8

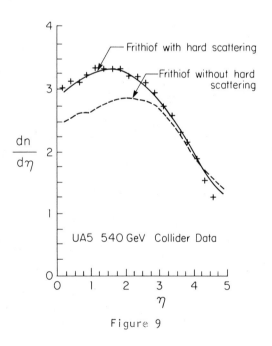

Figure 9

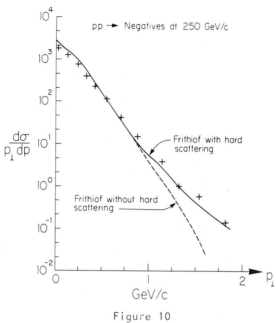

Figure 10

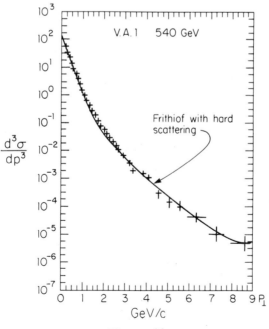

Figure 11

3. On the Structure of the Lund Fragmentation Model

There are several earlier reviews [9] on the details of the model and I will therefore be satisfied with a few remarks on some selected parts.

Firstly, the model contains a unique probability distribution for the production of a set of final state hadrons with energy momenta $p_j(p_j^2 = m_j^2 \ j = 1,...,n)$ from a string with total energy-momentum $P_{tot}(P_{tot}^2 = s)$.

$$d\Gamma_n = \prod_{j=1}^{n} Ndp_j\delta(p_j^2 - m_j^2) \ \delta(\Sigma p_j - P_{tot}) \times \exp{-bA} \qquad (6)$$

Here A is the connected energy-momentum space area of the string before the breakup (cf fig. 12).

The original $q\bar{q}$-pair $(q_0\bar{q}_0)$ produced at the vertex $0 = (0,0)$ carries the energy-momentum (W_+, W_-) and will each drag out the forcefield to the space-time points 0_+ and 0_- with $0_+ = (W_+/K,0)$, $0_- = (0,W_-/K)$. New $q\bar{q}$-pairs are produced at different vertex points v_1, ..., v_n along the the string history. The pairs are dragged apart by the string tension κ and accelerated together with a neighbouring q-$(\bar{q}$-) particle into a final state "yo-yo" hadron. There is evidently a one to one correspondence between the space-time and the energy-momentum vectors with the fixed "exchange rate" $\kappa \approx 1$ GeV/fm.

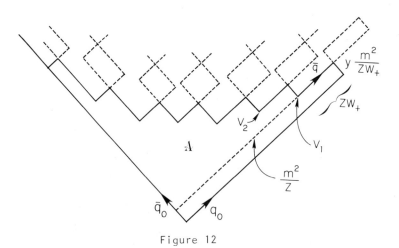

Figure 12

The distribution in eq. (1) has a set of properties [10]. I would like to mention those in order to clarify the relationship to older phenomenology, i.e. Regge-Mueller theory etc.

If we sum over all n and integrate over all momenta, we obtain

$$\sum_n \int d\Gamma_n \approx s^a$$

This is a power law result like in general multiperipheral Regge models with a parameter a which is a certain function of N and bm^2 (m^2 is some mean hadronic mass-square).

The parameter a also occurs in two different other connections. The single particle inclusive distribution, i.e. the distribution in $p_{+1} = ZW_+$ (note that $p_{-1} = m_1^2/p + 1$ in this simple picture) is given by

$$\frac{dP}{dz} = \frac{N}{z} (1-z)^a \exp\left(-\frac{bm_1^2}{z}\right)$$

Here the argument in the exponent is the area related to the production of the first rank particle. In an intuitively appealing way one can say that this is the region inside which the colour of q_0 must be remembered so that the state $(\bar{q}, q_0)$ becomes a colour singlet (cf below).

The phase-space factor $(1-z)^a$ evidently tells us that also the remaining particles want a share of the total W_+.

Finally for the vertex points there is a distribution in proper time τ, in particular for

$$\Gamma = \kappa^2 \tau^2$$

which is

$$\frac{dP}{d\Gamma} = C \Gamma^a \exp\text{-}b\Gamma \qquad (7)$$

It is useful to note that the quantity Γ can be interpreted as the momentum transfer between the group of particles going "to the right" and the other group going "to the left" due to the production vertex in question. If the total mass square of the rightmovers is s_1 and of the leftmovers s_2, it is easy to show that

$$\Gamma \approx \frac{s_1 s_2}{s}$$

Thus

$$\Gamma \approx \exp{-\Delta y}$$

with $\Delta y = \ell n \; s/m^2 - \ell n \; s_1/m^2 - \ell n \; s_2/m^2$, i.e. Δy is the rapidity gap between the right- and left-movers. Consequently eq. (7) tells us that the rapidity gap distribution is for large Δy

$$\frac{dP}{d\Delta y} \sim \exp{-(a+1)\Delta y}$$

like Regge-Mueller theory would tell us.

The eq. (6) has a structure reminiscent to Fermi's Golden Rule, i.e. the phase-space of the final state multiplied by a quantity characteristic of that state, i.e. the negative exponent of the connected string area. The result in eq. (6) is, however, obtained, from a consistency requirement on the stochastic process of string fragmentation [10]. Nevertheless the area law is in agreement with our expectations from a confined gauge theory and I would like to make a few remarks along these lines.

What is actually going on is the production of $q\bar{q}$-pairs, i.e. in field theoretical language the application of $\bar{\psi}\psi$-operators in different space-time points along the string history. Afterwards two adjacent such particles, i.e. one ψ from the vertex (j) and one $\bar{\psi}$ from the vertex (j + 1) are projected onto a final state hadron wave function. In order to obtain a gauge invariant result it is necessary to introduce the gauge field connection

$$\exp ig \int_{(j)}^{(j+1)} A^\mu d\xi_\mu$$

and consequently we expect that a matrix element for the transition to the final state must contain the factor

$$I = \exp i g \oint A^\mu d\xi_\mu$$

with $\oint$ symbolizing an integral around the border of the string field.

The Wilson criterion for confinement is just the requirement that this factor I should behave as

$$I = \exp i \xi S$$

with S the area and with $\text{Re}(\xi) = \kappa$, the string constant. S is evidently related to our earlier area A by $\kappa^2 S = A$.

The appearance of an $\text{Im}(\bar{\xi})$ is related to the production of $q\bar{q}$-pairs in the field presumably in accordance with a Kramers-Konig argument on a di-electricity constant for the vacuum.

Schwinger [11] has given an argument along these lines and shown that the squared decay matrix element of the no-particle vacuum in the presence of an external field is given by

$$\exp\text{-}c\kappa^2(VT)$$

Here (VT) is the space-time region over which the field is spanned and c is a constant which for spin 1/2 particles of mass m is

$$c = \frac{1}{4\pi^2} \sum_{n=1}^{\infty} \frac{1}{n^2} \exp\text{-}\left(\frac{n\pi m^2}{\kappa}\right) \approx \frac{1}{24\pi}$$

The last line is for massless $q\bar{q}$-pairs.

The space-time volume in our case is

$$\kappa^2(VT) = A \cdot S_T$$

with S_T the transverse space-time area of the string flux-tube and A again the characteristic breakup-area in energy-momentum space.

Thus the factor I in eq. (7) can in this interpretation be written as

$$I = \exp i A\left[\frac{1}{\kappa} + \frac{icS_T}{2}\right] \qquad (8)$$

In this way we obtain an interpretation of the area law in eq. (6) as the square of a matrix element given by I and then also

$$b = cS_T$$

Phenomenologically $b \approx 1\text{ GeV}^{-2}$ which would imply a radius for the transverse string flux tube

$$R_T \sim 0.9\text{ F}$$

Let me end this section with a few comments on the transverse momentum and quark mass dependence in the Lund String Model.

Consider a simple Hamiltonian model [12] for the motion of a q-particle in a linearly rising potential:

$$H = -\kappa z + \sqrt{k_z^2 + k_T^2 + \mu^2} \qquad (9)$$

For the energy $E = 0$ it defines the orbit of a q-particle with mass μ passing through the point $(x_T = 0, z = \sqrt{k_T + \mu^2}/\kappa)$ at time $t = 0$:

$$k_z(t) = \kappa t$$

$$z(t) = \frac{1}{\kappa} \sqrt{(\kappa t)^2 + k_T^2 + \mu^2}$$

$$x_T(t) = \frac{k_T}{\kappa} \ell n \left(\frac{\kappa t + \sqrt{(\kappa t)^2 + k_T^2 + \mu^2}}{\sqrt{k_T^2 + \mu^2}} \right)$$

The corresponding quantum mechanical wave-function ψ_q (defined by a wavepacket in the energy around $E = 0$) is easily constructed but unfortunately the space-time function cannot be expressed in terms of elementary functions. The corresponding $\bar{q}$-particle wavefunction ψ_q is obtained by the change $\kappa \rightarrow -\kappa$.

In ref. [12] a model for the production matrix element of a pair with the vertex situated at the origin was taken as the overlap of the product wavefunctions with a constant corresponding to the "flat" field:

$$M = \int dx\, dt\, \psi_q(x,t)\psi_{\bar{q}}(x,t) \qquad (10)$$

It is easily found that the integral in eq. (10) becomes

$$M \approx \exp\left(-\frac{(k_T^2 + \mu^2)\pi}{2\kappa}\right) \cdot \gamma$$

with γ the overlap of the wavepacket distributions and multiplied by a slowly varying function of k_T^2/κ

This result is the same as the one obtained from a WKB-approximation of the wavefunction from the point $z_0 = \frac{1}{\kappa} \sqrt{k_T^2 + \mu^2}$ to $z = 0$:

$$\psi(0) = \psi(z_0) \exp i \int_{z_0}^{0} dz \sqrt{(\kappa z)^2 - k_T^2 - \mu^2} =$$

$$= \psi(z_0) \exp - \frac{(k_T^2 + \mu^2)\pi}{4\kappa}$$

and we interpret

$$|M|^2 \propto \exp - \frac{(k_T^2 + \mu^2)\pi}{\kappa}$$

as the probability to produce a $(q\bar{q})$-pair with masses μ and transverse momenta $\vec{k}_T$ and $-\vec{k}_T$ in a constant forcefield with energy per unit length equal to κ. It is easily seen that the slope of the Gaussian distribution π/κ is very close to our estimate from the b-parameter of R_T. The final state meson stemming from $(q_j\bar{q}_{j+1})$ in this way obtains q transverse momentum

$$\vec{P}_{Tj} = \vec{k}_{Tj+1} - \vec{k}_{Tj}$$

4. Bose-Einstein Correlations

In this section I would like to discuss Bose-Einstein correlations in connection with the fragmentation of a string [13].

It is well know that there is an enhancement for two identical boson particles to be produced with small invariant energy-momentum difference $Q^2 = -(p_1-p_2)^2 = -q^2$. This is usually measured by the two-particle correlation coefficient R:

$$R = \sigma d\sigma/dp_1 dp_2 / \frac{d\sigma}{dp_1} \frac{d\sigma}{dp_2} \qquad (11)$$

The ordinary interpretation is that (as in astrophysical investigations with photon sources) $R = 1 + |\rho(Q)|^2$ with ρ the Fourier transform

$$\rho(q) = \int d^4x \rho(x) \exp^{iqx}$$

of the space-time distribution $\rho(x)$ of production points.

We note that in particle production at high energies we expect (cf fig. 12) that the particle sources move with relativistic velocities with respect to each other. Thus particle production points in the Lund model scatter around a hyperbola with proper time determined by the distribution in eq. (7). Although this implies a large longitudinal width of the particle sources ($\sim \sqrt{s/\kappa}$) one should notice that there is a strong correlation between the production points and the corresponding particle momenta. Thus particles from distant sources will typically exhibit large momentum differences.

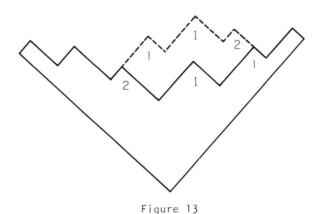

Figure 13

In this way the typical scale measured by Bose-Einstein correlations is the region inside which the momentum distributions overlap. As this is of the order of less that a unit in rapidity we expect that the B-E correlation length in q^2 should be of the order of $1/\tau^2$.

We may further expect a basically "round" distribution, i.e. not elongated along the direction of the event axis and also an energy-independent correlation.

Now suppose that we consider the production of the identical pair (1,2) in a string with the matrix-element as given by eq. (8) above:

$$M = \frac{1}{\sqrt{2}}(M_{12} + M_{21}) = \frac{1}{\sqrt{2}}(\exp i\,\xi A_{12} + \exp i\,\xi A_{21})$$

Here A_{12} is the area of the colour field for a given production order of the pair (12) while A_{21} is the area resulting from an exchange of the two particles (cf fig. 13).

There is evidently an area-difference and thus a phase-difference between the matrix elements M_{12} and M_{21}:

$$\Delta A = |A_{12}-A_{21}| = |\vec{p}_1 E_2 - \vec{p}_2 E_1 + (\vec{p}_1 - \vec{p}_2)E_I - (E_1 - E_2)\vec{p}_I|$$

which is easily seen to vanish if $p_1 = p_2$. If $p_1 \neq p_2$, ΔA also depends upon the momentum of the intermediate state I which must be there always if we have like-sign pions.

Using the expression for the matrix element we obtain for the ratio R in eq. (11)

$$R \sim 1 + \frac{\cos(\frac{\Delta A}{\kappa})}{\cosh(\frac{b\Delta A}{2})} \qquad (12)$$

where the second term should be averaged over all intermediate states I.

It should be noted that the string-mechanism we have exhibited for particle production is not equivalent to the Schwinger model [14] of coherent production. In this latter case the field couples directly and locally to a boson, while in the string mechanism the $q\bar{q}$-pair ending up in a particular boson are produced in different space-time points.

Before using eq. (12), which only applies to a $1+1$ dimensional world of massless quarks one needs to introduce transverse momentum and different flavours in accordance with the tunneling process [12,13].

The resulting prediction from the model is shown together with data from the PEP-TCP group in fig. 14. The full line shows the prediction with $\kappa = 0.2$ GeV². There is essentially no dependence of this curve on b. In this curve we have ignored the fact that some particles stem from resonance decays. It does not matter whether we only produce charged and neutral pions or if we produce the usual mixture of stable and unstable particles but evaluate the matrix elements for the final stable particles ignoring their production through resonances.

I feel that the result of this simple model with its implications from colour dynamics is very good. It would, of course, be even better if our experimental friends could measure Q-values down to $\sim$ 30-50 MeV where the main increase in our predicted values for R would be really noticeable.

Acknowledgements

I would like to thank all my present and former collaborators in the Lund group for a pleasant and continuously amusing atmosphere where lots of dynamically interesting ideas are produced and handled.

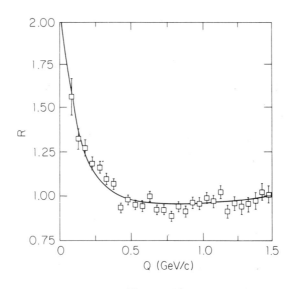

Figure 14

References

1. B. Andersson, G. Gustafson, G. Ingelman, T. Sjöstrand, Phys. Rep. 97 (1983) 33

2. B. Andersson, Invited talk at the Tashkent Multiparticle Symposium 1987, LU TP 87-20

3. K. Gottfried, F.E. Low, Phys. Rev. D17 (1978) 2487

4. B. Andersson, G. Gustafson, B. Nilsson-Almqvist, Nucl. Phys. B281 (1987)
   289, LU TP 87-6, LU TP 87-7
   B. Nilsson-Almqvist, E. Stenlund, Comp. Phys. Comm. 43 (1987) 387

5. B. Andersson, G. Gustafson, O. Månsson, LU TP 82-13 (1982)

6. G. Marchesini, B.R. Webber, Nucl. Phys. B238 (1984) 1
    B.R. Webber, Nucl. Phys. B238 (1984) 492

7. G. Gustafson, Phys. Lett. 175B (1985) 453
    G. Gustafson, U. Pettersson, LU TP 87-9 (1987)
    U. Pettersson, (to be published)

8. H.-U. Bengtsson, T. Sjöstrand, Comp. Phys. Comm. 46 (1987) 43
    T. Sjöstrand, M. van Zijl, LU TP 87-5 (1987)
    H.-U. Bengtsson, Comp. Phys. Comm. 31 (1984) 323

9. Cf ref. 1. B. Andersson, Invited talk at the Physics in Collision 2, 1982 (Plenum
    Press, NY)
    The Symposia for Multiple Particle Dynamics 1980-83
    G. Gustafson, Review talk at the Uppsala Conference, 1987
    T. Sjöstrand, LU TP 87-18 (1987), review article to be published in the International
    Journal of Modern Physics A

10. B. Andersson, G. Gustafson, B. Söderberg, Z. Physik C20 (1983) 317

11. W. Heisenberg and H. Euler, Z. Physik 98 (1936) 714
    J. Schwinger, Phys. Rev. 82 (1951) 664

12. B. Andersson, G. Gustafson, LU TP 82-5 (1982)

13. B. Andersson, W. Hofmann, Phys. Lett. 169B (1986) 364

14. J. Schwinger, Phys. Rev. 128 (1962) 2425
    A. Casher, J. Kojut, L. Susskind, Phys. Rev. C10 (1974) 732

15. M.G. Bowler, Z. Physik C29 (1985) 617
    Phys. Lett. 180B (1986) 299
    Phys. Lett. 185B (1987) 205

PARTON DISTRIBUTIONS IN NUCLEI AND POLARISED NUCLEONS

F.E. Close

Dept of Physics, University of Tennessee
Knoxville, TN37996, USA

ABSTRACT

I review the emerging information on the way quark, antiquark and gluon distributions are modified in nuclei relative to free nucleons. Recent data claiming that quarks carry a vanishingly small fraction of the proton's spin are criticised and the need for better understanding of $x \to 0$ behaviour is emphasised.

If we are to identify the formation of quark-gluon plasma in heavy ion collisions by changes in the production rates for ψ relative to Drell-Yan lepton pairs then it is important that we first understand the "intrinsic" changes in parton distributions in nuclei relative to free nucleons. So, in the first part of this talk I will review briefly our emerging knowledge on how quark, antiquark and gluon distributions are modified in nuclei relative to free nucleons.

In the second part of the talk I will review the status of quark distributions in polarised nucleons relative to those in unpolarised nucleons. In Bjorken's presentation[1] we heard about claims made at the Uppsala conference by the EMC collaboration[2] that, taken at face value, seem to imply that valence quarks may carry a vanishingly small percentage of the proton's spin. If this is true then it could be the most significant clue that our understanding of the proton is all at sea.

PARTONS IN NUCLEI

The best known nuclear distortion is that of the EMC effect which reveals a modification of the valence quark distributions in nuclei relative to those in free nucleons.

All experiments now show broad agreement[3]. The rise in F_A/F_N at $x \gtrsim 0.7$ is due to Fermi motion causing the structure function F_A to leak out to $x > 1$; dramatic as this appears it occurs where $F_{A,N} \to 0$ and is in fact a very minor contributor to the overall phenomenon. Indeed, overall the effect is a subtle 10% affair and we don't need to rewrite the nuclear physics textbooks. As $x \to 0$ we are beginning to see evidence for shadowing, a subject on which theory is now also starting to develop[4].

QUARKS IN NUCLEI

In the "intermediate" region $0.2 \leqslant x \leqslant 0.6$ the ratio falls below unity as (valence) quarks lose momentum due to nuclear binding. The A dependence was successfully predicted in advance of data[5] and is rather well understood. Significant fluctuations are predicted at small A which have yet to be studied. At large A the behaviour is smooth and it is safe to interpolate. Thus one can infer the $F_A(x)$ when A = tungsten, say, and use this as input to $\pi W \to \mu\bar{\mu} \ldots$ analyses for example.

A common feature of models is that the degradation of the valence quarks transfers energy-momentum to some other component (gluons and $q\bar{q}$ in rescaled QCD[6] or the partons in the π that are responsible for nuclear binding[7,8]). Thus they generate an increased sea in nuclei relative to that measured in free nucleons. In turn, this implies that $F_A/F_N > 1$ as $x \to 0$. However, this predicted enhancement will probably be blacked out by shadowing (which has not been incorporated in these models so far). Mueller and Qiu[4] have begun to illuminate us about the x and Q^2 dependence of nuclear shadowing; the quantitative combination of their work with "soft π"[7,8] or rescaled QCD[5] remains to be completed.

What impact does this have on Drell-Yan?

If $x_{1,2}$ refer to the beam and target partons, $x_F \equiv x_1 - x_2$ and $X_1 X_2 = Q^2/s$ then the ratio of cross sections for some Q^2, s fixed,

$$\frac{\sigma^{bA}}{\sigma^{bN}} \sim \frac{\bar{q}(x_1)q^A(x_2) + q(x_1)\bar{q}^A(x_2)}{\bar{q}(x_1)q^N(x_2) + q(x_1)\bar{q}^N(x_2)}$$

where sum over flavours weighted by their squared charge is understood. In the case of π^- beams, if $x_2 > 0.2$ so that $\bar{q} \ll q$, the DY process is dominantly due to $\bar{q}^\pi$ annihilating with $q^{A,N}$. Thus, in this kinematic regime

$$\frac{\sigma^{\pi^- A}}{\sigma^{\pi^- N}} \sim \frac{u^A(x_2)}{u^N(x_2)} \sim \frac{F^A(x_2)}{F^N(x_2)}$$

This is the same ratio as measured in inelastic lepton scattering ("EMC effect") and must obtain here too if factorisation is valid. Thus we should not be surprised by the results from NA10[9] who study $\frac{\sigma(\pi^- W \to \mu^+ \mu^- \ldots)}{\sigma(\pi^- D \to \mu^+ \mu^- \ldots)}$ and by varying Q^2 and x_F can separate both the pion and target structure functions. In ref 9 they exhibit the resulting ratio of $q^A/q^N(x_2)$ and $\bar{q}^{\pi(A)}/\bar{q}^{\pi(N)}(x_1)$.

The latter should be unity and is within errors, when one combines data from two energies, 140 GeV and 286 GeV incident π beams. (However if one restricts attention to the lower energy sample the situation is more messy and the pion distributions do not seem to factorise. Why this should be is unclear to me but bear it in mind as an empirical observation for later reference).

ANTIQUARKS IN NUCLEI

The Drell-Yan process with incident nucleons can probe $\bar{q}$ in the target if suitable kinematics are chosen, e.g. $x_1 \simeq 0.7$ and $x_2 \ll x_1$. An investigation of this in various models has been made by Bickerstaffe et al[10]. As an example, in fig 1, I show the predictions for $\frac{\bar{q}^A}{\bar{q}^N}(x)$ in iron in three models compared with information gleaned from CDHS[11]. The dramatic rise in the Berger-Coester model at $x \gtrsim 0.3$ is

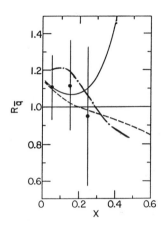

Fig.1 Data from ref.11 on the ratio $q^{-Fe}/q^{-D}(x)$ from neutrino scattering.
Here $\bar{q} = (\bar{u}+\bar{d}+2\bar{s})$. The solid curve illustrates predictions of pion
exchange model (ref.8, 11), the dot-dash is the pion model of ref.7
and the dashed curve is the rescaling model, ref.6.

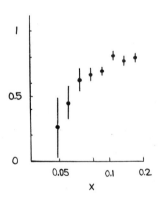

Fig.2 Ratio of ψ production in $\pi^- W/\pi^- Be$ from E537 (ref.14) which is
indicative of $g^W/g^{Be}(x)$, plotted against x-Bjorken for the nuclear
target (x_2).

due to their prediction that $\bar{q}$ leak out to moderate x values in nuclei. However it is illusory to some degree as both $\bar{q}^A$ and $\bar{q}^N$ are vanishingly small; even so experiment E772 may be able[23] to test this. Independent of specific models it is an interesting question whether $\bar{q}$ leak to "large" x in nuclei as this will bear on the Q^2 shape of Drell-Yan pairs in nuclei which may differ from the ψ production (produced by gluons) and potentially provide a background to the plasma signal sought in heavy ion collisions.

Recently WA25 and WA59 in collaboration have studied the EMC effect using ν and $\bar{\nu}$ interactions in neon and deuterium. The x,y distributions allow separation of quark and antiquark distributions and there is some indication that the sea <u>decreases</u> in going from deuterium to neon. Depending upon the model assumed for σ_L/σ_T the fractions of sea in neon is 7 ± 6% or 18 ± 10% less than in deuterium. Presumably these data are being dominated by nuclear shadowing, like the electromagnetic data for $x \lesssim 0.1$, again highlighting the need to see how shadowing modifies the curves in fig 1 at small x.

GLUONS IN NUCLEI

Insofar as inelastic $\gamma A \rightarrow \psi + \ldots$ proceeds via photon-gluon fusion and $NA \rightarrow \psi + \ldots$ involves gluon-gluon fusion, these processes probe $g^A(x)$. There have been early claims that $g^A > g^N$ ($x \sim 0.05$), this based on the EMC data[12] for $(\gamma Fe \rightarrow \psi)/(\gamma D \rightarrow \psi \ldots)$. Expressed as A^α this gave $\alpha = (1.10 \pm 0.03 \pm 0.04$. However it is now less clear whether <u>coherent</u> ψ production (for which $\alpha \simeq 4/3$) has been entirely removed from the data. Indeed E691 (Sokoloff et al) report[13] that (at $Q^2 = 0$)

$$\alpha_{coherent} = 1.40 \pm 0.06 \pm 0.04$$
$$\alpha_{incoherent} = 0.94 \pm 0.02 \pm 0.03$$

This suggests that gluons are shadowed in nuclei.

There are also confusing signals[14] coming from E537 who probe the gluon distribution with $\pi^- W/Be \rightarrow J/\psi \ldots$ at 125 GeV, measuring the x_F distributions and thereby enabling $g^{W/Be}$ to be measured for $x_{Bj} \lesssim 0.2$. If $x_F = x_1 - x_2$, then for J/ψ production at 125 GeV

$$x_F \simeq \frac{1}{25X_2} - x_2$$

Thus we can replot the data from E537 against X_2 (fig 2). This is equivalent to $g^W(x)/G^{Be}(x)$ only if $g^{\pi/(W)}/g^\pi(Be)$ cancels out. However, we have no immediate way of knowing if this is true empirically as $Q^2(= m_\psi^2)$ is fixed. Prima facie one may justifiably be worried. First, which is theoretical prejudice, if fig (2) is interpreted as $g^{W/Be}(x)$, it implies that gluons are significantly shadowed for x as large as 0.2. Our understanding of shadowing is still rather primitive, but such behaviour would be against all current models.

The NA(10) data[9] on $\pi^- W/D \rightarrow \mu^+ \mu^-$... may hint that we should be wary of E537's interpretation. Recall their extracted quark ratio which is in line with the EMC data from inelastic muon scattering – "EMC effect"[3]. This is fine for the 140 GeV and 286 GeV data combined but when one looks at the NA(10) data 140 GeV sample alone, things are less clear. The x_F distribution from NA(10) (which is a convolution of beam and target and thus "nearest" to the E537) for 140 GeV matches smoothly onto E537 at $x \simeq 0.2$ – and the reason is that $q/\pi(W) \neq q/\pi(Be)$ in the 140 GeV data sample. I have no idea why this should be so, but if it is true for gluons too that $g/\pi(W) \neq g/\pi(Be)$ at 125 GeV, then it raises a question about extracting $g^{W/Be}$ from E537. If we take the NA(10) data on quarks as a guide, then it is possible that the $g^{W/Be}(x)$ is, in effect, to be renormalised upwards by 20%. (More legitimately, I don't know why there is such an energy dependence, or even if it is real, but I would be happier to see the E537 experiment with 300 GeV incident beams or comparison with Υ production so as to get some lever on the $x_1 x_2$ separation directly).

If one renormalises the ratio in fig 2 upwards by 20% then there is no shadowing at $x \simeq 0.2$ and, furthermore, the data then look quantitatively as expected in Mueller-Qiu theory of shadowing[4]. It is important that this problem with separation be better understood before we can conclude very much on the $g^{W/Be}(x)$ ratio. (Another reason why one might regard this renormalisation as reasonable is that "infinite" shadowing leads to an $A^{2/3}$ behaviour and for W/Be ratio this is 0.37. The trend of the E537 data looks set to violate this whereas a 20% increase would bring this into line).

We have heard a progress report[15] from E672 who are measuring

$\pi^-A \rightarrow \psi$ on four nuclei at 530 GeV. I await their "high energy" extraction of $g^{A/N}(x)$. Until the conundrum of energy dependence (i.e the non-factorisation of the partons in the incident beam) is settled, I conclude that $g^{A/N}(x)$ probably falls as $x \rightarrow 0$, in qualitative agreement with the shadowing phenomenon, but the quantitative measure in unclear.

Thus, with the exception of valence quarks for $x > 0.2$, there is little or no evidence for non-trivial behaviour for $\tilde{q}^{A/N}$ and $g^{A/N}$. It is imperative to know these quantities much better, and to understand the anomalous energy dependence manifested by NA(10), and implicitly hinted at by E537. Until we do, then we cannot with any confidence use J/ψ relative to Drell-Yan production in AA collisions as a signal for quark-gluon plasma formation.

SPIN DISTRIBUTIONS IN POLARISED PROTONS

EMC have measured the structure function $g_1(x)$ for polarised protons. If their presentation at Uppsala[16] is taken at face value it seems to imply that only a small fraction of a polarised proton's spin is carried by quarks. If true, then we must significantly reassess our understanding of hadron structure. I will review here some of the hidden assumptions and argue that it is premature to claim such a dramatic result. These remarks will be reported in more detail elsewhere[17].

Bjorken derived a sum rule for the <u>difference</u> of proton and neutron polarised structure functions $g_1^{p,n}(x)$[18]

$$\int_0^1 dx(g_1^p(x) - g_1^n(x)) = \frac{1}{3} \left| \frac{g_A}{g_V} \right| . \tag{1}$$

Many people have written sum rules for the p and n separately[18,19]. Ellis and Jaffe in particular write[19]

$$\int_0^1 dx \ g_1^{p,n} = \frac{1}{12} \left| \frac{g_A}{g_V} \right| \left\{ \pm 1 + \frac{5}{3} \frac{3F - D}{F + D} \right\} \tag{2}$$

which involves the F,D values for the octet of β-decays. EMC choose to analyse this and then to translate the result in terms of the quark spins. However the introduction of F,D tends to obscure the relation to quark spins, which can be obtained rather directly[17,20], and has been misused in some analyses (see later). First I will show where the small value for the percentage quark spin comes from and highlight the sensitivity of this conclusion to small changes in data and analysis.

In the quark parton model one has[20]

$$\int dx \, g_1^P(x) = \int \frac{1}{2} \left[\frac{4}{9} \Delta u + \frac{1}{9} \Delta d + \frac{1}{9} \Delta s \right] dx \tag{3}$$

where $\Delta q(x) \equiv q\uparrow(x) - q\downarrow(x)$ refers to the distribution of quarks polarised parallel or antiparallel to the parent proton. (For a neutron target replace Δu by Δd and vice versa). We can decompose (eq 3) into the 3,8 components of octet and a singlet combination

$$\int dx \, g_1^{P,n}(x) = \pm \frac{1}{12} \int dx \Delta (u-d) + \frac{1}{36} \int dx \Delta (u+d-2s) + \frac{1}{9} \int dx \Delta (u+d+s) \tag{4}$$

The first two integrals on the R.H.S. are related to the (measured) $\dfrac{g_A}{g_V}$ in hadron beta-decays, thus

$$\int dx \Delta (u-d) = \left| \frac{g_A}{g_V} \right|_{n \to p} \quad (\equiv F+D) = 1.254 \pm 0.005 \tag{5}$$

$$\int dx \Delta (u+d-2s) = 3 \left| \frac{g_A}{g_V} \right|_{\Xi \to \Lambda} \quad (\equiv 3F-D) = 0.25 \pm 0.05 \quad . \tag{6}$$

Alternatively one can sum the 3,8 contributions and, for the proton target, write

$$\frac{1}{18} \int d\Delta (2u-d-s) = \frac{1}{6} \left| \frac{g_A}{g_V} \right|_{\Lambda \to p} \quad (\equiv \frac{1}{6}(F+ \frac{1}{3}D)) = 0.116 \pm 0.004 \tag{7}$$

(note that if one assumes that $\Delta s = 0$ then the relations (5,6) between the parton sum rules and F,D combined with eq (4) yields the Ellis-Jaffe form of the sum rule eq (2). But notice also that there is a constraint, viz that $(\frac{g_A}{g_V})_{np}$ / (F+D)=1 which is not satisfied in ref. 16*).

---

*This is because the derivation of F,D in ref. 21 used as input the neutron lifetime which was believed to be 925 sec. This gave $g_A = 1.233$ and disagreed with the decay asymmetries which imply $g_A = 1.25$. The modern value of the lifetime is 898 sec. which implies $g_A = 1.25$, in agreement with the asymmetry. A new fit to F,D is awaited.

Using only the well determined constraint (eq 5) yields from eq (4)

$$\int dx g_1^{p,n}(x) = \pm \frac{1}{12} \left|\frac{g_A}{g_V}\right|_{np} + \frac{5}{36} <S_Z^{ud}> + \frac{2}{36} <S_Z^s> \qquad (8)$$

If we also impose the constraint (eq 6) we can write

$$\int dx g_1^p(x) \simeq \frac{1}{10} \left|\frac{g_A}{g_V}\right|_{np} + \frac{1}{9} \int dx \Delta (u+d+s) \qquad (9)$$

(depending on which data on g_A are input ie eq 6 or 7, the coefficient of $\frac{g_A}{g_V}$ in eq 9 ranges between $\frac{1}{10}$ and $\frac{1}{11}$ but does not alter the essential point of our argument). Thus the $\frac{1}{10} \left|\frac{g_A}{g_V}\right| \simeq 0.125$ or 0.116 ± 0.004 (eq 7) nearly saturate the claimed integral of ref 16 ($\int dx g_1^p \simeq 0.113 \pm 0.012 \pm 0.026$). Thus we see immediately that the values of $<S_Z^q>$ are rather sensitive to the value of the integral. A 10% increase in the left hand side gives a much larger, even 50%, increase in the $<S_Z>$. There are several possible ways that 10% effects can accumulate on the L.H.S. The QCD corrections $\sim (1 + \alpha/\pi)$ are of this order and while small in the integral, they lead to large percentage changes in the inferred S_Z.

By far the most serious worry is the assumed extrapolation of $g(x)$ into the $x \to 0$ region. First we should realise that what is measured is a spin asymmetry, $A(x)$, which is falling as $x \to 0$. This has to be multiplied by $F_1(x)$ (which could be rising rapidly as $x \to 0$) in order to obtain $g_1(x) = A(x)F_1(x)$. The behaviour of $g_1(x)$ thus depends on whether the fall in A wins or loses against the rise in F_1. Small changes in the assumed behaviour as $x \to 0$ can lead to sizeable changes in the inferred integral and S_Z.

There are some indirect hints in unpolarised structure functions that convergence as $x \to 0$ may be slow. The $F_2^n(x)/F_2^p(x)$ should $\to 1$ as $x \to 0$, but the trend of the ratio seems to be below this. The reason is that one tends to think that x=0 is near to x=0.1 whereas physics tends to be logarithmic; $\log(x \to 0)$ is infinitely remote. This slow convergence of $F_2^n(x) - F_2^p(x)$ may also cause the convergence of the Gottfried sum rule to be slow

$$\int dx (F_1^p - F_1^n) = 1/6 \quad . \qquad (10)$$

Naive extrapolations of $F_1^p(x) - F_1^n(x)$ seem to imply that this sum rule is not saturated. Perhaps the $x \to 0$ behaviour of $F_1(x)$ is more interesting than we have suspected; bear in mind that $g_1 = A(x)F_1(x)$ and the similarity of (10) to the polarised sum rules is obvious. It may be incorrect assumptions on the unpolarised structure functions that are misleading us as to the behaviour of the $g_1(x)$. Paradoxically it may be data on $F(x \to 0)$ from HERA that will illuminate some features of the $g_1(x \to 0)$.

In ref (17) some explicit examples are presented of reasonable extrapolations of $F(x)$, designed to satisfy the Gottfried sum rule (10), and which may give values for $\int g_1(x)dx$ that are rather larger than claimed in the EMC analysis[2]. Whether or not things turn out this way remains to be seen but there is nothing to prove the contrary. Therefore one need not conclude (yet!) that the quarks carry a vanishingly small percentage of a polarised proton's spin. Indeed they can easily carry over 50% of the spin of the target, commensurate with their share of the total momentum. Notwithstanding these remarks, it is an interesting question as to how the proton's spin polarisation is shared among quarks, antiquarks and gluons[22] and varies with Q^2. Polarisation experiments probe some fine details of hadronic structure and need more study, both experimentally and theoretically.

REFERENCES

1.   J.D. Bjorken, these proceedings
2.   T. Sloan, Proc of Uppsala Int. Con. on HEP (1987).
3.   EMC Collaboration: Phys. Lett. 123B, 275 (1983)
      BCDMS: Phys. Lett. 163B, 282 (1985)
      A. Bodek et al, Phys. Rev. Lett. 50, 1431, and 51, 534 (1983)
      Phys. Rev. Lett. 52, 727 (1984)
      R. Arnold et al
      T. Sloan, ref 2.
4.   A. Mueller and J. Qiu, Nucl. Phys. B268, 427 (1986)
      J. Qiu, Nucl. Phys. B291, 746 (1987)
      Early qualitative ideas are in N.N. Nikolaev & V. Zakharov, Phys. Lett. 55B, 397 (1975).
5.   F.E. Close, R.L. Jaffe, R.G. Roberts and G.G. Ross, Phys. Letters 134B, 449 (1984).
6.   F.E. Close, R.G. Roberts and G.G. Ross, Phys. Letters 129B, 346 (1983).
7.   C.H. Llewellyn Smith, Phys. Lett. 128B, 107 (1983)
      M. Ericson and A. Thomas, Phys. Lett. 128B, 112 (1983).
8.   E. Berger and F. Coester, Phys. Rev. D32, 1071 (1985)
      E. Berger, F. Coester and R. Wiringa, Phys. Rev. D29, 398 (1984).
9.   NA(10): Physics Letters 193B, 368 (1987).

10. R. Bickerstaffe, M. Birse and G. Miller, Phys. Rev. Lett. $\underline{52}$, 2532 (1984).

11. CDMS: H. Abramowicz et al, Z. Phys. $\underline{C25}$, 29 (1987)
    E. Berger and F. Coester, Argonne report ANL-HEP-PR-87-13 (to appear in Ann. Rev. of Nuclear and Particle Science).

12. EMC: J. Aubert et al, Phys. Letters $\underline{152B}$, 433 (1985); Nucl. Phys. $\underline{B213}$, 1 (1983)
    F.E. Close, R.G. Roberts and G.G. Ross, Z. Phys. $\underline{C26}$, 515 (1985).

13. E691, M.D. Sokoloff et al, Flab 86/120E

14. E537, S. Katsanevas et al.

15. E672, reported by A. Zieminski in these proceedings.

16. T. Sloan (ref 2) and private communication.

17. F.E. Close and R.G. Roberts, RAL report in preparation.

18. J.D. Bjorken, Phys. Rev. $\underline{148}$, 1467 (1966).

19. J. Ellis and R.L. Jaffe, Phys. Rev. $\underline{D9}$, 1444 (1974).

20. F.E. Close, Introduction to Quarks & Partons (Academic Press, 1978)p

21. M. Bourquin et al, Z. Phys. $\underline{C21}$, 27 (1983).

22. P. Ratcliffe, Queen Mary College, London Report

23. G. Garvey et al., E772.

HEAVY QUARK PRODUCTION IN QCD

R. K. Ellis

Fermi National Accelerator Laboratory
P. O. Box 500, Batavia, Illinois 60510

1. Introduction

In this paper I review the status of the theory of heavy quark production. The production of heavy quarks in hadron-hadron and photon-hadron collisions continues to be a topic of great theoretical interest. The reasons for this enthusiasm are the existence of experimental data on charm production and the recent publication of data on bottom quark production in hadronic reactions. For a review of the data on the hadroproduction and photoproduction of heavy quarks I refer the reader to refs. [1] and [2] respectively. An additional motivation is provided by the need to relate the results of the search for the top quark[3] to a value of the heavy quark mass.

In the hadroproduction of heavy quarks there are two incoming strongly interacting particles. The produced quarks are coloured objects which subsequently fragment into the heavy mesons and baryons observed in the laboratory. Therefore heavy quark production is an important check of the QCD improved parton model in a complex hadronic environment. Because of their semi-leptonic decays, heavy quarks give rise to electrons, muons and neutrinos. Prompt leptons and missing energy are often used as signals for new phenomena. Therefore an accurate understanding of heavy quark production cross-sections is necessary to calculate background rates in the search for new physics.

The standard perturbative QCD formula for the inclusive hadroproduction of a heavy quark Q of momentum p and energy E,

$$H_A(P_1) + H_B(P_2) \to Q(p) + X \tag{1.1}$$

determines the invariant cross-section as follows,

$$\frac{E\, d^3\sigma}{d^3p} = \sum_{i,j} \int dx_1\, dx_2 \left[\frac{E\, d^3\hat{\sigma}_{ij}(x_1 P_1, x_2 P_2, p, m, \mu)}{d^3p}\right] F_i^A(x_1, \mu)\, F_j^B(x_2, \mu). \qquad (1.2)$$

The functions F_i are the number densities of light partons (gluons, light quarks and antiquarks) evaluated at a scale μ. The symbol $\hat{\sigma}$ denotes the short distance cross-section from which the mass singularities have been factored. Since the sensitivity to momentum scales below the heavy quark mass has been removed, $\hat{\sigma}$ is calculable as a perturbation series in $\alpha_S(\mu^2)$. The scale μ is *a priori* only determined to be of the order of the mass m of the produced heavy quark. Variations in the scale μ lead to a considerable uncertainty in the predicted value of the cross-section. This uncertainty is diminished when higher order corrections are included. The photoproduction of heavy quarks is described by a similar formula.

The theoretical justification for the use of Eq. (1.2) in heavy quark production has been discussed in refs. [4, 5]. Although there is no proof of factorisation in heavy quark production, arguments based on the examination of low order graphs suggest that factorisation will hold. No flavour excitation contributions are included in Eq. (1.2), since the sum over partons runs only over light partons. Graphs having the same structure as flavour excitation are included as higher order corrections. Interactions with spectator partons give rise to terms not shown in Eq. (1.2) which are suppressed by powers of the heavy quark mass. If factorisation holds the power corrections should be suppressed relative to the leading order by $(\Lambda/m)^2$. Note however that in ref. [6] a non-relativistic calculation has been performed which finds terms which are only suppressed by (Λ/m). It would be useful (particularly in the interpretation of charm data) to establish that the power corrections in the full relativistic theory are in fact $(\Lambda/m)^2$.

The lowest order parton processes leading to the hadroproduction of a heavy quark Q are,

$$(a) \quad q(p_1) + \bar{q}(p_2) \quad \rightarrow \quad Q(p_3) + \bar{Q}(p_4)$$

$$(b) \quad g(p_1) + g(p_2) \quad \rightarrow \quad Q(p_3) + \bar{Q}(p_4). \qquad (1.3)$$

The four momenta of the partons are given in brackets. The invariant matrix elements squared for processes (a) and (b) have been available in the literature for some time[7,8,9] and are given by,

$$\overline{\sum} \left|M^{(a)}\right|^2 = \frac{g^4 V}{2N^2}\left(\tau_1^2 + \tau_2^2 + \frac{\rho}{2}\right) \qquad (1.4)$$

$$\overline{\sum} \left|M^{(b)}\right|^2 = \frac{g^4}{2VN}\left(\frac{V}{\tau_1\tau_2} - 2N^2\right)\left(\tau_1^2 + \tau_2^2 + \rho - \frac{\rho^2}{4\tau_1\tau_2}\right) \qquad (1.5)$$

where the dependence on the $SU(N)$ colour group is shown explicitly, ($V = N^2 - 1$, $N = 3$) and m is the mass of the produced heavy quark Q. The matrix elements squared in Eqs. 1.4 and 1.5 have been summed and averaged over initial and final colours and spins. For brevity we have introduced the following notation,

$$\tau_1 = \frac{p_1 \cdot p_3}{p_1 \cdot p_2}, \quad \tau_2 = \frac{p_2 \cdot p_3}{p_1 \cdot p_2}, \quad \rho = \frac{2m^2}{p_1 \cdot p_2}. \tag{1.6}$$

For photoproduction of heavy quark the lowest order reaction is

$$(c) \quad \gamma(p_1) + g(p_2) \rightarrow Q(p_3) + \overline{Q}(p_4). \tag{1.7}$$

The matrix element squared for this reaction[10] is given in terms of the charge e_Q of the heavy quark as,

$$\overline{\sum} \left| M^{(c)} \right|^2 = \frac{g^2 e_Q^2}{\tau_1 \tau_2} \left(\tau_1^2 + \tau_2^2 + \rho - \frac{\rho^2}{4\tau_1 \tau_2} \right). \tag{1.8}$$

In terms of these lowest order matrix elements the invariant cross-section can be written as,

$$\frac{E_3 \, d^3 \hat{\sigma}}{d^3 p_3} = \frac{1}{16\pi^2 s^2} \overline{\sum} |M|^2 \tag{1.9}$$

where s is the square of the total parton centre of mass energy.

The phenomenological consequences of the lowest order formulae can be summarised as follows. The average transverse momentum of the heavy quark or antiquark is of the order of its mass and the p_T distribution falls rapidly to zero as p_T becomes larger than the heavy quark mass. The rapidity difference between the produced quark and antiquark is predicted to be of order one. The theoretical arguments summarised above do not address the issue of whether the charmed quark is sufficiently heavy that the hadroproduction of charmed hadrons in all regions of phase space is described by the QCD parton model, neglecting terms suppressed by powers of the charmed quark mass. For the application of these formulae to heavy quark production at fixed target energy see ref. [11] and references therein.

There are arguments[12] which suggest that higher order corrections to heavy quark production could be large. These are mostly due to the observation that the fragmentation process,

$$\begin{array}{c} g + g \quad \rightarrow \quad g + g \\ \quad \quad \quad \hookrightarrow Q + \overline{Q} \end{array} \tag{1.10}$$

although formally of order α_S^3, can be numerically as important as the lowest order $O(\alpha_S^2)$ cross section. This happens because the lowest order cross section for the process $gg \rightarrow q\overline{q}$ is about a hundred times smaller than the cross section for $gg \rightarrow gg$. A gluon jet will fragment into a pair of heavy quarks only a fraction $\alpha_S(m^2)/2\pi$ of the time. Because of the large cross-section for the production of gluons, the gluon fragmentation production process is still competitive with the production mechanisms of Eq. (1.3). The description of heavy quark production by the gluon fragmentation mechanism alone is appropriate only when the produced heavy quark is embedded in a high energy jet[13].

The matrix elements squared for the hadroproduction of a heavy quark pair plus a light parton have all been calculated[12,9,14,15]. By themselves, they have physical significance only when the jet associated with the light parton has a large transverse momentum. When the produced light parton has small transverse momentum the matrix elements contain collinear and soft divergences, which cancel only when the virtual corrections are included, and the factorisation procedure is carried out. Corresponding results for the photoproduction of heavy quarks are given in ref. [16].

A partial $O(\alpha_S^3)$ calculation involving the quark gluon fusion process which is free from soft gluon singularities, but contains collinear singularities has been presented in ref. [9]. This calculation provides a concrete example of the factorisation scheme. However this calculation is valid when the quark gluon process dominates over the competing processes. In other regions one cannot use any partial calculation of higher order effects; both real and virtual diagrams contribute. They separately contain divergences which cancel in a complete calculation.

A full calculation of the inclusive cross section for heavy quark production to order α_S^3 is described in ref. [17]. Some aspects of this calculation are discussed in section 2. Corresponding results for photoproduction to order $\alpha_S^2 \alpha_{em}$ are given in ref [18]. They are discussed briefly in section 3.

2.   The hadroproduction of heavy quarks

The basic quantities calculated in ref. [17] are the short distance cross sections $\hat{\sigma}$ for the inclusive production of a heavy quark of transverse momentum p_T and rapidity y. This requires the calculation of the cross-sections for the following parton inclusive processes,

$$g + g \to Q + X, \quad q + \bar{q} \to Q + X, \quad g + q \to Q + X, \quad g + \bar{q} \to Q + X$$

$$g + g \to \overline{Q} + X, \quad q + \bar{q} \to \overline{Q} + X, \quad g + \bar{q} \to \overline{Q} + X, \quad g + q \to \overline{Q} + X. \quad (2.1)$$

The inclusive cross-sections for the production of an anti-quark $\overline{Q}$ differ from those for the production of a quark Q at a given y and p_T. This interference effect, which first arises in $O(\alpha_S^3)$, is small in most kinematic regions. From these results and Eq. (1.2) we can calculate the distributions in rapidity and transverse momentum of produced heavy quarks correct through $O(\alpha_S^3)$. At this point we list the parton sub-processes which contribute to the inclusive cross-sections.

$$q + \bar{q} \to \quad Q + \overline{Q}, \quad \alpha_S^2, \alpha_S^3$$

$$g + g \to \quad Q + \overline{Q}, \quad \alpha_S^2, \alpha_S^3$$

$$q + \bar{q} \to \quad Q + \overline{Q} + g, \quad \alpha_S^3$$

$$g + g \to Q + \overline{Q} + g, \quad \alpha_S^3$$

$$g + q \to Q + \overline{Q} + q, \quad \alpha_S^3$$

$$g + \overline{q} \to Q + \overline{Q} + \overline{q}, \quad \alpha_S^3. \qquad (2.2)$$

Note the necessity of including both real and virtual gluon emission diagrams in order to calculate the full $O(\alpha_S^3)$ cross-section.

In order to describe the results in a relatively concise way, I concentrate on the calculation of the total cross section for the inclusive production of a heavy quark pair. Integrating Eq. (1.2) over the momentum p we obtain the total cross section for the production of a heavy quark pair,

$$\sigma(S) = \sum_{i,j} \int dx_1 dx_2 \; \hat{\sigma}_{ij}(x_1 x_2 S, m^2, \mu^2) \; F_i^A(x_1, \mu) F_j^B(x_2, \mu) \qquad (2.3)$$

where S is the square of the centre of mass energy of the colliding hadrons A and B. The total short distance cross section $\hat{\sigma}$ for the inclusive production of a heavy quark from partons i, j can be written as,

$$\hat{\sigma}_{ij}(s, m^2, \mu^2) = \frac{\alpha_S^2(\mu^2)}{m^2} f_{ij}\left(\rho, \frac{\mu^2}{m^2}\right) \qquad (2.4)$$

with $\rho = 4m^2/s$, and s the square of the partonic centre of mass energy. μ is the renormalisation and factorisation scale. In ref. [17] a complete description of the functions f_{ij} including the first non-leading correction was presented. These may be used to calculate heavy quark production at any energy and heavy quark mass.

Eq. (2.4) completely describes the short distance cross-section for the production of a heavy quark of mass m in terms of the functions f_{ij}, where the indices i and j specify the types of the annihilating partons. The dimensionless functions f_{ij} have the following perturbative expansion,

$$f_{ij}\left(\rho, \frac{\mu^2}{m^2}\right) = f_{ij}^{(0)}(\rho) + g^2(\mu^2)\left[f_{ij}^{(1)}(\rho) + \overline{f}_{ij}^{(1)}(\rho)\ln(\frac{\mu^2}{m^2})\right] + O(g^4). \qquad (2.5)$$

In order to calculate the f_{ij} in perturbation theory we must perform both renormalisation and factorisation of mass singularities. The subtractions required for renormalisation and factorisation are done at mass scale μ. The dependence on μ is shown explicitly in Eq. (2.5). The energy dependence of the cross-section is given in terms of the ratio ρ,

$$\rho = \frac{4m^2}{s}, \quad \beta = \sqrt{1 - \rho}. \qquad (2.6)$$

The running of the coupling constant α_S is determined by the renormalisation group,

$$\frac{d\alpha_S(\mu^2)}{d\ln\mu^2} = -b_0\alpha_S^2 - b_1\alpha_S^3 + O(\alpha_S^4), \quad \alpha_S = \frac{g^2}{4\pi}, \quad b_0 = \frac{(33 - 2n_{lf})}{12\pi}, \quad b_1 = \frac{(153 - 19n_{lf})}{24\pi^2} \qquad (2.7)$$

269

where n_{lf} is the number of light flavours.

The quantities $f^{(1)}$ depend on the scheme used for renormalisation and factorisation. Our results are obtained in an extension of the $\overline{MS}$ renormalisation and factorisation scheme. Full details are given in ref. [17]. In this scheme heavy quarks are decoupled at low energy. The light partons continue to obey the same renormalisation group equation as they would have done in the absence of the heavy quarks. Thus our results should be used in conjunction with the running coupling as defined in Eq.(2.7) and together with light parton densities evolved using the two loop $\overline{MS}$ evolution equations.

The functions $f_{ij}^{(0)}$ defined in Eqs. (2.4,2.5) are,

$$f_{q\bar{q}}^{(0)}(\rho) = \frac{\pi\beta\rho}{27}\left[2+\rho\right]$$

$$f_{gg}^{(0)}(\rho) = \frac{\pi\beta\rho}{192}\left[\frac{1}{\beta}(\rho^2 + 16\rho + 16)\ln\left(\frac{1+\beta}{1-\beta}\right) - 28 - 31\rho\right]$$

$$f_{gq}^{(0)}(\rho) = f_{g\bar{q}}^{(0)}(\rho) = 0. \tag{2.8}$$

We now turn to the higher order corrections in Eq.(2.5) which are separated into two terms. The $\overline{f}^{(1)}(\rho)$ terms are the coefficients of $\ln(\mu^2/m^2)$ and are determined by renormalisation group arguments from the lowest order cross-sections,

$$\overline{f}_{ij}^{(1)}(\rho) = \frac{1}{8\pi^2}\left[4\pi\, b_0\, f_{ij}^{(0)}(\rho) - \int_\rho^1 dz_1 f_{kj}^{(0)}\left(\frac{\rho}{z_1}\right)P_{ki}(z_1) - \int_\rho^1 dz_2 f_{ik}^{(0)}\left(\frac{\rho}{z_2}\right)P_{kj}(z_2)\right]. \tag{2.9}$$

P_{ij} are the lowest order Altarelli-Parisi kernels. The quantities $f^{(1)}$ in Eq.(2.5) can only be obtained by performing a complete $O(\alpha_S^3)$ calculation. We do not have exact analytical results for the quantities $f^{(1)}$. In ref. [17] a physically motivated fit to the numerically integrated result is given. The fit agrees with the numerically integrated result to better than 1%. The functions $f^{(0)}$, $f^{(1)}$ and $\overline{f}^{(1)}$ are shown plotted in Figs. 1, 2 and 3 for the cases of quark-antiquark, gluon-gluon and gluon-quark fusion respectively. Notice the strikingly different behaviour of the gluon-gluon and gluon-quark higher order terms in the high energy limit, $\rho \to 0$. These latter processes allow the exchange of a spin one gluon in the t-channel and are therefore dominant in the high energy limit. These cross sections tend to a constant at high energy. The lowest order terms involve fermion t-channel exchange and therefore fall off at large s as can be seen from Figs. 1 and 2. At high energy we find that,

$$f_{gg}^{(1)} \to 2Nk_{gg} + O(\rho\ln^2\rho), \qquad \overline{f}_{gg}^{(1)} \to 2N\overline{k}_{gg} + O(\rho\ln^2\rho)$$

$$f_{gq}^{(1)} \to \frac{V}{2N}k_{gg} + O(\rho\ln^2\rho), \qquad \overline{f}_{gq}^{(1)} \to \frac{V}{2N}\overline{k}_{gg} + O(\rho\ln^2\rho) \tag{2.10}$$

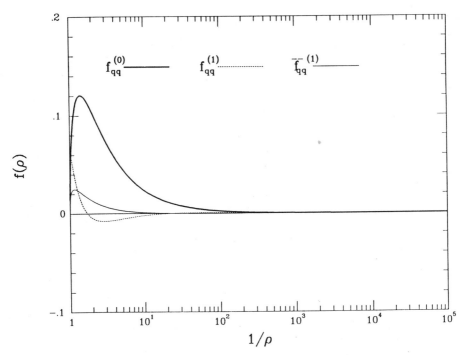

Figure 1. The quark-antiquark contributions to the parton cross section

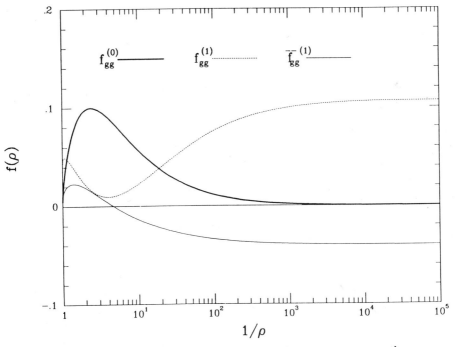

Figure 2. The gluon-gluon contributions to the parton cross section

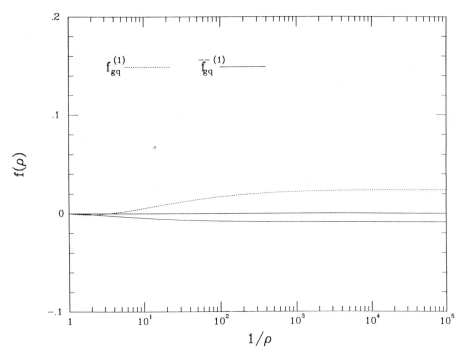

Figure 3. The gluon-quark contributions to the parton cross section

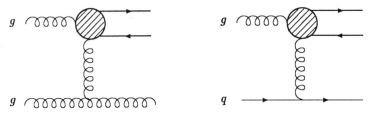

Figure 4. The diagrams responsible for the constant behaviour of the cross-section.

where the constants k_{gg} and $\bar{k}_{gg}$ are,

$$k_{gg} = \frac{1}{\pi V}\left(\frac{V}{2N}\frac{41}{108} - N\frac{793}{43200}\right) \quad \approx 0.018$$

$$\bar{k}_{gg} = -\frac{1}{\pi V}\left(\frac{V}{2N}\frac{7}{36} - N\frac{11}{360}\right) \quad \approx -0.0067 \qquad (2.11)$$

The colour factors V and N are defined after Eq. 1.5. The dominant diagrams are shown in Fig. 4. The asymptotic values of $f_{gg}^{(1)}$ and $f_{gq}^{(1)}$ are proportional to the colour charge of the line which provides the exchanged gluon, since in this limit the upper blob in Fig. 4 is the same for both diagrams and the lower vertex can be approximated by the eikonal form. In the gluon-gluon sub-process the exchanged spin one gluon can come from either incoming gluon, whereas in the gluon quark subprocess it can only come

272

from the incoming quark line. This, together with the ratio of the gluon and quark charges, explains the relative factor of $9/2$, shown in Eq.(2.10) and evident in Figs. 2 and 3.

A preliminary idea of the size of the corrections can be obtained from Figs. 1, 2 and 3 even before folding with the parton distribution functions. Taking a typical value for $g^2 \approx 2$, we see that the radiative corrections are large, particularly in the vicinity of the threshold. The significance of the constant cross-section region (gg, gq) at high energy will depend on the rate of fall-off of the structure functions with which the partonic cross-section must be convoluted.

Near threshold,$(\beta \to 0)$, we have,

$$f_{q\bar{q}}^{(1)} \to \mathcal{N}_{q\bar{q}} \left[-\frac{\pi^2}{6} + \beta \left(\frac{16}{3} \ln^2 (8\beta^2) - \frac{82}{3} \ln(8\beta^2) \right) + O(\beta) \right]$$

$$f_{gg}^{(1)} \to \mathcal{N}_{gg} \left[\frac{11\pi^2}{42} + \beta \left(12 \ln^2 (8\beta^2) - \frac{366}{7} \ln(8\beta^2) \right) + O(\beta) \right]$$

$$f_{gq}^{(1)} \to O(\beta). \tag{2.12}$$

The normalisation, $\mathcal{N}_{ij}$ of the expressions in Eq.(2.12) is determined as follows,

$$\mathcal{N}_{ij} = \frac{1}{8\pi^2} \left. \frac{f_{ij}^{(0)}(\rho)}{\beta} \right|_{\beta=0}, \quad \mathcal{N}_{q\bar{q}} = \frac{1}{72\pi}, \quad \mathcal{N}_{gg} = \frac{7}{1536\pi}. \tag{2.13}$$

Notice that in this order in perturbation theory the cross-section is finite at threshold. This is due to the $1/\beta$ singularity which is responsible for the binding in a coulomb system. The coulomb attraction tends to increase the cross-section when the incoming partons are in a singlet state (gg), and decrease the cross-section when the incoming partons are in an octet state $(gg, q\bar{q})$. This results in a net positive term for the gg case. Note that the numerical importance of the term due to the coulomb singularity is quite small.

We now examine the region near threshold in more detail. The terms in Eq.(2.12) which are finite at threshold have already been explained. The $\ln^2(\beta^2)$ terms in Eq. (2.12) have a general origin. They are due to terms of the form

$$f^{(1)}(\rho) \sim \frac{1}{8\pi^2} \int_\rho^1 dz \left[\frac{\ln(1 - z)}{1 - z} \right]_+ f^{(0)}(\rho/z) \tag{2.14}$$

For attempts to resum these terms of this form in Drell-Yan processes we refer the reader to ref. [19].

The phenomenological consequences of the complete $O(\alpha_S^3)$ formulae have been investigated in ref. [17], and to greater extent in ref. [20]. An attempt has been made to estimate the uncertainties due to the form of the gluon distribution and the value of Λ, the value of the heavy quark mass and the choice made for the scale μ. Predictions are given for charm, bottom and top production at fixed target and collider energies.

3. The photoproduction of heavy quarks

In this section I review the results which describe the total parton cross-section for the photo-production of a heavy quark pair[18]. The treatment closely parallels the hadroproduction results given in the previous section. The interaction of the point-like component of the photon with a parton gives the cross-section,

$$\hat{\sigma}_{\gamma j}(s, m^2, \mu^2) = \frac{\alpha_S(\mu^2)\alpha_{\text{em}}}{m^2} f_{\gamma j}\left(\rho, \frac{\mu^2}{m^2}\right) \tag{3.1}$$

α_{em} is the fine structure constant. Knowledge of the functions $f_{\gamma j}$ (and the hadronic functions f_{ij}) gives a complete description of the short distance cross-section for the production of a heavy quark of mass m by a photon γ. The index j specifies the type of the incoming parton. The dimensionless functions $f_{\gamma j}$ have the following perturbative expansion,

$$f_{\gamma j}\left(\rho, \frac{\mu^2}{m^2}\right) = f_{\gamma j}^{(0)}(\rho) + g^2(\mu^2)\left[f_{\gamma j}^{(1)}(\rho) + \overline{f}_{\gamma j}^{(1)}(\rho)\ln\left(\frac{\mu^2}{m^2}\right)\right] + \mathcal{O}(g^4) \tag{3.2}$$

The functions $f_{\gamma j}$ defined in eqs. (3.2) depend on the charge of the quark which interacts with the photon. In order to make these dependences explicit we define the quantities,

$$f_{\gamma g}(\rho, \frac{\mu^2}{m^2}) = e_Q^2 c_{\gamma g}(\rho, \frac{\mu^2}{m^2})$$

$$f_{\gamma q}(\rho, \frac{\mu^2}{m^2}) = e_Q^2 c_{\gamma q}(\rho, \frac{\mu^2}{m^2}) + e_q^2 d_{\gamma q}(\rho, \frac{\mu^2}{m^2}) \tag{3.3}$$

The charges of the heavy and light quarks are denoted by e_Q and e_q respectively. The interference term proportional to $e_Q e_q$ makes no contribution to the total cross-section. The perturbative expansions of c and d are defined by a formula with exactly the same structure as eq. (3.2). For the lowest order terms we find,

$$c_{\gamma g}^{(0)}(\rho) = \frac{\pi\beta\rho}{4}\left[\frac{1}{\beta}(3 - \beta^4)\ln\left(\frac{1+\beta}{1-\beta}\right) - 4 + 2\beta^2\right] \tag{3.4}$$

$$c_{\gamma q}^{(0)}(\rho) = d_{\gamma q}^{(0)}(\rho) = c_{\gamma \bar{q}}^{(0)}(\rho) = d_{\gamma \bar{q}}^{(0)}(\rho) = 0. \tag{3.5}$$

We now turn to the higher order corrections in eq. (3.2) which are separated into two terms. The $\overline{f}^{(1)}(\rho)$ terms are the coefficients of $\ln(\mu^2/m^2)$ and are determined by renormalisation group arguments from the lowest order cross-sections,

$$\overline{f}_{\gamma j}^{(1)}(\rho) = \frac{1}{8\pi^2}\left[2\pi\, b_0\, f_{\gamma j}^{(0)}(\rho) - \sum_k \int_\rho^1 dz_1 f_{kj}^{(0)}(\frac{\rho}{z_1})P_{k\gamma}(z_1) - \sum_k \int_\rho^1 dz_2 f_{\gamma k}^{(0)}(\frac{\rho}{z_2})P_{kj}(z_2)\right]. \tag{3.6}$$

The quantities $f_{\gamma j}^{(1)}$ in eq. (3.2) can only be obtained by performing a complete $\mathcal{O}(\alpha_{\text{em}}\alpha_S^2)$ calculation. The results of such a calculation are given in ref. [18].

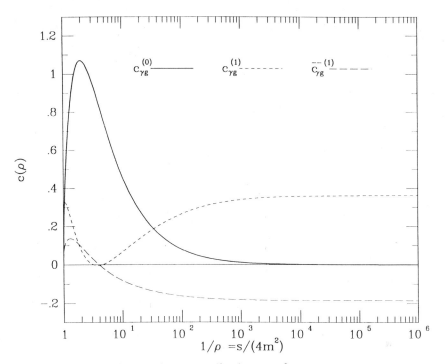

Figure 5. The photon gluon contributions to the parton cross-section

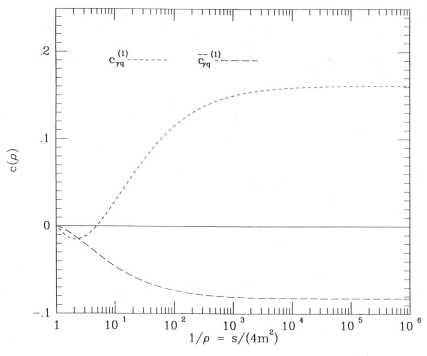

Figure 6. The photon quark contributions proportional to the square of the heavy quark charge.

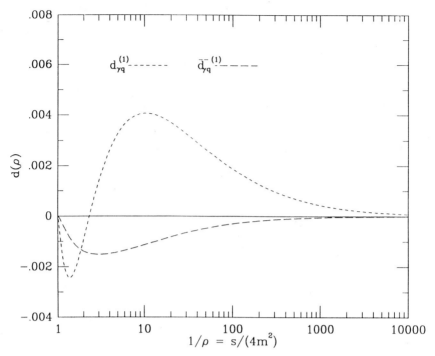

Figure 7. The photon quark contributions proportional to the square of the light quark charge.

The functions $c^{(0)}, c^{(1)}$ and $\bar{c}^{(1)}$ are shown plotted in fig. 5 for the case of photon-gluon scattering. In figs. 6 and 7 we plot $c^{(1)}, \bar{c}^{(1)}, d^{(1)}$ and $\bar{d}^{(1)}$ for the case photon-quark scattering. Notice that the term proportional to the square of the charge of the light quark is quite small.

At high energy we find that,

$$c^{(1)}_{\gamma g} \to N k_{\gamma g} + O(\rho \ln^2 \rho), \qquad \bar{c}^{(1)}_{\gamma g} \to N \bar{k}_{\gamma g} + O(\rho \ln^2 \rho)$$

$$c^{(1)}_{\gamma q} \to \frac{V}{2N} k_{\gamma g} + O(\rho \ln^2 \rho), \qquad \bar{c}^{(1)}_{\gamma q} \to \frac{V}{2N} \bar{k}_{\gamma g} + O(\rho \ln^2 \rho) \qquad (3.7)$$

where $k_{\gamma g}$ and $\bar{k}_{\gamma g}$ are,

$$k_{\gamma g} = \frac{41}{108\pi} \approx 0.121$$

$$\bar{k}_{\gamma g} = -\frac{7}{36\pi} \approx -0.0619 \qquad (3.8)$$

Thus the photon-parton cross-section goes to a constant as $s \to \infty$, in complete analogy with Eqs. 2.10, 2.11. Note that the constants $k_{\gamma g}$ and $\bar{k}_{\gamma g}$ are related to the 'Non-Abelian' parts of k_{gg} and $\bar{k}_{gg}$.

Near threshold we can extract the exact analytic behaviour,

276

$$c_{\gamma g}^{(1)} \to \mathcal{N}_{\gamma g}\left[-\frac{\pi^2}{6} + \beta\left(6\ln^2\left(8\beta^2\right) - 30\ln(8\beta^2)\right) + \mathcal{O}(\beta)\right]$$

$$c_{\gamma q}^{(1)} \to \mathcal{O}(\beta)$$

$$d_{\gamma q}^{(1)} \to \mathcal{O}(\beta). \tag{3.9}$$

The normalisation, $\mathcal{N}_{\gamma g}$ of the expression in eq. (3.9) is determined as follows,

$$\mathcal{N}_{\gamma g} = \frac{1}{8\pi^2}\left.\frac{c_{\gamma g}^{(0)}(\rho)}{\beta}\right|_{\beta=0} = \frac{1}{16\pi}. \tag{3.10}$$

The origin of these large terms near threshold has been explained in the previous section.

The phenomenological consequences of these formulae have been described in ref. [18]. It is interesting to compare the higher order corrections to hadroproduction with the corrections to photoproduction. Once allowance has been made for the fact that two incoming gluons can participate in the hadroproduction of heavy quarks, whereas at most one gluon can participate in the photoproduction reaction, the corrections to the two processes are remarkably similar. This suggests that a much tighter theoretical prediction can be obtained by comparing the two processes.

References

[1]   S. P. K. Tavernier, *Rep. Prog. Phys.* **50** (1987) 1439 .

[2]   S. D. Holmes, W. Lee and J. E. Wiss, *Ann. Rev. Nucl. Part. Sci.* **35** (1985) 397 .

[3]   C. Albajar *et al.*, Cern preprint, CERN-EP 87/190 (October 1987).

[4]   J. C. Collins, D. E. Soper and G. Sterman, *Nucl. Phys.* **B263** (1986) 37 .

[5]   S. J. Brodsky, J. C. Collins, S.D. Ellis, J. F. Gunion and A. H. Mueller, in *Proc. 1984 Summer Study on the Design and Utilization of the Superconducting Super Collider*, R. Donaldson and J. Morfín, (eds.), Fermilab, Batavia, Illinois, 1984, p. 227.

[6]   S. J. Brodsky, J. F. Gunion and D. E. Soper, *Phys. Rev.* **36** (1987) 2710 .

[7]   M. Gluck, J. F. Owens and E. Reya, *Phys. Rev.* **17** (1978) 2324 .

[8]   B. L. Combridge, *Nucl. Phys.* **B151** (1979) 429 .

[9]   R. K. Ellis, in *Strong Interactions and Gauge Theories*, edited by J. Tran Thanh Van, Editions Frontières, Gif-sur-Yvette, 1986, p. 339.

[10] L. M. Jones and H. W. Wyld, *Phys. Rev.* **D17** (1978) 759 .

[11] R. K. Ellis and C. Quigg, Fermilab preprint FN-445 (1987).

[12] Z. Kunszt and E. Pietarinen, *Nucl. Phys.* **B164** (1980) 45 .

[13] A. H. Mueller and P. Nason, *Phys. Lett.* **156B** (1985) 226 ;
Nucl. Phys. **B266** (1986) 265 .

[14] R. K. Ellis and J. C. Sexton, *Nucl. Phys.* **B282** (1987) 642 .

[15] J. F. Gunion and Z. Kunszt, *Phys. Lett.* **178B** (1986) 296 .

[16] R. K. Ellis and Z. Kunszt, Fermilab-Pub-87/226-T, (December 1987). Nuclear Physics B (to be published).

[17] P. Nason, S. Dawson and R. K. Ellis, Fermilab-Pub-87/222-T, (December 1987), Nuclear Physics B (to be published).

[18] R. K. Ellis and P. Nason, Fermilab-Pub-88/54-T, (May 1988). Nuclear Physics B (submitted).

[19] D. Appel, G. Sterman and P. Mackenzie, Stonybrook preprint-ITP-SB-88-6 (1988).

[20] G. Altarelli, M. Diemoz, G. Martinelli and P. Nason, CERN-TH-4978/88 (1988).

SOME RECENT DEVELOPMENTS IN THE DETERMINATION OF PARTON DISTRIBUTIONS*

J.F. Owens

Physics Department
Florida State University
Tallahassee, Florida 32306

ABSTRACT

Recent data relevant for determining parton distributions are discussed. Particular attention is paid to the question of how to improve our knowledge of the gluon distribution. The role of photon-jet and photon-hadron correlation data is discussed in conjunction with a new technique for performing next-to-leading order two-particle inclusive calculations.

1. INTRODUCTION

Calculations of short distance hadron induced processes require knowledge of the parton momentum distributions within the incoming hadrons. Such distributions are measurable in various deep-inelastic lepton scattering processes, for example. When effects due to gluon bremsstrahlung and quark-antiquark pair production are taken into account, the resulting distributions change with varying Q^2, the invariant mass squared of the exchanged vector boson. Such scaling violations are easily calculated by solving the coupled Altarelli-Parisi equations [1]. Having obtained such solutions, it is often convenient to develop parametrizations of the distributions, thereby avoiding the need to solve the equations repeatedly.

Many parametrizations of parton distributions are available in the literature. They are repeatedly refined as new data become available and, therefore, some of the older sets are rather out of date. Two relatively recent parametrizations are those of Eichten, Hinchliffe, Lane, and Quigg [2] and Duke and Owens [3]. Throughout this talk I will use the latter set of distributions in order to have a convenient reference when looking at new data.

The parametrizations in references [2] and [3] have many feature in common. Both sets are based upon fits using the leading logarithm approximation. Furthermore, both contain two sets of parametrizations based on differing choices for the gluon distribution - one relatively soft and one somewhat harder. The corresponding values of the QCD scale parameter Λ are different, as a result of the well-known Λ-gluon correlation - a harder gluon distribution gives a larger value of Λ and vice versa.

The analysis of reference [3] was performed in the summer of 1983. Since that

---

* Work supported in part by the U.S. Department of Energy.

time a number of experiments have published new data sets and QCD-based analyses. These have had an impact on a variety of topics, some of which will be discussed below. The outline of this talk is as follows. In the next section I will discuss some recent determinations of $\Lambda_{\overline{MS}}$. Section 3 contains a discussion of the determination of the ratio of the d and u quark distributions, and the measurement of the gluon distribution is discussed in Section 4. In Section 5 the outlook for future measurements is discussed while Section 6 contains a summary and some conclusions.

2. MEASUREMENTS OF Λ (LO, $\overline{MS}$,...)

The Altarelli-Parisi equations [1] for the evolution of quark and gluon distribution functions can be written as

$$\frac{dG_{q_i/A}(x,Q^2)}{dt} = \frac{\alpha_s(Q^2)}{2\pi} \int_x^1 \frac{dy}{y} \left[P_{qq}(x/y)G_{q_i/A}(y,Q^2) + P_{qg}(x/y)G_{g/A}(y,Q^2) \right] \quad (1a)$$

and

$$\frac{dG_{g/A}(x,Q^2)}{dt} = \frac{\alpha_s(Q^2)}{2\pi} \int_x^1 \frac{dy}{y} \left[\sum_{i=1}^{2f} P_{gq}(x/y)G_{q_i/A}(y,Q^2) + P_{gg}(x/y)G_{g/A}(y,Q^2) \right]. \quad (1b)$$

Here t is defined as $\ln(Q^2/\Lambda^2)$ and the splitting functions, P_{ij}, can be found, for example, in reference [1]. In order to solve this set of coupled equations, one must first specify the boundary conditions in the form of a set of parton distributions at initial value of $Q^2 = Q_0^2$. The running coupling, $\alpha_s(Q^2)$, must also be specified via the scale parameter Λ. If one is working beyond the leading logarithm approximation, then the two-loop splitting functions and running coupling must be used, as is discussed in reference [4], for example. Having obtained the evolved distribution functions, they must be convoluted with the appropriate hard scattering cross section in order to compare with the data. The appropriate expressions may also be found in reference [4].

The usual procedure followed when analyzing deep inelastic data in the framework of QCD is to vary the value of Λ and the forms of the input parton distributions in order to obtain an optimum description of the data. It is important to note that the output value of Λ can be affected by the choices made for the parton distribution parametrization and the cuts placed on the data, as well as a variety of other factors, as discussed in reference [5], for example.

Two techniques are widely used when extracting information from deep inelastic data sets. These are referred to as non-singlet and singlet analyses. In the non-singlet analysis, combinations of structure functions which involve only valence quarks are used. The relevant evolution equation is then much simpler, as the gluon distribution does not appear:

$$\frac{dG_{q_i/A}(x,Q^2)}{dt} = \frac{\alpha_s(Q^2)}{2\pi} \int_x^1 \frac{dy}{y} P_{qq}(x/y)G_{q_i/A}(y,Q^2). \quad (2)$$

An example of such an observable is the structure function $xF_3(x,Q^2)$ measured in neutrino experiments. Often, data for $F_2(x,Q^2)$ are used in non-singlet analyses with x

280

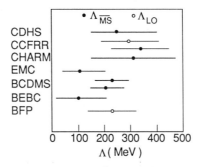

Figure 1. Summary of recent determinations of Λ. The data are from references [6-13].

large enough that the contributions from sea quarks and gluons will be negligible (typically $x \geq 0.3$ or so). The advantage of the non-singlet analysis is that a value for Λ can be determined independently of the gluon distribution. The major disadvantage has usually been that the data for the non-singlet observables have lower statistics than in the singlet case.

The singlet analyses utilize the full range of observables and, as a result, the contributions from the sea and gluon distributions must be included. The data generally generally have higher statistics than in the non-singlet case, but the correlation between the shape of the gluon distribution and the value of Λ makes the interpretation of the fits to the data somewhat harder.

A summary of recent determinations of Λ is depicted in Fig. 1. The values are from the following references: CDHS [6], CCFRR [7], CHARM [8], EMC [9], BCDMS [10,11], BEBC [12], and BFP[13]. Determinations of both Λ_{LO} and $\Lambda_{\overline{MS}}$ are shown. It is not possible to average the values shown in Figure 1, since the various analyses have utilized different kinematic cuts, gluon distributions (in the singlet analyses), and choices for the forms of the input parametrizations. Nevertheless, it is clear that a value for $\Lambda_{\overline{MS}}$ of approximately 200 MeV is compatible with all of the determinations. Bearing in mind that fitted values for Λ_{LO} and $\Lambda_{\overline{MS}}$ tend to be close to each other in deep inelastic scattering, the results in Figure 1 support the softer gluon distibutions of the Duke-Owens Set 1 or the EHLQ Set 1.

3. d/u RATIO

In deep inelastic scattering experiments, the (u+d) combination of quark distributions is well-measured in high statistics experiments on isoscalar targets. In the case of heavy targets such as iron, it is possible to incorporate corrections for various nuclear effects (EMC effect, fermi-motion corrections, etc.) into the analysis. This has been discussed in detail in reference [14] where a global fit to a wide variety of data has been performed. On the other hand, the determination of the u- and d-quark distributions separately requires the use of data taken with a hydrogen target. Until recently, the highest statistics data on the ratio of F_2 from neutron and proton targets were those from SLAC [15]. These data were used to constrain the d-quark distribution in the fits of reference [3]. Since then, data have been published from the EMC Collaboration [16] and preliminary results have been obtained by the BCDMS Collaboration [17]. These newer results lie systematically lower than the earlier SLAC results as is shown in Figure 2. Also shown in Figure 2 is the result of the Duke-Owens Set 1 fit which utilized only the SLAC data. Repeating that fit, but without the SLAC ratio data, yields the dashed curve. Note that in each case the d-quark distribution is normalized to give one d-quark in the proton, so that the two curves eventually cross at a very small value of x.

Several neutrino groups have reported measurements of the ratio of the d- and u-quark valence distributions. Measurements from the CDHS [18] and BEBC [19] Collaborations are shown in Figure 3. The solid and dashed curves correspond to those in Figure 2. It appears that the lower d/u ratio is favored by the neutrino data.

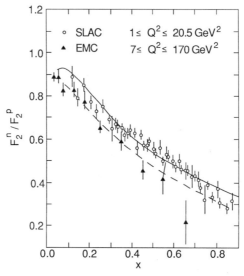

Figure 2. Comparison of several fits to data for the ratio of F_2 on neutrons and protons. The data are from reference [15] (SLAC) and reference [16] (EMC). The solid curve is from Duke-Owens Set 1 whereas the dashed curve was obtained by refitting after removal of the SLAC ratio data.

The apparent discrepancies noted above may be due to the different ranges of Q^2 covered by the data. The SLAC experiment utilized data in the region from 1.0 to 20.5 GeV^2. In order to present the F_2 ratio data it was assumed that there was no dependence on Q^2 for this ratio over the indicated range. On the other hand, the EMC data cover the region of 7 to 170 GeV^2 while the BCDMS data have average values of Q^2 from 18 to 83 GeV^2. The two neutrino results shown in Figure 3 have $Q^2 = 66x GeV^2$ for the CDHS data and Q^2 from 1 to 60 GeV^2 for the BEBC data.

For many large transverse momentum processes the precise value of the ratio of the d and u quark distributions is not critical. Recently, however, attention has been focussed on an exception to this. The ratio $\sigma(\bar{p}p \rightarrow W + X)/\sigma(\bar{p}p \rightarrow Z + X)$ can be used to place limits on the t quark mass [20]. The major theoretical uncertainty is due

282

to the variations in the d/u ratio in different parton distribution sets. This is due to the fact that W production involves the product $\bar{u}d + \bar{d}u$ whereas Z production involves the product $\bar{u}u + \bar{d}d$. Thus, the cross section ratio is proportional to the d/u ratio and the theoretical uncertainty is merely a reflection of the variation between the various experimental results. If only the high Q^2 data for the d/u ratio are utilized in the determination of the parton distributions, the theoretical uncertainty in the cross section ratio will be decreased. This point has been discussed in reference [14]. In reference [21] a complete set of input parton distributions have been extracted from deep inelastic data. A thorough discussion of the various uncertainties is presented, in addition to a variety of predictions for high energy processes.

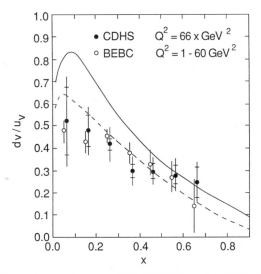

Figure 3. Data from references [18] (CDHS) and [19] (BEBC) for the ratio of the d and u valence distributions are compared to the results from the two fits discussed in the text. The meaning of the curves is the same as for Figure 2.

4. GLUON DETERMINATIONS

During the past several years a number of experimental groups have published gluon distributions determined from analyses of deep inelastic scattering data. The CHARM Collaboration has presented the results of a next-to-leading order analysis in reference [8]. They obtain $\Lambda_{\overline{MS}} = 310 \pm 140 \text{MeV}$ and their fits yield a band of allowed values for the gluon distribution. Their results at $Q^2 = 20 \text{GeV}^2$ are shown in Figure 4 by the dashed curve. The tick marks indicate the width of the allowed band determined in their analysis. The solid curve corresponds to the gluon distribution of the Duke-Owens Set 1 evolved up to the same value of Q^2. The CHARM results favor a somewhat softer gluon than that used in Set 1.

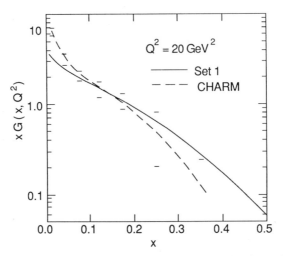

Figure 4. The dashed curve and the tick marks show the CHARM determination [8] of the gluon distribution at $Q^2 = 20\text{GeV}^2$. The solid curve is the Duke-Owens Set 1 result evolved to the same value of Q^2.

The CDHS Collaboration has reported several gluon distributions in reference [6]. At $Q_0^2 = 5\text{GeV}^2$ in a leading order fit they find

$$xG(x, Q^2) = 1.75(1 + 8.9x)(1 - x)^{6.03}$$

with $\Lambda_{LO} = 290 \pm 20\text{MeV}$. This result is nearly identical to that used in the Set 1 Duke-Owens distributions.

Recently, the BCDMS Collaboration has presented preliminary results [11] of an analysis of their hydrogen data. Two fitting programs were used and the input gluon distribution at $Q_0^2 = 5\text{GeV}^2$ was parametrized as

$$xG(x, Q_0^2) = \frac{0.45}{\eta + 1}(1 - x)^{\eta}.$$

Their results are summarized in the following table.

table 1

Fit	Λ_{LO}	η_{LO}	$\Lambda_{\overline{MS}}$	$\eta_{\overline{MS}}$
1	196 ± 19	5.2 ± 1.5	214 ± 19	10.3 ± 1.5
2	183 ± 25	5.4 ± 1.3	195 ± 20	8.9 ± 1.5

The leading order results are compatible with those from other analyses. However, the next-to-leading order results favor a very much softer gluon, at least in the region $x \leq 0.3$, which is the region in which their data are sensitive to the gluon contribution. A similar analysis of their carbon data favors a value of $\eta_{\overline{MS}}$ greater than 6.

It is, at this point, worth examining in some detail how the gluon distribution enters into the QCD predictions for deep inelastic scattering. In leading order, we can use equation (1) to write an expression for the Q^2 dependence for F_2:

$$\frac{dF_2(x,Q^2)}{dt} = \frac{\alpha_s(Q^2)}{2\pi} \int_x^1 dy \left[F_2(\frac{x}{y},Q^2)P_{qq}(y) + \sum_{i=1}^{2f} e_i^2 \frac{x}{y}G(\frac{x}{y},Q^2)P_{qG}(y) \right]. \quad (3)$$

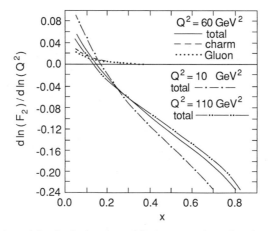

Figure 5. The logarithmic derivative of $F_2(\mu p)$ as given by the Duke-Owens Set 1 distributions. The results are shown for three values of Q^2. The gluon and charm contributions at $Q^2 = 60\text{GeV}^2$ are shown separately.

After substituting the expressions for the splitting functions into equation (3), the expression for the slope may be written as:

$$\frac{dF_2(x,Q^2)}{dt} = \frac{\alpha_s(Q^2)}{2\pi} \left\{ \frac{4}{3}\left[\frac{3}{2} + 2\ln(1-x)\right]F_2(x,Q^2) \right.$$
$$+ \frac{4}{3}\int_x^1 dy \frac{(1+y^2)F_2(\frac{x}{y},Q^2) - 2F_2(x,Q^2)}{1-y}$$
$$\left. + \sum_{i=1}^{2f} e_i^2 \int_x^1 dy \frac{x}{y}G(\frac{x}{y},Q^2)\frac{y^2+(1-y)^2}{2} \right\}. \quad (4)$$

285

The first term in equation (4) is positive for $x \leq 1 - e^{-3/4} \approx 0.53$, whereas the second term is negative for all values of x between zero and one. The last term, which is where the gluon distribution enters, is positive for all x in the same interval. The net result is that the slope of F_2 with respect to $\ln Q^2$ will be positive at low x and negative in the high x region. The precise cross over point is controlled, in part, by the gluon distribution. In addition, the contributions from heavy quarks appear predominantly in the low x region. They depend, in part, on the gluon distribution through the subprocess $\gamma^* g \rightarrow Q\bar{Q}$. They also depend on the model used for the threshold dependence of the heavy quark distributions. Finally, it should be pointed out that the slope depends strongly on the value of Q^2 at which it is measured. This makes it difficult to compare the results of different experiments unless the Q^2 range used for measuring the slope is quoted for each value at which the slope is given.

As an illustrative example, Figure 5 shows the logarithmic derivative of F_2 with respect to $\ln Q^2$ as given by the Duke-Owens Set 1 distributions. Results are shown for three values of Q^2 and the differences indicate the need to know the correct value of Q^2 for each x value, before comparing with experimental results. Also shown separately at $Q^2 = 60\text{GeV}^2$ are the contributions from the charm quark and gluon distributions. The gluon contribution, in particular, is rather small even in the low x region and it is negligible above $x \approx 0.25$.

The preceeding discussion shows that deep inelastic scattering is a hard place to measure the gluon distribution. It is, at best, an indirect measurement, and there are other effects which are comparable in size, thereby further obscuring the gluon contribution. Ideally, what is needed is a hard scattering process where the dominant contribution is proportional to the gluon distribution. An example is direct photon production in proton-proton collisions where the dominant subprocess is $gq \rightarrow \gamma q$. The increased sensitivity to the gluon distribution of direct photon production as compared to deep inelastic scattering was nicely illustrated in the talk by R. Baier at this conference. The data base for inclusive single photon production has now reached the point where we can start to learn something quantitative about the gluon distribution. Even more can be learned when data are available in which the recoiling jet or leading hadron is measured in addition to the trigger photon. An analysis of this type has been presented by the AFS Collaboration [22] who measured the process $pp \rightarrow \gamma + jet + X$ at $\sqrt{s} = 62\text{GeV}$. They have extracted the gluon distribution in a leading logarithm analysis using a K-factor of two. The results correspond to $Q^2 = \frac{4}{3}p_T^2\text{GeV}^2 = 1315x^2\text{GeV}^2$. The AFS results are shown in Figure 6, along with the Duke-Owens Set 1 gluon (solid curve) and the CHARM gluon (dashed curve). In each case, the distributions have been evaluated at the Q^2 scale mentioned above. The data favor a somewhat softer gluon than that contained in the Set 1 distributions; they agree rather nicely with the CHARM result.

The overall lesson at this point is that the gluon may well be somewhat softer than previously thought, but the question has not yet been definitively settled. In this regard it is interesting to note that J. Collins has suggested [23] that the gluon distribution may increase more rapidly than x^{-1} as x approaches zero. More precisely, one usually chooses an x^{-1} behavior at Q_0^2. The evolution equations then build up a low-x spike at higher values of Q^2. Unfortunately, the evolution equations are notoriously unstable if one evolves backwards from Q_0^2 with this input. This can be taken as a suggestion that a steeper power of x is needed. Studies have shown that using $x^{-\eta}$ with $1.0 < \eta \leq 1.5$ helps stabilize the evolution. This could result in a softer gluon distribution while still retaining a high x tail of sufficient magnitude to describe the observed high-p_T phenomena. Such an ansatz has been utilized in the analysis of reference [14].

5. FUTURE DEVELOPMENT

When next-to-leading order calculations are performed for hard scattering processes, it is necessary to have a reference process for the definition of the required parton distributions. Deep inelastic scattering serves the purpose for quark distributions, but the gluon distribution first enters there in the next-to-leading order. We need a next-to-leading order calculation for a hard scattering process where the gluon enters in the leading order. Such a process is direct photon production. The required calculation for the inclusive single photon cross section has been reported in reference [24]. The results of this calculation are in remarkably good agreement with the data which has been reported at this conference. Nevertheless, the inclusive single photon cross section still involves a convolution of the parton distributions with the hard scattering subprocesses. A change in the shape of the gluon distribution essentially just gives an overall shift in the normalization of the predictions. Typically, predictions based on the Duke-Owens Set 1 and Set2 distributions are very nearly parallel, with one curve just shifted from the other. As was discussed in the previous Section, what is needed is data for double-arm observables, e.g., γ + jet or γ + leading hadron. Such data allow one to have more control over the underlying parton-level kinematics and, thereby, to obtain information on the *shape* of the gluon distribution.

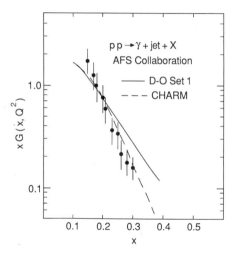

Figure 6. The gluon distribution as determined by the AFS Collaboration [22] compared to that from the CHARM Collaboration [8] (dashed curve) and the results of the Duke-Owens Set 1 distributions (solid curve).

In order to maximize the information gained from such data, it will be necessary to have a next-to-leading order calculation for the correlation observables. This, however, is made complicated by the need to be able to implement experimental cuts, match detector acceptances, change the definitions of the variables used to calculate distributions, etc. To date, no such calculations for direct photon processes exist which have this flexibility.

The requirements noted in the preceeding paragraph are most easily met by utilizing Monte Carlo techniques. However, in order to use the Monte Carlo approach for a next-to-leading order calculation, some method of handling the singularities associated with the two- and three-body final states must be developed. This has recently been accomplished for the case of high-p_T dihadron production [25]. Let $\sigma(\epsilon_1, \epsilon_2)$ denote the differential cross section for producing a dihadron final state where the hadrons have charges ϵ_1 and ϵ_2. The combination $\sigma(+, +) + \sigma(-, -) - \sigma(+, -)$ involves only non-singlet fragmentation. Specifically, we have constructed a Monte Carlo program which calculates observables for the process $pp \to (h^+ - h^-) + (h^+ - h^-) + X$ where $(h^+ - h^-)$ denotes non-singlet fragmentation and all species of charged hadrons are summed over. The original motivation for this calculation was the Fermilab experiment E-711 which will measure such charge combinations in an effort to perform a high statistics measurement of the angular distribution of quark-quark scattering.

To order α_s^3, the following subprocesses contribute:

$$\alpha_s^2 + \alpha_s^3 \qquad \begin{aligned} qq &\to qq \\ qq' &\to qq' \\ gg &\to q\bar{q} \end{aligned}$$

$$\alpha_s^3 \qquad \begin{aligned} qq &\to qqg \\ qq' &\to qq'g \\ gg &\to q\bar{q}q \\ gq' &\to q\bar{q}q' \\ gg &\to q\bar{q}g. \end{aligned}$$

In the three-body final states only the quarks and antiquarks contribute to the fragmentation. Dimensional regularization is used to regulate the infinities and the ultra-violet divergences are subtracted in the $\overline{MS}$ scheme. The three-body subprocesses have both soft and collinear singularities. In the $qq \to qqg$ subprocess, for example, we define the soft region to be that for which the gluon has an energy in the subprocess center-of-momentum frame which is less than $\delta_s\sqrt{\hat{s}}$ where $\sqrt{\hat{s}}$ is the total subprocess energy. The collinear region consists of the sum of all regions in phase space where the gluon and another parton have an invariant mass squared or momentum transfer which is smaller than $\delta_c\hat{s}$ in magnitude. The three-body contribution is thus partitioned into three pieces:

1) soft α_s^3 singular contribution
2) hard collinear α_s^3 singular contribution
3) finite three-body contribution.

Contributions 1) and 2) are added to the α_s^2 and α_s^3 two-body terms and the collinear singularities are factorized, thereby yielding a finite result. At this point we have finite two- and three-body contributions which can be handled with standard Monte Carlo techniques. The two types of final states are generated separately and combined in the histogramming program. This can therefore be considered as a "two pass" Monte Carlo program. We have varied the cut-offs δ_s and δ_c and found regions where the numerical results are stable with respect to variations of them. We are currently studying the dependence of the results on the choices made for the renormalization and factorization scales and expect to publish the final results soon [25].

The method outlined above is reasonably straightforward to apply to any two-body process, e.g., γ + jet, γ + hadron, or two jet final states, etc. Such calculations can be expected to yield important constraints or even outright measurements of the gluon distribution once sufficient data become available.

6. SUMMARY AND CONCLUSIONS

A survey of the recent additions to the data base for deep inelastic scattering shows that data of high statistical precision are continuing to be added. Accordingly, there has been a significant improvement in certain details of our knowledge of parton distributions. Three items, in particular, stand out as being significant developments.

1. It appears that the values for $\Lambda_{\overline{MS}}$ are settling down in the vicnity of 200 MeV and the errors continue to shrink. This is consistent with the observation that the Duke-Owens Set 2 distributions (with $\Lambda = 400$MeV) tend to overestimate the data in direct photon production when optimized next-to-leading order calculations are used.

2. Recent data relevant for the d/u ratio show a somewhat smaller value than previously thought. Precise data on the ratio of neutron and proton structure functions will help to reduce the errors on the theoretical predictions for the ratio of W and Z production.

3. Both deep-inelastic and $\gamma + $ jet data suggest a softer gluon distribution than has been commonly used, but there is still some uncertainty as to the precise shape. Ongoing direct photon experiments will undoubtedly add to our knowledge in this area in the near future.

REFERENCES

1. G. Altarelli and G. Parisi, Nucl. Phys. B126:298 (1977).

2. E. Eichten, I. Hinchliffe, K. Lane, and C. Quigg, Rev. Mod. Phys. 56:599 (1984).

3. D.W. Duke and J.F. Owens, Phys. Rev. D30:49 (1984).

4. R.T. Herrod, S. Wada, and B.R. Webber, Z. Phys. C9:351 (1981).

5. A. Devoto, D.W. Duke, J.F. Owens, and R.G. Roberts, Phys. Rev. D27:508 (1983).

6. H. Abramowicz et al., CDHS Collaboration, Z. Phys. C17:283 (1983).

7. D.B. MacFarlane et al., CCFRR Collaboration, Z. Phys. C26:1 (1984).

8. F. Bergsma et al., CHARM Collaboration, Phys. Lett. 153B:111 (1985).

9. J.J. Aubert et al., EMC Collaboration, Nucl. Phys. B259:189 (1985).

10. A.C. Benvenuti et al., BCDMS Collaboration, Phys. Lett. 195B:97 (1987).

11. A.C. Benvenuti et al., BCDMS Collaboration, Contribution to the International Europhysics Conference on High Energy Physics, Uppsala, 1987.

12. K. Varvell et al., BEBC WA59 Collaboration, CERN/EP 87-46 (1987).

13. P.D. Meyers et al., BFP Collaboration, Phys. Rev. D34:1265 (1986).

14. A.D. Martin, R.G. Roberts, and W.J. Sterling, Rutherford Appleton Laboratory Report Ral-87-052 (1987).

15. A. Bodek et al., Phys. Rev. D20:1471 (1979).

16. J.J. Aubert et al., EMC Collaboration, Phys. Lett. 123B:123 (1983).

17. BCDMS Collaboration, private communication.

18. H. Abramowicz et al., Z Phys. C25:29 (1984).

19. D. Allasia et al., Phys. Lett. 135B:231 (1984).

20. F. Halzen, Phys. Lett. 182B:388 (1986).

21. M. Diemoz, F. Ferroni, and E. Longo, CERN report CERN-TH.4751/87 (1987).

22. T.Åkesson et al., Z. Phys. C34:293 (1987).

23. J.C. Collins, Semi-hard Processes and QCD in:"Physics Simulations at High Energy," V. Barger, T. Gottschalk, and F. Halzen eds., World Scientific, Singapore.

24. P. Aurenche, A. Douiri, R. Baier, M. Fontannaz, and D. Schiff, Phys. Lett. 140B:87 (1987).

25. L. Bergmann and J.F. Owens, in preparation.

DIRECT PHOTON STUDIES

CURRENT STATUS OF EXPERIMENT E706 (FERMILAB)

George K. Fanourakis
(For E706 Collaboration*)

University of Rochester
Rochester, NY 14627

1 INTRODUCTION

Experiment E706 is designed to study interactions of protons, pions and kaons in a variety of targets, triggering on high transverse momentum (greater than 4 GeV/c) electromagnetic showers. This is a second generation fixed target experiment, using high energy beams (800 GeV primary protons from the Fermilab Tevatron), and a wide acceptance spectrometer, intended to reach 10-11 GeV/c in direct photon transverse momenta.

The first run of E706 is in progress, and the various parts of the detector are being debugged. The purpose of this presentation is to describe the status of our spectrometer, and to display the ability of the experiment to observe neutral pions, which is the most important step for the separation of direct photons from background.

2 THE E706 APPARATUS

The apparatus consists of a set of Silicon Strip Detector (SSD) planes in front of and behind the target, a set of Proportional Wire Chambers (PWC's) downstream of an analysis magnet, a Liquid Argon Calorimeter (LAC) and a Forward Calorimeter (FC). Downstream of E706 is experiment E672, which utilizes a toroidal magnet and drift chambers to study muon pairs. The beam is defined through beam counters, and tagged using a differential Cherenkov counter. The muon flux in the beam line is reduced by utilizing spoiler magnets in the beam line. The hadronic halo is absorbed in an iron hadron shield in front of the experiment. Figure 1 details the lay out of the E706 and E672 experiments in the Mwest Experimental Hall.

2.1 The Cherenkov Counter

The Cherenkov counter is about 33m long (focal length = 32.3m), and has a radius of 24.4cm. It uses helium gas in the pressure range of 4-7

TABLE I

Expected yield of Direct Photons

P_T (GeV/c)	π⁻C (500 hrs at 530 GeV/c)		π⁺C (700 hrs at 530 GeV/c)		pC (400 hrs at 800 GeV/c)	
	[γ]	[γγ]	[γ]	[γγ]	[γ]	[γγ]
5	34,000	330	34,000	120	29,000	90
6	9,500	90	9,000	40	6,000	20
7	2,500	26	2,100	11	850	3
8	650	8	450	3	160	
9	180	3	120	1	40	
10	50	1	25		7	
11	12		5		1	
5-11	46,892	458	45,700	175	36,058	113

The one or two photon yields in this table were obtained using our acceptance (Ref. 7). We assumed 10^7 pps for incident pions and 1.5×10^7 pps for incident protons. A 10% interaction length target was assumed, and a 25% duty factor for the Tevatron. The columns labeled [γ] are for a single photon in the indicated P_T range; the columns labeled [γγ] are for two photons and at least one is within the given P_T range.

M WEST

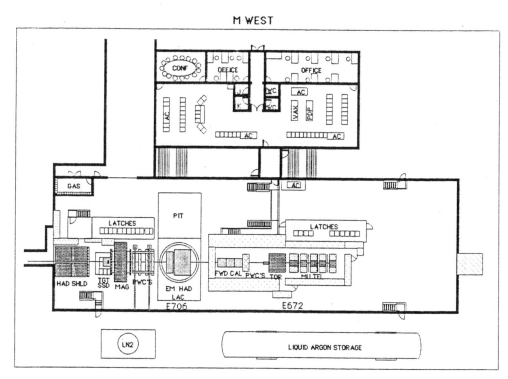

Figure 1

Plan view of experiments E706 and E672 in the MWEST experimental hall.

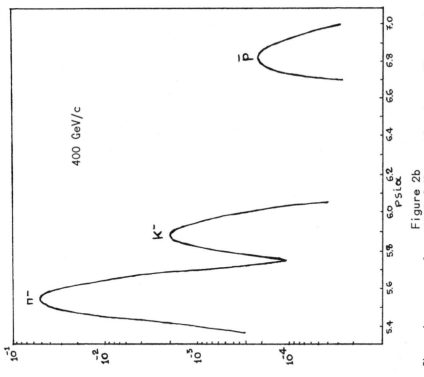

Figure 2a

Cherenkov curve using a 530 GeV negative beam. The vertical axis is the efficiency using OR'ed threefold coincidences in the coincidence ring, and two or less hits allowed in the anticoincidence ring. The horizontal axis is the helium gas pressure.

Figure 2b

Cherenkov curve for a 400 GeV negative beam. The vertical axis is the efficiency using a sixfold coincidence and only one hit allowed in the anticoincidence ring. The horizontal axis is the helium gas pressure.

psia; it has one coincidence and one anticoincidence ring, each containing 6 phototubes. The Cherenkov angle for the coincidence ring is 5 mrad. The typical efficiency for twofold coincidences is 96%. Figures 2a and 2b show the pressure curves for 400 GeV and 530 GeV negative beam.

2.2 The Silicon Strip Detectors

The SSD's consist of 6 X-Y modules (12 planes), with a total of 6600 active strips (Ref. 1). Two X-Y modules, consisting of 3cmx3cm wafers, are positioned before the target, and serve to define the trajectory of the beam particle. Two similar modules follow the target, and another two, made up from 5cmx5cm wafers, are used to measure trajectories of the produced particles. All SSD's have a 50 microns pitch, and are about 300 microns thick.

A minimum ionizing particle deposits about 2×10^4 electron-hole pairs, and the charge is collected in about 20ns; consequently a low noise wide band preamplifier must be used for signal processing. We use the RelLab IO-323-C charge sensitive preamplifier. The Nanometric (Ref. 2) system has been adapted to further amplify, discriminate and convert the analog signals to logic levels for storage as digital data. Figure 3 shows the X and Y beam-profiles obtained with the SSD's.

The target consists of about 2mm-thick segments, separated by about 1.5mm gaps, to aid in vertex reconstruction. We are planning to use carbon, copper, aluminum and beryllium targets of about 0.1 radiation lengths thickness. The Monte Carlo estimated primary vertex resolution is 10 microns in the transverse direction, and 350 microns along z (the longitudinal direction).

2.3 The Proportional Wire Chambers

The PWC's consist of 16 planes (4 sets of XYVU modules). They have a 2.54mm wire spacing. The module dimensions are one 1.22mx1.63m, two 2.03mx2.03m, and one 2.44mx2.44m. The gas used is an argon, isobutane, isopropyl alcohol and freon mixture.

The readout system relies on the same Nanometric amplifier, discriminator and latching system used for the SSD's. Figure 4 shows the reconstructed tracks from an interaction in which the analysis magnet was off.

2.4 The Liquid Argon Calorimeter

The LAC consists of an electromagnetic and a hadronic part, each mechanically independent of the other (Ref. 3). The two are suspended individually via suspension rods from a gantry structure that is able to move perpendicularly to the beam direction (see Fig. 5). The tranverse sizes of these detectors are 3 and 3.6 meters respectively. A beam pipe, about 40cm in diameter, passes through the centers of both structures. This is filled with helium gas to reduce the amount of material in the beam region. A thin-walled vessel, filled with low density foam (Rohacell), is fastened in front of the detector, again, to reduce the amount of passive material.

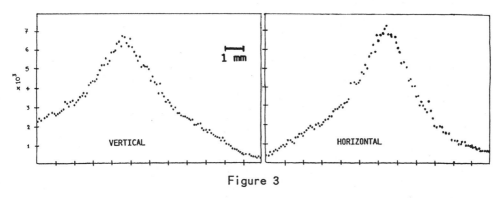

Figure 3

X and Y beam profiles obtained with the SSD's.

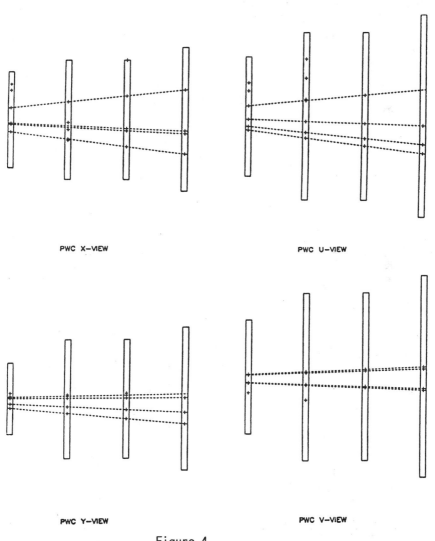

Figure 4

Reconstucted tracks from a low multiplicity event as seen by the PWC's.

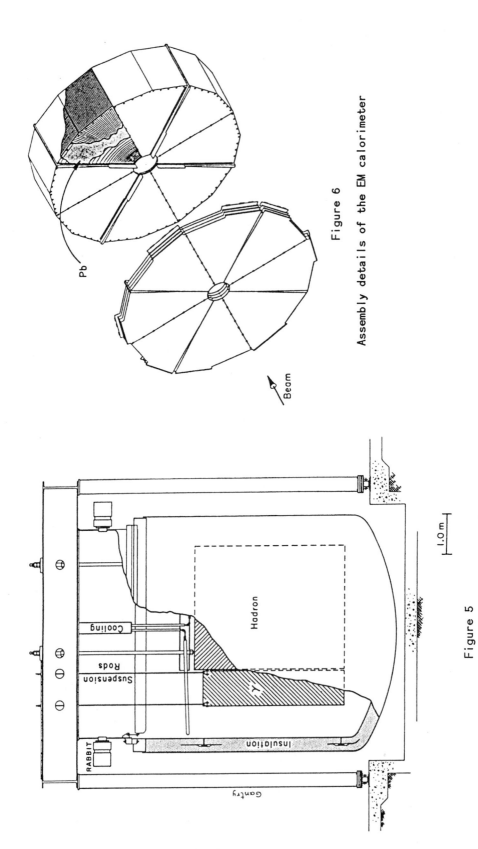

Figure 6

Assembly details of the EM calorimeter

Figure 5

E706 liquid argon calorimeter; suspension and cooling details.

A 5.18 meter diameter, polyurethane-foam-insulated, cylindrical cryostat, attached via a flange to the cryostat's top hat, contains the liquid argon and encloses the detector. The cryostat can be lowered into a specially built pit, when it is necessary to access the detector.

The detector components were produced and assembled in a clean environment in order to minimize any argon contaminants in the system. The cryostat and its contents were thoroughly degassed by an elaborate pumping, purging and heating procedure. The liquid argon used was the finest commercial grade produced in the USA. Its oxygen content was measured by an oxygen analyzer to be less than 0.2ppm. The purity of the delivered liquid argon was checked using a liquid argon test cell, employing a radioactive Ruthenium beta source, prior to acceptance of the liquid from the vendor. The average equivalent oxygen contamination was measured to be less than 0.4ppm. After the cryostat was filled with liquid argon, the measurement was repeated, (this time using an electron beam), and the dependence of pulse height on high voltage was fitted to the form:

$$Q = Q_0 - \{1 - \frac{s}{d} - \frac{s}{d}(1 - e^{-d/s})\}$$

Where d is the argon gap size (2.5mm), Q_0 is the total electronic free charge deposited in the argon, and s=aE/p, where a=0.06 cm ppmO_2/kV/cm (Ref. 4), E is the applied electric field, and p is the molecular impurity concentration (parts per million). The fit gave a result for p of 0.6ppm. This corresponds to a reduction of the electron signal by less than 5%, when compared to absolutely pure argon.

A set of four copper-finned cooling coils are suspended from the top of the cryostat. Liquid nitrogen flows inside these coils for cooling down the detector and for keeping it cold. The detector was cooled down by intoducing gaseous argon into the cryostat and adjusting the nitrogen flow through the cooling coils. The cooling was monitored by about 240 thermocouples, attached at mechanically critical areas of the detector. The temperature differentials during the cooldown were kept to under $30°K$ in order to keep the mechanical stresses low. The stresses on the hanging rods, at the warm top-part of the cryostat (above the cooling coils), were monitored via strain gauges during cooldown.

2.4.1   The Electromagnetic (EM) Calorimeter

The electromagnetic calorimeter consists (see Fig. 6) of four mechanically independent quadrants, fastened together by special ring sections. Each quadrant contains 66 layers of 2mm thick lead plates, interleaved with 1.6mm thick G10 radial (r) or azimuthal (phi) readout boards. The r-coordinate readout boards are further subdivided into octants, each containing 256 r-strips, each about 5.5mm wide. The strips are focussed in a tower-like way to the target. The phi coordinate strips are subdivided into inner and outer segments. Each inner strip subtends an azimuthal angle of 16.4 mrad, whereas each outer phi strip covers 8.2 mrad. The separation between inner and outer phi strips corresponds to a laboratory angle of 45 mrad. All argon gaps are 2.5mm.

This detector is read out in two sections along the beam direction. The first 1/3 (22 layers, 10 radiation lengths) is summed strip by strip, and the signal routed to three special interface boards in front of the detector. The corresponding interface boards for the back half (44 layers, 20 radiation lengths) are located at the back of this calorimeter.

These interface boards also provide connectors for external cables, which carry the signals to the amplifiers and digital electronics, which are part of the RABBIT (Redundant Analog Bus Based Information Transfer) data acquisition system (Ref. 5).

Each amplifier card has 16 channels; each channel includes a low noise integrating amplifier, a fast output used for trigger purposes, a 800 nsec delay line and a dual ADC sample and hold circuit to sample the pulse train before and after the event of interest. Also, four neighboring channels are added and fed into a TDC sample and hold circuit to obtain timing information to use for excluding energy depositions that are out of time.

The position resolution of electrons and photons in the calorimeter is less than 1mm. The electron energy resolution anticipated from EGS Monte Carlo, not taking into account the electrical noise or any gain variations, is $\sigma/E = 10\%/\sqrt{E}$. At present, prior to tuning and calibration, the resolution obtained is about $20\%/\sqrt{E}$ per view (see Fig. 7a). Typical transverse profiles in phi and r are shown in Figs 7b and c.

2.4.2 The Hadronic Calorimeter

The hadron detector has 52 3.7mx3.7mx2.5cm stainless steel plates, interleaved with 53 readout units ("cookies"). All the plates have a roughly octagonal shape, except two "superplates", which are constructed from a better grade of cryogenic stainless steel, and are used to support the weight of the rest of the hadron calorimeter.

The readout cookies consist of two complementary halves, which combine to give a full and uniform coverage of the cross section of the calorimeter (fig. 8). The readout pads are triangular, and can be combined logically to form hexagons, larger area triangles, etc. Each half of the readout is a sandwich of two 0.8mm thick G10 boards separated by 3mm thick G10 strips. The outer boards are copper clad on both sides, with the outside copper held at ground and the inside at high voltage. Facing the high voltage side are the readout boards with triangular pads in horizontal rows, and gaps for the readout leads (covered by protective strips). The back sides of the readout boards are glued to long vertical 3mm thick G10 ribs, which are attached to a header from which a cookie is suspended. The cookies were assembled using a cryogenic epoxy glue and a pneumatic press.

The pads of successive cookies are ganged to form towers which focus on the interaction target. This pattern requires that the readout boards of all cookies are of different size. Manufacturing the boards by conventional photographic etching techniques would have been very expensive. A computer controlled commercial plotter (made by Gerber Scientific) was therefore converted to a mechanical etching, cutting, and drilling machine to produce the readout and high-voltage boards. This technique is much faster than chemical etching, and is especially useful in producing large area readout boards.

The same readout system is used for the hadron calorimeter as was described for the EM calorimeter. The readout is also subdivided in two longitudinal sections, the front being two and the back eight interaction lengths. The pad energies are summed in towers using special interface boards which also contain the cables leading to the electronics.

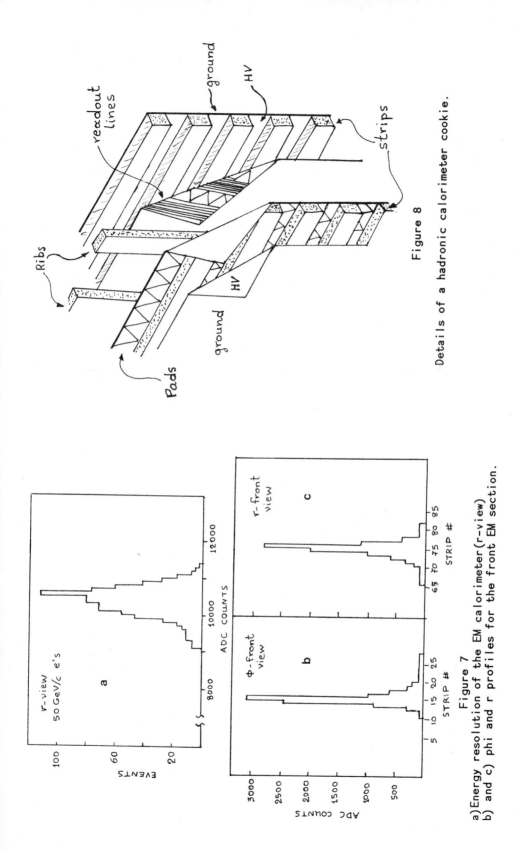

Figure 8

Details of a hadronic calorimeter cookie.

Figure 7

a)Energy resolution of the EM calorimeter(r-view)
b) and c) phi and r profiles for the front EM section.

299

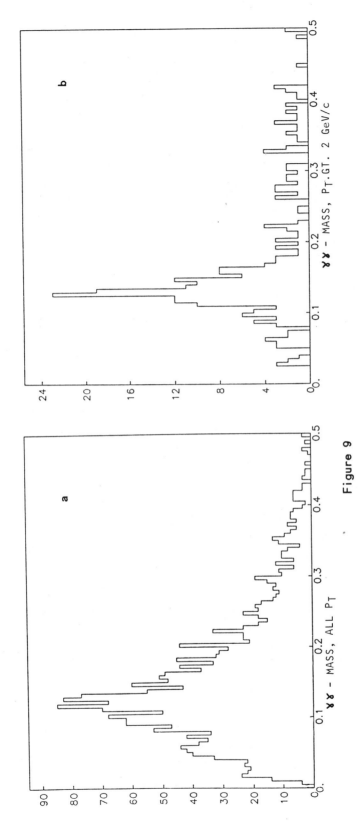

Figure 9

The distribution of the mass of any two isolated energy deposition ("photon") combinations: a) All P_T's and b) of the combinations with P_T above 2 GeV/c.

The position resolution of the hadron calorimeter is expected to be a few centimeters. The expected energy resolution from the GEANT (GHEISHA) Monte Carlo is $\sigma/E = 78\%/\sqrt{E}$ (ignoring amplifier noise).

2.5 The Forward Calorimeter

The forward calorimeter is designed to intercept the forward jet region not covered by the LAC (because of the central hole). It is made of three longitudinal identical modules, each consisting of alternating layers of steel and acrylic scintillator. Each module contains 32 steel plates 1.9cm thick, 114cm in diameter, separated by 6.3mm gaps, where 4.6mm thick plexipop is inserted. A central 3.2cm diameter hole allows the passage of the non-interacting beam.

The readout is accomplished via BBQ wavelength-shifter rods (doped with a UV absorbing agent) that run parallel to the beam direction. The signals from the photomultiplier tubes coupled to each rod are digitized by a 100 MHz flash ADC system.

3 THE DATA AQUISITION AND ANALYSIS

We use VAXONLINE, the data aquisition system developed by the Fermilab Computing department (Ref. 6). Four PDP11's (3 for E706, 1 for E672) collect the event fragments from the different parts of the apparatus and send them to a MicroVAX, where they are concatenated to the final event and written to magnetic tape.

At present, we have aquired only a small amount of 530 GeV pion data, on a Carbon target. The trigger required a local deposition of transverse momentum of >1.5 GeV/c in any eight neighboring r-strips, and a global requirement (octant) of >3 GeV/c.

The distribution of the mass of any two gamma combinations, each of energy greater than 5 GeV, is shown in Fig 9a. The presence of a pion peak can be discerned above the combinatorial background. The signal can be enhanced by requiring >2 GeV/c transverse momentum for the two photon systems, as shown in Fig 9b. Considering the status of the analysis (prior to calibration and tuning of the detector and trigger), the result is remarkably good!

Table I shows the expected yields of direct photons for a total of 1600 hours of running. Although this amount of beam time will not be available during this running period, it is clear that the detector is working well and the experiment should eventually reach its stated goals.

Acknowledgements

We would like to acknowledge the support of the DOE, NSF and the UGC (India) for this research program.

* THE E706 COLLABORATION

G. BALLOCCHI, L. DE BARBARO, W. DESOI, G. FANOURAKIS, T. FERBEL, G. GINTHER, P. GUTIERREZ, A. LANARO, F. LOBKOWICZ, J. MANSOUR, G. PEDEVILLE, E. PREBYS, D. SKOW, P. SLATTERY, N. VARELAS and M. ZIELINSKI
Department of Physics and Astronomy, University of Rochester, New York 14627, USA.

T.CHAND, B.C. CHOUDHARY, V. KAPOOR, S. MATHUR, B.M. RAJARAM and R. SHIVPURI
Department of Physics and Astrophysics, Delhi University, Delhi 11 00 07, India.

W. BAKER, D. BERG, D. CAREY, C. JOHNSTONE and C. NELSON
Fermilab, P.O. Box 500, Batavia, Illinois 60510, USA.

C. BROMBERG, D. BROWN, J. HUSTON and R. MILLER
Department of Physics, Michigan State University, East Lansing, Michigan 48824, USA.

R. BENSON, P. LUKENS and K. RUDDICK
School of Physics, University of Minnesota, Minneapolis, Minnesota 55455, USA.

G. ALVERSON, W. FAISSLER, D. GARELICK, G. GLASS, M. GLAUBMAN, I. KOURBANIS, C. LIRAKIS, E. POTHIER, A. SINANIDIS, G.-H. WU, T. YASUDA and C. YOSEF
Department of Physics, Northeastern University, Boston, Massachussets 02115, USA.

S. EASO, K. HARTMAN, B.Y. OH, W. TOOTHACKER and J. WHITMORE
Department of Physics, Penn State University, University Park, Pennsylvania 16802, USA.

E. ENGELS Jr, S. MANI, P.D.D.S. WEERASUNDARA, and P.F. SHEPARD
Department of Physics and Astronomy, University of Pittsburgh, Pennsylvania 15600, USA.

References

1) E. Engels et al, NIM 226(1984) 59-62.
   E. Engels et al, NIM A253(1987) 523-529.
2) Multi Wire Proportional Chamber Readout System, MPWC-1, Nanometric Systems, Inc., 451 S. Boulevard Oak Park, Il. 60302.
3) F. Lobkowicz et al, NIM A235(1985) 332-337.
4) Private communication from G. Blazey.
5) T.F. Droege, K.J. Turner, T.K. Oshka, RABBIT System Spesifications, Fermilab note PIN521/CDF119 (1982).
6) VAXONLINE, Fermilab Computing Department, Data Aquisition Software Group, notes: PN252, PN271.2, PN272.2, PN278, PN280.
   Also, IEEE transactions on Nuclear Science, NS-34 No.4(1987) 763-767.
7) W. Baker et al, Fermilab P695 (Revised)/P706 Proposal, October, 1981.

EXPECTATIONS FOR DIRECT PHOTON PHYSICS

FROM FERMILAB EXPERIMENT E705

D. E. Wagoner[8], M. Arenton[1], T. Y. Chen[1], S. Conetti[6], B. Cox[4],
S. Delchamps[4], B. Etemadi[5], L. Fortney[3], K. Guffey[5], M. Haire[6],
P. Ioannu[2], C. M. Jenkins[4], D. J. Judd[8], C. Kourkoumelis[2],
I. Koutentakis[2], J. Kuzminski[6], K. W. Lai[1],
A. Manousakis-Katsikakis[2], H. Mao[9], A. Marchionni[6],
P. O. Mazur[4], C. T. Murphy[4], T. Pramantiotis[2], R. Rameika[4],
L. K. Resvanis[2], M. Rosati[6], J. Rosen[7], C. H. Shen[9], Q. Shen[3],
A. Simard[6], R. Smith[4], L. Spiegel[7], D. Stairs[6], R. Tesarek[3],
W. Tucker[5], T. Turkington[3], F. Turkot[4], L. Turnbull[6],
S. Tzamarias[7], M. Vassiliou[2], G. Voulgaris[2], C. Wang[9], W. Yang[4],
N. Yao[1], N. Zhang[9], X. Zhang[9], and G. Zioulas[6]

[1] University of Arizona, Tucson, Arizona
[2] University of Athens, Athens, Greece
[3] Duke University, Durham, North Carolina
[4] Fermilab, Batavia, Illinois
[5] Florida A&M University, Tallahassee, Florida
[6] McGill University, Montreal, Quebec, Canada
[7] Northwestern University, Evanston, Illinois
[8] Prairie View A&M University, Prairie View, Texas
[9] Shandong University, Jinan, Shandong, People's Republic of China

ABSTRACT

The E705 scintillation glass/lead glass electromagnetic calorimeter is described. The trigger used for recording high transverse momentum direct photon signals from 300 GeV/c π^-, π^+, $\bar{p}$, p interactions in a Li^7 target is explained. Preliminary results on the response of this direct photon trigger and electromagnetic calorimeter and expected event sensitivities are presented.

INTRODUCTION

Experiment E705[1] at Fermilab is a fixed target experiment to study charmonium and direct photon production with 300 GeV/c beams of π^-, π^+, $\bar{p}$, and p on a Li^7 target. This experiment is currently in the middle of its data taking run so any results reported here are strictly preliminary. This paper will describe the direct photon goals of E705 and the electromagnetic calorimeter and direct photon trigger used in the experiment. The charmonium part of the experiment is described in these proceedings in Ref. 2.

Direct Photon Physics Goals of E705

Production of high transverse momentum direct photons provides a clean method to probe the parton nature of hadrons and to perform tests of the theory of QCD. Refs. 3 and 4 are excellent reviews on direct photon production in QCD and a theoretical review of this topic is also presented in these proceedings in Ref. 5. A goal of E705 is to study the contributions of the annihilation and Compton processes to direct photon production at $\sqrt{s} = 23.7$ GeV/c by performing comparisons of π^- vs π^+ and $\bar{p}$ vs p direct photon production for $p_T > 4$ GeV/c. E705 also plans to study leading particle distributions from the recoil quark or gluon fragmentation and measure the nucleon and pion structure functions in the accessible kinematic range. The study of high mass diphoton production is also another goal of E705.

In particular by studying the difference of direct photon cross sections between the beam types, i.e., $\sigma_\gamma(\pi^- \text{ Li}) - \sigma_\gamma(\pi^+ \text{ Li})$ and $\sigma_\gamma(\bar{p} \text{ Li}) - \sigma_\gamma(p \text{ Li})$, one can cleanly isolate the quark-antiquark annihilation term. In this way the annihilation and Compton contributions can be separated and valence quark and gluon structure functions for the pion and nucleon can be extracted. Also by studying the difference distributions for the characteristics of the recoil leading particles one can attempt to determine properties of gluon and quark fragmentation.

E705 EXPERIMENTAL APPARATUS

E705 is a large aperture open geometry fixed target experiment. Fig. 1 shows the layout of the experiment. The beam used is a 300 GeV/c 98.5% π^-/1.5% $\bar{p}$ or 40% π^+/60% p beam. The beam particle types are tagged with two gas threshold Cherenkov counters and the beam trajectory is measured at three beam PWC stations. The target is 33 cm of Li^7.

Between the target and the analysis magnet are three PWC chamber sets and three drift chamber sets. These chamber sets are deadened in the beam region to allow the non-interacted beam to pass through without producing hits. To detect particles in the beam region there are three PWC chamber sets with finely spaced wires which are sensitive only in the beam region. The aperture covered by the combination of these nine chamber sets matches the outer aperture of the analysis magnet which is ±142 mrad $\times$ ±81 mrad. The analysis magnet has an equivalent p_T kick of 0.766 GeV/c. Following the analysis magnet are three large drift chamber stations with deadened beam regions. These sets of upstream and downstream chambers will permit E705 to perform tracking on leading particles from the direct photon recoil fragmentation. Some level of K_S^0 reconstruction can also be done.

Following the downstream drift chambers are two sets of finely segmented scintillation hodoscope planes one oriented horizontally and the other vertically. Then follows the electromagnetic calorimeter which will be described more fully later. Finally there is the steel muon absorber and tungsten beam dump. The muon absorber has a stopping power of 6 GeV/c and contains within it the muon scintillator hodoscopes. There are three vertically oriented planes of muon hodoscope and one horizontally oriented plane.

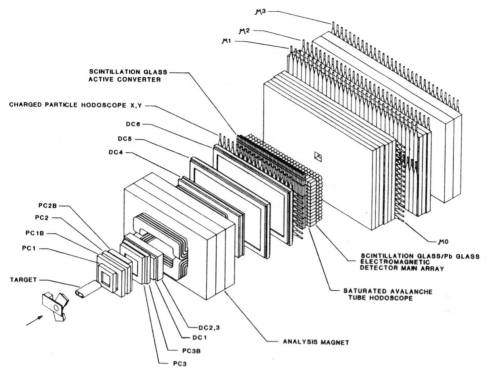

Figure 1. Experimental Layout of E705.

Figure 2. Side view of E705 electromagnetic calorimeter. The beam enters from the left. Shown are the two active converter layers and the main array.

These muon hodoscopes are to define the dimuon trigger used in the charmonium part of E705.

E705 ELECTROMAGNETIC CALORIMETER

Fig. 2 shows a side view of the E705 electromagnetic calorimeter. It is located 10 m from the target and is composed of three parts: i) the main array (the downstream part on the right), ii) the active converter/lead-gas converter (the upstream part on the left), and iii) the gas tube hodoscope (which is located in the gap between the active converters and main array). The active converter/lead-gas converter/gas tube hodoscope is used to start the electromagnetic shower and determine its position. They are also used to provide rejection against hadron showers. The main array is used to determine the energy of the electromagnetic shower. These devices are now described more fully.

Main Array

Fig. 3 shows a front view of the main array of the E705 electromagnetic calorimeter. The area inside the dashed lines contain SCG1-C scintillation glass blocks 21 radiation lengths (X_0) long and the blocks outside are SF5 lead glass 18 X_0 long. The small SCG1-C blocks in the center are 7.5 cm × 7.5 cm in transverse size and are read out with RCA 6342A photomultipliers. The large SCG1-C and SF5 blocks are 15 cm × 15 cm and are read out with EMI 9791KB photomultipliers. There is an empty beam hole of size ±15 cm × ±7.5 cm.

SCG1-C is a heavy Barium glass containing an inorganic scintillator that is produced by Ohara Optical Glass Manufacturing Co., Ltd. SCG1-C has been tested extensively by E705[6] and some of its properties compared to SF5 are given in Table 1. The two main advantages of SCG1-C are that it produces over 5 times as much light as SF5 (which implies that its energy resolution should be at least a factor of 2 better than that for SF5) and that it is $\approx$ 150 more resistant to radiation darkening than SF5. Fig. 4 shows the energy resolution of SCG1-C as measured in Ref. 6. The energy resolution for E705 conditions is expected to be: $\sigma_E/E \approx 1.2\% + 2.0\%/\sqrt{E}$ (with E in GeV). Fig. 5 shows the increased radiation resistance of SCG1-C compared to that of SF5 as determined in Ref. 6.

All 392 channels of the main array go to a precision ADC (with 15 effective bits) specially built for E705. The signals of all main array channels also go to LeCroy 4290 TDC's to determine the time of the various energy deposits. Each block has mounted on its upstream face an optical fiber coming from the central LED pulsing system used to monitor the gain of each block between electron calibration runs.

All channels of the main array also go the fast analog cluster finder which is used to form the high p_T direct photon trigger and diphoton trigger used in E705. The outermost layer of SF5 blocks and innermost layer of small SCG1-C blocks are not in the trigger fiducial volume and are not used to produce direct photon or diphoton triggers; they are used only to contribute energy as neighbors to clusters within the trigger fiducial volume.

Table 1. Properties of SCG1-C scintillation glass and SF5 lead glass[6].

	SCG1-C		SF5	
Composition:	BaO	43.4%	PbO	55%
(by weight)	SiO_2	42.5%	SiO_2	38%
	Li_2O	4.0%	K_2O	5%
	MgO	3.3%	Na_2O	1%
	K_2O	3.3%		
	Al_2O_3	2.0%		
(inorganic scintillator →)	Ce_2O_3	1.5%		
Density:	3.36 g/cm^3		4.08 g/cm^3	
Radiation Length:	4.25 cm		2.47 cm	
Refractive Index:	1.603		1.67270	
Critical Energy:	21.9 MeV		15.8 MeV	
dE/dx_{min}	5.16 MeV/cm		5.77 MeV/cm	
Nuclear Absorption Length: (for 30–200 GeV π's)	45.6 cm		42.0 cm	
Flourescent Wavelength:	429 $\pm$ 37.8 cm (isotropic)			
Flourescent Decay Time:	$\approx$ 80 ns (with fast leading edge)			
PMT Light Output:	5.1 × SF5 (85% scintillation, 15% Cherenkov)			
Radiation Resistance:	$\approx$ 150 times more resistant than SF5			

Active Converter and Lead-Gas Converter

Fig. 6 shows a front view of the SCG1-C active converters and the central lead-gas converter (LGC). The LGC is an 8 layer lead/proportional tube sandwich occupying the central $\pm$52.5 cm × $\pm$97.5 cm region of the calorimeter. The LGC is a total of 3.8 X_0 deep and has a beam hole matching that of the main array. The anode segmentation is 1 cm with similar segmentation for cathode strip readout in the other dimension. The 8 longitudinal samples are ganged together in both the anode and cathode and the pulse height of each ganged signal is read out using a 12 bit LeCroy 2280 ADC system. This device will enable E705 to improve the position resolution and two photon separation over that of the SCG1-C active converter/gas tube hodoscope system and should not substantially degrade the energy resolution of the main array to the high energy photons expected in the central part of the calorimeter.

Two longitudinal layers of active converter blocks of SCG1-C lie to either side of the central LGC. Each layer of SCG1-C is 1.75 X_0 deep and there is a pre-converter of 1 X_0 of steel upstream of the first layer. The transverse size of the active converter is 7.5 cm × 97.5 cm and the blocks are read out with RCA 6342A photomultipliers. The two longitudinal measurements of the shower development in the active converter will be used in identifying photons and electrons and rejecting hadrons. Each active converter block has attached to it an optical fiber from the E705 central LED pulsing system which is used to monitor the gain. Each channel of the active converter is digitized with the E705 precision ADC system that is also used for the main array; the

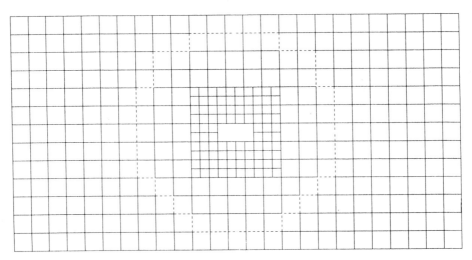

Figure 3. Front view of E705 Main Array.

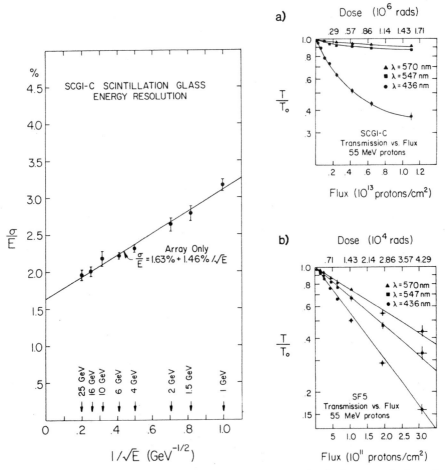

Figure 4. Energy Resolution for an SCG1-C array.

Figure 5. Optical transmission along the beam axis at the specified wavelength for 1 cm thick samples of (a) SCG1-C and (b) SF5 as a function of dose.

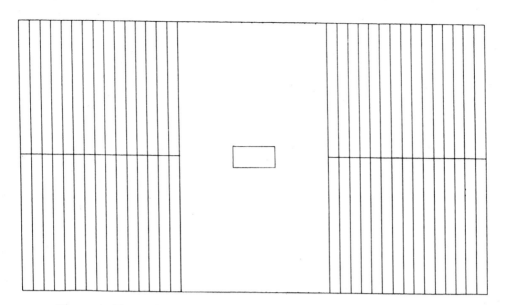

Figure 6. Front view of E705 active converter and lead-gas converter.

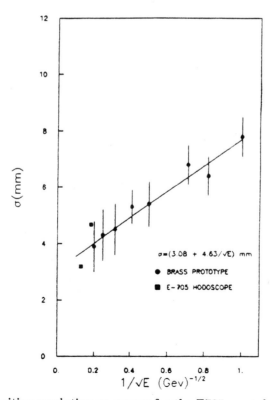

$$\sigma = (3.08 + 4.63/\sqrt{E}) \text{ mm}$$

● BRASS PROTOTYPE

■ E-705 HODOSCOPE

Figure 7. Position resolution vs energy for the E705 gas tube hodoscope.

active converter is not used in the formation of the direct photon or diphoton trigger.

Gas Tube Hodoscope

The gas tube hodoscope contains two planes of conducting plastic proportional tubes with induced cathode strip readout. It is placed between the active converter and main array and covers the same area as the active converter. The proportional tubes have a 0.7 cm spacing. This device is used to determine the electromagnetic shower position for showers originating in the active converter. The properties and resolution of the gas tube hodoscope have been studied in test beams and are reported in Ref. 7. The anode and cathode signals from the two longitudinal layers are ganged together and digitized with a 12 bit LeCroy 2280 ADC system. The measured position resolution from Ref. 7 of this device is shown in Fig. 7.

Electromagnetic Calorimeter—General

The entire E705 electromagnetic calorimeter contains 25 tons of glass and sits inside an enclosed temperature controlled house in order to control temperature dependent gain variations. The temperature is controlled though banks of thermoelectric coolers and can be kept stable to $\pm 0.06°$ C. The house containing the calorimeter is moveable and can be driven vertically and horizontally under computer control to ± 0.5 mm. This enables each main array block to be positioned into a calibration electron beam for determining absolute calorimeter calibration constants. The calibration beam is an electron or positron beam with a tunable energy from 2–100 GeV. Electrons are Cherenkov identified and can be momentum tagged.

For monitoring the gain between calibration runs, each glass block has attached to it an optical fiber coming from a central LED pulser[8]. 92 Hewlett-Packard HLMP-3950 LED's are driven by fast triggered transistor avalanche pulsers and the light from these LED's are mixed into a single light source. This single LED light source then passes through a computer controlled filter wheel and is viewed by a bundle of quartz optical fibers which are fanned out and attached to the glass blocks of the calorimeter. The LED light source is temperature controlled by the same system used for the calorimeter and is monitored by 3 PIN photodiodes attached to 3 of the optical fibers. The light level of the LED pulser is stable to $\pm 1\%$. With this system the gain of the calorimeter glass blocks can be accurately tracked between calibration runs.

E705 DIRECT PHOTON AND DIPHOTON TRIGGER

The high p_T direct photon trigger in E705 is formed from a specially built system[9]. In general, the photon trigger forms analog energy sums from adjacent main array blocks to find energy clusters. The peak block and energy sum of each cluster are found in parallel for all clusters. The cluster energies are converted to transverse momentum and this p_T level can then be triggered on. The operation of this trigger will now be explained in more detail.

The signal from each main array block goes into front end electronics. The front end is effectively a charge integrator with pole-zero correction to compensate for the various input signal shapes coming from the various block and PMT types. In order to correct for pile-up and permit high rate operation, baseline restoration is performed

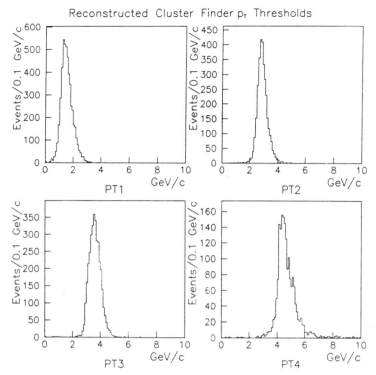

Figure 8. Raw uncorrected p_T sums for p_T trigger thresholds of:
PT1=1.7 GeV/c, PT2=2.5 GeV/c, PT3=3.5 GeV/c,
PT4=4.5 GeV/c.

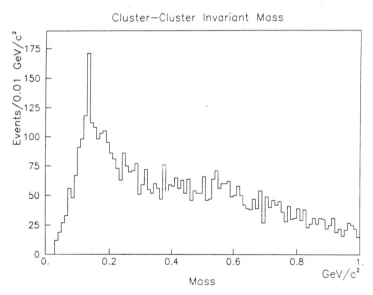

Figure 9. Preliminary invariant mass for all combinations of clusters outside of
±52.5 cm for PT1 and PT2 single photon triggers. No other cuts on
the mass combinations are made.

using a using a system of two track and holds to perform a "before-trigger" level subtraction. During read out this "before" level is also digitized with a 3 bit ADC.

The baseline corrected signal then goes to the precision ADC. This ADC is an auto-ranging 12 bit ADC with 15 bit sensitivity. The signal is sampled early and a decision is made whether or not to turn on an 8X amplifier. In this way a fast 12 bit ADC can be used which will effectively have 15 bit sensitivity. This permits sensitivity to a large dynamic range in energy deposition of from 6 MeV to 200 GeV.

The front end also forms the analog output signal which is used by the cluster finder and high p_T trigger. This analog signal is formed by the difference of signals from before and within a tapped delay line. The high p_T photon cluster finder finds clusters of energy deposition by forming analog energy sums with each block and its nearest neighbors. These cluster sums are formed through a custom wire-wrapped backplane corresponding to the E705 main array block layout. If a block's energy is larger than that of all its neighbors and larger than a computer setable minimum threshold (currently 4.2 GeV), then an energy cluster is defined having the position of the peak block and the energy of the analog cluster sum. This analog energy cluster sum is then converted to an analog p_T by attenuating the energy of the cluster by an amount proportional to the value of $\sin \theta$ of the peak block, i.e., $p_T \approx E_{\text{cluster}} \sin \theta_{\text{peak}}$. This analog p_T signal is then available for triggering.

To form the high p_T photon trigger, the main array is divided into quadrants and each quadrant is handled independently for trigger formation. Each quadrant can produce p_T triggers at 4 different computer setable p_T levels which can each be separately prescaled. The 4 p_T levels (labeled PT1–PT4) are presently set in E705 to 1.7, 2.5, 3.5, and 4.5 GeV/c where all PT4 ($p_T > 4.5$ GeV/c) triggers are read out and the others are appropriately prescaled. Fig. 8 shows preliminary plots of the raw uncorrected p_T sums (for the main array only) for each of the 4 p_T trigger levels A very good correspondence between the p_T levels set and triggered on can be seen. A diphoton trigger is implemented by requiring two high p_T clusters in opposite quadrants. The lowest p_T trigger level is 1.7 GeV/c, so this requires that the triggered diphoton mass be ≥ 3.4 GeV/c². Since in the E705 cluster finder the cluster summing is analog and all channels are done in parallel, the cluster finder trigger is quite fast and is capable of operating at a 2 MHz interaction rate.

A preliminary search for π^0's has also been performed. Clusters from the two lower p_T triggers (PT1 and PT2) were reconstructed in the region outside of the LGC (± 52.5 cm). Fig. 9 shows a preliminary plot of the invariant mass of all combinations of all these reconstructed clusters. No other cuts were made on the invariant mass combinations. A π^0 peak is clearly evident in this plot at the expected mass.

DIRECT PHOTON EVENT SENSITIVITY

E705 has a fast CAMAC/VME-based data acquisition system[10] utilizing smart crate controllers and large buffered dual port memory. The E705 trigger and data acquisition system allow a trigger rate of 200 events/sec of spill (with 4600 bytes/event) to be written to tape. Over half of these triggers are high mass dimuon triggers[11] used for the charmonium part of the experiment; the remainder are devoted to the single photon and diphoton triggers. To achieve this trigger rate, E705 is currently running

Table 2. Direct photon event sensitivities for various beams in units of
events / pb/nucleon. E705 values are based on extrapolated running
through Feb. 15, 1988.

Experiment	π^-	$\bar{p}$	π^+	p	p_{lab}	Target
E705	9.1	0.32	5.8	6.1	300 GeV/c	Li^7, 33 cm
NA24	1.33	—	0.19	0.45	300 GeV/c	LH_2, 1 m
WA70	3.3	—	1.3	5.2	280 GeV/c	LH_2, 1 m
UA6	—	1.0	—	1.6	315 GeV/c	H_2 gas jet

at an interaction rate of 0.7–1.0 MHz. The duty factor for beam delivery at Fermilab is very favorable with a 23 second long spill once a minute. The E705 target of 33 cm of Li^7 is $\approx$ 13% of an interaction length. With the enhancement due to the A dependence for direct photon processes ($\approx A^1$) compared to that for total cross section processes ($\approx A^{0.72}$), E705 expects to achieve a good sensitivity for direct photon production. Table 2 shows the direct photon event sensitivity expected for E705 at the end of the present fixed target run. The E705 sensitivities are determined from the current data set and extrapolations based on it for running through Feb. 15, 1988. These sensitivities should allow E705 to probe direct photon production up to $p_T \approx 8$ GeV/c for $\pi^\pm$ and p. Also shown are the event sensitivities reported at this workshop by the CERN direct photon experiments in the same energy range as E705: NA24[12], WA70[13], and UA6[14]. Compared to WA70, E705 should have 3–4 times better sensitivity for $\pi^\pm$ production and roughly the same sensitivity for p production. UA6 has clearly the best sensitivity for $\bar{p}$ production at these energies.

CONCLUSION

The goal of experiment E705 is to study the contribution of the annihilation and Compton processes to direct photon production at $\sqrt{s} = 23.7$ GeV and to extract quark and gluon structure functions for the pion and nucleon through the comparison of π^-/π^+ and $\bar{p}/p$ high p_T photon production. E705 is presently collecting data and by the end of the data run should have the sensitivity to be able to make a significant statement about direct photon production for p_T's up to 8 GeV/c.

REFERENCES

1. Fermilab Proposal E-705 (1981).
2. S. Conetti, these proceedings.
3. J. F. Owens, Rev. Mod. Phys. **59**, 465 (1987).
4. T. Ferbel and W. R. Molson, Rev. Mod. Phys. **56**, 181 (1984).
5. R. Baier, these proceedings.
6. D. E. Wagoner, et al., Nucl. Inst. and Meth. **A238** (1985) 315; B. Cox, et al., Nucl. Inst. and Meth. **A238** (1985) 321; B. Cox, et al., Nucl. Inst. and Meth. **219** (1984) 487.
7. C. M. Jenkins, et al., Proceedings of the Gas Sampling Calorimetry Workshop II, Fermilab, 201 (1985); R. Rameika, et al., Nucl. Inst. and Meth. **A236** (1985) 42; B. Cox, et al., Nucl. Inst. and Meth. **219** (1984) 491.

8. L. Fortney, Proceedings of the Calorimeter Calibration Workshop, Fermilab, 1 (1983).

9. R. Rameika, Proceedings of the Workshop on Triggering, Data Acquisition and Computing for High Energy/High Luminosity Hadron-Hadron Colliders, Fermilab, 111 (1985).

10. S. Conetti, M. Haire, and K. Kuchela, IEEE Trans. Nucl. Sci. **NS-32** (1985) 1318; S. Conetti and K. Kuchela, IEEE Trans. Nucl. Sci. **NS-32** (1985) 1326.

11. H. Areti, et al., Nucl. Inst. and Meth. **212** (1983) 135.

12. P. Seyboth, these proceedings; C. De Marzo, et al., Phys. Rev. D **36**, 8 (1987).

13. M. Martin, these proceedings; M. Bonesini, et al., CERN-EP/87-185 (1987).

14. P. T. Cox, these proceedings.

JET PHYSICS FROM THE AXIAL FIELD SPECTROMETER

Hans Bøggild

Niels Bohr Institute
Copenhagen, Denmark

1. Introduction

One of the central physics issues of the Axial Field Spectrometer (AFS) at the CERN ISR was the study of jet-events in terms of the Parton model and QCD.

The part of the apparatus which is relevant for jet physics is shown on Fig.1 . The transverse energy ΣE_T for all particles with c.m.s. rapidity less than 0.9 was measured using the box shaped uranium scintillator calorimeter and the momenta of charged particles were measured in the central drift chamber in the 0.5 Tesla axial magnetic field provided by a big "horseshoe" magnet (not shown). Details can be found in ref. 1.

A combination of triggers selected events with one, two or three high E_T jets and also events with very high ΣE_T without jet requirements. The many other triggers included a special selection of events with high p_T photons, where jets were found off-line. The analysis of these events, with a photon plus a jet has been discussed at this Workshop by Julia A. Thompson - the main result was a measurement of the gluon structure function in the proton, indicating a rather soft distribution.

In this talk I will discuss the AFS results on jet production for proton-proton collisions at $\sqrt{s}=63$ GeV in the central region, i.e. $|\eta| < 0.9$.

First I will shortly summarize the results on two jet
production (ref.2), which is found to be well described by
QCD and observed to be dominated by quark-quark collisions
(rather than quark-gluon or gluon-gluon). Then I will discuss the production of three jets (ref.3), which is well
described in terms of gluon bremsstrahlung and where a
value for the strong coupling constant α_s is extracted.
Finally I will talk about events where four jets were found
(ref 4): A part of these events are interpreted as resulting from double parton scattering and a transverse size of
the valence quark system is extracted.

At this point I would like to comment on the experimental problems in studying jets at the ISR. The c.m.s.
energy range covered by the ISR is actually the range where
jets start to manifest themselves in a clear and unbiased way.
At the lower ISR energies $\sqrt{s}$ = 23 and 31 GeV jets can only be
inferred in a statistical way and a reliable Monte Carlo
model is a necessary tool. At the highest energies $\sqrt{s}$ = 53
and 63 GeV the "unbiased" trigger requiring only high ΣE_T
leads to events where back to back jets are visible by eye
in single events (ref 5). Still only ∿50% of the events seem
to be clear jet events and it remains a problem how to account for the other 50%.

When three of four jets are required, or whenever quantitative results are wanted we are in a situation where a reliable Monte Carlo model, in terms of both physics and detector description, is necessary. A jet is in our case defined
by a so-called jet window algorithm as a maximum E_T found
within a particular size region in $\Delta\eta$ and $\Delta\varphi$. After a maximum E_T (jet) has been found a second non-overlapping E_T(jet)
can be found etc. In the results I am going to present here
a great care has been put into the construction and test of
a Monte Carlo model - we have found that using ISAJET (ref 6)
with some modification leads to a consistent description of
our data. Using the Duke-Owens Structure Functions (ref 7)
and increasing the contribution from the so-called underlying event (or beam jets) by roughly 50% gives a good result. For the 3-jet analysis where gluons play an important
role, it is furthermore found that the transverse momentum

in the gluon fragmentation has to be increased by 33% compared to quark fragmentation.

2. Two jet events

The di-jet mass spectrum is shown on Fig.2 and illustrates the good qualitative agreement with leading order QCD, but also shows the need for the so-called K-factor - a constant multiplied on the cross section to account for the higher order corrections that have not yet been calculated.

The longitudinal jet fragmentation function $D(z)$, where $z = P_z/P_{max}$ along the jet axis, is found to be in good agreement with results from e^+e^- when the di-jet mass is similar to the e^+e^- cms energy. The ratio of the fragmentation functions for positive and negative particles is seen in Fig.3. The three shaded bands show the expectation if the source of the high E_T jets were qq, qg and gg scattering. It is seen that at these values of di-jet mass $\gtrsim$ 20 GeV/c² corresponding to constituents having $\gtrsim$ 1/3 of the proton momentum we are entirely dominated by quark-quark scattering (valence quarks). This dominance is actually more pronounced than expected from the model with Duke-Owens Structure functions, which in turn agrees with the previously mentioned result on the gluon structure function (see J.A.Thompson's talk).

3. Three jet events

In QCD, gluon bremsstrahlung can give rise to a third jet, a fourth jet, etc. The rate of this compared to the production of two jets is controlled by the strong coupling $\alpha_s(Q^2)$.

Fig.4 shows two examples from the AFS of three-jet candidates. We now define $x = 2 E_T(\text{jet})/\sum_{123} E_T(\text{jets})$ where $x_1 > x_2 > x_3$. Since $x_1+x_2+x_3=2$ we can now plot x_2 versus x_1 in a "Dalitz plot" where all the points have to be within a triangle. Fig 5a shows the data for the so-called "star 3-jets", i.e. particularly clear cases where it has been required that the smallest x, x_3 is bigger than 0.25 and that the angle, ω , between the two lowest energy jets is bigger than 40°. Fig 5b shows the configurations corresponding to different areas in the "Dalitz plot". Fig 6a shows what is expected from the Monte Carlo model with both 2-jets and multi-

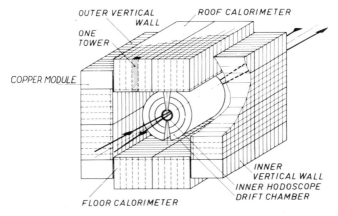

Fig.1 The AFS

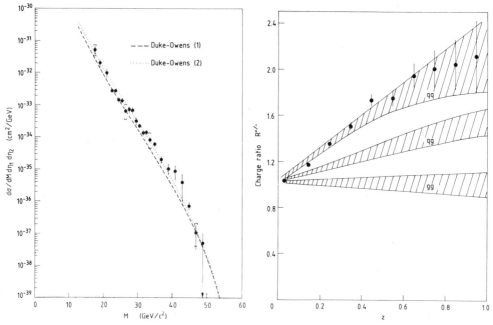

Fig.2 The dijet cross-section at 90°
as a function of the constituent--
constituent centre-of-mass energy,
M. The outer error bars show the
systematic errors originating from
uncertainties in the energy scale
and the Monte Carlo correction of M.
Shown are also the results of leading-order QCD calculations based on
the structure functions of [7], taking $Q^2=M^2/3$ and four flavours

Fig.3 Ratio between the numbers of positive and negative particles as a function of the corrected fractional momentum z. For comparison are shown predictions by the ISAJET Monte Carlo for
qq, qg, and gg scattering. The bands
reflect the statistical errors in the
Monte Carlo calculations

jets, and Fig 6b shows the situation when gluon bremsstrahlung is switched off. It is clear that the 3-jet event sample (star 3-jets) are genuine 3-jets and not artifacts created by the event selection.

The uncorrected ΣE_T distribution for star 3-jet events and 2-jet events are compared in Fig 7. From the ratio of these spectra α_s can in principle be extracted after the necessary corrections for instrumental effects and imperfections of the jet finding algorithm.

The good agreement between the QCD model (see also remarks in the introduction) and the data is illustrated on Fig 8a, for the angle, ω, between the second and third jet and in Fig 8b for the distribution of the fractional momentum of the third jet, x_3 .

From the ratio between 3-jet to 2-jet events at the same total jet mass, the value of α_s is extracted:

$$(K_3/K_2) \; \alpha_s \; = \; 0.18 \; \pm \; 0.03(\text{stat}) \; \pm 0.04(\text{syst})$$

The factors K_3 and K_2 come from the fact that the QCD calculations as previously stressed are limited in order and that 3- and 2-jet events of course have different sets of higher order corrections.

4. Four jet events

The study of four-jet events (central jets) at the ISR energies is tricky because the minimum energy per jet becomes quite low (~4 GeV). At such energies jets are neither clearly seen nor well measured, and the analysis necessarily relies on the Monte Carlo simulation as discussed in the introduction. In this case (as for the three jet study) the Monte Carlo program has been carefully checked to reproduce all the main features of the events both inside and outside the jet regions.

In the four jet analysis the pairwise transverse momentum balance is studied with the aim of distinguishing between double parton scattering DPS (i.e. two independent scatters) and double bremsstrahlung DBS. A total of $\Sigma E_T > 25$ GeV is required and where four jets, each with more than 4 GeV, are found we boost to the rest system of the four jets and define an "imbalance variable", $I_b = p_T^2(\text{min})$ where $p_T(\text{min})$ is the vector transverse momentum of a pair, taking the minimum of

319

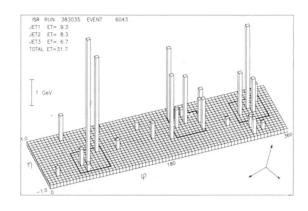

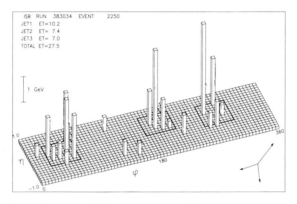

Fig.4 Two examples of star 3-jet events. The transverse energies of the clusters (particles) are represented by towers on a pseudorapidity-azimuth grid with cells $\Delta\phi \times \Delta\eta = 5° \times 0.10$. The windows defining the jets are indicated. The insets give the jet configuration in the transverse plane.

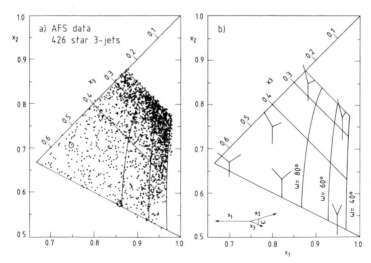

Fig.5 (a) and (b) 'Dalitz plot'; x_2 versus x_1 for 3-jet events in the transverse plane. (a) AFS experimental data. The circles indicate the two events shown in Fig.4. (b) The geometry of 3-jet events in the 'Dalitz plot'. The inset gives the definition of the inter-jet angle ω. Lines of constant x_3 or ω are shown. The small diagrams are examples of the momentum vectors of the jets in 3-jet events.

320

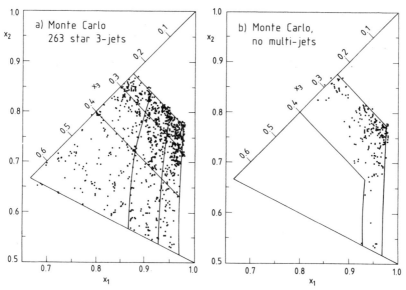

Fig.6 (a) and (b). 'Dalitz plot' for Monte Carlo events. (a) The event generator produces both 2-jet and multi-jet events, which are passed through the detector simulation. (b) The event generator produces 2-jet events only. The region of star 3-jet events is indicated. The Monte Carlo data have been normalized to the same number of 2-jet events as the experimental data in Fig.2a.

the three possible combinations. For independent pairwise balanced DPS events I_b will be small, while for DBS events it should have a wider distribution. Fig 9a,b shows two typical events. The imbalance distribution, Fig 10, shows a strong excess of events with small imbalance compared to the bremsstrahlung Monte Carlo when $E_T(jets)$ is low (25 to 28 GeV). However, when $\Sigma E_T(jets)$ is larger than 31 GeV the shape is as expected from bremsstrahlung. This is interpreted as a DPS component dying away when ΣE_T is increased.

The relative importance of DPS and DBS depends on the ratio of the structure functions at different x-values, and turns out to lead to a strong s-dependence and E_T-dependence. This is illustrated by Fig 11, where the ratio of four-jet production to two-jet production is shown as a function of $\Sigma E_T(jets)$ (at different values of s) for events where each jet has more than 1/6 of $\Sigma E_T(jets)$, and where all jets are separated with more than 60°. The dashed curves show the ratio for bremsstrahlung processes. The overall DBS 4-jet to 2-jet ratio will have a weak s-dependence because of the different abundance of gluons vs quarks at different parton x. The expected s- and ΣE_T dependence is, however, much stronger for double parton scattering (solid lines), and in this calculation a turnover at $\Sigma E_T \sim 31$ GeV seems reproduced.

The calculation uses a particular typical transverse size of the parton distribution in the proton as a parameter (i.e. the more tightly packed the partons are the more DPS). The reason for the strong energy dependence at fixed $\Sigma E_T(jets)$ is that it turns out to be more economical to pick out two pairs of partons at one half the x-value needed for a single scatter. In the AFS analysis we find a typical transverse size of ~5mb, corresponding to a Gaussian distribution of dn $\propto \exp(-R^2/2D^2)d^3R$ where D = 0.14 fm or to a proton radius of 0.3 fm. It should be stressed that this rather low value corresponds to contituents having a high x (>0.2) and that a larger radius seen at lower x is in no contradiction. The picture would be of a hard tightly packed core of valence quarks with a more fuzzy cloud of partons around!

It is interesting to remark that a small dimension of the valence quark system naturally leads to a high correspond-

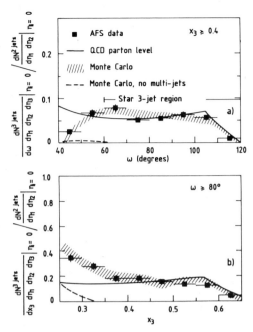

Fig.7(a) Distribution of the angle ω in the transverse plane between the third and the second jet in 3-jet events with $x_3 \geq 0.40$. (b) Distribution of the fractional momentum x_3 of the third jet in 3-jet events with $\omega \geq 80°$. The data are corrected for trigger and filter inefficiencies and for the finite size of the rapidity interval for jets. The solid lines 'QCD parton level' represent the lowest-order QCD matrix elements. The shaded bands represent the Monte Carlo model, i.e. including fragmentation effects and detector resolution. The dashed lines represent the Monte Carlo model with multi-jet events excluded from the event generator. The star 3-jet region is indicated in (a).

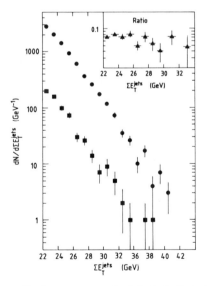

Fig.8 The distribution of the uncorrected total 2-jet E_T and total 3-jet E_T for 2-jet and star 3-jet events. The inset gives the ratio of the two spectra. These spectra are uncorrected for instrumental effects and imperfections of the jet algorithm, so the variation of the ratio with E_T cannot be interpreted directly as the variation of the strong coupling α_s.

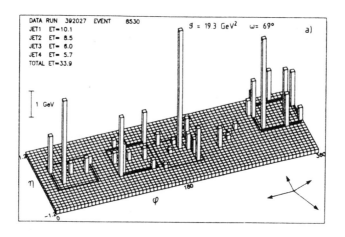

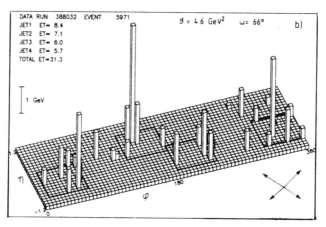

Fig.9(a),(b) Two examples of 4-jet events. The imbalance g (see text and Fig.3) and the smallest interjet angle ω are given. The insets give the momenta of the jets in the transverse plane. Events a and b are likely to be due to DBS and DPS, respectively. Notice that only a little E_T (typically 3-4GeV) falls outside the indicated jet windows, although these cover only 1/3 of the total acceptance region.

324

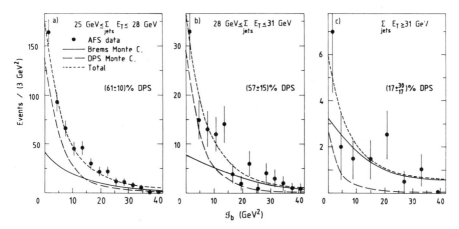

Fig.10(a-c) The distribution of the 'boosted imbalance' g_b (see text) in three different intervals of the 4-jet E_T; $E_T^4 \geq 4$GeV and $\omega \geq 55°$ are required.

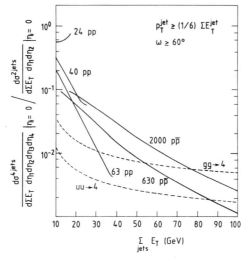

Fig.11 The 4-jet cross-section relative to the 2-jet cross-section. The DPS cross-section (solid lines) for pp or $p\bar{p}$ collisions at various values of $\sqrt{s}$ (in GeV) are expressed for jets produced at 90° to the beam. Each parton in the DPS events is required to have at least 1/6 of the total jet E_T. The scale is set to 5 mb to agree with our measurement. The expected DBS cross-section with the same cuts is shown as dashed lines.

325

ing value of the so-called "intrinsic k_T" in Drell-Yan production. Our value of D = 0.14 fm thus leads to $\langle q_T \rangle \sim 1.2$ GeV/c.

According to this analysis (and Fig 11) double parton scattering should be very abundant at $\sqrt{s} \sim 24$ GeV; this might explain why experiments in that energy regime (e.g. NA5) using full azimuth large E_T triggers have had such a hard time.

5.   Conclusion

Since I am running out of jets, I had better conclude by thanking the organizers of this Workshop for giving me an opportunity to visit my "old home land".

References

1)   T.Åkesson et al: Nucl Instrum Methods 241 (1985) 17

2)   T.Åkesson et al: Z. Phys C.   30 (1986) 27

3)   T.Åkesson et al: Z. Phys C.   32 (1987) 317

4)   T.Åkesson et al: Z. Phys C.   34 (1987) 163

5)   T.Åkesson et al: Phys Lett   128B (1983) 354

6)   F.Paige, S.Protopopescu: ISAJET.   BNL 31987 (1982)

7)   D.Duke & J.Owens: Phys Rev   D30 (1984) 49

MEASUREMENTS ON W AND Z PRODUCTION AND DECAY FROM THE UA2 EXPERIMENT AT THE CERN p̄p COLLIDER

THE UA2 COLLABORATION
BERN-CERN-COPENHAGEN (NBI)-HEIDELBERG-ORSAY (LAL)-PAVIA-PERUGIA-PISA-SACLAY (CEN)

Presented by Allan G. Clark[*]

ABSTRACT

A summary is presented of results from the UA2 experiment on the production and decay of the W and Z bosons. A comparison is made with expectations of the standard electroweak model, and the QCD model for strong interactions. Experimental uncertainties relevant to past and future measurements in the UA2 detector are briefly discussed.

1. INTRODUCTION

The UA2 detector at the CERN p̄p Collider collected data over a period of four years from 1981; the total integrated luminosity corresponded to 142 nb^{-1} at $\sqrt{s}$=546 GeV and 768 nb^{-1} at $\sqrt{s}$=630 GeV. In this talk we summarize aspects of these data relevant to the production of W and Z bosons in p̄p collisions, and their subsequent decay properties. These data have been published [1-4] and this summary serves as a 'route map' highlighting recent results. In section 2, we note the existing W and Z event samples, and then in sections 3 and 4 discuss the data in the context of existing models. An important recent result from the UA2 experiment is the observation of decays W/Z $\rightarrow$ q̄q and we present this in section 5. In section 6, the ratio of the W and Z cross-sections is used to place a limit on the number of light neutrino species. In section 7, limits are placed on the rate of selected exotic W and Z decays, and on the production cross-section for additional W and Z bosons. Finally, the expected accuracy of future measurements is discussed in section 8.

2. W AND Z EVENT SAMPLES

As described elsewhere [5-6] the UA2 apparatus allows the identification and energy measurement of electrons and hadrons (or hadronic jets) as in Table 1. The existence of noninteracting particles (for example ν_e in the decay W $\rightarrow$ eν_e) can also be inferred.

---

[*] Present address: Fermi National Accelerator Laboratory, P. O. Box 500, Batavia, Illinois, 60510.

TABLE 1.

Acceptance of the UA2 Detector

Polar Angle θ°	Azimuthal angle ϕ°	Particles identified	Particles with measured energy
$40^\circ < \theta < 140^\circ$	$0 < \phi < 360^\circ$	Electrons	Electrons
		Hadrons	Hadrons
$20^\circ < \theta < 37.5^\circ$	$0 < \phi < 360^\circ$	Electrons	Electrons
$142.5^\circ < \theta < 160^\circ$		Hadrons	

Electrons and hadrons are identified as clusters of energy deposition in the highly segmented calorimeters of the detector [5-6]. In the case of electrons, the pattern of deposition must be consistent with the expectation for electrons, and must furthermore match spatially with a charged track. Further identification power is obtainable from the signal of a preshower detector preceding the calorimeter. Of particular relevance for W and Z analyses is the accuracy of the measured energy deposition in the calorimeter. The response of the calorimeter to electromagnetic and hadronic showers has been extensively studied at the CERN PS and SPS machines using electron, muon and hadron beams of momenta in the range 1 to 70 GeV. Continuous monitoring of the calibration is performed using radioactive sources and calibrated light flashers. Electromagnetic showers are measured with an energy resolution of $0.14/\sqrt{E}$(E in GeV), whereas the energy resolution for single pions varies from 32% at 1 GeV to 11% at 70 GeV, approximately like $E^{-1/4}$. The systematic uncertainty on the energy calibration is estimated to be $\pm1.6\%$ for the central electromagnetic calorimeter and less than $\pm6\%$ for the hadronic one. They combine to result in a global systematic error of less than $\pm4\%$ on the jet energy scale. For the forward calorimeter the absolute energy scale is known to within $\pm2.5\%$. In typical $\bar{p}p$ interactions a large fraction of the total collision energy is carried by particles at small angle that remain undetected. Consequently, only the component, $\vec{p}_T^{\,m}$, of missing momentum transverse to the beam axis can be reliably measured. Depending on the interpreted physics process, $\vec{p}_T^{\,m}$ may be ascribed to one or more neutrinos, or to some other non-interacting particle. In particular, decays $W \rightarrow e\nu_e$ are characterized by an identified charged electron with significant missing energy $(|\vec{p}_T^{\,m}|$ or $|\vec{p}_T^{\,\nu}|)$ at approximately opposite azimuth. The UA2 detector is not equipped to identify muons and furthermore there is no detection at angles $\theta < 20^\circ$ to the beam, and only a poor measurement of hadron energies in the region $20^\circ < |\theta| < 40^\circ$. The accuracy of $|\vec{p}_T^{\,\nu}|$ is therefore deteriorated, and more importantly has non-gaussian tails.

Figure 1 shows the p_T^e vs. p_T^ν distribution for all identified electrons satisfying $p_T^e > 11$ GeV. At large p_T^e values, signals from the W $(p_T^\nu \sim p_T^e)$ and Z $(p_T^\nu \sim 0)$ are visible, above a background of misidentified low p_T hadrons.

Two hardware triggers were used for electron studies at the UA2 detector. The W trigger selects electrons with a deposited transverse energy $E_T^e \gtrsim 11$ GeV. Electron pairs were selected from the Z trigger, which required two energy depositions in the calorimeter of transverse energy $E_T^e \gtrsim 6$ GeV and an azimuthal separation exceeding 60°. These triggers are constructed from

328

calorimeter signals, and so are dominated by hadronic final states. Nevertheless, their efficiency for electron detection is close to 100%. The region of low p_T^e is dominated by hadronic background and we restrict the W sample to electron candidates of $p_T^e > 20$ GeV. The resulting m_T^{ev} distribution is shown in Fig. 2a. Background contributions from misidentified hadrons or hadronic jets, from $W \to \tau v_\tau$ decays, and from $Z \to e^+e^-$ decays are superimposed. Of all the $W \to ev_e$ decays within the acceptance of the apparatus, 80% are expected to satisfy $p_T^e > 20$ GeV and $m_T^{ev} > 50$ GeV. These selections have been applied to the $W \to ev_e$ sample of 251 events. The p_T^e distribution of this sample is shown in Fig. 2b, again with superimposed background estimates.

In the distributions of Fig. 2b, one event contains an electron candidate of transverse momentum $p_T^e = 77.4$ GeV. The event has no associated jet activity, and a neutrino with $p_T^e \cong 80$ GeV is interpreted, opposite in azimuth to the electron. Background contributions to this event are negligible. However, events of large m_T^{ev} are expected from W exchange in the s-channel but off the mass shell, and we estimate that 0.07 such events should be observed with $p_T^e > 70$ GeV in the W sample.

Figure 1 includes an event containing an isolated electron candidate of $p_T^e = 90.1$ GeV. The electron is balanced by a hadronic jet of $E_T = 77 \pm 8$ GeV. A careful examination of the jet topology suggests the overlap of an electron having $p_T^e \cong 22$ GeV with a hadronic jet having $E_T \cong 52$ GeV. The natural interpretation of this event is therefore the production of a Z having transverse momentum $p_T^Z \cong 70$ GeV, with an associated jet that overlaps one of the electrons from $Z \to e^+e^-$ decay. This event does not pass the calorimetric selections of the Z analysis, and is excluded from the Z sample.

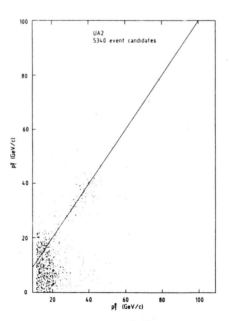

Figure 1. The distribution in the (p_T^e, p_T^v) plane of 5340 events containing at least one electron candidate of $p_T^e > 11$ GeV. If more than one electron candidate is identified, that of highest p_T is selected. The superimposed line represents $p_T^v = p_T^e$.

A similar analysis can be made for decays $Z^\circ \to e^+e^-$, but if the selection criteria used for the W analysis are used on both clusters of the $Z \to e^+e^-$ candidate, the detection efficiency is low (~50%). However, because of the increased rejection resulting from the requirement of two clusters of energy with lateral and longitudinal profiles consistent with those of isolated electrons, less stringent selection criteria can be applied while maintaining good Z identification efficiency and good rejection of hadronic background. Details of the analysis are listed in reference [1]. As shown in figure 3, a total of 39 events satisfy $m_{ee} > 76$ GeV and are attributed to Z production, with an estimated background of 1.3 events.

The process of internal bremsstrahlung can produce events $Z \to e^+e^-\gamma$; one such event is included in the final event sample. The probability of seeing at least one $(ee\gamma)$ event having an $(ee\gamma)$ configuration less likely than that observed in the sample of 39 Z decays is estimated to be $\cong 0.4$.

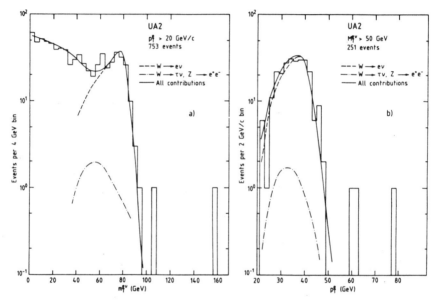

Figure 2. Transverse mass and transverse momentum spectra for the final event sample. The expectation for W decay is evaluated using $m_W = 80.2$ GeV. (a)The m_T^{ev} spectrum of 753 events satisfying $p_T^e > 20$ GeV. The expected signal from $W \to ev_e$ decay is superimposed, together with $W \to \tau v$ decays (5.2±0.5 events) and Z decays with one electron outside the UA2 acceptance (9.6±1.6 events). The solid line shows the total of all expected contributions to the $m_T^{ev} > 50$ GeV). (b) The distribution of p_T^e for 251 events satisfying $m_T^{ev} > 50$ GeV. The superimposed background contributions are as for (a).

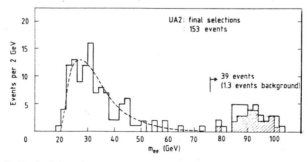

Figure 3. The distribution of m_{ee} in events for which at least two electron candidates are identified. The contribution of QCD background processes is superimposed (----). The hatched region shows the m_{ee} distribution of the sample of 25 events used in the evaluation of the Z° mass.

The data used for studies of the $W \to q\bar{q}$ and $Z \to q\bar{q}$ decay modes were recorded by requiring large transverse energy depositions in opposite azimuthal calorimeter wedges. The threshold for each transverse energy deposition was initially set at $\cong 20$ GeV (for an integrated luminosity of 0.31pb^{-1}). Later, thresholds of 15 GeV (0.27pb^{-1}) and 12.5 GeV (0.15pb^{-1}) were used. The subsequent analysis is described in Reference [4] and at this conference [7], and results are summarized in section 5.

Finally, for specific studies of low-p_T electrons and low-mass electron pairs (see section 8), more stringent electron criteria were used to minimize the hadronic background, at the expense of the electron detection efficiency. Details are given in Reference [3].

3. PRODUCTION OF W AND Z BOSONS

The measured cross-section for W and Z production followed by their decays $W \to ev_e$ and $Z \to e^+e^-$ respectively, is shown in Table 2. The experimental measurement uncertainty is dominated by systematics and in particular uncertainties of the efficiency and integrated luminosity measurement. However, theoretical uncertainties are almost as large because of uncertainties in the QCD evaluation, (for example the choice of momentum scale, the evolution of lower energy structure function measurements, etc.). Using the predicted total production cross-section [8], the leptonic branching fraction can be inferred. The results are shown in Table 2, and are compared with predictions of the Standard Model (assuming 3 lepton and quark doublets, and a top quark mass in the range 40 to 100 GeV). The need for better experimental measurements and improved theoretical evaluations is evident.

TABLE 2.

Cross-section Measurements

	Experiment	Prediction [8]
σ_Z^e(nb) $\sqrt{s} = 546$ GeV	$.116 \pm .039 \pm .011$	
σ_Z^e(nb) $\sqrt{s} = 630$ GeV	$.073 \pm .014 \pm .007$	
$B(Z \to ee)$	$.046 \pm .008(\text{exp}) ^{+.008}_{-.004}(\text{th})$	$.031 - .035$
$R = \sigma_Z^e(630)/\sigma z^e(546)$	0.6 ± 0.3	1.25
σ_w^e(nb) $\sqrt{s}=546$GeV	$.61 \pm .10 \pm .07$	
σ_w^e(nb) $\sqrt{s}=630$GeV	$.57 \pm .04 \pm .07$	
$B(W \to ev_e)$	$.10 \pm .01(\text{exp}) ^{+.02}_{-.03}(\text{th})$	$.086 - 0.11$
$R = \sigma_w^e(630)/\sigma_w^e(546)$	$.93 \pm .17$	1.23

At CERN energies, the partons involved in the W and Z production are dominantly valence quarks; therefore, a measurement of the fractional longitudinal momentum x_w (or x_Z) provides a test of leading order QCD. The experimental distributions dn/dx_w and dn/dx_Z are compared with QCD expectations in Figure 4. The agreement is excellent.

A comparison of the transverse momentum carried by the W and Z bosons with expectations of QCD evaluations, provides a sensitive test of higher order QCD processes. Experimentally, p_T^w cannot be accurately measured and the most accurate measurement is that of p_T^Z where both decay

331

electrons are directly measured. The measurement error of ~±2GeV is mainly the result of uncertainties of the energy measurement. Consequently the distribution of p_η^Z (the component of p_T^Z normal to the e^+e^- bisector) has an improved accuracy (~±0.23GeV) and can be compared directly with theory as in Figure 5. The agreement is excellent.

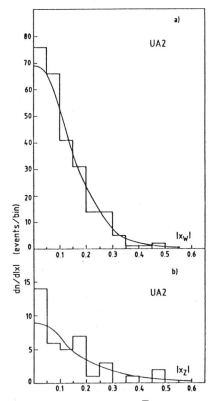

Figure 4. Distributions of the variable $y = 2p_L/\sqrt{s}$ for W production (a) and Z production (b). The curves are predictions using the structure functions of Ref. [9], with n=0.2 GeV.

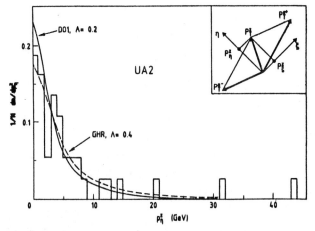

Figure 5. The distribution of the η-component of the Z transverse momentum compared with theoretical predictions from Ref. [10] using two different sets of structure functions [9,11].

Real non-leading QCD contributions generate final states involving at least one quark or gluon jet, at a rate determined by the strong coupling constant α_S. In the UA2 analysis, associated jets of transverse energy exceeding 10 GeV are compared with QCD expectations; the W sample of 251 events contains 29 events with one associated jet and one event with two associated jets. The transverse jet energy distribution is shown in Figure 6, and is compared with pertubative QCD calculations [12] assuming $\alpha_S=0.15$. The agreement is excellent. Two events of Figure 5 having very large E_T values are discussed in detail elsewhere [13]. There is no evidence of deviations from QCD expectations. The Z sample contains 4 events with one jet, and no event with 2 or more jets; again, the agreement with QCD expectations is good.

As shown in Figure 7, there is no evidence of differences between the transverse energy and summed (vector) transverse momentum distributions of 'spectator' hadrons associated with W and Z production. Superimposed is the distribution obtained when minimum bias events are selected, and all momenta are 'inflated' by a factor 1.45 before a transverse boost according to the theoretical p_T^W and p_T^Z distributions. The agreement is impressive, and suggests at least qualitatively a larger 'spectator' event activity for W and Z production.

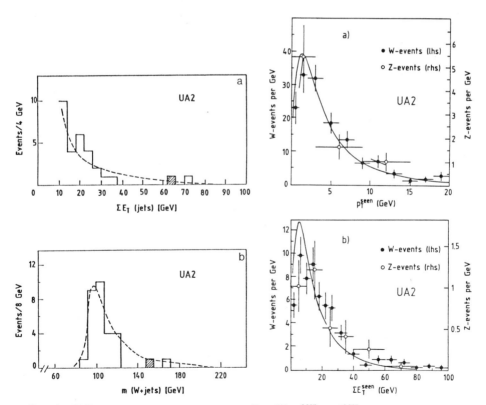

Figure 6. (a) Transverse energy distribution of W-associated hadronic jets. The dotted line is obtained using the QCD model of Ref. [12]. The shaded event is the only one containing two jets. (b) Invariant mass distribution of the W + jet(s) system. The shaded event is the only one containing two jets.

Figure 7. The p_T^{seen} and ΣE_T^{seen} distributions of hadrons observed in association with W-bosons (black circles) and Z-bosons (open circles) are compared with:. (a) The transverse momentum distribution of 'inflated' minimum bias events (see text) boosted according to the W-transverse momentum distribution [8, 10], (b) The ΣE_T distribution of 'inflated' minimum bias events.

4. THE STATUS OF STANDARD MODEL PARAMETERS

The distribution of transverse mass, m_T^{ev}, is used to evaluate the W mass. A Monte Carlo simulation which takes account of the expected W longitudinal and transverse motion, and expected detector measurement characteristics, is compared with the data. The result of the comparison is shown in Table 3. Using a fit to the m_T^{ev} distributions that includes the W width Γ_W as a free parameter, limits on Γ_W are also quoted in Table 3. Systematic uncertainties limit the accuracy of the data. These include theoretical uncertainties of the generated distributions and the dominant uncertainty of the calorimeter energy scale. Since in measurements of $\frac{m_W}{m_Z}$ the scale uncertainty nearly cancels, this error is quoted separately in Table 3.

The UA2 experiment has an essentially background-free sample of the decay $Z \rightarrow e^+e^-$. The mass enhancement of Figure 3 is fitted to a relativistic Breit-Wigner shape, distorted by the experimental mass resolution, and results from the fit as shown in Table 3.

A first estimate of the Z width, Γ_Z, can be obtained by a direct measurement of the width of the mass peak. However, given the limited statistics and a measurement accuracy of the same order as the width itself, the determination of Γ_Z depends critically on a precise knowledge of the measurement accuracy. Resultant limits on Γ_Z are given in Table 3.

Ignoring the Fermion and Higgs scalar masses, and the elements of the Kobayashi-Maskawa matrix [14], the minimal Standard Model is characterized by three parameters, taken here to be α_{EM} (the fine structure constant), and the masses m_W and m_Z. To compare the UA2 measurements with predictions from the Standard Model, suitably renormalized and radiatively corrected theoretical quantities must be used [15]. The definition for which

$$\sin^2\theta_W = 1 - [\frac{m_W}{m_Z}]^2$$

is used [16], leading to the following predictions:

$$m_W^2 = A^2/[(1-\Delta r)\sin^2\theta_W]$$
$$m_Z^2 = 4A^2/[(1-\Delta r)\sin^2 2\theta_W]$$

TABLE 3.

Measurements of the W and Z mass and width

m_W (GeV)	$80.2 \pm 0.6(\text{stat})(\pm 0.5(\text{syst})^1 \pm 1.3(\text{syst})^2$
Γ_W (GeV) (90%CL)	< 7.0
m_Z (GeV)	$91.5 \pm 1.2(\text{stat}) \pm 1.7(\text{syst})$
Γ_Z (GeV) (90%CL)	< 5.6

$(\text{syst})^1$: Systematic error resulting from uncertainties in Γ_W (± 0.2GeV) and in the measurement of p_T^v (± 0.3 LGeV), and cell-to-cell calibration uncertainties of ± 0.3 GeV.

$(\text{syst})^2$: Systematic error resulting from energy scale uncertainty.

where $A = [\frac{\pi\alpha_{EM}}{\sqrt{2}G_F}]^{1/2} = (37.2810 \pm 0.0003)$ GeV using the measured values of α_{EM} and G_F. In the above equations, the value of Δr reflects the effect of one-loop radiative corrections on the W and Z masses and has been computed to be [16]

$$\Delta r = .0711 \pm .0013$$

assuming $m_t = 45$ GeV and a Higgs boson mass, $M_H = 100$ GeV. Although the quoted theoretical error of Δr is small, Δr can be significantly changed [16, 17] in the case of, for example, a very heavy t-quark ($\Delta r \sim 0$ for $m_t = 240$ GeV), or if a new fermion family exists with a large mass splitting between the two members of an SU(2) doublet.

From the equations above, and assuming the theoretical value of Δr, a model-dependent estimate of $\sin^2\theta_W$ can be evaluated. The results are shown in Table 4 and are in agreement with a recent fit to existing data for deep inelastic scattering [18], which gives $\sin^2\theta_W=.233 \pm .006$ with radiative corrections appropriate to deep inelastic scattering.).

In the above discussion, the ρ parameter [19]

$$\rho = m_W^2/(m_Z^2\cos^2\theta_W)$$

is defined to be $\rho = 1$. An estimate of ρ is obtainable and the results listed in Table 4 are compatible with the expectation $\rho = 1$ of the minimal Standard Model with scalar Higgs particles.

A measurement of the radiative correction Δr is also possible from these data. Eliminating $\sin^2\theta_W$ from the equations above, the results (called Δr in Table 4) are obtained. Alternatively, by using low-energy measurements of $\sin^2\theta_W$ and the measured values of m_W and m_Z, the results $\Delta r^{(a)}$ of Table 4 are obtained. Within the present statistical and systematic limits, the presence of radiative corrections cannot be demonstrated. This conclusion is very evident from Figure 8 where the 68% confidence level contours of (m_Z-m_W) vs. m_Z are compared with the Standard Model predictions and the range allowed by the low-energy data. Moreover, the contribution of the interesting weak radiative contribution to Δr (for example from fermion loops or Higgs corrections) is small («1%) for values of $m_{top} \leq 100$ GeV, and is beyond the capability of present Collider measurements.

TABLE 4.

Measurements of the Standard Model Parameters

Parameter	Measurement
$\sin^2\theta_W = 1 - m_W^2/m_Z^2$	$0.232 \pm 0.025 \pm 0.010$
$\sin^2\theta_W = A^2/(1-\Delta r)m_W^2$	$0.232 \pm 0.003 \pm 0.008$
ρ	$1.001 \pm 0.028 \pm 0.006$
Δr	$0.068 \pm 0.087 \pm 0.030$
$\Delta r^{(a)}$	$0.068 \pm 0.022 \pm 0.032$

(a)The first value of Δr is extracted using only collider data, whilst the second uses the best value of $\sin^2\theta_W$ as input.

5. HADRONIC DECAYS OF W AND Z BOSONS

A first indication of the validity of the Standard Model expression for the coupling between quarks and W or Z comes from the measurement of the W and Z production cross-sections at the CERN Collider, in good agreement with the QCD predictions. Another consequence of the Standard Model is that the W and Z bosons should decay into quark-antiquark pairs with probabilities

$$\frac{\Gamma(W \to q\bar{q})}{\Gamma(W \to ev_e)} \cong 6 \quad \text{and} \quad \frac{\Gamma(Z \to q\bar{q})}{\Gamma(Z \to ee)} \cong 20.$$

UA2 data also show direct evidence for this decay channel, the signal appearing in the jet-jet invariant mass spectrum around the W and Z mass values as a small excess of events over the copious QCD background. This result demonstrates that jets can be identified with their parent partons and may under some circumstances be used for jet 'spectroscopy'. These data have been treated in some detail at this conference [7], and only the final result is presented here.

Because of the measured mass-resolution of the detector near the W and Z mass ($\Delta m = 7.9 \pm 0.5$ GeV), and because of the small mass difference, the two mass peaks cannot be separated and the event excess is expected in a wide mass range, $65 < m < 105$ GeV. A signal is defined as an excess of events in this region, above the background which is obtained by interpolating between continuum samples outside the same region. The shape of the background regions is parametrised in several different ways, and all give a good description of the data. Typically, χ^2 values of 20 with 12 degrees of freedom are obtained.

Figure 9 shows the mass distribution of the events which pass all the cuts. A structure is visible in the region of the W and Z masses. To evaluate the number of events, the data are fitted with a superposition of a signal shape and of the interpolated background shape. The

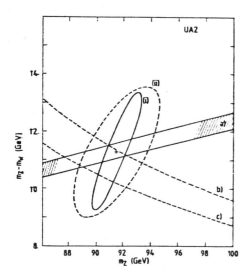

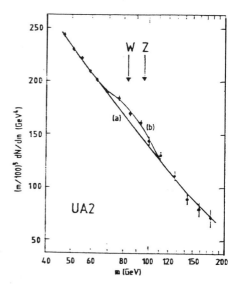

Figure 8. Confidence contours (68% level) in the (m_Z - m_W, m_Z) plane taking into account the statistical error only (i), and with systematic errors added in quadrature(ii). The region (a) is allowed by the average of recent low-energy measurements. Curve (b) is the Standard Model prediction for $\rho = 1$ with known radiative corrections, and curve (c) is the expectation without radiative corrections.

Figure 9. The jet-pair mass distribution of events passing the selection criteria. The smooth curves are the results of the best fits to the strong interaction background alone (curve a) or including two gaussians describing W,Z decays (curve b).

336

signal shape is the sum of two gaussians, representing the W and Z peaks; the W mass is a free parameter, whereas the ratio m_Z/m_W is fixed. From the fit we find 686 ± 210 events, with $m_W = 82 \pm 3$ GeV. The Monte Carlo expectation is 340 ± 80 events. The result is insensitive to the parametrization of the background. No systematic effect capable of creating the observed signal has been identified.

The evidence for a signal at the level of 3 standard deviations is in reasonable agreement with the Standard Model expectations for W and Z decaying into two quark jets. Continuing evidence and a quantitative measurement of the branching ratio awaits the collection of a significantly larger data sample using the UA2' detector.

6. NEUTRINO LIMITS

The expected value of the Z width is sensitive to the input parameters of the Standard Model, and in particular on the number of 'light neutrinos' ($m_v < m_Z/2$). The full Z width is given by

$$\Gamma_Z = \Sigma\Gamma_Z(l\bar{l}) + \Sigma\Gamma_Z(q\bar{q}) + N_\upsilon\Gamma_Z(\nu\bar{\nu})$$

where the first sum is over all leptons with masses of less than $m_Z/2$, the second term over all quark flavours, and the third term over all 'light neutrinos'. It is assumed that the charged members of any new families are too heavy to contribute. Unfortunately, given the present statistics and mass resolution, a direct measurement of Γ_Z is not sufficiently accurate. Instead, the UA2 experiment measures the ratio [20]

$$\sigma_W^e/\sigma_Z^e = [\sigma(\bar{p}p \to W + X)/\sigma(\bar{p}p \to Z + X)][\Gamma_W(e\nu)/\Gamma_Z(ee)]\Gamma_Z/\Gamma_W = R_{exp}.$$

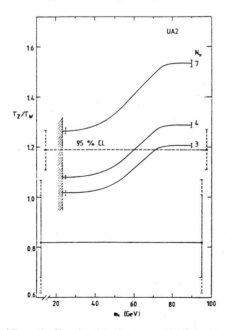

Figure 10. The value of Γ_Z/Γ_W measured by UA2, with superimposed statistical and theoretical uncertainties (full and dashed lines respectively). The 95% confidence limit is also shown (----). The expected variation with m_t is shown as a solid line for 3, 4, and 7 types of light neutrino. The superimposed error bars represent uncertainties of the theoretical evaluation. The shaded region marks the limit on m_t as measured at PETRA.

This quantity is well measured, since the errors on luminosity and efficiencies almost completely cancel. The extraction of Γ_Z/Γ_W from this measurement requires that the Standard Model couplings are assumed. Then the total cross-section ratio can be calculated, with an error resulting from the uncertainty on the input structure functions [21]. This error does not cancel because the u- and d-quark structure functions enter differently for Z and W production. Both the cross-section ratio and the ratio of the leptonic widths depend on $\sin^2\theta_W$, but the product of the two is insensitive to the value chosen.

The UA2 experiment then measures $R_{exp} = 7.2 \, ^{+1.7}_{-1.2}$, and therefore,
$$\Gamma_Z/\Gamma_W = .82 \, ^{+0.19}_{-0.14} \pm 0.6 \text{ (theory)} < 1.19 \pm 0.8 \text{ (95\%cl)},$$
where for Γ_Z/Γ_W, statistical uncertainties and the error due to structure functions uncertainties are evaluated separately.

Even after the assumptions given above, the ratio is affected by the existence of a t-quark, since no contribution from $Z \to t\bar{t}$ occurs for $m_t < m_Z/2$, while the process $W \to t\bar{b}$ can occur for $m_t > (m_W - m_b)$ [22]. The expected variation of Γ_Z/Γ_W as a function of the t-quark mass is shown in Figure 10 for various assumed numbers of light neutrino species. The uncertainty of the predictions results from the error on $\sin^2\theta_W$.

The UA2 experiment measures a limit at 95% cl of $n_\nu < 7$ independently of m_t, and $n_\nu < 3$ if $m_t > 74$ GeV. Contingent on model limitations, these values will be significantly improved when the Z sample is increased, and when the top quark mass is known.

7. DEVIATIONS FROM THE STANDARD MODEL

A study has recently been published [3] giving limits for various hypothetical processes. In this section we briefly summarize the processes relevant to deviations from the minimal Standard Model.

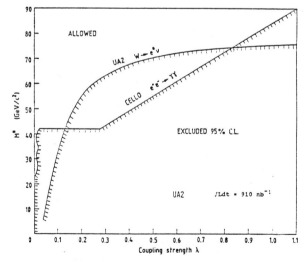

Figure 11. Limits on excited electrons shown as 95% confidence level curves in the λ - M* plane, where λ is the coupling strength and M* the excited electron mass. Also shown are recent limits from e^+e^- experiments.

Excited lepton states are expected in models where leptons are composite particles [23]. The UA2 experiment searched for electrons through the decay sequence $W \to e^* \nu_e \to e \gamma \nu_e$, using the inclusive sample of 5340 events containing at least one electron candidate shown in Figure 1. Only 24 events contain an additional photon candidate with transverse momentum above 10 GeV. These events are compatible with the expected background from jet pairs, where one jet fakes an electron and the other jet fakes a photon. In addition, if we require that the missing transverse momentum be larger than $(M_W - M^*)/4$ (M^* the invariant electron-photon mass), none of these events survives, whereas more than 95% of the events originating from the above process should survive (the exact number depending upon the mass of the excited electron). A Monte Carlo simulation of this process was performed in the UA2 detector, assuming that the W couples to the transition magnetic moment of the e^*-ν_e current. Taking into account the detector acceptance and energy resolution and in particular the experimental requirement of a 30° opening angle between the electron and photon, we obtain a 95% confidence limit as shown in Figure 11.

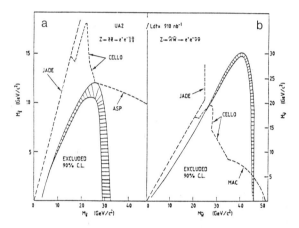

Figure 12. (a)Limits on the $\tilde{e}$ and $\tilde{\gamma}$ masses shown as 90% confidence level contours in the case of a stable photino. The dashed area illustrates the theoretical uncertainty on Drell-Yan production arising from higher-order terms. Also shown (dashed curves) are the recent limits from e^+e^- experiments. (b)Same as a) for the $\tilde{W}$ and $\tilde{\nu}$ masses, with the assumptions quoted in the text.

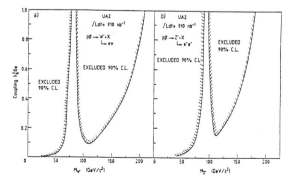

Figure 13. (a)Limits on an additional charged vector boson W', obtained from an analysis of single electron candidates and shown as 90% confidence level contours in $M_{W'}$ - $\lambda_q^2 B_e$ space. (b)Same as a) for an additional neutral vector boson Z' from an analysis of electron pairs.

TABLE 5.

Accuracy of Standard Model parameters in the improved UA2 detector

Quantity	Value	Expected accuracy (fraction)	
		Statistical	Systematic
σ_W^e	570 pb	~0.028	~0.085
σ_Z^e	73 pb	~0.071	~0.085
$\dfrac{\sigma_Z^e}{\sigma_W^e}$	0.13	~0.11	small
m_W	80.2	~0.002	~0.003 (analysis)
			~0.01 (energy)
m_Z	91.5	~0.004	~0.01
m_W/m_Z	~0.88	~0.004	≤ 0.002

Derived Quantities	Value	Approximate expected accuracy (absolute)	
		Statistical	Systematic
$\sin^2\theta_W$ [a]	0.232	~0 006	~0 .004
$\sin^2\theta_W$ [b]	0.232	~0.001	~0.006
Δr	0.068	~0.025	~0.025
Δr a	0.068	~0.017	~0.02
ρ	1.001	~0.01	~0.003

a - b : see Table 3

The hypothetical processes, [$\bar{p}p \rightarrow Z + X; Z \rightarrow \tilde{e}\tilde{e}; \tilde{e} \rightarrow e\tilde{\gamma}$] where the $\tilde{\gamma}$ escapes detection (stable photino), and [$\bar{p}p \rightarrow Z + X; Z \rightarrow \tilde{W}\tilde{W}; \tilde{W} \rightarrow e\tilde{\gamma}_e$] where the $\tilde{v}$ escapes detection (a stable $\tilde{v}$ or $\tilde{v} \rightarrow v\tilde{\gamma}$), are studied. The final states are characterized by large -p_T electrons pairs. The limits, obtained taking into account the expected contributions from background and from the Drell-Yan continuum, are shown in Figure 12 as 90% confidence level contours. The limits of Figure 12 are valid for a stable photino and assume $\tilde{e}_L$ and $\tilde{e}_R$ to be degenerate in mass. The limits of Figure 12a assume that the branching fraction for $\tilde{W} \rightarrow e\tilde{\gamma}$ is 33%. The hatched areas show how the limits vary if the expected Drell- Yan contribution is increased by 50%.

If any additional vector bosons [24] exist, limits for their production can be estimated as a function of their mass, $M_{W',Z'}$, their coupling to quarks, λ_q, and their branching ratio to electrons, B_e (λ_q and B_e are normalized to the Standard Model values). The cross-section, σ, for the production of an additional vector boson and its subsequent electronic decay, is $\sigma = \sigma_o\lambda_q^2 B_e$ (σ_o being the cross-section for standard couplings). Assuming $\Gamma_{W',Z'} = \Gamma_{W,Z} M_{W',Z}$, Figure 13 shows the 90% confidence level contours in M_W—$\lambda_q^2 B_e$ and M_Z—$\lambda_q^2 B_e$ space. Additional W' bosons of mass less than 209 GeV and additional Z' bosons of mass less than 180 GeV are excluded if their couplings to quarks and leptons are the same as for the standard vector bosons, unless the mass is within ± 5 GeV of the standard W or Z value.

8. FUTURE COLLIDER MEASUREMENTS

Results collected from the $\bar{p}p$ Collider during the past four years have exceeded all expectations. In this period the UA2 Collaboration identified some limitations of the existing detector design, and has now completed modifications to their detector to take advantage of corresponding improvements to the CERN Collider.

From results of this presentation, one can conclude that for measurements related to the Standard Model:

i.   The energy-scale uncertainty of calorimeter calibrations was a major limitation. In the upgraded UA2 detector, an energy-scale calibration uncertainty of ±1% is anticipated.

ii.   The p_T^ν- resolution of the UA2 detector was inadequate. It has been improved by adding full calorimetric coverage to within ~5° of the beam line. The accuracy of Standard Model measurements will be largely unaffected by this change, but the analysis will be simplified, background contributions to the data sample will be reduced, and possible deviations from the Standard Model will be more accessible.

iii.   The rejection power of the electron identification criteria to hadronic background was marginal in the UA2 detector for low cross-section processes such as

$$\bar{p}p \rightarrow W + X; W \rightarrow t + b; t \rightarrow e\nu_e b.$$

The construction of a new central tracking detector, with a transition-radiation detector to reject overlaps of $\pi°$'s with charged particles and an array of silicon counters to reject $\pi°$ conversions, is expected to improve the rejection of background in the UA2 detector by at least one order of magnitude.

Given these improvements, the expected accuracy of some Standard Model and cross-section parameters is listed in Table 10, assuming an integrated luminosity of 10 pb^{-1}.

The following comments are appropriate.

i.   The present (σ_Z/σ_W) measurement is limited by statistics; an improved measurement will place a better constraint on the number of additional neutrinos.

ii.   Given the accurate data on the decay $Z \rightarrow e^+e^-$ expected from future e^+e^- machines, an accurate measurement of the ratio (m_W/m_Z) is of importance.

iii.   Using Collider measurements alone, expected deviations $\Delta r \neq 0$ will be accessible at approximately the 3 s.d. level, but a measurement of the weak radiative contribution will remain inaccessible, at least if the top quark mass is ~50-100 GeV.

iv.   Measurements of $\sin^2\theta_W$ at the $\bar{p}p$ Collider will be much improved, but will remain limited by analysis systematics or by energy-scale uncertainties.

So far, there is no evidence for any deviation from the minimal Standard Model of electroweak or strong interactions. The thrust of future measurements will be to identify deviations if they exist.

Acknowledgements

Thanks are due to Professor B. Cox and his colleagues for the organization of a superb meeting. I wish to thank my colleagues in the UA2 Collaboration for their support since the start of $\overline{p}p$ Collider operation. The transformation of this script into legibility is due to L. Rauch, to whom I am grateful.

References

1. R. Ansari et al.,Phys. Lett. 186B:440 (1987).
2. R. Ansari et al., Phys. Lett. 194B:158 (1987).
3. R. Ansari et al., CERN-EP/87:117 (July 1987), *to be published* in Phys. Lett.
4. R. Ansari et al., Phys. Lett. 186B:452 (1987).
5. A. Beer et al., Nucl. Instr. Meth. 224:360 (1984).
6. B. Mansoulie, 1983, The UA2 Apparatus at the CERN $\overline{p}p$ Collider, in "Proc. 3rd Moriond Workshop on $\overline{p}p$ Physics, La Plagne," J. Tran Thanh Van, ed., Editions Frontieres, Paris.
7. UA2 Collaboration, presented by K-H. Meier, *these proceedings*.
8. G. Altarelli, et. al, Z. Phys. C-Particles and Fields 27:617 (1985).
9. D. W. Duke and J.F. Owens, Phys. Rev. D30:491 (1984).
10. G. Altarelli et al., Nucl. Phys. B246:12 (1984).
11. M. Gluck et al., Z. Phys. C-Particles and Fields 13:119 (1982).
12. S. D. Ellis, R. Kleiss and W.J. Sterling, Phys. Lett. 154B:435 (1985).
13. P. Bagnaia, et al., Phys. Lett. 139B:105 (1984)
    J. Appel et al., Z. Phys. C-Particles and Fields 30:1 (1986).
14. M. Kobayushi, et al., Prog. Theor. Phys. 49:652 (1973).
15. A. Sirlin, Phys. Rev. D22:971 (1980).
    W. Marciano, Phys. Rev. D20 :274(1979).
    M. Veltman, Phys. Lett. 91B:95 (1980).
16. W. Marciano and A. Sirlin, Phys. Rev. D29:945 (1984);
    F. Jerlehner, Bielefeld preprint BI-TP 1986/8 (1986);
    W. Hollick, DESY preprint DESY 86-049 (1986).
17. M. Veltman, Nucl. Phys. B123:89 (1977).
    Z. Hioki, Nucl. Phys. B229:284 (1983).
18. U. Amaldi et al., Phys. Rev. D26:1385 (1987).
19. A. Ross and M. Veltman, Nucl Phys. B95:135 (1975).
    P. Hung and J. Sakurai, Nucl Phys. B143:81 (1978).
20. F. Halzen and K. Murula, Phys. Rev. Lett. 51: 857 (1987)
    K. Hikasa, Phys. Rev. D29:1939 (1984).
21. N.G. Deshpande et al., Phys. Rev. Lett. 54:1757, (1985).
22. J.H. Kuhn et al., Nucl. Phys. B272:560 (1986).
    D. Yu Bardin et al., Z. Phys. C-Particles and Fields - 32:121 (1986).
23. H. Terazawa et al., Phys. Lett. 112B:387 (1982)
    A. de Rujula et al., Phys. Lett. 140B:253 (1984).
24. For example:
    P. Langacher et al., Phys. Rev. D30:1470 (1984).
    E. Cohen et al., Phys. Lett. 165B:76 (1985).
    D. London and J. Rosner, Phys. Rev. D34:1530 (1986).

THEORETICAL REVIEW OF JET PRODUCTION IN HADRONIC COLLISIONS

C.J. Maxwell

Department of Physics
University of Durham
Durham, England

ABSTRACT

Recent theoretical progress in several different areas of jet production in high-energy hadron collisions is reviewed and discussed.

1. Introduction

One of the great experimental successes of the CERN pp collider has been the observation of large-p_T hadronic jets with angular distributions and cross-sections in good agreement with the expectations of QCD hard scattering dynamics. QCD jets will provide a daunting background to many potentially interesting physics signals at future colliders - top production, Higgs, heavy W's and Z's ..., and so in addition to the inherent QCD tests provided by jet production, there is an added interest in studying and reliably predicting jet backgrounds.

In this review I shall first discuss the QCD predictions for two and three jet production, indicating that at tree-level the predictions have some surprising simplifying features. I will also briefly review some two and three jet experimental results from the CERN $\bar{p}p$ collider. These predictions and tests are all based on tree-level QCD calculations. Ellis and Sexton have now completed an $O(\alpha_s^3)$ calculation of the $2 \to 2$ subprocess and I shall briefly discuss this. I shall then outline the impressive recent theoretical progress in developing techniques for calculating multi-jet production, in particular duality properties and their use in calculating multi-gluon scattering. There are some related attempts to estimate $2 \to n$ scattering processes.

2. Two-jet Production

Figure 1 represents the standard hard-scattering ansatz. The cross-section for producing two jets at large p_T is factorized into a convolution of structure functions $f_{a|h}(x,Q^2)$, for finding parton constituent 'a' in hadron h with momentum fraction x, and a hard scattering cross-section $\hat{\sigma}_{ab \to cd}$ to be computed in perturbative QCD for the $2 \to 2$ parton-parton scattering subprocess ab $\to$ cd.

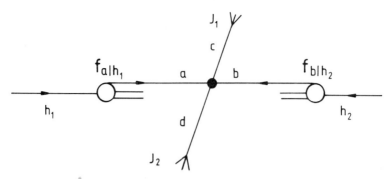

Figure 1.   Hard-scattering ansatz

$$\sigma_{2-j} = \sum_{a,b} \int dx_a \, dx_b \, f_{a|h_1}(x_a, Q^2) \, f_{b|h_2}(x_b, Q^2) \, \hat{\sigma}_{ab \to cd} \tag{1}$$

The sum is taken over all nine subprocesses, $gg \to gg$, $qg \to qg$ ... etc. The tree-level $\hat{\sigma}_{ab \to cd}$ were first correctly calculated by Combridge, Kripfganz and Rauft [1]. One might have expected that with nine subprocesses there would be many striking tests of QCD, with different subprocesses, having distinctive angular distributions, becoming dominant in different kinematical regions. Various features of tree-level QCD conspire, however, to make the result surprisingly simple. It is useful in understanding this to introduce a single kinematical variable [2]

$$\chi \equiv e^{y_1 - y_2} = \frac{1 + \cos\theta^*}{1 - \cos\theta^*} = \frac{\hat{u}}{\hat{t}} \tag{2}$$

y_1, y_2 are the rapidities of the jets, θ^* represents the c.m. scattering angle. In Table 1 we display $\frac{d\sigma}{d\hat{t}}$ (suitably normalized) for the nine subprocesses, expressed in terms of χ.

The first six subprocesses, in which the initial and final particles are the same, all have a χ^2 behaviour. The remaining three subprocesses go like χ. The presence of the χ^2 terms is traceable to $\hat{t}$-channel gluon exchange. This means that for χ large (θ^* small) the first six subprocesses dominate and have a common χ^2 angular dependence. Furthermore the coefficients of χ^2 are in geometrical progression,

$$qq \to qq : qg \to qg : gg \to gg = 1 : \frac{C_A}{C_F} : \left(\frac{C_A}{C_F}\right)^2$$

where C_A, C_F are group Casimirs and $\frac{C_A}{C_F} = \frac{9}{4}$ for SU(3). The dominance of $\hat{t}$-channel gluon exchange and the colour structure of QCD thus conspire to enable the cross-section to be written in the form

$$d\sigma_{2-j} \sim F(x_a, Q^2) \, F(x_b, Q^2) \, d\hat{\sigma}_{gg \to gg} \tag{3}$$

with $F(x, Q^2) \equiv g(x, Q^2) + \frac{4}{9}[q(x, Q^2) + \bar{q}(x, Q^2)]$ an effective structure

Table 1. $\dfrac{d\hat{\sigma}}{d\hat{t}}$ for the nine $2 \to 2$ subprocesses expressed in terms of χ.

Subprocess	$\dfrac{\hat{S}^2}{\pi\alpha_S^2}\ \dfrac{N_g}{4C_F^2}\ \dfrac{d\hat{\sigma}}{d\hat{t}}$
$\left.\begin{array}{l} q_1 q_2 \to q_1 q_2 \\ q_1 \bar{q}_2 \quad q_1 \bar{q}_2 \end{array}\right\}$	$\chi^2 + \chi + \chi^{-1} + \chi^{-2} + 1$
$q_1 q_1 \to q_1 q_1$	$\chi^2 + \chi + \chi^{-1} + \chi^{-2} - \dfrac{1}{N}(\chi+2+\chi^{-1})$
$q_1 q_1 \to q_1 \bar{q}_1$	$\chi^2 + \chi + \chi^{-1} + \chi^{-2} + \dfrac{1}{N}(\chi-1+\chi^{-1}) - \dfrac{\chi^2+1}{(\chi+1)^2}$
$qg \to qg$	$\dfrac{C_A}{C_F}\left[\chi^2 + \chi + \chi^{-1} + \chi^{-2} + \dfrac{C_F}{C_A}\left(\dfrac{1}{2}\chi + \dfrac{3}{2} + \dfrac{1}{2}\chi^{-1}\right)\right]$
$gg \to gg$	$\left(\dfrac{C_A}{C_F}\right)^2\left[\chi^2 + \chi + \chi^{-1} + \chi^{-2} + 2 - \dfrac{\chi}{(1+\chi)^2}\right]$
$q_1 \bar{q}_1 \to q_2 \bar{q}_2$	$\dfrac{(\chi^2+1)}{(\chi+1)^2}$
$q\bar{q} \to gg$	$\dfrac{N_g C_A}{2 C_F}\left[\dfrac{C_F}{C_A}(\chi+\chi^{-1}) - \dfrac{(\chi^2+1)}{(\chi+1)^2}\right]$
$gg \to q\bar{q}$	$\dfrac{4 C_F^2 C_A N}{N_g^3}\left[\dfrac{C_F}{C_A}(\chi+\chi^{-1}) - \dfrac{(\chi^2+1)}{(\chi+1)^2}\right]$

function. The common angular dependence has been taken to be that of the dominant $gg \to gg$ reaction in eq. (3), although any of the first six subprocesses with χ^2 behaviour could have been used. Even at moderate values of χ ($\chi = 1$, $\theta^* = 90°$ for instance) this "single effective subprocess approximation" works well [2,3]. In the absence of scaling violations in α_S and the structure functions, χ has the useful property that

$$\frac{d\sigma}{d\chi} \simeq \frac{\chi^2+\chi^{-2}+\dots}{(\chi+1)^2} \sim \text{ constant (large } \chi) \tag{4}$$

Thus deviations of $\dfrac{d\sigma}{d\chi}$ from constancy may be used to examine scaling violations and Q^2 choice.

An important caveat to stress is that we are using an $O(\alpha_S^2)$ tree-level calculation so that Q^2 is quite arbitrary and no information on Λ can be properly inferred. We shall discuss a recent $O(\alpha_S^3)$ calculation in Sec. 4, but this has yet to be implemented using an experimental jet definition, to actually predict measured jet cross-sections.

We shall now briefly discuss some UA1, UA2 results on two-jet production. For a fuller discussion see the talks of Scott and Meier [4,5].

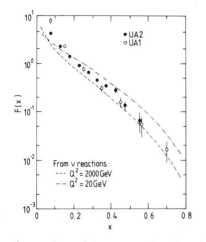

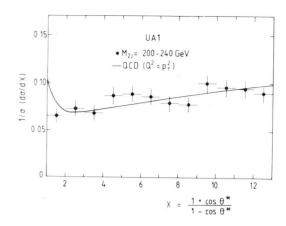

Figure 2. The measured effective structure function of eq. (3) compared with extrapolated CDHS structure function.

Figure 3. The observed χ distribution compared with the lowest-order QCD prediction with $Q^2 = p_T^2$.

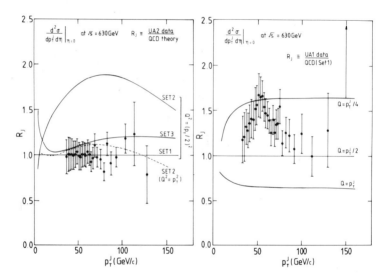

Figure 4. UA1, UA2 p_T spectra compared with lowest order QCD predictions using the structure functions of ref. 6.

In Fig. 2 we show the effective structure function of eq. (3) extracted from UA1 and UA2 data. Both agree well with the corresponding combination of CDHS structure functions measured in neutrino reactions, evolved in Q^2 from 20 to 2000 GeV². Fig. 3 shows $d\sigma/d\chi$ for UA1 data in a narrow bin of two-jet mass between 200 and 240 GeV. There is a clear deviation from the lowest-order scaling prediction of a horizontal straight line for $\chi > 2$. The curve is a lowest order prediction using $Q^2 = p_T^2$, notice that $Q^2 = \hat{s}$ is disfavoured since this is independent of χ for the binned data. Fig. 4 shows the experimental UA1 and UA2 two-jet p_T spectrum ($p_T > 20$ GeV) normalized to the lowest order QCD prediction, using a new structure function fit by Martin, Roberts and Stirling [6]. Set 1 with $Q = p_T/2$ provides an excellent fit of the UA2 data. Of course

346

Q may be varied, and Set 2 with $Q = p_T$ provides an acceptable fit. Set 3 has an $x \to 0$ behaviour for the gluon structure function of $x^{-1.5}$ in accordance with a suggestion of Collins [7]. One can see that it would be necessary to go to very low p_T's, where the jets are intrinsically ill-defined, in order to see the rapid increase due to this low-x behaviour.

3. Three-Jet Production

There are eleven $\hat{\sigma}_{ab \to cde}$ $2 \to 3$ scattering subprocesses. All have been computed at tree-level $O(\alpha_s^3)$ [8], but no higher order corrections have been calculated or are in prospect. Just as in the two-jet case there are some astonishing simplifications. The eleven subprocesses have remarkably similar angular distributions and in fact

$$\sigma_{3-j} \sim F(x_a, Q^2) F(x_b, Q^2) \hat{\sigma}_{gg \to ggg} \text{(angles)}$$

$$F(x) = g(x) + \lambda(q(x) + \bar{q}(x))$$

$$\lambda \to \frac{4}{9} \text{ for forward jets.} \tag{5}$$

To highlight the common angular distributions we show in Table 2 the lowest order QCD $2 \to 3$ subprocess cross-sections (summed over all possible assignments of partons to jets) normalized to the dominant $gg \to ggg$ (10 ratios) for transverse three-jet production (all three jets are at 90° to the beam in the subprocess c.m.) [9]. We consider four kinematical

Table 2. Lowest order $2 \to 3$ QCD subprocess cross-sections, normalized to $gg \to ggg$, for various transverse 3-jet configurations.

	{.9,.9,.2}	{.9,.8,.5}	{.8,.7,.5}	$\{\frac{2}{3},\frac{2}{3},\frac{2}{3}\}$	COLL
$q_1 q_1 \to q_1 q_1 g$	.05	.05	.06	.07	.05
$q_1 q_2 \to q_1 q_2 g$	.08	.08	.09	.09	.07
$q_1 \bar{q}_1 \to q_1 \bar{q}_1 g$	.13	.13	.14	.14	.08
$q_1 \bar{q}_2 \to q_1 \bar{q}_2 g$	.12	.12	.13	.13	.07
$qg \to qgg$	.29	.29	.28	.27	.29
$q\bar{q} \to ggg$	.03	.03	.02	.02	.03
$gg \to q\bar{q}g$	.11	.13	.13	.12	
$q_1 g \to q_1 q_2 \bar{q}_2$	.02	.02	.02	.02	
$q_1 g \to q_1 q_1 \bar{q}_1$	.00	.01	.01	.01	
$q_1 \bar{q}_1 \to q_2 \bar{q}_2 g$	.02	.03	.03	.03	.03

configurations ranging from Mercedes Benz ($x_1 = x_2 = x_3 = \frac{2}{3}$) at the centre of the transverse Dalitz plot, to the edges of the Dalitz plot. The ratios remain constant within $\sim 10\%$ whilst the absolute rates are varying by

several orders of magnitude. Furthermore there is an approximate geometrical progression

$$gg \to ggg \; : \; qg \to qgg \; : \; qq \to qqg \; \simeq \; 1 \; : \; 28 \; : \; (.28)^2 \tag{6}$$

To try and understand these features better [10] we can study the
variation of the ratios around the boundary of the Dalitz plot - that is,
consider configurations where one jet takes half the energy and the
remaining two jets are collinear, with fractional energies Z, $(1-Z)$. We
tabulate the Z-dependence of the ratios in Table 3. The subprocess
ratios depend on two ratios of Altarelli-Parisi splitting functions

$$f(Z) \; \equiv \; \frac{P_{qq}(Z) + P_{qq}(1-Z)}{P_{gg}(Z)} \tag{7}$$

and

$$g(Z) \; \equiv \; P_{qg}(Z)/P_{gg}(Z) \; .$$

Table 3. Z-dependence of $2 \to 3$ subprocess ratios around the boundary of
the transverse Dalitz plot.

$q_1 q_1 \to q_1 q_1 g$	$\dfrac{704}{6561} f(Z)$
$q_1 q_2 \to q_1 q_2 g$	$\dfrac{320}{2187} f(Z)$
$q_1 \bar{q}_1 \to q_1 \bar{q}_1 g$	$\dfrac{448}{6561} g(Z) + \dfrac{1120}{6561} f(Z)$
$q_1 \bar{q}_2 \to q_1 \bar{q}_2 g$	$\dfrac{320}{2187} f(Z)$
$qg \to qgg$	$\dfrac{440}{2187} (1 + f(Z))$
$q\bar{q} \to ggg$	$\dfrac{224}{6561}$
$^*gg \to q\bar{q}g$	$N_f \left[2g(Z) + \dfrac{7}{729} f(Z) \right]$
$^*q_1 g \to q_1 q_2 \bar{q}_2$	$(N_f-1) \dfrac{880}{2187} g(Z)$
$^*q_1 g \quad q_1 q_1 \bar{q}_1$	$\dfrac{880}{2187} g(Z)$
$q_1 \bar{q}_1 \to q_2 \bar{q}_2 g$	$N_f \left[\dfrac{448}{6561} g(Z) + \dfrac{32}{2187} f(Z) \right]$

With the exception of the processes indicated with an asterisk $f(Z)$
dominates the Z-dependence, since $g(Z) \ll f(Z)$. A curious result of QCD
is that $P_{gg}(Z) = \frac{9}{4}[P_{qq}(Z) + P_{qq}(1-Z)] - 6 P_{qg}(Z)$, an exact relation between

splitting kernels first noted by Dokshitzer [11]. This means that
$f(Z) = \frac{4}{9} + \frac{24}{9} g(Z)$, and since $g(Z) << f(Z)$, $f(Z) \approx \frac{4}{9}$ for all Z (about a
10% variation).

Thus the eight unstarred subprocess ratios will be constant within
$\sim$ 10% around the boundary of the Dalitz plot. Their values on the boundary
(at $Z = \frac{1}{2}$ for definiteness) are indicated in the column labelled "COLL" in
Table 2, are are close to those throughout the interior of the Dalitz plot.
Around the boundary one has

$$qq \rightarrow qqg : qg \rightarrow qgg : gg \rightarrow ggg \simeq \left(\frac{4}{9}\right)^3 : \frac{1}{2}\frac{4}{9}\left(1 + \frac{4}{9}\right) : 1 \qquad (8)$$

which is $(.30)^2 : .32 : 1$, an approximate geometrical progression which
appears to be a numerical accident. These considerations do not change
greatly if one moves out of the transverse plane.

Experimentally, the angular dependence in 3-jet UA1/UA2 data agrees
reasonably with that of $gg \rightarrow ggg$ and the ratio of three-jet to two-jet
cross-sections has been used to estimate $\alpha_s K_3/K_2$, where k_3, k_2 are
unknown K-factors associated with two-jet and three-jet production. The
experiments quote [12]

$$\alpha_s \frac{K_3}{K_2} = \begin{array}{ll} .24 \pm .04 \pm .04 & UA2 \\ .23 \pm .04 \pm .04 & UA1 \end{array} \qquad (9)$$

With no k-factors one would have anticipated $\alpha_s \simeq .15$ at the
$\langle Q^2 \rangle \simeq 4000$ GeV² relevant to this determination.

4. Higher Order Corrections

Ellis and Sexton [13] have performed a calculation of the next-to-
leading $O(\alpha_s^3)$ corrections to all nine $2 \rightarrow 2$ subprocesses, and all $2 \rightarrow 3$
subprocesses at tree-level have been computed in n-dimensions. The one
remaining step which has yet to be completed is to combine the real and
virtual corrections using dimensional regularization, to produce a finite
cross-section. For this to be done one needs to implement a jet algorithm,
preferably the one used by the experiment which the theoretical calculation
is to be compared with. It is hoped that results will be available within
the next six months [14].

In the meantime one can note in passing an amusing feature of the
n-dimensional tree-level $gg \rightarrow ggg$ result. Labelling the momenta $12 \rightarrow 345$
and denoting $(p_i \cdot p_j)$ by the shorthand (ij) one has

$$|M_{gg \rightarrow ggg}|^2 = g^6 N^3 (N^2-1) \sum_P \frac{F(1,2,3,4,5)}{(12)(23)(34)(45)(51)}$$

where

$$F(1,2,3,4,5) = \frac{(n-2)^2}{8} \sum_{i<j} (ij)^4 - \frac{3}{4}(n-4)(n-10)\{(12)^2(23)^2 + \ldots\}$$

$$(10)$$

The first term in F is just the familiar compact CALKUL result for
$2 \rightarrow 3$ gluon scattering [15]. The extra term vanishes in both four and
ten dimensions. Thus the five gluon scattering amplitude in four-

dimensions is proportional to that in ten dimensions! No explanation for this curious fact is known.

5. n-jet Production

QCD multi-jet production will be an important background to many physics processes that we will wish to study at future hadron colliders. Higgs production, heavy quark (top) pairs, vector boson pairs, all have multi-jet final states as the dominant decay mode. To reliably estimate the background one will need the cross-section for hard, well-separated jets characteristic of the new physics signals. In such configurations jet fragmentation Monte Carlos based on leading pole implementations of QCD will not be valid. One will rather need exact tree-level matrix elements for $2 \to n$ parton-parton subprocesses.

All $2 \to 4$ subprocesses at tree-level have now been calculated [16]. Until recently the dominant $gg \to gggg$ six-gluon scattering subprocess was only known as a lengthy computer code [16] taking $\sim .1$ sec CPU time per four-jet event. New techniques now provide a "compact" analytic form involving several hundred terms for the 6g amplitude, and hold out the promise of exact calculations of $7g, 8g \ldots$, $2 \to 5$, $2 \to 6 \ldots$ subprocesses. In the interim there are suggestions for estimating $2 \to n$.

The new technique, proposed by Mangano, Parke and Xu [17], is to define a set of sub-amplitudes of multi-gluon scattering which have duality properties characteristic of string amplitudes. Specifically one writes the tree-level vector particle scattering amplitude for colours $a_1, a_2, \ldots a_n$, momenta $p_1, p_2, \ldots p_n$ and helicities $\varepsilon_1, \varepsilon_2, \ldots \varepsilon_n$ in the form

$$M_n = \sum_{\text{"perms"}} \text{tr}(\lambda^{a_1} \lambda^{a_2} \lambda^{a_3} \ldots \lambda^{a_n}) m(p_1, \varepsilon_1; \, p_2, \varepsilon_2; \ldots; \, p_n, \varepsilon_n) \tag{11}$$

The sum is over non-cyclic permutations of $1, 2, \ldots, n$. Abbreviating the sub-amplitudes by $m(1, 2, 3, \ldots, n)$, for some particular set of helicities, one finds they satisfy the following properties.

(1) $m(1, 2, 3, \ldots n)$ is gauge invariant.

(2) $m(1, 2, 3, \ldots n)$ is invariant under cyclic permutations of $1, 2, \ldots, n$.

(3) $m(n, n-1, \ldots, 1) = (-1)^n m(1, 2, 3, \ldots, n)$

(4) $\sum m(1, 2, 3, \ldots n) = 0$ (sum over cyclic perms of $1, 2, \ldots n$).

(5) $m(1, 2, 3, \ldots n)$ factorizes on multi-gluon poles.

(6) $|M_n|^2 = \dfrac{N^{n-2}(N^2-1)}{2^n} \sum_{\text{"perms"}} \{ |M(1, 2, \ldots, n)|^2 + O(N^{-2}) \}$

These 'duality' properties are automatically satisfied by string amplitudes in an SU(N) open bosonic string theory. The sub-amplitudes of eq. (11) are the zero-slope limit of the string sub-amplitudes and so the same properties hold. The traces of λ-matrices are just the Chan-Paton factors of the string theory. Mangano, Parke and Xu use the duality properties (1)-(6) to simplify the calculation of sub-amplitudes using Feynman diagrams and obtain relatively compact expressions for the 6g case.

A more radical approach is that of Kosower, Lee and Nair [18] who have explicitly taken the zero-slope limit of a string calculation to obtain the well-known gg → ggg result. They claim that using algebraical tricks larger numbers of gluons will be practicable with this method.

Let us now summarize what is known in general about the n-gluon scattering amplitude. We can decompose the overall squared amplitude as

$$|M_n|^2 = \sum_s |M_n^{(s)}|^2 \tag{12}$$

where $s = |\sum \lambda_i|$ labels the total helicity and the sum runs over $s = n, n-2,$ $n-4, \ldots 1(0)$. For general n

$$|M_n^{(n)}|^2 = |M_n^{(n-2)}|^2 = 0 \tag{13}$$

follows from helicity conservation arguments. Also for general n

$$|M_n^{(n-4)}|^2 = |M_n^{PT}|^2 + 0\left(\frac{1}{N^2}\right) \tag{14}$$

where

$$|M_n^{PT}|^2 = \frac{g^{2n-4}N^{n-2}2^{4-n}}{(N^2-1)} \sum_{i<j} (ij)^4 \sum_{"P"} \frac{1}{(12)(23)(34)\ldots(n1)} \tag{15}$$

is a generalization of the 2 → 3 compact CALKUL formula, proposed by Parke and Taylor [19]. The sum is over the $(n-1)!/2$ distinct permutations of $1,2,\ldots n$. The $0(1/N^2)$ terms are known for general n, since m(1+,2+,3-, 4-,...n-) is known [17] and the $1/N^2$ terms are just the interference terms between different orderings (c.f. duality property (6)). In the 2 → 4 case the $0\left(\frac{1}{N^2}\right)$ corrections are typically small ∿ 1% corrections. For 2 → 2 and 2 → 3 the formula of (15) reproduces the known results

$$|M_4|^2 = \frac{1}{2}|M_4^{PT}|^2$$

$$|M_5|^2 = |M_5^{PT}|^2 \tag{16}$$

For soft gluon emission (a string of eikonal factors)

$$|M_n|^2 \rightarrow |M_n^{PT}|^2 + 0\left(\frac{1}{N^2}\right) \tag{17}$$

This is a result of Bassetto, Ciafaloni and Marchesini [20] and implies that as n increases $|M_n^{(n-4)}|^2$ should remain a reasonable fraction of the overall $|M_n|^2$, offsetting to some extent the increasing number of helicity combinations.

There are various proposals for estimating 2 → n multi-gluon cross-sections. Kunszt and Stirling [21] have estimated 2 → (n-2) QCD jet production rates and compared these QCD backgrounds with the expected signal for top production (2 → 6 background) at SSC, Higgs decay etc. Their conclusion is that even with cunning and drastic cuts the QCD jet background typically swamps the signal by orders of magnitude. To make these estimates for n > 6 they assumed that all the helicity combinations contribute equally and corrected the PT result of eq. (15) by a combinatoric factor, i.e.

$$|M_n|^2 \simeq \text{COMBINATORIC FACTOR} \times |M_n^{PT}|^2 \tag{18}$$

For example, for n = 6 there are 30 helicity combinations of the form (--++++) or (++----) contributing to the s = 2 term of eq. (12) which is given by $|M_6^{PT}|^2$ (neglecting $\frac{1}{N^2}$ corrections), and 20 of the form (---+++) contributing to the s = 0 term. Assuming that all these combinations contribute equally the relevant combinatoric factor in (18) will be 50/30 = 1.67 ... The combinatoric factors up to n = 10 are tabulated in Table 4.

The result of eq. (17) implies that these naive combinatoric factors will overestimate $|M_n|^2$, and that $|M_n^{PT}|^2$ will remain a much larger fraction of $|M_n|^2$ than implied by Table 4.

Table 4. Combinatoric factors in eq. (18) up to n = 10.

n	combinatoric factor
6	1.67
7	2.67
8	4.25
9	6.83
10	11.13

I have attempted in [22] to obtain more accurate estimates. The trick is to approximate $|M_n|^2$ for some n-2 jet configuration by

$$|M_n|^2 \simeq \frac{|M_n|^2}{|M_n|^2}\bigg|_{IR} \times |M_n^{PT}|^2 \tag{19}$$

where $\big|_{IR}$ denotes that the ratio of amplitudes is evaluated at a 'nearby' configuration where two of the jets i,j, having the smallest invariant mass, have been replaced by a collinear pair of jets in the direction of $p_i + p_j$ with energy fractions Z, (1-Z), Z = $E_i/(E_i+E_j)$, and rescaled four momenta to make the particles on-shell. We shall term this procedure "infra-red reduction". Since the ratio contains no kinematical poles it will be reasonably slowly-varying and its value at the 'nearby' configuration should approximate the ratio for the original (n-2) jets. The approximate ratio is then multiplied by $|M_n^{PT}|^2$ evaluated at the original (n-2) jet configuration. One has

$$\frac{|M_n|^2}{|M_n^{PT}|^2}\bigg|_{IR} = F(R,Z) \frac{|M_{n-1}|^2}{|M_{n-1}^{PT}|^2} \tag{20}$$

352

$F(R,Z)$ is a simple rational function of R,Z, R being itself a function of the kinematics of the (n-3) jets remaining after i,j have been combined. Since $|M_5|^2 = |M_5^{PT}|^2$ eq. (20) gives

$$|M_6|^2 \simeq F(R,Z) \ |M_6^{PT}|^2 \tag{21}$$

we can repeat the infra-red reduction several times and the naive combinatoric factor is replaced by a product of F's.

$$|M_n|^2 \simeq \prod_{i=1}^{n-5} F(R_i,Z_i) \ |M_n^{PT}|^2 \tag{22}$$

Equation (21) gives excellent accuracy for the 6-gluon 2 → 4 process when compared with the exact result [16]. Using jet cuts typical of UA1/UA2 at fixed $\sqrt{\hat{s}} = 100$ GeV one finds (generating events uniformly in phase space with RAMBO [23]) that for 95% of events the approximation of (21) is within 20% of the exact result, and 90% within 10%. For similar cuts we display in Table 5 the product of F's, using the approximation of (22), compared with the combinatoric factor of Table 4 for n = 6,7,8,9,10.

Table 5. Growth of $|M_n|^2$. $\langle \Pi F_i \rangle$ entering in eq. (22) compared with the naive combinatoric factor of Table 4.

n	combinatoric factor	$\langle \Pi F_i \rangle$
6	1.67	1.31
7	2.67	1.56
8	4.25	1.86
9	6.83	2.16
10	11.13	2.90

As can be seen the combinatoric factor does over-estimate the cross-section as compared to our more detailed approximation. We stress, however, that this overestimation by perhaps a factor of 2 or so for n = 8, does nothing to alter the conclusions of Ref. [21] concerning QCD backgrounds.

6. Summary and Conclusions

In two-jet production we have seen that θ^* and p_T distributions are well-reproduced by lowest-order QCD with $Q^2 = p_T^2$ (not $\hat{s}$). It seems that higher order corrections leave the simplifying features present at tree-level intact. In order to say more we need an implementation of the Ellis and Sexton $O(\alpha_s^3)$ calculation. For three-jet production we showed that the universal angular distribution of the 2 → 3 processes was connected with the infra-red behaviour of the amplitudes, and in particular with a curious relationship between the Altarelli-Parisi splitting kernels. Experimentally there is reasonable agreement with the angular distribution for gg → ggg, and the ratio of 3-jet rate to 2-jet rate has been used to infer $\frac{K_3}{K_2}\alpha_s = .24$, where K_3, K_2 are unknown K-factors.

There has been much recent progress in calculating and estimating multi-jet production rates. Knowledge of these will be crucial for reliable determinations of QCD backgrounds in forthcoming hadron collider experiments.

ACKNOWLEDGEMENT

It is a pleasure to thank the organisers for the invitation to participate in this enjoyable workshop, in such an idyllic setting.

REFERENCES

[1] B.L. Combridge, J. Kripfganz and J. Rauft, Phys. Lett. 70B (1977) 234.
[2] B.L. Combridge and C.J. Maxwell, Nucl. Phys. B239 (1984) 429.
[3] F. Halzen and P. Hoyer, Phys. Lett. 130B (1983) 326.
[4] W. Scott, these proceedings.
[5] K.H. Meier, these proceedings.
[6] A.D. Martin, R.G. Roberts and W.J. Stirling, Rutherford Lab. preprint
       RAL-87-052 (1987).
[7] J.C. Collins, IIT preprint 86-0298 (1986).
[8] T. Gottschalk and D. Sivers, Phys. Rev. D21 (1980) 102;
       Z. Kunszt and E. Pietarinen, Nucl. Phys. B164 (1980) 45.
[9] B.L. Combridge and C.J. Maxwell, Phys. Lett. 151B (1985) 299.
[10] K. Makhshoush and C.J. Maxwell, in preparation.
[11] Yu.L. Dokshitzer, Sov. Phys. JETP 46 (1977) 641.
[12] UA1 Collab., G. Arnison et al, Phys. Lett. 158B (1985) 494;
       UA2 Collab., J.A. Appel et al, Z. Phys. C30 (1986) 341.
[13] R.K. Ellis and J.C. Sexton, Nucl. Phys. B269 (1986) 445.
[14] Famous phenomenologist, private communication.
[15] F.A. Berends, R. Kleiss, P. de Causmaecker, R. Gustmans and T.T. Wu,
       Phys. Lett. 103B (1981) 124.
[16] S.J. Parke and T.R. Taylor, Nucl. Phys. B269 (1986), 410;
       Z. Kunszt, Nucl. Phys. B271 (1986), 333;
       J. Gunion and J. Kalinowski, Phys. Rev. D34 (1986), 2119.
[17] Michelangelo Mangano, Stephen Parke and Zhan Xu, FERMILAB Pub-
       87/52-T (1987).
[18] David Kosower, Bum-Hoon Lee and V.P. Nair, Columbia University
       preprint CU-TP-378 (1987).
[19] S.J. Parke and T.R. Taylor, Phys. Rev. Lett. 56 (1986) 2459.
[20] A. Bassetto, M. Ciafaloni and G. Marchesini, Phys. Rep. 100 (1983)
       201.
[21] Z. Kunszt and W.J. Stirling, Durham University Preprint DTP/87/16
       (1987).
[22] C.J. Maxwell, Phys. Lett. 192B (1987) 190.
[23] S.D. Ellis, R. Kleiss and W.J. Stirling, Comput. Phys. Commun. 40
       (1986) 359.

RESULTS ON JET PRODUCTION FROM THE UA2 EXPERIMENT AT THE

CERN PROTON-ANTIPROTON COLLIDER

Karlheinz Meier

CERN
Geneva, Switzerland

INTRODUCTION

The UA2 experiment operating at the CERN proton-antiproton collider has accumulated an integrated luminosity of about 750 nb-1 at $\sqrt{s}$ = 630 GeV during the 1984/85 run periods. This review summarizes the results obtained for multijet production (Section 2) and the first successful attempt to perform particle spectroscopy using hadronic jets originating from the decay of W/Z bosons into jet pairs (Section 3). Special emphasis is put on a detailed discussion of jet detection algorithms (Section 1).

1. JET DETECTION TECHNIQUES IN UA2

The UA2 experiment is instrumented with a highly segmented electromagnetic and hadronic central calorimeter covering the region of polar angles 40° < θ < 140° over the full azimuth Φ. Transversely the calorimeter is segmented into 240 cells of 10° × 15° size. In longitudinal direction each cell is divided into an electromagnetic compartment (17.5 radiation lengths deep) and two hadronic compartments (each 2 absorption lengths deep). A detailed description of the system can be found in Ref. 1. The following describes techniques and features of jet identification algorithms optimized and adapted to the physics subjects adressed here.

The UA2 collaboration has recently presented a study of energy flow in hadronic events at $\sqrt{s}$ = 630 GeV (Ref. 2). The strong energy dependence of the means of event shape variables (Fig. 1) reflects the onset of final states dominated by two jet configurations. The typical proton-antiproton collision at high transverse energy (E_t > 70 GeV) is therefore jet-like. This observation motivates the identification of the localized energy depositions in the calorimeter with "jets". Jets are in turn related with their parent partons. The study of jet dynamics is therefore a direct test of the underlying production mechanisms.

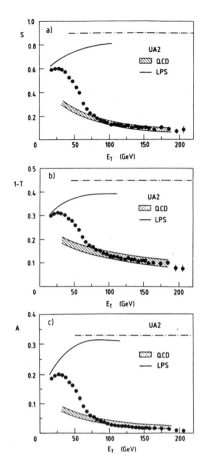

Fig. 1.
Mean values of event shape
variables as a function of
the total transverse energy
a) sphericity
b) 1-thrust
c) aplanarity
solid curve: phase space model
dashed band: QCD calculation
dashed line: instrumental limit

In the analysis of multijet events clustering methods
have been used for jet identification. Clusters are
structures of adjacent calorimeter cells having an energy in
excess of a 400 MeV threshold. A momentum vector is assigned
to each cluster joining the center of the UA2 detector with
the cluster centroid and measuring its length by the cluster
energy. The cluster energy is obtained from the weighted sum
of the three calorimeter compartment energies. The weights
are determined to reflect the different response of the
compartments to hadrons and photons and the mean particle
composition of hadronic jets (Ref. 1).

This basic cluster method provides a good two jet
resolution in space. The energy collection is however limited
to the central core of the jet (depending on the threshold
cell energy). For a jet with a transverse energy E_t = 40 GeV
typically 10% of the energy is not seen in the clustering
procedure. To refine the algorithm according to the special
requirements of the multijet analysis the basic method has
been tuned in two ways.

The study of bremsstrahlungs-type angular distributions
needs to resolve closeby jets whereas efficient energy
collection is less important in this case. To achieve this
the primary clusters are reprocessed using the same
algorithm, but applying a higher threshold, set at 5% of the

total energy of the primary cluster. The content of a primary cluster cell whose energy is below the new threshold is redistributed among secondary clusters according to their relative location and energy content. The angular resolving power associated with this algorithm has a distribution with an approximately energy independent cut-off at 30° ± 10°.

For a quantitative analysis of multijet production (measurement of the strong coupling constant described in section 2) a good understanding of the energy collection in a cluster as well as a uniquely defined angular resolving power are essential. The clustering algorithm described before is therefore followed by one additional step in which cluster pairs, having an opening angle below 50° and a transverse energy in excess of 5 GeV, are merged into a single cluster. After merging the effective resolving power exceeds 99% above 50° and cuts off sharply below this angle.

In jet spectroscopy with low (= 2) jet multiplicities the primary concern is to arrive at the best possible mass resolution at the correct mass scale (Section 3). This is achieved with a cone algorithm. Here the energies of all calorimeter cells having their center within a cone of angle ω around the jet (cluster) axis are summed. The value of ω is adjusted in order to minimize energy measurement errors. The optimization procedure requires a variable sensitive to the jet energy resolution. For two jet events the difference between the average values of $p_t{}^\zeta$ and $p_t{}^\eta$, the components of the transverse momentum of the jet pair projected on the bisectors of the tranverse momenta is such a variable. $p_t{}^\zeta$ describes the energy imbalance of a two jet event. This variable receives contributions from soft initial state bremsstrahlung and the energy measurement error. $p_t{}^\eta$ describing non back-to-back two jet configurations is dominated by the bremsstrahlung contribution because the error in the position measurement of jets is negligible. Therefore the difference is mainly affected by the jet energy resolution. If the cone is chosen too large, more energy from the underlying event activity is incorrectly assigned to the jet. If the cone is chosen too small, the energy of the jet is not sufficiently contained. A value $\cos\omega = 0.6$ is used for jet spectroscopy in the UA2 calorimeter, as that corresponding to the narrowest cone compatible with an optimum energy resolution.

2. THE PHYSICS OF MULTIJET EVENTS

Hard proton-antiproton collisions sample the short distance interactions between nuclear constituents (quarks and gluons). A comparison with the predictions of perturbative quantum chromodynamics (QCD) is possible by identifying hadron jets with the partons produced in the hard scattering process. The analysis presented in this section gives quantitative comparisons with QCD calculations of order $\alpha_s{}^3$ and $\alpha_s{}^4$. The possibility of 4 jet events being produced by two independent scattering processes of order $\alpha_s{}^2$ within the same proton-antiproton collision is also investigated. This production mechanism is referred to as "multiparton scattering".

Experimentally exclusive n-jet topologies are selected with the following criteria:

- $E_{t,jet} > 10$ (15) GeV
  for n clusters (the value in brackets is used in the quantitative three jet analysis)

- $-0.8 < \eta < 0.8$
  $|\sum \vec{p}_{t,jet}| < 20$ GeV
  to select well contained events

- $\sum E_{t,jet} > 70$ GeV
  to remove trigger biases

- special cuts used in the search for multiparton scattering:
  energy collection in 30° cones
  $\sum E_{t,cones} > 60$ GeV

The event numbers obtained are 25087, 12158, 2458 and 278 for the 2, 3, 4 and >4 jet topologies respectively.

THREE JET PHYSICS

The aim of the three jet studies presented here is twofold . First the characteristic QCD behavior of the observed events will be shown by comparing the Dalitz diagram with a QCD calculation and a 3-body phase space model. Motivated by the success of the QCD model a value for the strong coupling constant α_s is then extracted from the data. A discussion on the relevance of such a measurement is given.

The model calculations contain structure functions, the relevant QCD calculation describing the hard scattering process, final state parton fragmentation and the response of the calorimeter to the fragmentation products. The QCD calculations available in α_s^3 (Ref. 3) and α_s^4 (Ref. 4) include all leading order terms contributing to the $2 \to 2,3,4$ parton cross sections. The difference between these leading order calculations $\sigma_{n,LO}$ and the full topological cross section σ_n is usually expressed in form of a K-factor $K_n = \sigma_n/\sigma_{n,LO}$. All comparisons are therefore proportional to $\alpha_s^n \cdot K_n$ and not simply to α_s^n.

A large variety of distributions has been compared to the QCD model (Ref. 5). As an example figure 2 shows the Dalitz plot, constructed by defining scaled variables

$$x_{12} = m_{12}^2/M^2 \quad \text{and} \quad x_{23} = m_{23}^2/M^2$$

where M is the mass of the three jet system. An increased event density in the region of small x_{23} (corresponding to final state bremsstrahlung of a soft third jet) is clearly visible and well explained by the QCD calculation. A simple 3-body phase space model (indicated in figure 2) would predict a uniform distribution in the Dalitz plot (except for acceptance effects) and is therefore excluded.

It has been shown that QCD gives a good description of the three jet data sample. From former studies (Ref. 6) it is

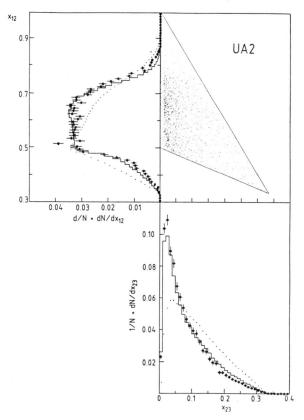

Fig. 2. Dalitz plot for three jet events
histogram: QCD calculation
dotted line: 3-body phase space

known that it also provides a good description of the two jet
sample. This motivates an evaluation of the quantity
$\alpha_s \cdot K_3/K_2$ by adjusting α_s in the QCD model until the
theoretical value of the ratio between the three jet cross-
section and the two jet cross-section (R_{QCD}) is equal to its
experimental value (R_{exp}). The value obtained is

$$\alpha_s \cdot K_3/K_2 = 0.236 \pm 0.004 \text{ (stat.)} \pm 0.04 \text{ (syst.)}$$

The large systematic error for this measurement is mainly
caused by uncertainties in the parton fragmentation model and
the contribution from the underlying event activity to the
jet energy . Both uncertainties affect the experimentally
observed jet energy scale. Althought these uncertainties
cancel largely in the ratio R_{exp}, they cause an error in
$\alpha_s \cdot K_3/K_2$ of $\sim$ 15%. Smaller contributions come from the
uncertainties in the structure functions, the calorimeter
response and the leakage of four jet events into the three
jet topology.

   It is instructive to investigate the dependence of
$\alpha_s \cdot K_3/K_2$ on variables related to the momentum transfer Q^2
of the QCD scattering process. Figure 3a shows $\alpha_s \cdot K_3/K_2$
evaluated in bins of the effective center-of-mass energy in
the parton-parton scattering system. The dashed line

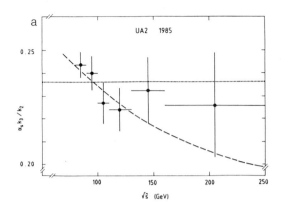

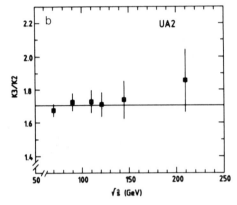

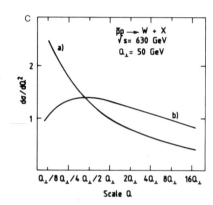

Fig. 3. a) $\alpha_S \cdot K_3/K_2$ evaluated as a function of the parton-parton center-of-mass energy

b) fitted values of K_3/K_2 evaluated as a function of the parton-parton center-of-mass energy (assuming Λ = 250 MeV)

c) scale dependence of W-boson production for
   a) 1st order QCD calculation
   b) 1st and 2nd order QCD calculation

indicates the expected variation ("running") of the strong coupling constant α_S. It is evident that the accurary of the data does not allow a conclusive statement about the running of α_S. The upgraded UA2 detector with its larger angular coverage and a ten times increased statistics at ACOL will improve the experimental points especially at high Q^2 values.

The following comment concerns the relevance of this collider measurement in view of the many different measurements of the strength of the strong interaction coming from other observed processes (Ref. 7). Those data suggest a value for the QCD scale parameter Λ of order 250 MeV. The UA2 measurement was performed at large Q^2 values with a mean of 1700 GeV^2 taking the maximum transverse energy per jet as momentum transfer. As a consequence the corresponding value for Λ would be as large as 2.8 GeV! This is clearly excluded by the wide range of experimental numbers available for this parameter. One can on the other hand use the UA2 measurement to learn about the non-completeness of the used LO QCD calculations (K-factors). Assuming the correct value of Λ to be 250 MeV figure 3b shows the best fit for the ratio K_3/K_2

over the measured range of parton-parton center-of-mass energies. A mean value of about 1.7 is required to describe UA2 data. This number represents a firm measurement of a higher order QCD contribution to jet production at the collider and awaits confirmation by a theoretical calculation.

Another theoretical uncertainty connected to the α_s measurement presented here concerns the scale of momentum transfers involved. Uncertainties in the values used for Q^2 affect the α_s-measurement only if they are different for the two jet and the three jet events. Using for example the mean jet transverse energy instead of the maximum jet transverse energy would leave the two jet cross-section unchanged. However the three jet calculation would predict a larger value and consequently the measured value of α_s. K_3/K_2 would drop by about 25%. The effect of uncertainties in the Q^2 scale is demonstrated in figure 3c (Ref. 7). Here the scale dependence is shown for the process of W-boson production in proton-antiproton collisions where both, LO and next-to-LO calculation are completed. The predicted cross-section shows a much reduced scale dependence for the combined 1st and 2nd order calculation. This stresses again the need for next-to-LO calculations of jet production at the collider.

FOUR JET PHYSICS

Four jet events are the subject of the following study. Their yield and phase-space density provide additional tests of perturbative QCD. Here, however a new production mechanism is expected to play a role: the production of two independent pairs of incoming partons (Ref. 8). Recent ISR data have been claimed by the AFS collaboration (Ref. 9) to provide evidence for such a signal.

Initially the observed four jet events will be compared to the predictions of a leading order QCD calculation in which four jet events exclusively result from elementary subprocesses having two incoming and four outgoing partons with cross-sections taken from Ref. 4. The four jet system can be described in terms of nine variables: three describing the orientation of the jet system and six describing the internal configuration of the jets. This large number of variables makes the description of this final state rather complex. The jets do not lie in a plane, nor can they be described by simple Dalitz plots. A natural set of variables are the space angles between the jets in their center-of-mass, $\cos\omega_{ij}^*$. Only five of these variables are independent. The sixth degree of freedom corresponds to the total mass of the jet system. Figure 4a compares four $\cos\omega_{ij}^*$ distributions of data and simulated events. For this purpose jets are sorted in order of decreasing energy in the center-of-mass system with the result that jets 3 and 4 are generally radiated gluons, one from jet 2, the other from jet 1 or again from jet 2. This is visible from the enhancements observed in the vicinity of $\cos\omega_{ij}^* = 1$ with respect to phase space distributions obtained by setting the subprocess matrix elements equal to unity in the QCD model calculation. The agreement between data and 4th order QCD calculation is remarkable. It is instructive to compare two equivalent

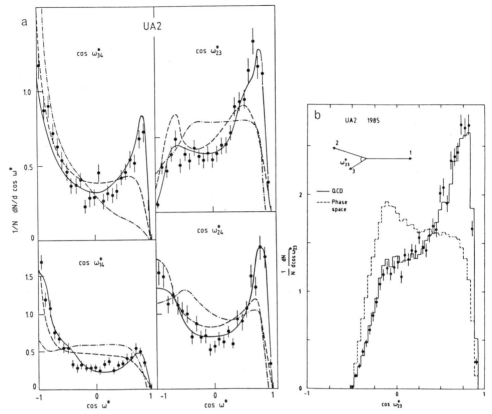

Fig. 4. a) Four $\cos\omega_{ij}{}^*$ distributions of four jet events
compared to the leading order QCD model (solid
line), the 4-body phase-space (dashed line) and the
multiparton scattering model described in the text
(dash-dot line)

b) $\cos\omega_{23}{}^*$ of three jet events compared to the leading
order QCD model (solid histogram) and the 3-body
phase-space (dashed histogram)

angular distributions from three and four jet events. Figure
4b showing $\cos\omega_{23}{}^*$ of three jet events exhibits the same
features as the corresponding distribution of four jet
events. Both topologies prefer small jet opening angles
characteristic for gluon bremsstrahlung.

The success of 4th order QCD in describing the features
of the observed four jet events already precludes a dominant
contribution from multiparton scattering. The prediction of a
multiparton calculation (see next chapter for details)
describes the observed angular distributions (figure 4a) much
less well. This observation is now quantified in an explicit
search for multiparton scattering events. The cross-section
to produce two jet pairs via the multiparton mechanism σ_{MP}
can be expressed relating it to the single two jet production
cross-section σ_2:

$$\sigma_{MP} = \sigma_2{}^2/\sigma_{eff} \quad , \quad \sigma_{eff} = \pi \cdot R_{eff}{}^2$$

The motivation for this expression is that in a naive picture

assuming complete independence of the two jet pairs σ_{eff} is
given by the inelastic non-diffractive proton-antiproton
cross-section of approximately 40 mb. In practice deviations
from this picture are expected from trivial kinematic
constraints and from the dynamics proper. The equation given
for σ_{MP} suggests a comparison between real four jet events
and a multiparton model obtained by merging pairs of
experimentally observed two jet events. Such a simple
multiparton model includes automatically many instrumental
factors (e.g. geometrical acceptance, calorimeter energy
response) and some dynamical factors (e.g. soft gluon
radiation from the incoming interacting partons). The
articifical multiparton events are ordered in azimuth angle
Φ. It turns out that after the ordering procedure jet1/jet3
and jet2/jet4 form the correctly associated pairs in 98% of
all cases. Using this information a separation variable Δ is
being constructed:

$$
\Delta \quad = \quad \left[\frac{(\pi - \Delta\Phi_{13})^2 + (\pi - \Delta\Phi_{24})^2}{\sigma^2(\Delta\Phi)} + \frac{\Delta E_{t\,13}^2 + \Delta E_{t\,24}^2}{\sigma^2(\Delta E_t)} \right]^{\frac{1}{2}}
$$

The construction of this variable is motivated by the fact
that multiparton scattering events should be composed out of
two two jet events each balanced in the azimuthal angular
difference $\Delta\Phi$ and in the transverse energy difference ΔE_t.
The distribution of Δ is shown in figure 5 and can be used to
place limits on a possible multiparton signal. Without any
specific assumption on the nature of the observed events a
limit of $\sigma_{eff} > 7$ mb (90% C.L.) is obtained by simply
assigning all events in the low Δ region ($\Delta < 1$) to the
multiparton mechanism. This limit, which corresponds to

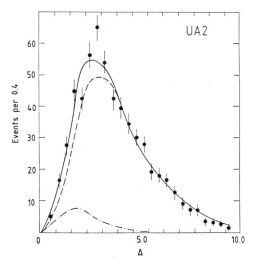

Fig. 5. Δ distribution compared to the best fit (solid line)
to a combination of a bremsstrahlung contribution
(dashed line) and a multiparton contribution (dash-
dot line)

$R_{eff} > 0.5$ fm, can be improved by fitting the Δ distribution to a linear combination of multiparton events (from the simple multiparton model) and of bremsstrahlung events (from the QCD model of Ref. 4). The best fit (shown in figure 5) gives $\sigma_{eff} > 10$ mb (90% C.L.) corresponding to $R_{eff} > 0.6$ fm. This limit is expected to be improved by experiments using a lower transverse energy threshold and accumulating a significantly higher integrated luminosity as the new UA2 detector in the near future.

3. JET SPECTROSCOPY IN THE HADRONIC DECAY OF W/Z BOSONS

The intermediate vector bosons W/Z are expected to decay into quark-antiquark pairs with well defined branching fractions (Ref. 10) :

$$\Gamma(W \to q\bar{q})/\Gamma(W \to e\nu) \approx 6$$

$$\Gamma(Z \to q\bar{q})/\Gamma(Z \to ee) \approx 20$$

(excluding modes with a top quark)

The observation of such decays is an important check of the standard model. An additional motivation is to provide the first test-case of the ability of future collider experiments to perform spectroscopy with hadron jets identified with their parent partons (Ref. 11).

The experimental requirements are the selection of well measured jets with optimized mass resolution and a good control of the overwhelming background from QCD two jet production. Jets have been selected using the cone algorithm described in section 1. Only jets well transversely and longitudinally contained in the calorimeter have been considered. The longitudinal containment was assured by selecting jets with a small fraction (< 40%) in the second hadronic compartement and a large fraction (> 24%) in the electromagnetic compartement of the UA2 calorimeter. Together with additional cuts on the transverse momentum balance and the quality of jet reconstruction the overall efficiency of the selection procedure is 66%. The mass resolution of the two jet system was determined to be ≈ 8 GeV in the W/Z mass region. A detailed discussion on the determination of this number can be found in Ref. 12. Figure 6 shows the two jet mass spectrum for the data taken at $\sqrt{s} = 630$ GeV. A fit to a background + signal shape (displayed in figure 6) results in a signal size of 686 ± 210 events (3.3 standard deviations) in the W/Z mass region. The fit was performed using two gaussians for the W and Z respectively. The ratio m_Z/m_W, the expected position for the W peak, the instrumental mass resolution and the ratio between the production rates of W and Z bosons where fixed to their known values. The expected signal is 340 ± 80 events. The observed effect and the expectation are compatible within 1.4 standard deviations.

Stronger evidence for a signal and a significant quantitative measurement of the W/Z branching fractions into two jets require the collection of a much larger data sample.

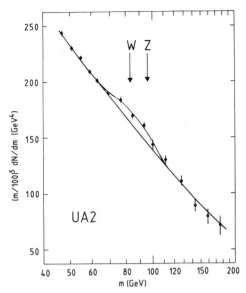

Fig. 6. Mass spectrum of two jet events with best fit to QCD
background alone (a) and including two gaussians for
W/Z decays (b)

SUMMARY

The UA2 experiment at the CERN proton-antiproton
collider has studied the characteristics of a large sample of
multijet events.

Three jet events show a good qualitative agreement with
a leading order QCD calculation. In a quantitative study of
those events a value for $\alpha_S \cdot K_3/K_2 = 0.234 \pm 0.004$ (stat.) $\pm$
0.04 (syst.) was derived. A deeper understanding of the
numerical value in the framework of other α_S measurements
needs the results of higher order QCD calculations.

Four jet events are well described by a leading order
QCD calculation of order α_S^4. A significant contribution from
multiparton scattering is not required. A limit has been
placed on the effective proton cross-section $\sigma_{eff} > 10$ mb
(90% C.L.).

A signal of 3 standard deviations significance has been
observed in the two jet mass spectrum. Its properties are
compatible with the hadronic decays of the W/Z bosons.

The upgraded UA2 detector at the improved CERN proton-
antiproton collider complex will be able to repeat these
studies with improved quality and significance.

Acknowledgements

I would like to thank Brad Cox for organizing this very
pleasant workshop at St. Croix. The results presented here
are a collaborative effort of the whole UA2 group. I wish to
thank all my colleagues for their support. Special thanks go
to Annette for the careful preparation of the manuscript.

REFERENCES

1. A. Beer et al., Nucl. Instr. and Methods 224 (1984) 360.
2. UA2 Collaboration, R. Ansari et al., Z. Phys. C36 (1987) 175.
3. Z. Kunszt and E. Pietarinen, Nucl. Phys. B164 (1980) 45;
   T. Gottschalk and D. Sivers, Phys. Rev. D21 (1980) 102;
   F. A. Berends et al., Phys. Lett. 118B (1981) 124.
4. Z. Kunszt and W. J. Stirling, Phys. Lett. 171B (1986) 307.
5. UA2 Collaboration, J. A. Appel et al., Z. Phys. C30 (1986) 341.
6. UA2 Collaboration, J. A. Appel et al., Phys. Lett. 160B (1985) 349.
7. W. J. Stirling, talk given at the 1987 International Symposium on Lepton and Photon Interactions at High Energies, Hamburg, 1987.
8. AFS Collaboration, T. Akesson et al., Z. Phys. C34 (1987) 163.
9. B. Humpert, Phys. Lett. 131B (1983) 461;
   N. Paver and D. Treleani, Phys. Lett. 146B (1984) 252;
   B. Humpert and R. Odorico, Phys. Lett. 154B (1985) 211;
   F. Halzen, U. of Wisconsin-Madison preprint MAD/PH/340 (1987).
10. J. Ellis et al., Ann. Rev. Nucl. Part. Sc. 32 (1982) 443.
11. T. Akesson et al., Proceedings of the Workshop on Physics at Future Accelerators, La Thuile (Italy) and Geneva (Switzerland), 1987, CERN 87-07, 174.
12. UA2 Collaboration, R. Ansari et al., Phys. Lett. 186B (1987) 452.

RESULTS ON JETS FROM THE UA1 EXPERIMENT

W.G. Scott

Oliver Lodge Laboratory
University of Liverpool

INTRODUCTION

This contribution reviews some of the results on jets from the UA1 experiment at the CERN SPS $p\bar{p}$ collider. The results presented are based on data taking corresponding to an integrated luminosity $LT \simeq 120nb^{-1}$ at $\sqrt{s} = 546GeV$ during $1982-1983$ and data taking corresponding to an integrated luminosity $LT \simeq 600nb^{-1}$ at $\sqrt{s} = 630GeV$ during $1984-1985$. In addition interesting results were obtained from the 1985 ramping run in which the energy of the SPS was ramped continuously over a range of energies $\sqrt{s} = 200 - 900GeV$ at lower luminosities, $LT \simeq 5\mu b^{-1}$.

The bulk of the data was obtained using the UA1 jet-trigger which requires a localised transverse energy deposition in the calorimetry exceeding a specified E_T-threshold ,typically $E_T > 20, 30GeV$ depending on the running conditions. Small samples of data were also taken using lower jet-trigger thresholds $E_T > 5 - 10GeV$. Finally substantial data samples were recorded under minimum-bias conditions for which no trigger requirement is imposed. All of the data taken during the ramping run were recorded using the minimum-bias trigger.

JET CROSS-SECTIONS

Data from the ramping run have been used to investigate the energy dependence of jet production [1]. Figure 1 shows the cross-section for events with one (or more) jets exeeding a nominal transverse energy threshold $E_T > 5GeV$ for various values of of $\sqrt{s}$ as measured by UA1 in the ramping run. The jets are defined using the UA1 jet algorithm [2] which combines calorimeter hits in a cone $\Delta R = \sqrt{(\Delta\eta^2 + \Delta\phi^2)}$ around a jet initiator which satisfies $E_T > 1.5GeV$. The threshold $E_T > 5GeV$ is nominal to the extent that no corrections have been applied to at this stage to account for energy uncorrelated with the jet, which "splashes" randomly into the jet-cone from the rest of the event. The "splash-in" effect leads to a lower effective threshold which is probably in the range $E_T > 2 - 4GeV$. From Figure 1 we may conclude that the jet cross-section increases rapidly with increasing $\sqrt{s}$. The behaviour of the total cross-section is also shown for comparison. From an experimental point of view it is an

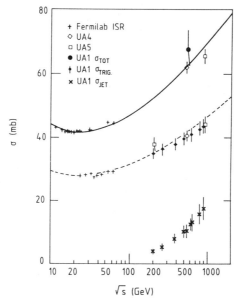

Figure 1. The cross-section for events with one or more jets exceeding a nominal threshold $E_T > 5GeV$ as a function of $\sqrt{s}$. The total cross-section is also shown.

open question whether or not minijet production may be regarded as the cause of the increase in the total cross-section in this energy range. In any event it is apparent that minijet production represents an appreciable fraction of the total cross-section at energies $\sqrt{s} \simeq 1TeV$.

A rapid increase in the jet production cross-section for a fixed transverse momentum threshold is predicted in QCD. The increase reflects the rapid growth in the number of contributing parton-pairs The subprocess cross-section $\sigma(p_T) \simeq \alpha_s^2/p_T^2$ is constant independent of $\sqrt{s}$. If the partons have a bremsstrahlung distribution then the number of contributing parton-pairs grows like ln^2s. More detailed QCD calculations [3] predict essentially the same behaviour, and in fact agree rather well with the data if the effective threshold is suitably chosen ($p_T \simeq 2GeV$).

The p_T dependence of the jet cross-section at $\sqrt{s} = 630GeV$ is shown in Figure 2 [1,4]. At this stage of the analysis the jet energies have been corrected to account for jet energy which falls outside the the jet-cone ($\simeq 10\%$) and for "splash-in" ($2-4GeV$). Corrections have also been applied to the measured rates to correct for energy resolution smearing and for trigger inefficienies. In Figure 2 the solid curve represents the leading order QCD prediction based on Duke and Owens structure functions assssuming $\Lambda_{QCD} = 0.2GeV$. The prediction is absolute and no normalisation adjustment has been made to fit the data

The comparison of Figure 2 is among the more spectacular and convincing of QCD tests. The data and the theoretical curve track each other closely over about nine orders of magnitude in rate. More critically, over most of the p_T-range, the data tend to lie above the theorteical curve by about a factor of two. This tendency is reminiscent of the famous K-factor effect observed in Drell-Yan processes. Unfortunately systematic errors in the jet energy and scale, due to uncertainties in calorimeter calibrations and in the jet energy corrections, lead to systematic errors in the jet cross-sections at the level of a factor of two, and rule out a significant measurement of the K-factor in this case. In the theory the K-factor represents the effect of of the next to leading order

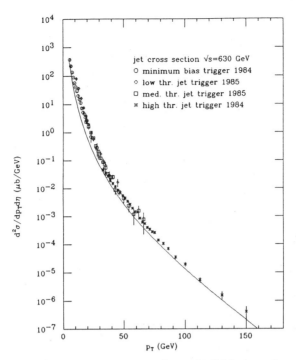

Figure 2. The inclusive jet cross-section at $\sqrt{s} = 630 GeV$ plotted as a function of p_T. The solid curve is the QCD prediction based on Duke and Owens structure functions assuming $\Lambda_{QCD} = 0.2 GeV$.

corrections;these are much more complicated here than for the Drell-Yan process and will depend sensitively on the jet definition ,i.e. the jet-cone size $\Delta R < 1$ etc.

In Figure 3 the jet cross-section is plotted as a function of p_T for several values of $\sqrt{s}$ in the range $\sqrt{s} = 200 - 900 GeV$ based on the ramping run data. The data are fully corrected as described above. All the remarks regarding the systematic errors apply here except that even more caution should be excercised regarding the low p_T points $p_T = 5 - 10 GeV$,where jet energy corrections are very large. The solid curves represent the QCD predictions which in this case have been scaled up (all three curves together) to match the data better. Although qualitatively correct, the predictions seem to fail quantitatively: the 900GeV data tend to lie above the theoretical curve for $p_T > 10 GeV$ suggesting that the energy dependence is actually somewhat faster than predicted by the theory. Note that the errors shown in Figure 3 are statistical only. In view of the uncertain systematics which are necessarily associated with the low p_T jets, the true significance of the discrepancy is difficult to asssess.

A partial cancellation of systematic errors is acheived by computing the ratio of cross-sections for different energies. Figure 4 shows the cross-section ratios $\sqrt{s} = 900 GeV / \sqrt{s} = 500 GeV$ and $\sqrt{s} = 200 GeV / \sqrt{s} = 500 GeV$ plotted as a function of p_T. In this case the errors shown include estimates for the systematic errors. It is interesting to that the behaviour of these ratios in the limit $p_T \to 0$ is a sensitive measure of the behaviour of the gluon structure function at small x. A bremsstrahlung $(1/x)$ distribution leads to a ratio of unity as indicated by the solid curves in Figure 4. A $(1/x^{3/2})$ distribution as suggested by Collins [5] leads to a ratio equal to the ratio of beam energies as indicated by the solid curves. The present data seem to favour a $1/x$ behaviour, though higher energies are needed to settle the question definitively.

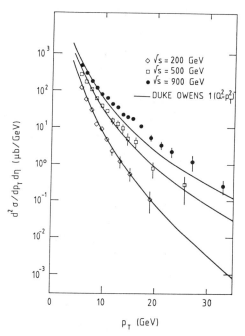

Figure 3. The inclusive jet cross-section at $\sqrt{s} = 200, 500, 900 GeV$ plotted as a function of p_T. The solid curves are the QCD predictions which have been scaled up by a common factor to fit the data better

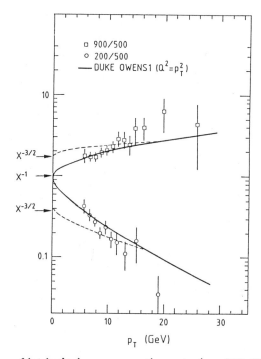

Figure 4. The ratios of jet inclusive cross-sections at $\sqrt{s} = 200, 500$ and 900GeV. The curves indicate the possible behaviour as $p_T \to 0$ depending on the behaviour of gluon structure function at small x

TWO-JET AND THREE-JET CROSS-SECTIONS

Two-jet and three-jet production have been studied extensively in the UA1 experiment with the aim of developing quantitative QCD tests using hadron beams. The values of the space-like momentum transfer have been extended far beyond those explored present day lepton-hadron experiments. In hadron-hadron experiments two-jet production results direcly from parton-parton scattering processes which proceed by purely strong interactions. Three-jet production is of interest as a strong radiatative correction to parton-parton scattering.

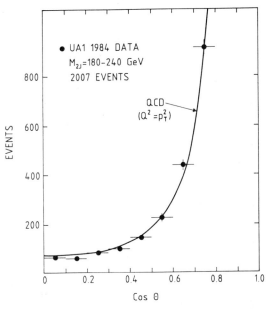

Figure 5. The cms angular distribution of jet-pairs for two-jet mass $m_{2J} = 180 - 240 GeV$. The curve is the QCD prediction which incorporates various scale-breaking effects.

In the UA1 experiment considerable emphasis has been placed on the measurement of the two-jet cms angular distribution [6,7,8] In QCD the cms angular distribution for parton-parton scattering is expected to show a characteristic $(1 - cos\theta)^{-2}$ dependence on cms scattering angle θ, quite analogous, for example to the case of Bhabha-scattering in QED. Representative results from UA1 based on the 1984 running are shown in Figure 5. In passing we note that a jet-trigger based on a fixed E_T-threshold requirement is not ideal for the study of the two-jet angular distribution: a very restrictive mass cut has to be applied to ensure uniform angular acceptance over the full angular range and a large fraction of the events selected by the trigger are not used in the final analysis. A better procedure would be to implement two-jet mass cuts and cms angle cuts from the start, if possible, extending the analysis to lower values of the two-jet mass. In Figure 5 the data are plotted over the angular range $cos\theta = 0 - 0.8$ and for two-jet mass $m_{2J} = 180 - 240 GeV$.

In Figure 5 the solid curve shows the QCD prediction normalised to the data. The prediction takes account of scale-breaking effects in the strong coupling constant and

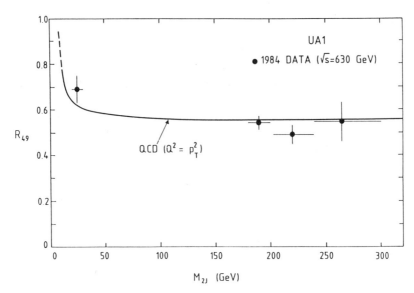

Figure 6. The wide-angle to small-angle ratio R (see text) plotted as a function of two-jet invariant mass. The curve is the QCD prediction assuming $Q^2 = p_T^2$

in the structure functions. These effects modify the scaling curve by suppressing the wide-angle scattering relative to the small-angle scattering and are required to fit the data. Clearly, in principal, precise measurements of the shape of the two-jet angular distribution can yield information on the value of Λ_{QCD} in full analogy with the case of deep inelastic scattering.

A sensitive measure of the shape of the two-jet angular distribution is obtained by computing the ratio of wide-angle to small-angle scattering: $R = \sigma(\cos\theta = 0 - 0.6)/\sigma(\cos\theta = 0.6 - 0.8)$ The dependence of this ratio on two-jet mass is shown in Figure 6. Over most of the range the wide-angle to small-angle ratio is essentially constant independent of two-jet mass. The average value of R for $m_{2J} > 180 GeV$ is given by $R = 0.521 \pm 0.025$. It will be interesting to see if this ratio remains constant at the higher values of two-jet mass which will be available at the FNAL Tevatron collider.

Full α_S^3 corrections to parton-parton scattering exist [9] but have not yet been incorporated into a phenomenological analysis. As is well known, in the absence of higher order corrections, the definition of the Q^2-scale parameter appearing in the theoretical formulae remains arbitrary. Figure 7 shows the theoretical prediction for R plotted as a function of the Q^2-scale choice, assuming $\Lambda_{QCD} = 0.2 GeV$. As a consequence of the Q^2-scale ambiguity it is not yet possible to convert the measured value of R into a serious measurement of Λ_{QCD}. This will be possible however when the Q^2-scale question is resolved: for example if $Q^2 = p_T^2/20$ then $ln(m_{proton}/\Lambda_{QCD}) = 1.12 \pm 0.78$. Note that the error in $ln(m_{proton}/\Lambda_{QCD})$ is symmetric and independent of the Q^2-scale ambiguity.

Figure 8 shows the two-jet mass distribution, based on the combined 1984 and 1985 data samples, for two different angular ranges $\cos\theta = 0 - 0.2$ and $\cos\theta = 0.6 - 0.8$. The shape of the mass distribution is well reproduced by the QCD curve in each case. In particular there is no evidence for new resonances, in this mass range, decaying to two-jets. In QCD the shape of the mass distribution is determined by the shape of the effective structure function $F(x) = G(x) + 4/9(Q(x) + \bar{Q}(x))$ [10].

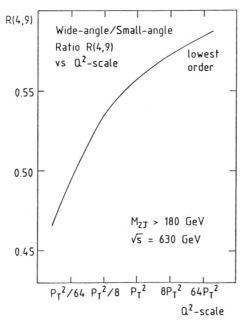

Figure 7. The theoretical prediction for the wide-angle to small-angle ratio R (see text) plotted as a function of the Q^2-scale choice.

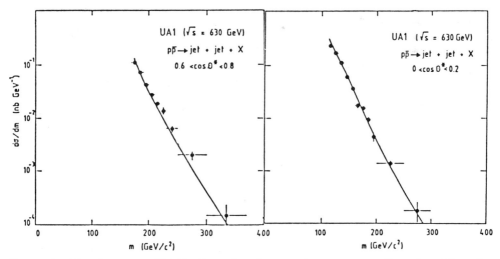

Figure 8. Two-jet mass distributions for the angular ranges $cos\theta = 0 - 0.2$ and $cos\theta = 0.6 - 0.8$ The curve is the QCD prediction which has been scaled up to fit the data.

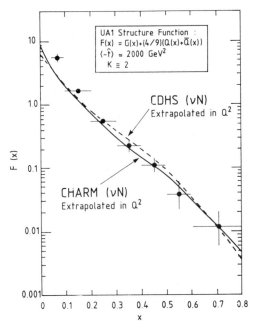

Figure 9. The effective structure function $F(x) = G(x) + 4/9(Q(x) + \bar{Q}(x))$

In the UA1 two-jet analysis [6] the effective structure function has been extracted from the data assuming factorisation in x_1 and x_2: i.e. $S(x_1, x_2) = F(x_1)F(x_2)$, where x_1 and x_2 are the scaled longitudinal momenta of the incoming partons and $S(x_1, x_2)$ is the event density, appropriately weighted, in the x_1, x_2-plane . if the x_1, x_2-plane is divided into n by n square bins with coordinates $x_1(i), x_2(j), i, j = 1, n$ then the structure function may be projected-out using the expression:

$$F(x(k)) = \sqrt{(\sum_{i=1}^{n} \sum_{j=1}^{n} S(x_1(i), x_2(j)))} - \sqrt{(\sum_{i \neq k}^{n} \sum_{j \neq k}^{n} S(x_1(i), x_2(j)))}$$

The measured structure function , based on the 1982 data is shown in Figure 9. More precise data based on the combined 1984/1985 sample have been presented recently [11].

To a good approximation the same effective structure function is also relevant for three-jet production[12]. In the three-jet case, for fixed cms energy, there are four independent variables at the subprocess level. In the UA1 three-jet analysis [7] the data are plotted as a function of dimensionless phase-space variables: the Dalitz-plot variables x_3, x_4 ,which determine the sharing of the available energy between the three jets and the angular variables θ and ψ, which specify the orientation of the three-jet system with respect to the incoming parton axis. The distributions in these variables, based on the combined 1983/1984 data, are shown in Figure 10 compared to leading-order (α_S^3) QCD predictions [13]. The data are consistent with the QCD predictions and demonstrate very clearly initial and final-state bremsstrahlung effects.

The ratio of three-jet to two-jet production is shown in Figure 11 as a function of the three-jet (two-jet) mass. The three-jet to two-jet ratio is consistent with a constant over the mass range shown ($m_{2J} = 150 - 300 GeV$). Very naively the three-jet to two-jet ratio measures the value of α_S, and for the central acceptance region UA1

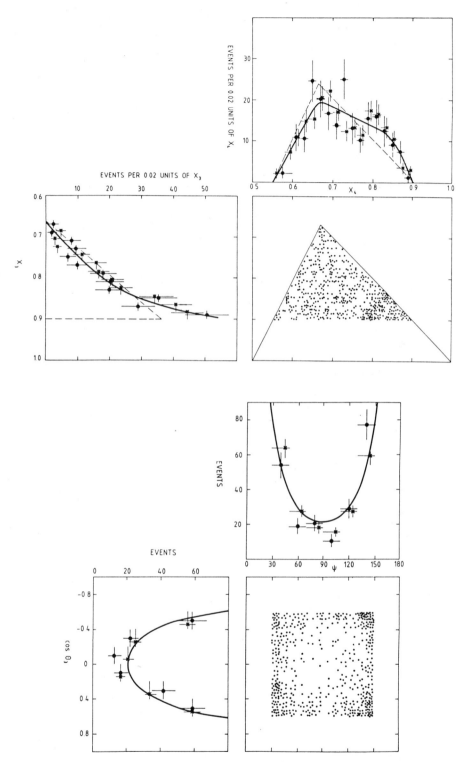

Figure 10. Differential cross-sections for three jet events plotted as a function of the Dalitz-plot variables x_3 and x_4 and as a function of the angular variables θ and ϕ (see text). The curves show the leading-order QCD predictions.

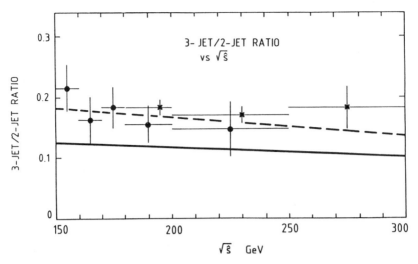

Figure 11. The ratio of three-jet to two-jet cross-sections plotted as a function of the three-jet (two-jet) mass. The curves show the QCD predictions for two different choices of the three-jet Q^2-scale.

obtains the value "α_S" $= 0.23 \pm 0.02 \pm 0.03$. This result is in excellent agreement with a corresponding result obtained by the UA2 collaboration [14]. However such a large value of α_S would be in contradiction with existing low energy measurements from e^+e^- experiments and from deep inelastic scattering; any extrapolation of existing low energy measurements up to collider energies predicts $\alpha_S \simeq 0.12 - 0.14$. It is very likely, that at least part of the problem here is due to a mismatch in the Q^2-scales for three-jet and two-jet events. In Figure 11 the solid curve shows the predicted three-jet to two-jet ratio plotted on the assumption that the Q^2-scales for three-jet and two-jet production are matched i.e. $Q_{3J} = Q_{2J}$. The broken curve, which is in better agreement with the data, has been calculated on the assumption $Q_{3J} = 2/3 Q_{2J}$. In the latter case there is an imperfect cancellation of the common factor α_S^2 (and of the effective structure function squared) which are now evaluated at different Q^2-scales in the three-jet and two-jet cross-sections. Since complete higher order (α_s^4) corrections to three-jet cross-sections are unlikely to become available in the near future the problem of the ambiguous Q^2-scale for three-jet cross-sections is likely to be with us for some time to come. In fact neither UA1 nor UA2 make any claim to measure α_S itself, based on the measurement of the three-jet to two-jet ratio.

FRAGMENTATION

In a recent publication [15], UA1 have presented results on the ratio of quark and gluon fragmentation functions. The analysis is based on the measurement of charged hadrons within jets using the UA1 central detector. A statistical separation of quark and gluon jets is acheived,exploiting the difference in shape of the quark and gluon structure functions in the proton to compare quark and gluon jets at the same $< Q > \simeq < p_T > \simeq 45 GeV$. In Figure 12 the open squares represent the UA1 data for the ratio of gluon to quark fragmentation functions plotted as a function of the fragmentation variable z. Within large errors the UA1 data suggest a softer fragmentation function for gluon jets relative to quark jets.

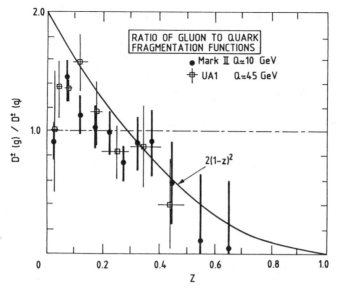

Figure 12. The ratio of gluon to quark fragmentation functions plotted as a function of the fragmentation variable z

Remarkably these data are in excellent agreement with recent data from the MARK II collaboration [16]. In the MARK II analysis three-jet events at $\sqrt{s} = 29 GeV$ are compared with two-jet events at $\sqrt{s} = 19.3 GeV$. The ratio of gluon to quark fragmentation functions at $Q = 10 GeV$ based on the MARK II data is shown in Figure 12. Clearly this comparison is strictly only valid in the scaling approximation; scale breaking effects tend to soften the gluon distribution further at the higher value of Q^2, but are evidently insignificant here. Taking both data sets together the conclusion has to be that the fragmentation function for gluons is indeed softer than the fragmentation function for quarks by at least a factor $(1 - z)^2$.

Acknowledgement

It is a pleasure to thank colleagues in the UA1 experiment for continued collaboration and Brad Cox for the opportunity to speak at such an enjoyable meeting.

References

[1]   C.Albajar et al. (UA1 Collab) to be published.

[2]   G.Arnison et al. (UA1 Collab) *Phys. Lett.* **123B** (1983) 115

[3]   W.J.Stirling. Proceedings of the International Workshop on $p\bar{p}$ physics; Aachen (1986);editor K.Eggert.

[4]   G.Arnison et al. (UA1 Collab) *Phys. Lett.* **172B** (1986) 481

[5]   J.Collins. Illinois Preprint ITT-86-0298 (1986)

[6]   G.Arnison et al. (UA1 Collab) *Phys. Lett.* **136B** (1984) 294

[7]   G.Arnison et al. (UA1 Collab) *Phys. Lett.* **158B** (1985) 494

[8]   G Arnison et al. (UA1 Collab) *Phys. Lett.* **177B** (1986) 244

[9]   R.K.Ellis et al. *Nucl. Phys.* **B269** (1986) 445

[10]   B.L.Combridge and C.J.Maxwell *Nucl. Phys.* **B239** (1984) 429

[11]   S.Li. Proceedings of the International Europhysics Conference on High Energy Physics; Uppsala (1987);editor O.Botner.

[12]   B.L.Combridge and C.J.Maxwell *Phys. Lett.* **151B** (1985) 299

[13]   F.Berends et al. *Phys. Lett.* **103B** (1981) 124

[14]   J.A.Appel et al. *Zeit. Phys.* **C30** (1986) 341

[15]   G.Arnison et al. (UA1 Collab) *Nucl. Phys.* **B276** (1986) 253

[16]   A.Petersen et al. *Phys. Rev. Lett.* **55** (1985) 1954

RECENT RESULTS FROM FERMILAB E557 AND E672 EXPERIMENTS

A. Zieminski

Indiana University
Bloomington, IN 47405

ABSTRACT

Recent results from two Fermilab experiments, E557 and E672 are reviewed. The E557 experiment is studying production of jets in pN collisions at c.m. energies 27 and 39 GeV. The A - dependence for jet production is well parametrized by an A^{α} behavior with $\alpha = 1.0$ - 1.2 . The beam jet fragmentation is broader than expected, strongly depends on the event configuration at 90° and does not show a leading proton signal at high E_T. The E672 experiment is aimed at studying hadronic final states produced in association with high mass dimuons. It has started to take data during the Fall 1987 run. We discuss the experimental setup, trigger rates and expected physics results on the hadroproduction of chi states and beauty.

1. Introduction

This paper summarizes recent results from two Fermilab experiments, E557 and E672. The E557 experiment has analysed production of jets in pN collisions at c.m. energies 27 and 39 GeV [1] . Beam energies of Fermilab fixed target experiments are somewhat low for jet studies. Nevertheless, results discussed here are not available from other laboratories and provide a unique information on:

— A - dependence of high transverse energy events and jet production [2] and

— beam jet fragmentation [3-4] .

The E672 experiment [5] is aimed at studying hadronic final states produced in association with high mass dimuons. It has started to take data during the Fall 1987 run. We discuss the experimental setup, performance of the trigger processor, trigger rates and expected physics results on the hadroproduction of chi states and beauty particles.

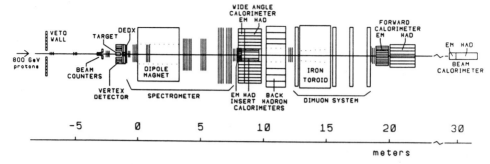

Fig. 1. Apparatus layout for the E557/E672 1984 run.

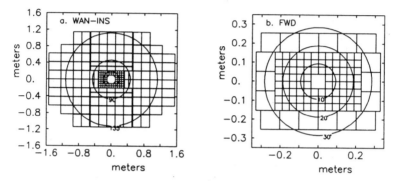

Fig. 2. Calorimeter segmentation.

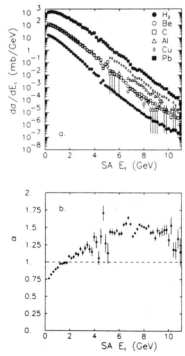

Fig. 3a. Cross section vs. E_T for pA collisions at 800 GeV/c. The SA trigger
   data with an aperture of 1.4 sr are plotted.
   3b. Exponent α from an A^α fit to the data of (a) vs. E_T. The pp data
   not included in the fit.

2. E557 experiment at Fermilab

This experiment studies production of high transverse energy events and jets in pp and p-nucleus (pA) at 400 GeV/c and 800 GeV/c. These studies expand the c.m. energy range for jet production, enable event structure analysis with an almost 4π coverage and , in the pA case, can measure the effect of interactions of secondaries produced in the initial scatter with the other nucleons in the nucleus.

The layout of the experiment for the Spring 1984 run is shown in Fig. 1. The magnetic spectrometer is followed by series highly segmented calorimeters,denoted by the names wide angle (WAN), insert (INS), forward (FWD) and beam. Each of these calorimeters consisted of an electromagnetic section followed immediately downstream by a fully absorptive hadronic section. The geometry and the granularity of the WAN,INS and FWD calorimeters are shown in Fig. 2. For pp data the beam was incident on a 45-cm liquid hydrogen target; for pA data nuclear targets of Be,C,Al,Cu and Pb replaced the hydrogen target and were constructed of three successive foils thin enough to avoid significant rescattering.

We triggered on events which deposit a large amount of transverse energy, E_T, within the aperture of a calorimeter located at $90°$ in the c.m.s. Two classes of triggers are discussed here: the GLOBAL trigger with 2π azimuthal and $45° - 135°$ c.m. polar angle coverage and small aperture trigger (SA) corresponding to a $\Delta\Phi = 90°$ and $60° < \theta^* < 110°$ c.m. coverage. At Fermilab energies GLOBAL triggers select predominantly non-jet events[6] , whereas SA triggers should select large fraction of hard scatters[7] .

· The E557 experiment has an almost 4π coverage and is able to observe on the average 92% of the incident 800 GeV energy. The total energy distribution (not shown) has an average value of 740 GeV with the FWHM of 150 GeV.

3. The A - dependence for high- E_T and jet event production

Two competing mechanisms are believed to be responsible for pA collisions that produce large amounts of transverse energy relative to the incident beam direction. They are:

— multiple soft scattering involving many nucleons within the nucleus and

— hard scattering of constituents that can produce high-pt jets which then propagate through the nucleus.

The separation of events produced by each of the mechanisms should lead to a better understanding of the space time development of the intermediate nuclear matter. Previous experiments[8] , including our results from the 1981 run, have found a strong nuclear dependence for the production of high E_T events. The new analysis expands these measurements and makes a first attempt to isolate the A-dependence of the hard scattering component.

The observed $d\sigma/dE_T$ distributions for the SA data obtained for pp and various pA collisions at 800 GeV/c are shown in Fig. 3. The A-dependence of $d\sigma/dE_T$ is well described by the A^α paramerization in the entire E_T range[2] . The asymptotic values of α at large E_T is 1.6 for the Global data (not shown) and 1.4 for the SA data. The values of α presented in Fig. 3 are for all events with a given E_T, irrespectively of the event structure in the transverse plane. The event structure in the transverse plane is usually characterized by the planarity variable[6] , which approaches 1 for 'jetty' events and is close to 0.4 for the isotropic distribution in the transverse plane. It has been shown before[7] that already at Fermilab energies high planarity events selected with 'jet' enhancing triggers are predominantly due to hard scattering. When the A

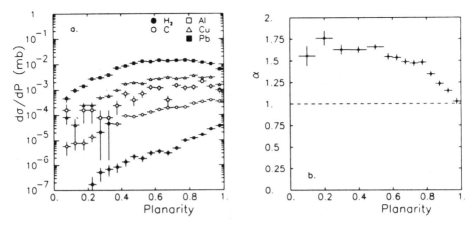

Fig. 4a. Planarity distributions for pA collisions. Same data as in Fig.3
with an additional cut of $E_T > 8.2$ GeV.
4b. α vs. planarity

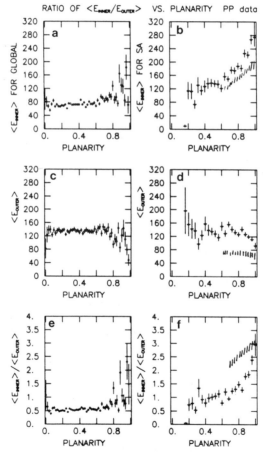

RATIO OF $\langle E_{INNER}/E_{OUTER} \rangle$ VS. PLANARITY PP data

Fig. 5. Average laboratory energies observed in the OUTER and INNER cones
(see text) and their ratio as a function of planarity for events
selected with the GLOBAL trigger ($E_T > 21$ GeV; a,c,e) and the SA
trigger ($E_T > 6.5$ GeV; b,d,f). pp data at 800 GeV/c are shown.

- dependence is examined as a function of planarity one notices that the exponent α reaches values of 1.1 - 1.2 at the highest planarities (Fig. 4). This effect has been also observed at 400 GeV/c by E609 Collaboration[9] , which used the so called "two-high" trigger[10] to enhance the jet signal.

These results are the first measurements of the A dependence of a hard scattering cross section in calorimeter jet experiments. However, we have to stress that values of α , as determined in this analysis are sensitive not only the A - dependence for the "jet" production rates but in addition reflect the possible change with A in the event structure as described by the planarity variable. The analysis to separate the two effects is in progress [11] .

The A - dependence of event structure for the GLOBAL trigger data is discussed in refs. [13-14] .

4. Beam Jet Fragmentation

Fixed target exeriments are a priori better equipped to study the beam jet fragmentation than the collider experiments which suffer from the beam pipe obstruction and limited spacial resolution in the forward direction. The available information on the beam fragmentation in hadronic jet experiments is rather sparce. The analysis of the 1984 E557 data is still in progress. Here we present some preliminary 1984 results and published results on charged particles distributions from the 1981 run.

We have studied the energy flows in cones corresponding to different c.m. polar angle regions for massless particles: $0° < \theta^* < 25°$: (INNER cone) and $25° < \theta^* < 45°$: (OUTER cone). The average laboratory energies in these cones are shown in Fig.5 as functions of planarity for the GLOBAL and SA triggers. We note that planarity was determined in the $45° < \theta^* < 135°$ region only. One observes a strong increase of the forward beam jet collimation for planar events, in particular for the SA data, which contain larger fraction of hard scatters than the GLOBAL data. The difference between the two data sets points to a dependence of the beam jet fragmentation on the mechanism leading to a large E_T event. Many partons from the incoming proton are likely to participate in the collision leading to a high E_T GLOBAL event. The resulting beam jet is relatively broad without strong leading particle effect. This point is better illustrated by the 400 GeV/c pp data on charged particle production in the forward direction (GLOBAL trigger)[3-4] . The data in Fig. 6 presented in terms of variable $x = 2 p_z/(\sqrt{s}-E_T)$, which takes into account depletion of the available c.m. energy by the E_T of the trigger. The difference in slope of x dσ/dx between negative and positive particles, visible at low E_T and related to a leading proton effect, completely disappears for $E_T > 8$ GeV. The ratio of positive and negative structure functions integrated over 0.5 $< x < 1.0$ decreases with E_T to an asymptotic value of 1.8+/-0.2 (Fig. 7). This result is consistent with the expectation for a single quark or diquark fragmentation into pions alone [15] , but does not leave room for a production of fast protons in the forward region. The laboratory polar angle charge distributions[4] presented in Fig. 8 further illustrate the disappearence of leading protons for high E_T events.

5. The Fermilab E672 experiment

(a) Physics goals

Experiment E672 is a fixed target experiment in the MW beam, the aim of which is to study hadronic processes yielding high mass dimuons (the trigger) and associated particles. The more specific goals are:

— the characteristics of ψ production and Drell-Yan processes including A-dependance,

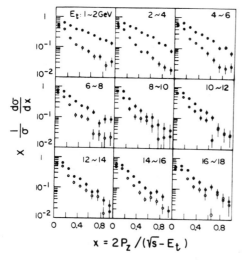

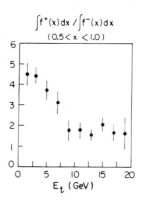

Fig. 6. Beam jet fragmentation function for positive and negative particles in nine E_T slices. Data for pp collisions at 400 GeV/c are shown. Variable x is defined in the text.

Fig. 7. The ratio of positive to negative beam jet fragmentation functions integrated over $0.5 < x < 1.0$ region vs. GLOBAL E_T. Same data as in Fig.6 were used.

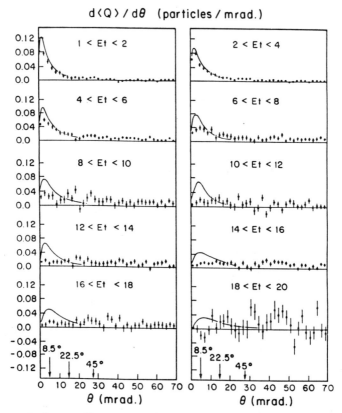

Fig. 8. Forward charge flows as a function of laboratory polar angles in different E_T bins.

multiplicities and momenta of associated particles (in almost the entire region of phase space), analysis of spectator jets

— the hadronic production of $\chi \rightarrow \psi + \gamma$ states and associated particles

— BB production where one of the Bs decays into $\psi + X$,or where the semileptonic decays of both B and B yield a $\mu\mu$ pair with large invariant mass.

We intend to run with negative and positive beams (π^-, K^-, π^+, p) at 530 GeV/c and with a proton beam of 800 GeV/c. The proposed targets include C, Al and W (2%-6% interaction lenghts thick). The bulk of the 1987 data will be taken with the negative pion beam of 530 GeV/c. The expected event rates for processes listed above are given in Sect.7.

(b) Apparatus

Experiment E672 studies dimuon production in an open-geometry configuration. The heart of the E672 experiment is a dimuon detector consisting of toroid magnet, proportional chambers, scintillator hodoscopes and a trigger processor. The E672 detector shares the MW beam line with E706 (see Fig. 9). The physics goal of E706 is a study of direct photon production. The E706 detector consists of a liquid argon calorimeter (LAC) and a spectrometer including a silicon strip detector (SSD; 14 x-y planes), 16 planes of PWC's and a forward calorimeter [16] . The dipole analysing magnet and three planes of the scintillating veto wall were assembled by the joint effort of E672 and E706. The two experiments take data simultaneously with the information from the entire apparatus being available to both groups. The data from individual detectors after being read by four independent PDP computers are send to a microVax where they are concatenated and routed to one of the two tape drives, depending which of the trigger bits (E672 or E706) was on.

The layout of the E672 dimuon detector, located 21 m to 33 m downstream of the target, is shown in Fig. 10. It consists of :

— a PWC station (D) consisting of 3 planes

— a tungsten and steel beam dump

— a toroid providing a nominal p_T kick of 1.3 GeV/c

— four PWC chambers (μ1 through μ4); three planes each

— two planes of scintillator, sixteen wedge-shaped counters each

— concrete and iron shielding and

— associated electronics and pretrigger and trigger processors.

Signals from the scintillator walls are used to form the dimuon pretrigger (2 or more hits in each of the walls). Data from the muon chambers are read via the LeCroy PCOS III system into a trigger processor (DMTP)[17] which: (a) finds space points at two of the chambers; (b) finds tracks; (c) computes momenta assuming the muons orginated in the target and (d) calculates effective masses. The calculations of the momenta and masses take into account correction for the magnetic field in the dipole magnet. The mass resolution of the processor is set by multiple scattering to about 700 MeV. The average trigger formation time has been measured to be less than 10 μs.

The pretrigger rate in the present configuration for πC interactions at 530 GeV/c is 10^{-4} interactions and is approximately 5 times higher than a genuine two muon tracks signal. The extra pretriggers are due to the leakage of muons from the beam dump thru the cracks around the toroid coils, hodoscope edge effects, bremsstrahlung off single tracks and the DMTP acceptance cuts. The trigger processor dimuon mass cutoff at 2 GeV reduces the trigger rate to $2*10^{-6}$ interactions.

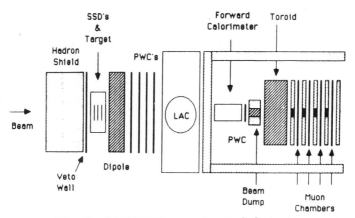

Fig. 9. E672/E706 experimental layout

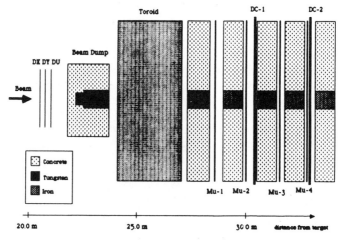

Fig. 10. E672 detector

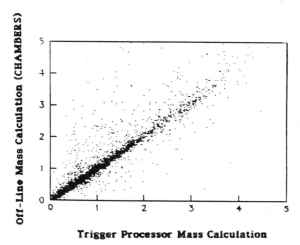

Fig. 11. Scatter plot of μ+μ⁻ masses calculated by the trigger processor vs. off-line mass calculations.

The $\mu^+\mu^-$ dimuon masses computed on-line by the trigger processor are compared with the off-line mass calculations in Fig. 11. The off-line analysis fitted tracks in the muon chambers only. The difference between the two mass values is due to more sofisticated algorithm used in the off-line analysis and higher computer precision. The sigma of the difference is consistent with the Monte Carlo predicted mass resolution for the DMTP of 650 MeV.

6. The A - dependence for the forward ψ production

During September 1987 data were taken with the E672 part of the apparatus only (no SSD or PWC data were available). The incident beam was 530 GeV/c π^-. We ran with a number of thick targets: 4" and 8" C, 5" and 6.5" Al, 4" Cu and 2" and 3" Pb. Events were selected using pretrigger only, however the trigger processor results were also recorded. The off-line track and mass reconstruction was based entirely on muon chambers hits. The $\mu^+\mu^-$ mass distribution for all targets combined is shown in Fig. 12A clear peak around mass of 3.0 GeV corresponds to ψ production. The fit of an exponential background and a gaussian yielded 1600 50 ψ s with the mass of 3.01 GeV and width of 500 MeV (the 0.1 GeV shift in the mass is consistent with the uncertaintities in the strength of the toroid field and corrections for the muon energy losses in various materials). The mass resolution improves considerabily when the D-station PWCs are included in the track fits (Fig. 13).

The mass fits were repeated for different samples of events to establish contributions from individual targets and to check effects of target thickness, backscatter of target debrits toward the muon veto wall, hello muons contamination etc. The preliminary results for the A - dependence of the cross sections are based on one third of the total event sample (the least biased). The cross sections presented in Fig. 14 were corrected for the $x_f > 0$ acceptance estimated to be 0.28 +/- 0.03. Other corrections: for the pretrigger and chambers efficiencies, software losses, Drell-Yan pairs, target thickness , non vertex interactions etc. were not applied. We have estimated that their net effect could increase the ψ production rate by atmost 20% and should be almost target independent. The observed rates are consistent with the extrapolated πp data [18] and indicate an $A^{0.9}$, rather than $A^{1.0}$ dependence for ψ forward production. Similar A - dependence has been recently reported by othet experiments [19] .

The observed and acceptance corrected x_f distribution for the ψ mass region are shown in Fig. 15. The x_f region accessible to this experiment extends from $x_f > 0.1$ to $x_f < 0.6$. The corrected x_f spectrum for the Al data is shifted toward $x_f = 0$ when compared to the πp data (taken at lower c.m. energies[19]).

Results presented here serve mainly as a benchmarking of the experiment. They confirm feasibility of triggering on high mass dimuons in an open geometry set up with more than 17 m decay path for secondaries. Let us also note that the expected ψ mass resolution , once the magnetic spectrometer data are available, is 110 MeV, sufficient for a clear, almost background free ψ selection.

The measured ψ production rate off Carbon target in our experiment is one ψ per 4 10^6 interactions. The pretrigger/interaction ratio is 10^4. The trigger processor cutoff of 2 GeV selects 2 events/10^6 interactions. Additional selection of $\mu^+\mu^-$ mass combination only reduces this ratio further by 50%. Under such triggering conditions 20% of selected events would contain a ψ particle. The actual running conditions are less selective yielding one ψ per 100 to 200 triggers.

7. Expected event rates for the 1987/1988 E672 run

At this writing the E672 and E706 are taking data reading out information from the entire apparatus. The expected event rates by the end of this run are given below.

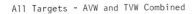
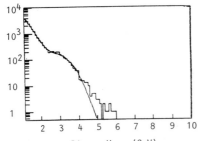

-Dimuon Mass (GeV)
Fitted to an Exponential Background + Gaussian

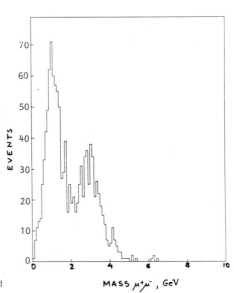

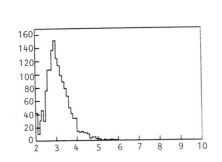

Dimuon Mass (GeV) - With Background Subtracted

MASS $\mu^+\mu^-$, GeV

Fig. 12. $\mu^+\mu^-$ off-line mass plot (muon chambers only).

Fig. 13. $\mu^+\mu^-$ off-line mass plot (muon detector (including D-station) only).

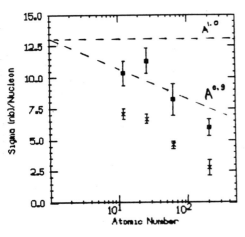

Fig. 14. A-dependence for ψ production ($x_F > 0$).

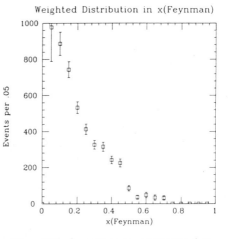

Fig. 15. Acceptance corrected x_f distributions for ψ s.

(a) Assumptions:

— negative pions at 530 GeV/c; $A^{0.9}$ dependence for hard processes.

— 500h of 2 MHz beam incident on a 6% C target; 23s long spills once a minute; The expected luminosity per nucleon: 5 pb^{-1}.

(b) ψ production: 18000 events

— σ (πp ; $x_f > 0.$) * BR($\psi \to \mu^+\mu^-$)= 13 nb[18] ; acceptance = 0.28.

(c) $\chi \to \psi + \gamma$ production: 3500 events; (1900 with E_γ >8 GeV).

— σ (π p $\to \chi \to \psi + \gamma$) * BR($\psi \to \mu^+\mu^-$) = 6.5 nb

— acceptance for γ s: 38% (all) ; 23% (E_γ >8 GeV).

χ mass resolution improves from 29 MeV (all) to 18 MeV (E_γ >8 GeV) assuming LAC energy resolution of 0.5% + 15%/$\sqrt{E}$ and 110 MeV ψ mass resolution (χ mass almost entierly depends on γ energy resolution). The expected χ mass resolution is mediocre, however our signal to background ratio after removing identified gammas from π° should be of the order 2:1.

(d) B B $\to \psi$ + X production: 10 events

— cross section = 10 nb [20] ; BR(B$\to \psi$)=1.2% ; BR($\psi \to \mu^+\mu^-$) = 7%

— acceptance = 11.5% assuming x_f , p_T distributions from[20] .

The SSD resolution for secondary vertices along the beam direction is approximately 1 mm. The resoulution in the transverse direction is better than 50 μm.

(e) two B semileptonic decays with $\mu\mu$ mass > 2.5 GeV; 21 events

— BR(B$\to \mu$ +X) = 11% [21] ; acceptance=3.5% ; (16% without the mass cut).

ACKNOWLEDGEMENTS

I thank my E672 Collaborators for prompt and efficient analysis of the data while they are being accumulated. Very stimulating discussions with R. Crittenden, A. Dzierba, T. Marshall, H. Martin, C. Stewart and T. Sulanke are highly appreciated.

REFERENCES

1) The members of the E557 Collaboration are: R. Gomez(Caltech), L. Dauwe, H. Haggerty, E. Malamud, M. Nikolic (Fermilab), S. Hagopian, B. Pifer (Florida State), R. Abrams, J. Ares, H. Goldberg, C. Halliwell S. Margulies, D. McLeod, A. Salminen, J. Solomon, G. Wu (UIC), S. Blessing, R. Crittenden, P. Draper, A. Dzierba, R. Heinz, J. Krider, T. Marshall, J. Martin, A. Sambamurti, P. Smith, A. Snyder, D. Stewart, T. Sulanke, A. Zieminski (Indiana), R. Ellsworth (George Mason),J. Goodman, S. Gupta, G. Yodh (Maryland), S. Ahn, T. Watts (Rutgers) V. Abramov, Yu. Antipov, B. Baldin, S. Denisov, V. Glebov, Yu. Gorin, V. Kryshkin, S. Petrukhin, S. Polovnikov, R. Sulyaev (Serpukhov)

2) Gomez, R., et al., Phys. Rev. D35, 2736 (1987)

3) Ahn, S., et al., Phys. Lett.B177, 233 (1986)

4) Ahn, S., et al., Phys. Lett.B183, 115 (1987)

5) The members of the E672 Collaboration are: R. Gomez, G. Ludlam (Caltech), J. Krider (Fermilab), H. Goldberg, S. Margulies, J. Solomon (UIC), T. Boyer, R. Crittenden, A. Dzierba,

S. Kartik, T. Marshall, J. Martin, C. Neyman, P. Smith, T. Sulanke, K. Welch, A. Zieminski (Indiana), C. Davis, H. Xiao (Louisville), L. Dauwe (Michigan at Flint), V. Abramov, Yu. Antipov, B. Baldin, S. Denisov, V. Glebov, Yu. Gorin, V. Koreshev, V. Kryshkin, S. Petrukhin, S. Polovnikov, V. Sirotenko, R. Sulyaev (Serpukhov)

6) De Marzo, C., et al., Nucl.Phys. B211, 375 (1983); Brown, B.C., et al., Phys. Rev. D29, 1895 (1984)

7) Akesson, T., et al., Phys. Lett. 128B, 354 (1983); Cormell, L., et al., Phys. Rev. Lett. 53, 1988 (1985)

8) Brown, B., et al., Phys. Rev. Lett. 50, 11 (1983)

9) Miettinen, H., et al., Nucl. Phys. A418, 315 (1984)

10) Arenton, M., et al., Phys. Rev. D31, 984 (1985)

11) Stewart, D., et al., 'Production of jets in pA collisions', to be submitted to Phys.Rev. D

12) Gomez, R., et al., Proceedings of the 2nd Inter. Workshop on Local Equilibrium in Strong Interaction Physics, Los Alamos,N.M., 1986; also: Fermilab preprint PUB-Conf.74/86

13) Halliwell, C., Proceedings of the Moriond Conf., 1987)

14) Ares, J., et al.,'Transverse Energy-Dependence of Event Structures' ,to be submitted to Phys.Rev.D

15) Allen, P., et al., Nucl. Phys. B214, 369 (1983)

16) Fanourakis, G., contribution to this Conference

17) Crittenden, R.,et al., E672 Memo No.141 (unpublished), 1986 also Crittenden, R. and Dzierba, A. in Proceedings from the Fermilab b-physics Workshop, Nov. 1987

18) Lemoigne, Y.,et al., Phys. Lett. 113B, 509(1982)

19) Badier, J.,et al., CERN-EP/83-86; also McDonald, K. and Conetti, S. contributions to this Conference

20) Berger, E. ,preprint ANL-HEP-PR-87-90 (1987)

SUMMARY AND ROUND TABLE DISCUSSION: HADRONIC JETS

Stephen D. Ellis

Department of Physics
University of Washington, Seattle, Washington 98195

Chairmen : S. D. Ellis, K. Pretzl

Round table members: L. Galtieri, A. Garfinkel and W. Scott

Hadronic jets play an essential role in the analysis of experimental data from both hadron machines and e^+e^- colliders. These jets consist of the collimated "spray" of hadronic debris generated by the fragmentation of a quark or a gluon which is either produced or scattered into an isolated region of phase space. While we believe that, due to the phenomenon of confinement, we can never observe truly isolated single quarks or gluons, these jets of hadrons are the clear "footprints" of individual quarks and gluons which, at least for a short time, were isolated in momentum space. However, since a colored object such as a quark or gluon cannot fragment in isolation into purely color singlet hadrons nor can such a process conserve energy and momentum, we know that the fragmentation process always involves some collaboration with the other partons in the event. Hence, even at the theoretical level, there is a fundamental and unavoidable ambiguity in the definition of a jet. This is an important but not essential limitation. While the earliest attempts to detect jets at hadron machines suffered from various forms of trigger bias difficulties, there are now clear jet signals at essentially all accelerators. At this meeting we heard about recent results on

- the fixed target physics of E-557/672[1] at Fermilab;

- jet physics in the AFS[2] data from the ISR;

- jets associated with $W^\pm/Z^o$ decay in the UA2 data[3,4];

- jet detection and multi-jet events at UA2[4];

- properties of jets and multi-jet events at UA1[5];

- the promise of the future at CDF[6] at the Tevatron;

- and a review[7] of recent theoretical progress in jet physics.

The following is not a comprehensive review of these presentations but rather a (biased) listing of the highlights of the jet session and the associated round table discussion.

FIXED TARGET

Using a spectrum of target materials E-557 has new results[1] on the A dependence of jet production. They trigger both on large global transverse energy, E_T, and on large E_T in a limited angular range with the latter sample being typically more "jetty" on average. Characterizing their results in the usual A^α form, they find α values only slightly larger than 1 (1.1 - 1.2) for the most "jetty" events while the more amorphous large E_T events exhibit α values near 1.6. This seems a nice confirmation that jets arise from single hard scatterings while large E_T events have a sizable multiple soft scattering component. Studies of the fragmentation of the beam particle into the so called beam jet suggest that this process is correlated with what is occurring in the transverse plane. Those events which are characterized by a single hard scattering exhibit a more narrow beam jet then those with large global E_T. Detailed studies contrasting how jets evolve in a nuclear environment versus the vacuum would also be very interesting.

COLLIDERS

While, as noted above, there is a fundamental ambiguity in the definition of a jet and while actual detectors only measure energies with finite accuracy, one of the most impressive features[2,4,5] of the collider jet data, is the agreement between theoretical predictions[7] and the jet production rate over a large range in the E_T of the jet and an incredible range in the magnitude of the cross section. The agreement between theory and experiment is at the 50% level and is as good as the agreement between the different experiments which implement somewhat differing jet definitions (the so called jet algorithms). Within these uncertainties this agreement constitutes a very strong indication that the jet structures being observed are those expected from QCD. The angular dependence observed in the jet data seems to strongly suggest[5] that an appropriate choice of characteristic hard scattering scale is $Q^2 = P_T^2$ as opposed to say $\hat{s}$ (the total parton-parton CM energy). It will be interesting to see how this result compares with forthcoming calculations[7,8] of higher order corrections to jet cross sections.

It was also reported[5] that UA1 is pushing their jet analysis to remarkably small E_T^{jet} (~ 5 GeV) where the perturbative analysis is suspect but the observed rate of so called "minijet" events is a finite fraction of the total rate. For that reason alone this topic deserves further study. While the energy dependence of the rate of these events seems to track the observed increase in the total cross section, it is far from clear that one should connect the two effects. The magnitude of the "minijet" rate is highly dependent on the minimum E_T^{jet} (although W. Scott argued that 5 GeV is a "natural" minimum jet energy) and the underlying soft interactions would likely be present (and contribute to σ_{TOT}) independent of the "minijets."

To perform more detailed studies, a typical technique is to define 2 and 3 jet samples via some jet definition and associated cuts and to consider the ratio of the corresponding rates. The resulting experimental numbers can be thought of as a measure of the quantity

$$R\left(\frac{3 \text{ jet}}{2 \text{ jet}}\right) = \alpha_s(Q^2)\frac{K_3(Q^2)}{K_2(Q^2)} \tag{1}$$

where the various sources of uncertainty (largely higher order corrections and Q^2 choice on the theory side and jet definition dependence on the experimental side – effects which are not uncorrelated) are all hidden in the "K- factors", K_3 and K_2. The resulting experimental numbers are

$$R = 0.23 \pm 0.02 \pm 0.03 \quad [\text{UA1, Ref.(5)}],$$

$$R = 0.236 \pm 0.004 \pm 0.04 \quad [\text{UA2, Ref.(4)}],$$

$$R = 0.18 \pm 0.03 \pm 0.04 \quad [\text{AFS, Ref.(2)}],$$

$$R = 0.19 \pm 0.02 \pm 0.04 \quad [\text{CMOR, Ref.(9)}]. \tag{2}$$

The appropriate value of Q^2 which should be associated with each of these numbers is somewhat uncertain but is order 2000 GeV^2 for the CERN collider data and order 300 GeV^2 for the ISR data. When compared to e^+e^- data, which suggest that α_s should be of 0.12 at the Q^2 of the CERN Collider, we must conclude that the two K factors are not equal and at least one is not too close to unity. It will interesting to have theoretical input on this issue but a full calculation of the first nontrivial term in K_3 seems unlikely. While the $2 \to 4$ processes are known at the tree level, an evaluation of the virtual corrections to the $2 \to 3$ processes is unlikely in the near future. On the other hand calculations of K_2 including the effects of the jet definition are in progress. [8] However, the conclusion must be that, for the moment, the quantity in Eq. (1) will not allow a precise test of QCD.

Another jet feature which has been studied[2,5] at colliders is the structure of the fragmentation function describing the distribution of hadrons within the jets. Moderately successful attempts have been made to separate quark and gluon jet samples on a statistical basis using our knowledge of the differing distributions of quarks and gluons in the incident protons. Within sizable systematic uncertainties the results are in good agreement with what is observed in e^+e^- collisions with the gluon jets exhibiting a "softer" distribution (more low E_T hadrons) than the quark jets as expected in QCD. (The comparison with the high quality data from MARKII[10] was especially emphasized by L. Galtieri during the round table discussion.)

Another interesting topic is the contribution to the observed 4 jet final states[2,4] from two independent $2 \to 2$ scattering processes (the double parton scattering process). The fact that the AFS reports such a contribution in their 63 GeV data while UA2 does not clearly see it in their 630 GeV data seems to be explained by the rather different kinematic regimes in which they look. UA2 looks at a sample characterized by a much larger value of $\sum_{\text{jets}} E_T$ than the AFS data. It would be very helpful to have more experimental information on this subject.

Finally it is important to highlight the recent success by UA2[3,4] in exhibiting the presence of the hadronic decays of the $W^\pm$ and Z° in their 2 jet sample. While this sort of analysis is in its infancy and for the moment relies on very detailed cuts (and knowing where to look), it will clearly be essential for much of the hoped for physics of the future. To identify the Higgs boson at the SSC or LHC through its decay into WW or ZZ (the bulk of the rate for a Higgs mass above ~ 200 GeV/c^2) it will be very helpful to be able to identify at least one of the W's or Z's via its $q\bar{q}$ or 2 jet decay. The lepton decay rates are an order of magnitude smaller than the hadronic rates. Thus this topic deserves considerable further study, both experimentally and theoretically.

THEORY

Recent progress on various issues of theoretical and phenomenological interest in jet physics were reviewed by C. Maxwell. [7] Of special interest are the essentially universal angular structures of the perturbatively evaluated 2 and 3 jet cross sections which seem to be in very good agreement with data. As noted earlier, there is an urgent need for a full implementation of the $2 \rightarrow 3$ and virtual $2 \rightarrow 2$ corrections calculated earlier by K. Ellis and J. Sexton[11] in a form applicable to data including the specific jet definition. A special focus was placed on recent theoretical work involving various creative attempts to reliably estimate the rate of production of final states with large (≤ 10) numbers of jets. This is not only a challenging and amusing theoretical problem but essential for the estimation of Standard Model backgrounds to signals of exciting new physics at the forthcoming SSC and LHC machines.

ROUND TABLE DISCUSSION

The round table discussion served to further highlight several of the points raised above. The fundamental ambiguities in the definition of jets were reviewed (S. Ellis) with a warning that this could be influencing the properties of jets as seen in hadron machines. The successes of the current data to indicate good agreement between jets seen in hadron and electron machines were reviewed by W. Scott and L. Galtieri. In response to a question from K. Pretzl, W. Scott confirmed that the separation of quark and gluon jets was not possible on an event by event basis with present techniques. At the same time there was optimism about what will be learned about jets at CDF.

As an appendix to the discussion of the possible determination of α_s in multi-jet studies at hadron machines, L.Galtieri gave a short review of the situation at e^+e^- machines. In this context the most reliable determinations of α_s involve measurement of the Energy-Energy Correlation (EEC)[12] and its asymmetry. This quantity is constructed from an energy weighted sum over all pairs of particles in the final state. It is a reliable quantity to use to measure α_s at least partially because, while it is very sensitive to contributions from 3 jet final states, it involves no explicit definition of a jet. While there were initially (as of 1986) results from different experiments which differed by as much as 30% in α_s, these seem to have been due largely to differences in fragmentation schemes and the way the perturbative calculation was implemented in the full Monte Carlo calculation. Note that the EEC is well defined in pure perturbation theory due to the cancellation of divergences in real and virtual contributions. However, these divergences must be handled with explicit cutoffs in Monte Carlo programs. Thus an extra ambiguity appears. In the most recent analyses (1987)[13] the differences are down to 12% with

$$\alpha_s = 0.14 \pm 0.02 \,. \tag{3}$$

This result is in good agreement with that found by analyzing[13] the ratio R of the fully hadronic to $\mu^+\mu^-$ rates. (There was also some discussion, led by M. Tannenbaum, as to whether the curve shown by L. Galtieri illustrated the claimed result for this latter quantity.)

There were general congratulations offered to the UA2 group for the work with the 2 jet W, Z signal. It is certainly reassuring that such a signal could be seen but one is still a long way from being able to use such a signal to trigger on W's and Z's as one would like for the Higgs search. It will be interesting to see how much the situation improves with the upgraded UA2 detector and whether UA1 and CDF can generate similar results. UA2 in particular hopes that a higher data rate at the upgraded CERN Collider will yield a better fit to the background rate.

The issues of "minijet" physics and "double parton scattering" were also discussed at some length at the round table. The general conclusion was that it is not obvious that either topic is under good control, theoretically or experimentally. However, there did seem to be a consensus that they both exhibit considerable potential interest and deserve further study.

SUMMARY

Recent advances in the experimental measurement of jets and their properties at hadron colliders are truly impressive. The quality of the data should serve to encourage theorists, phenomenologists and experimentalists to work toward the possibility of 10% comparisons of theory to data (to replace the present 50%). There are perturbative calculations of jet K-factors and Monte Carlo fragmentations studies which can be performed on realistic time scales to help accomplish this goal. On the experimental side, jet definitions based on finer grained measurements and more uniformity experiment to experiment will constitute significant improvements. The outlook for the next few years at UA1, UA2 and CDF is for considerably improved measurements of jet rates and properties. QCD tests involving the expected Q^2 dependence of these quantities should benefit from the larger energies available to CDF at the Tevatron. Progress in the theoretical understanding and numerical estimation of QCD backgrounds to new physics signals can be expected to improve. The exciting possibility of using two jet triggers in spectroscopy should be thoroughly studied in the laboratory.

ACKNOWLEDGMENTS

This work was supported in part by the US Department of Energy, Contract No. DE-AT06-88ER40423. The author would like to thank his many experimental and theoretical colleagues, especially K. Pretzl, for many informative and helpful discussions. B. Cox and P. Hale deserve special thanks for their unending efforts to organize this meeting at such a delightful location and for overseeing the production of these *Proceedings*.

REFERENCES

[1]   A. Zieminski, elsewhere in these *Proceedings*.

[2]   H. Boggild, elsewhere in these *Proceedings*.

[3]   A. Clark, elsewhere in these *Proceedings*.

[4]   K. Meier, elsewhere in these *Proceedings*.

[5]   W. Scott, elsewhere in these *Proceedings*.

[6]   M. Franklin, elsewhere in these *Proceedings*.

[7]   C. Maxwell, elsewhere in these *Proceedings*.

[8]   Private communications from various sources and S. D. Ellis, Z. Kunszt and D. E. Soper, in preparation.

[9]   A. L. S. Angelis, *et al.*, *preprint* CERN-EP/87-192.

[10] A. Petersen, *et al.*, *Phys. Rev. Lett.* **55** (1985) 1954.

[11] R. K. Ellis and J. C. Sexton, *Nucl. Phys.* **B269** (1986) 445.

[12] C. L. Basham, L. S. Brown, S. D. Ellis and S. T. Love, *Phys. Rev.* **D17** (1978) 2298.

[13] See the contributions from W. J. Stirling and S. L. Wu in the *Proc. of the International Symposium on Lepton and Photon Interactions at High Energies, Hamburg*, (1987).

QCD: PHOTO/HADROPRODUCTION OF HEAVY FLAVORS;

FERMILAB E691, E769 AND BEYOND

Jeffrey A. Appel

Computing Department
Fermi National Accelerator Laboratory
Batavia, IL. 60510

ABSTRACT

A series of heavy flavor experiments is being performed at the Fermilab Tagged Photon Laboratory. These experiments make use of accurate charged particle tracking, including precise vertex determinations, in an open geometry forward spectrometer. The global transverse energy trigger has high acceptance for all heavy flavor decay modes and allows a wide variety of questions to be addressed.

Some of the results for the completed charm photoproduction experiment (E691), especially production dynamics and relative branching ratios, have been selected for presentation. Their selection is due to their relevance to questions fundamental to QCD, the topic of this workshop. Other results (e.g., on charm particles lifetimes, mixing, Cabibbo single and doubly suppressed decays, etc.) have appeared[1] already or are about to appear.

The current experiment, charm hadroproduction (E769), is also described and its status presented. Finally, plans for future efforts, including beauty hadroproduction, are summarized.

THE TAGGED PHOTON SPECTROMETER

The Tagged Photon Spectrometer (Fig. 1) is by now a fairly standard forward open geometry multiparticle spectrometer.[2] The charged particles are momentum analyzed by a two-magnet system with 35 planes of drift chambers and are identified by use of two, one atmosphere, Cerenkov counters, the first filled with nitrogen and the second with a mixture of 80% helium and 20% nitrogen. An important part of the tracking is performed by 9 planes of $50\mu m$ pitch silicon microstrip detectors (SMDs) located just downstream of the target. Electron and photon identification and energy measurements are provided by the electromagnetic calorimeter which, in conjunction with the hadron calorimeter just downstream, measures the neutral and charged hadron energies and positions. One meter of steel absorber precedes a set of 15 muon defining scintillation counters at the downstream end of the experiment. The collaboration which performed E-691 included researchers from the CBPF and the University of Sao Paulo in Brazil, Carleton University, the University of Toronto and the NRC in

Canada and from the University of California at Santa Barbara, the University of Colorado and Fermilab in the U.S.[3]

The SMD system contained a total of 6840 readout strips with 50 micron spacing (pitch). The reconstructed interaction vertex had an uncertainty of 20 microns in each transverse direction and about 200 microns along the beam direction. The SMD system has been widely recognized as a crucial part of the improved capability of E691, reducing the background by as much as 2 orders of magnitude relative to previous experience. This reduction comes in two parts essentially. The first element is that by selecting events which have secondary vertices in the fiducial region, one greatly enhances the fraction of charm events in the data sample. Secondly, in calculating effective masses for identifying parent particles, one needs only consider combinations of tracks which come from common vertices downstream of the production vertex. This greatly reduces the number of combinations of tracks tried in searching for charm particles.

A rather beautiful illustration of this occurs in Fig. 2, where a D*D event is reconstructed. One sees the production vertex followed by the downstream decay of the two D^os in the event. The separation along the beam direction between the initial production point and the first downstream decay is labeled Δz and the ellipses indicate the 1σ resolution of each vertex. If one takes the z axis of the ellipses as uncertainties and divides the Δz by the uncertainty of the separation, one gets what is called $\Delta z/\sigma_z$. By cutting on this parameter, one can get signals for the mass peaks with very little background. Figure 3 shows this for D^os which decay to $K\pi$ and the even cleaner signals to background when one looks at the events which also have the low energy pion detected for those D^os which come from D*s.

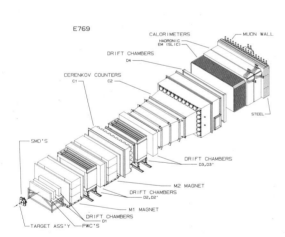

Fig. 1. *Schematic View of the Tagged Photon Spectrometer.*

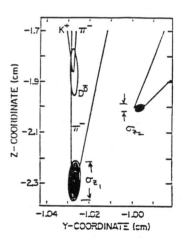

Fig. 2. *Reconstructed D*D Event*

The improved capability of E691 is a result of more than just the silicon microstrip detectors. Three other features were also important. The first of these is the large quantity of data collected and the possibility of rapidly analyzing the hundred million events. The ACP computer systems was used for a large fraction of the production charged track reconstruction and calorimetry. An equal amount of equivalent Cyber CPU power was also available for physics analysis and some of the production reconstruction. The total computing power applied to E691 was approximately 300,000 VAX 11/780 equivalent CPU hours.

A second additional feature of E691 was running at the Fermilab Tevatron. The accelerator duty factor for data taking was 22 seconds of spill out of a 57 second cycle time, better than 30%. In addition, the higher energy of the Tevatron beam allowed use of higher flux and higher energy secondary beams. Furthermore, the run itself was about 5 months long and the beam was available to the experiment throughout the entire period.

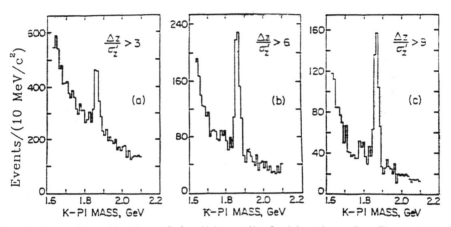

Fig. 3. *Mass Distribution of Candidate Kπ Combinations for Three Different Selections of Secondary (the Decay) Vertex Separation from the Production Point.*

Finally, the third additional feature in E691 was a global transverse energy trigger.[4] This trigger is based on the simple analog sum of signals from the two calorimeters and provided a rejection factor of about three for non-charm events. The trigger is very open to charm and proved to be about 80% efficient as seen in Fig. 4.

E691: PHOTOPRODUCTION OF CHARM

Production Dynamics

The distribution of charm events as a function of global transverse energy (E_t) in the forward hemisphere is the first feature of the production dynamics. This global transverse energy comes largely from the mass of the charm particles produced in pairs. However, it is seen that the forward transverse energy extends to quite high values and is not simply peaked at the threshold value. This is what makes the global E_t trigger work for charm production.

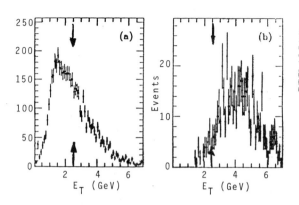

Fig. 4. Global E_T distribution of Events
Measured by the Spectrometer
(a) for all Hadronic Interactions
and (b) for charm events.

Fig. 5. Effective Mass
Distribution of
$K\pi\pi$ Combination
Candidates for D^+.

Figure 5 shows the sample of charged D^+ used for analysis of the production dynamics. Similar distributions exist for the D^0s and D^*s. The data shown here are taken from about 95% of the total data sample and all the results presented are considered preliminary. Distributions of the energy and x_F dependance of D^0 and D^+ production are shown in Figs. 6-8 as well as the average P_t^2 of the cross section. Table 1 gives some characterization to this data. We soon expect to have extracted the gluon structure function of the proton from this data assuming a photon gluon fusion model. Thus, there is some check on the internal consistency of the model and we will be reporting more than parameters for an arbitrary fit to the data.

TABLE 1

Characterization of Photoproduction of D^0s and D^+s*

Characteristic	D^0	D^+
$\sigma(200\ GeV)/\sigma(100\ GeV)$	1.82 ± 0.35	2.46 ± 0.54
n in $(1 - X_F)^n$	2.89 ± 0.12	2.57 ± 0.10
$\langle p_t^2 \rangle$	1.34 ± 0.04	1.43 ± 0.04

*The results shown are preliminary and only the statistical uncertainties are included.

In our range of 90-260 GeV photon energy, almost equal charm and anti-charm meson production is observed. The associated production seen at lower energies is less than 10% and shows no dependance on the incident photon energy or x_F in the range of 0.2 to 0.6.

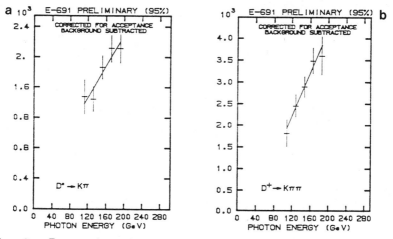

Fig. 6. *Energy dependence of photoproduced (a) D^0s and (b) D^+s.*

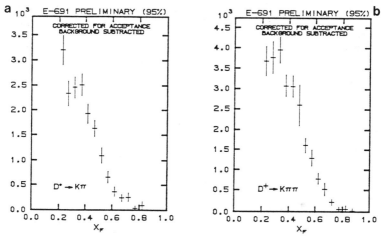

Fig. 7. *Feynman x distribution of photoproduced (a) D^0s and (b) D^+s.*

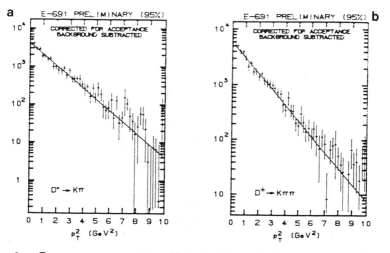

Fig. 8. *Transverse momentum distribution of photoproduced (a) D^0s and (b) D^+s.*

Hadronization and Charm Particle Branching Ratios

A fairly consistent picture of charm meson decays is beginning to emerge. This picture explains the unequal lifetimes of the charged and neutral Ds as well as relative rates of the D_S and Λ_c. Once one has isolated the decay diagrams, we have perhaps an even more isolated quark than in some of the hard QCD processes discussed at this conference. It is a challenge to the theory to describe the hadronization of these isolated quarks. What is the role of such effects as color cancellation, helicity conservation and final state interactions? What is the two body vs. many body and resonant vs. non-resonant nature of the hadronization process?

As an example of the quality of the data which should encourage additional efforts in this area, we examined the KKπ final state. In Fig. 9 is presented the Dalitz plot of decays corresponding to the charged D and D_S masses. One sees a vertical band corresponding to the ϕ meson and a horizontal band corresponding to K*. In both cases one also sees the effect of the vector-pseudoscalar nature of the resonant decays. This causes bunching in the relevant band near the edges of the Dalitz plot. All of this can be used to separate the resonant $\phi\pi$ and K*K decay modes and, after subtraction of the background, also determination of the non-resonant decay of the D meson. Similarly, in the 3π decay made shown in Fig. 10 and in the KK3π decay mode shown in Fig. 11, we can separate resonant and non-resonant contributions. The results of all of these data are summarized in Tables 2 and 3. Here we see precision on ratios of branching fractions at the 10-30% level.

TABLE 2

$D_S^\pm$ Branching Ratios (B.R) Relative to B($D_S^\pm$ -> $\phi\pi^+$)

Decay Mode	B.R. $\overline{B(D_S^\pm \to \phi\pi^+)}$
$D_S^\pm$ -> K*⁰K⁺	0.87 ± 0.13 ± 0.05
($D_S^\pm$ -> K⁻K⁺π^+)nr	0.25 ± 0.07 ± 0.05
$D_S^\pm$ -> $\phi\pi^+\pi^-\pi^+$	0.42 ± 0.13 ± 0.07
($D_S^\pm$ -> K⁺K⁻$\pi^+\pi^-\pi^+$)nr	< 0.32 (90% C.L.)

TABLE 3

D^+ Branching Ratios (B.R.) Relative to B(D^+ -> $^-\pi^+\pi^+$) and Absolute Branching Ratios*

Decay Mode	B.R. $\overline{B(D^+ \to K^-\pi^+\pi^+)}$	Absolute B.R. (%)
D^+ -> $\phi\pi^+$	0.075±0.008±0.007	0.68±0.07±0.12
D^+ -> K*⁰K⁺	0.061±0.009±0.006	0.45±0.08±0.10
(D^+ -> K⁻K⁺π^+)nr	0.052±0.008±0.006	0.47±0.07±0.09
D^+ -> $\phi\pi^+\pi^-\pi^+$	< 0.002 (90% C.L.)	< 0.02 (90% C.L.)
(D^+ -> K⁺K⁻$\pi^+\pi^-\pi^+$)nr	< 0.03 (90% C.L.)	< 0.24 (90% C.L.)

*Using $B(D^+ \to K^-\pi^+\pi^+) = 9.1 \pm 1.4\%$

404

The isolation of downstream decays allows one to study decay modes which were previously unmeasurable, even in electron-positron storage ring experiments. The 3π final state, in particular, has recently provided the same kind of results as just seen for the states with strange particles in them. Without the silicon microstrip vertexing, one would have had to take combinations of the large numbers of pions produced in final states and the charm states would be swamped by background. Figure 11 shows a significant charged D and D_s signal in the 3π decay mode and an ability to measure the resonant ρ decay state as well. Results are summarized in Table 4.

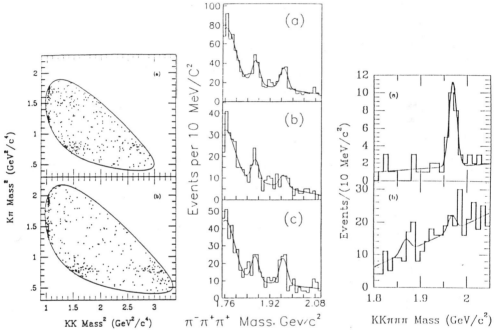

Fig. 9. The Dalitz plots for the (a) D^+ and (b) D_s^+ regions.

Fig. 10. (a) The $\pi^+\pi^-\pi^+$ mass spectrum for all events passing vertex cuts. The curve represents a fit with linear background, Gaussian peaks for the D^+ and D_s^+, and a peak for the misidentified $D^+ \rightarrow K^-\pi^+\pi^+$ events; (b) The same for events consistent with $\rho^0\pi^+$. (c) The same for events not included in the $\rho^0\pi^+$ sample.

Fig. 11. The $K^+K^-\pi^+\pi^-\pi^+$ mass spectra for the (a) $\phi\pi^+\pi^-\pi^+$ and (b) non-resonant $K^+K^-\pi^+\pi^-\pi^+$ final states.

405

E769: HADROPRODUCTION OF CHARM

A follow-on experiment, E769, is using a slightly modified tagged photon spectrometer and many of the same techniques to study the hadroproduction of charm particles. The goal of the experiment is to collect about 75,000,000 events for each of the plus and minus beam of tagged pions and kaons. With these, the experiment will study the flavor, A, x, and p_t dependance of charm and charm-strange production. The experimenters also hope to have larger statistics samples of D_s and Λc relative to E691. This should allow further improvement in lifetime and decay branching ratio information. The collaboration is again composed of physicists[5] from Brazil, Canada and the U.S., but has added Northeastern, Tufts, and Yale Universities and the University of Wisconsin at Madison to make up for the fraction of E691 collaborators who have concentrated their efforts on E691 data analysis and some of whom have moved to other experiments.

TABLE 4

E-691

$$\frac{B(D_s^\pm \to \pi^+\pi^-\pi^+)}{B(D_s^\pm \to \phi\pi^+)} \quad = \quad 0.29 \pm 0.07 \pm 0.05$$

$$\frac{B(D_s^\pm \to \rho\pi^+)}{B(D_s^\pm \to \phi\pi^+)} \quad < \quad 0.08 \ (90\% \ \text{C.L.})$$

$$\frac{B(D^+ \to \pi^+\pi^-\pi^+)}{B(D^+ \to K^-\pi^+\pi^+)} \quad = \quad 0.041 \pm 0.008 \pm 0.004$$

$$\frac{B(D^+ \to \rho\pi^+)}{B(D^+ \to K^-\pi^+\pi^+)} \quad = \quad 0.022 \pm 0.008 \pm 0.003$$

The collaboration has made four additions to the E691 apparatus. Given the incident charged beam, it is now possible to include information on this particle in defining vertices. To this end, a 25 micron pitch SMD pair and upstream PWCs of 1mm pitch have been added to the experiment. The target for the incident beam is a set of Be, Al, Cu and W foils with each foil about 200 microns thick. Therefore, primary interactions can be localized better than from the reconstructed tracks alone. The third set of additions is for the downstream tracking where a pair of crossed 25 micron pitch SMDs are added as well as two proportional wire chambers which measure the vertical position upstream of the first magnet. Finally, incident pions and kaons are tagged by Transition Radiation Detector (TRD) and a Differential Isochronous Self-focusing Cerenkov counter (DISC) respectively. The power of these devices to identify incident 250 GeV beam particles is illustrated in Fig. 12. The distributions shown there are only days old, the devices still being tuned for optimum performance. Nevertheless, one can see clean differences in the distributions of signals due to π's, K's and protons in the beam.

This higher beam energy (250 GeV) helps make up for the lower average x of the constituents in the scattering of hadrons relative to the photoproduction scattering. The average photon energy in E691 for charm events was 145 GeV. However, because of the smaller fractional charm production rate with incident hadrons, one can still ask whether E769 will be able to measure the hadroproduction of charm particles. By comparing the features of E769 to E691 we can get some feeling for the confidence felt by the experimenters in their eventual success in this effort. First among these are the extra background suppression tools available in E769. We have already mentioned the tracking of the incident particle and the use of foil targets which will allow a better determination of the interaction point in E769 relative to E691. Similarly, the 25 micron pitch of the central region of the 4 added SMD planes should help improve the resolution of vertices in the hadroproduction experiment. This will be especially useful in finding Λ_c particles where the lifetime had been measured to be about half that of neutral Ds, for example, and D_ss. The common knowledge on hadroproduction of charm states is that hadroproduction is much harder than photoproduction because of the extra quarks in the event. These eventually become additional particles in the final state which contaminate mass distributions due to the extra combinations in the higher multiplicity events. However, as already pointed out, the use of tracks coming from well defined secondary vertices removes these extra hadrons from the combinations included in mass distributions. Thus the use of SMDs should make hadroproduction much more like photoproduction than was the case before their use.

A major part of the success of E691 was the ability to take a much larger data sample than in previous experimental efforts. This capability has been extended for E769 by an additional factor of more than 3. Furthermore, the data acquisition system (shown in Fig. 13) provides a number of additional useful features. The readout of data in CAMAC has been sped up to 600 nanoseconds per word from the 3 μseconds per word of E691. The CAMAC data is read out in 7 parallel branches relative to 2 in E691. All of this allows more data to be taken into the system and the use of ACP memories allows buffering an entire spill's worth of data. This can then be written to magnetic tape essentially continuously. The tape drives themselves have been upgraded from 75 ips to 100 ips and are written directly from the memories of the ACP modules onto magnetic tape without traversing the internal bus of the general purpose host computer. The experiment is capable of recording greater than 300 events per second with 30% dead time. In E769, the use of ACP processor boards is strictly for event formating and storage, awaiting writing to magnetic tape. No filtering of data has yet been performed in this system although that capability exists.

The use of Smart Crate Controllers (SCCs) and readout buffers (RBUFs) follows an original concept due to Sergio Conetti of McGill University. His designs were re-engineered and extended at Fermilab by Ed Barsotti and Sten Hansen. The software for the system makes use of some general ACP and Computing Department software, but the overall system and most of the specific software in use was written by Steve Bracker and Colin Gay of E769 and Mark Bernett of the Computing Department. The success of this new data acquisition system is indicated by the fact that 1500 full tapes of data have been written in one month with a single day peak of 150 tapes.

BEYOND E769

Assuming that the run which is currently in progress continues to go well, the collaboration anticipates preparing a new proposal. Additional collaborators are also being sought for this effort which will aim at additional charm data and the hadroproduction of beauty. The same

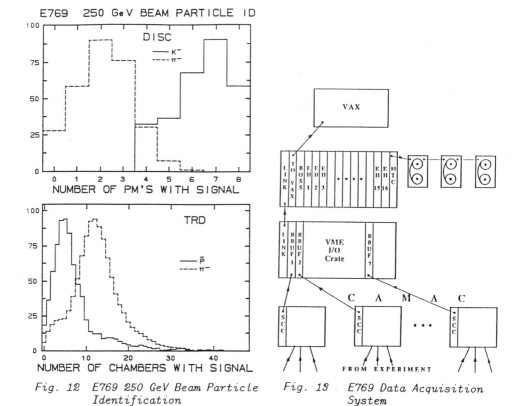

Fig. 12   E769 250 GeV Beam Particle   Fig. 13   E769 Data Acquisition
         Identification   System

apparatus and techniques will lie at the heart of the new proposal.
However, in order to get to beauty in sufficient quantities, one needs to
select events which are a thousand times more rare than the charm particle
events. For this, one will increase the incident beam energy and, with it,
the beauty cross section. Simply dialing up the threshold on the global
transverse energy trigger discriminator from 3 GeV to 10 GeV will provide
nearly an order of magnitude of additional rejection. The event readout at
that time can be reduced by an additional factor of 2 to 3 which would lead
to the need to filter data in the on-line system before writing data to
magnetic tape. These features should allow a 1 run experiment to collect
on the order of 100 recognizable beauty decays. The use of a hardware
trigger for selecting events with multiple decay vertices, high p_t leptons
or large multiplicity jumps near the target can increase this and lead to
the next generation of beauty physics capabilities.

ACKNOWLEDGMENTS

The author is grateful to all the collaborators on E691 and E769. It
is mostly the results of their work which is summarized here. Especially
worthy of note in this context are the analyses of Milind Purohit, Audrius
Stundzia, Marleigh Sheaff, Lee Lueking, Alice Bean, Paul Karchin, Greg
Punkar and Mike Witherell. The research is supported by the U.S.
Department of Energy and National Science Foundation, by the Natural
Science and Engineering Research Council of Canada, by the National
Research Council of Canada, and by the Brazilian Conselho Nacional de
Desenvolvimento Cientifico e Technologico.

REFERENCES

1. J.C. Anjos, et al., Phys. Rev. Lett. <u>58</u>, 311 (1987); Phys. Rev. Lett. <u>58</u>, 1818 (1987).

2. For details see J. Raab, et al., Fermilab-Pub-87-144, submitted to Phys. Rev. D (1987) and references therein.

3. E-691; University of California-Santa Barbara: A. Bean, T.E. Browder, P.E. Karchin, S. McHugh, R.J. Morrison, G. Punkar, J.R. Raab, M.S. Witherell; Carleton University: P. Estabrooks, J. Pinfold, J.S. Sidhu; Centro Brasileiro de Pesquisas Fisicas: J.C. Anjos, A.F.S. Santoro, M.H.G. Souza; University of Colorado at Boulder: L.E. Cremaldi, J.R. Elliott, M.C. Gibney, U. Nauenberg; Fermi National Accelerator Lab: J.A. Appel, P.M. Mantsch, T. Nash, M.V. Purohit, K. Sliwa, M.D. Sokoloff, W.J. Spalding, M.E. Streetman; National Research Council: M.J. Losty; Universidade de Sao Paulo: C.O. Escobar; University of Toronto: S.B. Bracker, G.F. Hartner, B.R. Kumar, G.J. Luste, J.F. Martin, S.R. Menary, P. Ong, A.B. Stundzia.

4. J.A. Appel, p. 555 in "The Search for Charm, Beauty, and Truth at High Energies," Proc. of a Europhysics Study Conference on High-Energy Physics held November 15-22, 1981, in Erice, Sicily, Italy, G. Bellini and S.C.C. Ting, editors, Plenum Press, New York (1984).

5. E-769; University of California-Santa Barbara: G. Punkar; Centro Brasileiro de Pesquisas Fiscas: G.A. Alves, J.C. Anjos, C. DeBarros, H. DeMotta Filho, A.F.S. Santoro, B. Schulze, M.H.G. Souza; University of Colorado at Boulder: L.M. Cremaldi; Fermilab: J.A. Appel, L. Chen-Tokarek, R. Dixon, H. Fenker, D. Green, L.Lueking, P.M. Mantsch, T. Nash, W.J. Spalding, M.E. Streetman, D. Summers; Institute of Theoretical Physics-Sao Paulo: A. D'Oliveira; Northeastern University: D. Kaplan, I. Leedom, S. Reucroft; University of Toronto: S.B. Bracker, C. Gay, R. Jedicke, G.J. Luste; Tufts University: J. Metheny, R. Milburn, A. Napier; University of Wisconsin at Madison: D. Errede, M. Sheaff; Yale University: P.E. Karchin, Z. Wu.

DIMUON EXPERIMENTS AT THE FERMILAB HIGH INTENSITY LABORATORY

Sergio Conetti

Institute for Particle Physics and McGill University
Montreal, Quebec, Canada H3A 2T8

INTRODUCTION

Over the last ten years, the High Intensity Area located at the Fermilab Proton-West beam has seen the development of a powerful large aperture spectrometer, utilised to study several aspects of the hard interactions of hadrons. The detector was first employed[1] to investigate the production of Drell-Yan high mass states as well as some properties of the hadronic production of J/ψ. This first experiment was performed in a "close geometry", where only the muons from the reaction under study were detected; as a natural continuation,[2] the study of virtual, "Drell-Yan" photons was extended to the measurement of real photons directly produced in hard parton-parton interactions. The investigation of the hadronic production of a bound state of heavy quarks was pursued in more detail by studying the production of charmonium χ states. Finally, the future plans for the FNAL P-West detector foresee[3] an extensive study of the production and decay of particles carrying the b-quark, detected via the presence of one or more muons with large transverse momentum. Apart from the direct photon measurement, whose detailed description appears in another contribution to these Proceedings[4], an essential element of all the other experiments is the dimuon trigger, developed over the course of the experiments to provide a three-level event selection. Before presenting more detailed performances and results from the various experiments, we will discuss the basic features of the dimuon trigger, since it plays a major role in the actual execution of the measurements.

THE THREE LEVEL DIMUON TRIGGER

The elements employed for the first level of dimuon event selection are visible in fig. 1a, showing the closed geometry version of the spectrometer utilised for E-537. At the back of the spectrometer one can see three planes of scintillation counters embedded in a massive steel and concrete shield, providing a total of 13 hadronic interaction lengths. In each plane, the counters are organised into two rows of vertical elements, covering respectively the upper and lower half of the spectrometer aperture. To form the fast trigger, each counter from the first plane is put in a triple coincidence with the OR of two counters from the second and two from the third plane. The geometrical arrangement of the AND/OR combinations guarantees that the direction of the μ candidate is consistent with a particle coming from the target, after accounting for magnetic bend and multiple scattering. A muon detected by the counters needs to possess an energy of at least 6 GeV, since this is the average energy loss of a particle originating from the target and traversing the absorber's thickness. In the open

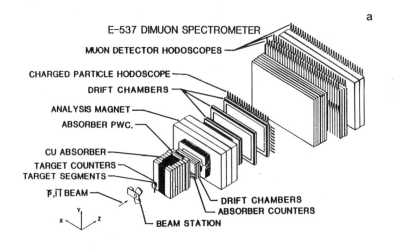

E-537 DIMUON SPECTROMETER

MUON DETECTOR HODOSCOPES

CHARGED PARTICLE HODOSCOPE

DRIFT CHAMBERS

ANALYSIS MAGNET

ABSORBER PWC.

CU ABSORBER

TARGET COUNTERS

TARGET SEGMENTS

$\bar{P}$,$\bar{\Pi}$ BEAM

DRIFT CHAMBERS

ABSORBER COUNTERS

BEAM STATION

a

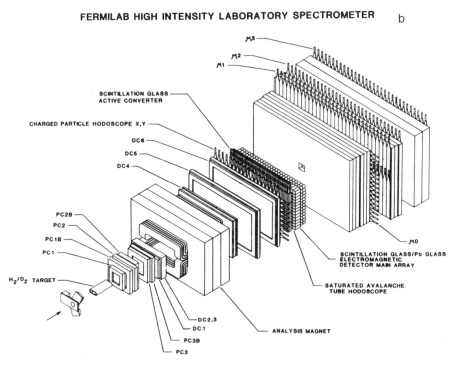

FERMILAB HIGH INTENSITY LABORATORY SPECTROMETER   b

M3

M2

M1

SCINTILLATION GLASS ACTIVE CONVERTER

CHARGED PARTICLE HODOSCOPE X,Y

DC6

DC5

DC4

PC2B

PC2

PC1B

PC1

H_2/D_2 TARGET

DC2,3

DC1

PC3B

PC3

M0

SCINTILLATION GLASS/Pb GLASS ELECTROMAGNETIC DETECTOR MAIN ARRAY

SATURATED AVALANCHE TUBE HODOSCOPE

ANALYSIS MAGNET

Fig. 1. The FNAL High Intensity Laboratory P-West Spectrometer. a) E–537 closed geometry configuration. b) E–705 open geometry configuration.

geometry version of the spectrometer (E-705, fig. 1b), the lack of the absorber positioned downstream of the target was compensated by increasing the thickness of the first muon wall to provide the same 6 GeV energy cutoff. In order to form the dimuon trigger, the outputs of the muon triple coincidences are grouped into four quadrants (see also ref. 5) and a trigger is defined as the coincidence of at least two quadrants. The individual counters are timed into the triple coincidences so that accidental coincidences between signals coming from different RF buckets are excluded. At the same time, the total timing spread between the signals from the global quadrant ORs is slightly larger than the 18.8 nsec separation between buckets, so that one is forced to accept a small contamination of accidental triggers. Such a background can nevertheless be resolved, at the analysis stage, by examining the TDC data recorded for the individual triple coincidences. This information was in fact used in the analysis of E-537 to study the accidental background to the Drell-Yan events.

The total time required to form the first level trigger, mainly due to the signal propagation delays, is of about 300 nsec, but since the trigger operation is fully pipelined, such a formation time does not introduce any dead-time. The overall trigger efficiency was determined to be of the order of 95%.

A comparison of the trigger's performance, in terms of rejection of the total interaction rate, for both the closed and open geometry configuration is shown in Table 1: the better figure for the closed geometry reflects the well known fact that, in an open geometry configuration, any attempt to trigger on prompt muons will be affected by the in flight decays of pions and kaons produced in the target. The situation is also summarised in fig. 2, showing, as a function of the particle's momentum, the probability for a muon to be recorded by the E-705 μ counters, either as the result of in flight decay or of the punch-through of a secondary particle into the muon absorber. The spectrum of fig. 2 was produced by generating minimum bias events with the Pythia Montecarlo,[6] propagating them through the E-705 spectrometer, while taking into proper account the decay and punch-through probability. Integrating the distribution of fig. 2 over all momenta should provide the single muon rate measured in the E-705 trigger counters: the result of such integration is .040/interaction, in very good agreement with the value of .034 measured in E-705. This figure represents the rate to be expected in the implementation of a single muon trigger. As mentioned before, the experiments presented here attempt to select events containing a high mass muon pair, originating from quarkonium decay or the Drell-Yan process. Fig. 3 shows the result of the same Montecarlo study, where events containing at least two decay or punch-through muons are selected, and the invariant mass of the dimuon system is computed. In this case, the integral of such a spectrum can be compared directly to the measured performance of the dimuon trigger: the integral value of 2×10^{-4} agrees fairly well with the value of 6×10^{-4}, as reported in table 1 for the first level trigger rejection factor, and the discrepancy between data and simulation can easily be explained in terms of the approximations inherent to the Monte Carlo.

The spectrum of fig.3 suggests that a considerable gain in the trigger selection power could be achieved by performing a fast estimation of the invariant mass associated with the dimuon system: such a procedure provides the cleanest selection of real high mass events, without introducing the bias attached to the more common procedure of imposing a threshold onto the transverse momenta of the triggering μ's. The task of the second and third levels of trigger was then chosen to be the reconstruction, with increasing degree of accuracy, of the invariant mass of the dimuon system: the devices employed for this purpose are respectively a fast processor, consisting of an assembly of general purpose arithmetic and logic modules, and an online micro-processor farm, running a filter program optimised for speed. The trade-off of accuracy versus speed of execution is quite crucial in the sequence of event selection: given that the background mass spectrum (fig. 3) is falling very fast, the performance of the online filters is very sensitive to the processor's mass resolution. As an illustration, the typical mass resolution of the second level processor is of the order of 10-15% : in order to have a two standard deviation acceptance of J/ψ events it is necessary to set the processor's mass threshold at 2.5 GeV; this implies that the processor will accept background events

TABLE 1

Performance of three-level dimuon trigger.

		E–537	E–705	E–771 (projected)
INTERACTIONS/SEC		1.5×10^7	10^6	$\geq 10^6$
LEVEL 1	Rejection Factor	3×10^4	1.6×10^3	$\geq 1.5 \times 10^3$
	Execution Time	300 nsec	300 nsec	300 nsec
LEVEL 2	Rejection Factor	10	4–5	≥ 10
	Execution Time	10 μsec	50 μsec	≤ 10 μsec
LEVEL 3	Rejection Factor		3	≥ 3
	Execution Time		50 msec	≤ 100 msec

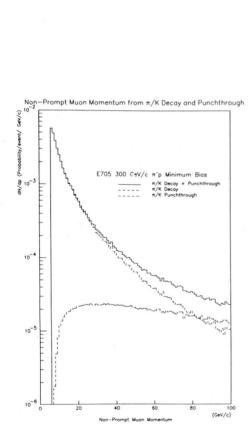

Fig. 2. Momentum spectrum of background decay and punch-through muons

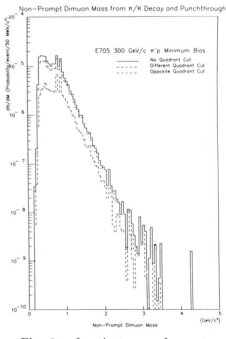

Fig. 3. Invariant mass of the background dimuons

having a mass as low as 2. GeV, where the dimuon rate is two orders of magnitude higher than at 3. GeV.

The second level processor employed in E-537, based on the Fermilab ECL-CAMAC general purpose set of modules, has been described in detail.[7] It exploits a set of fast encoders that prepare lists of hit scintillation counters and drift chamber cells. These lists are loaded into the stacks of the processor itself: loop controllers are responsible for fast track finding by looking for aligned sets of chamber and counter hits, and candidate tracks are combined two at the time to compute the invariant mass, based on the knowledge of the tracks direction and momentum. Complex calculation are performed very quickly ($<$ 100 nsec), by making use of pre-loaded memory look-up units, while arithmetic logic units are used for simpler calculations and for logic decisions. The exploitation of the processor for the open geometry environment of E-705 required some upgrading with respect to the E-537 version. The information of a new plane of horizontal muon counters was fed into the processor to provide a better definition of the μ's vertical position, and six planes (rather than three) of drift chambers were interrogated, allowing a more solid track definition. Even with these improvements, the processor's performance in the later experiment did not match the previous one (see table 1), due to the much higher hit multiplicity encountered in the open geometry and the poorer momentum resolution associated with the higher beam energy (300 GeV vs. 125). Even so, the combined rejection factor provided by the first and second level trigger approaches 10^4, allowing to run the experiment at interaction rates of the order of 1 MHz, subject to the Data Acquisition System capability of handling more than 100 events/sec. When running at 1 MHz and accounting for the level 1 performance, the average processing time of 50 μsec implies that the dead-time introduced by level 2 is around 3%. For the next experiment, beauty search E-771, there are plans to increase the interaction rate by up to a factor of ten: for this reason, as well as for the limitations mentioned before connected with hit multiplicity and momentum resolution, we are actively studying further upgrades of the second level processor.

The third level of online event selection relies on the fast Data Acquisition System specially developed for the High Intensity spectrometer. A set of intelligent CAMAC controllers[8] performs parallel reading of the event data into the memories of a microprocessor farm, whose task is to perform event building and online filtering of the events[9]. The system is capable of reading one event, several kilobytes in size, in less than a millisecond: as a rule of thumb, the dead-time is 10% for every unit of 100 events/sec. For the microprocessor farm, the final choice fell on the Fermilab ACP system[10], capable of providing a large amount of user friendly computer power at accessible price. A single-crate ACP system, containing up to 15 microprocessor boards, was installed for the E-705 data taking run and employed to collect all the data. In parallel, a Fortran program is under development to run in the ACP nodes and to perform online filtering of the events by computing, with greater accuracy than what is possible in the second level processor, the mass of the dimuon system. Preliminary tests have shown that a further rejection of at least a factor of three can be achieved, with an average processing time of less than 100 msec per event, so that a system of 10 processors could handle a rate of 100 events/sec. The ease of programming the third level filter in Fortran and the expected increase in the computing power of the ACP system suggest that an even higher rejection factor could be achieved in the future. When combined with the envisaged improvements of the level 2 processor, a total rejection factor of 10^5, allowing to run comfortably at 10^7 interactions/sec, seems to be quite realistic from the trigger and data acquisition point of view.

After the description of the basic triggering system we will now present the past and present achievements as well as the future plans for the Fermilab P-West High Intensity spectrometer.

EXPERIMENT 537 AND THE A-DEPENDENCE OF J/ψ HADRONIC PRODUCTION[11]

The primary goal of E-537 was to perform a low sistematics measurement of the production of Drell-Yan high mass dimuons by antiprotons and pions. In addition to these

TABLE 2

Total Cross Sections ($x_F > 0$) for J/ψ Production from
Be, Cu and W at 125 GeV/c (nbarns per nucleus)*

Target Beam	Be	Cu	W
π^-	560 ± 43 (2881)	3610 ± 210 (1958)	7900 ± 63 (33820)
$\bar{p}$	462 ± 18 (588)	2820 ± 110 (529)	6900 ± 89 (12530)

*Errors are statistical and include background subtraction. The number of J/ψ's in each data sample is given in () following the cross sections. Systematic errors are $\pm 5\%$ on all cross sections.

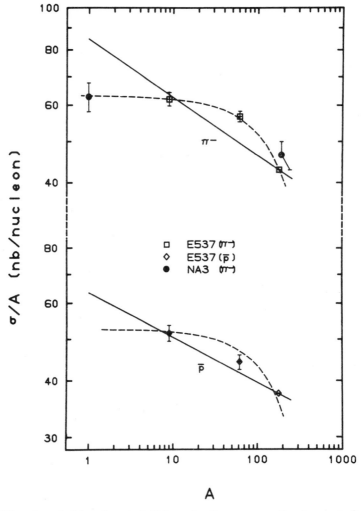

Fig. 4. A-dependence of J/ψ production cross section for $x_F > 0$.

results, already reported,[12] the dimuon trigger allowed to collect simultaneously a sample of J/ψ events, so that certain aspects of its production could be studied. In particular, data was collected with three different target materials, in order to investigate the presence of heavy target effects and their interpretation in terms of the behaviour of the parton structure functions in nuclei.[13] Among previous experiments, the richest set of ψ A-dependence data had come from the NA3 collaboration[14] and their sample of J/ψ produced on hydrogen and platinum. Following the usual convention of parametrising the A dependence of the total cross-section with the form:

$$\sigma(A) = \sigma(1)A^\alpha \qquad (1)$$

they used their two A values to otain $\alpha = .96\pm.01$, giving an indication that, at variance with some previous expectations, ψ production might not go like A^1. Moreover, NA3 observed that the ratio of H_2 over Pt (divided by A) cross section was not a constant as a function of x_F, but that in fact the normalised relative probability of producing a ψ from heavy nuclei was decreasing as x_F increased. Such effect, being prominent at large x_F, was interpreted as due to a diffractive component of the J/ψ production mechanism.

To study the same effect, E-537 collected data on J/ψ production by 125 GeV/c $\bar{p}$ and π interacting with beryllium, copper and tungsten targets. The total number of events collected from each target (for $x_F > 0$) is reported in table 2, together with the computed cross sections. The results are summarised in fig. 4, showing the π and $\bar{p}$ E-537 data on Be, Cu and W together with the NA3 H_2 and Pt data, averaged over the three energies measured by NA3 and extrapolated to 125 GeV/c. The data shows clearly that the parametrisation (1) does not describe correctly the A dependence of the J/ψ production cross section. Fitting with the purely empirical form

$$\sigma(A)/A = (a + b\,A),$$

one obtains $a = 63.17 \pm 2.0$ and $b = -0.11 \pm 0.01$, with $\chi^2/DF = 0.53$.

The data collected also allows to study the dependence of the nuclear effect on the x_F and p_t of the produced particles. The results are shown in fig. 5a and b, where the ratios R1 and R2 are plotted, both for π and $\bar{p}$ data, with the following definitions:

$$R1 = \frac{\frac{1}{A_1}\frac{d\sigma}{dx_F}|_{A_1}}{\frac{1}{A_2}\frac{d\sigma}{dx_F}|_{A_2}} \qquad R2 = \frac{\frac{1}{A_1}\frac{d\sigma}{dp_t}|_{A_1}}{\frac{1}{A_2}\frac{d\sigma}{dp_t}|_{A_2}}$$

In all cases it appears that the nucleons contained in a heavy target are less efficient in producing J/ψ's, and such effect is more pronounced at high x_F and low p_t, in agreement with the NA3 results. The phenomenon can be investigated further by examining the R1 (R2) ratio computed for different regions of $p_t(x_F)$, as shown in fig. 6a and b. In particular, one can see that the p_t effect is equally present for all regions of x_F, putting in doubt the diffractive interpretation.

In trying to understand the observed phenomena, it is natural to turn one's attention to the behaviour of the parton's distribution in a heavy nucleus rather than in a free nucleon. Much theoretical work has taken place in recent years, primarily as a result of the observation of the so called "EMC effect", to understand how the quark structure functions would have to be modified when dealing with a heavy nucleus. When referring to the production of J/ψ's, one can expect that, while the effects of the distortion of the quark distributions should still be present, one should also be able to observe the contribution of nuclear effects upon the gluon structure function, since ψ production, at high enough energy, is expected to be dominated by gluon fusion processes. Such gluon effects are not directly observable in deep inelastic type of experiments, where the EMC effect was discovered. Moreover, it appears that the trend observed in the E-537 data is not the one that would be expected from the EMC effect. The various models proposed to explain the EMC effect advocate a softening of the nuclear momentum distribution of quarks (the x_2 distribution), so that one would expect an enhancement of the cross section at high x_F for the heavy target (with the usual conventions, $x_F = x_1 - x_2$, and $x_1 x_2 = M^2/s$). Our data shows exactly the opposite,

417

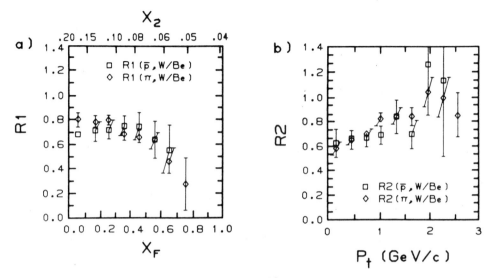

Fig. 5. Tungsten over Beryllium ratios of J/ψ differential cross sections. a) R1, x_F dependence. b) R2, $p_\perp$ dependence.

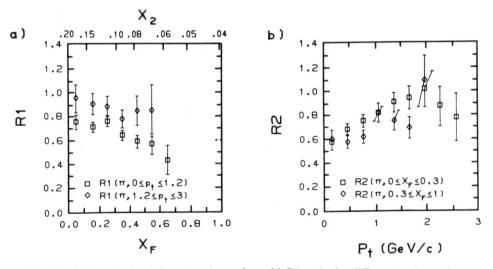

Fig. 6. a) R1 ratio for different regions of $p_\perp$. b) R2 ratio for different regions of x_F.

so that a stronger mechanism must be at work: a good candidate is the nuclear shadowing model,[13,15] where the soft gluon component of a nucleon in a heavy nucleus is considerably depleted by its absorption by the other nucleons. Such effect would dominate at low x_2 (large x_F), but it would be expected to vanish at large $x_2(x_F \rightarrow 0)$.[13] Another mechanism that should be taken into account is the rescattering of the incoming beam or the outgoing ψ in the target nucleus,[16] with the end effect of a relative decrease, for the heavy nucleus, of the number of J/ψ produced at high x_F. The situation is summarised in fig. 7, where the R1 ratio is compared with the predictions of the different models: as expected, the EMC type mechanisms do not reproduce the observed effects, while shadowing and rescattering go in the right direction. At low x_F, the rescattering alone can accomodate the data, while at large x_F the combination of the two mechanisms is in good agreement with the measured points: this is quite reasonable, since there is no reason why both mechanisms should not be present, and one could expect that their contributions factorise.

In conclusion, we have confirmed that the hadronic production of J/ψ is not linear in the atomic number, and moreover that the cross section does not follow the simple expression (1). When examining the x_F and p_t dependences, we observe a trend which is neither a diffraction nor an EMC-type process, but is likely due to the combination of nuclear shadowing and rescattering.

EXPERIMENT 705 AND THE HADRONIC PRODUCTION OF CHARMONIUM

The primary goal of the experiment currently running at the Fermilab P-West beam is to bring light to what still remains, more than ten years after the discovery of charm, an open question, that is what exact mechanism is responsible for the hadronic production of a bound state of heavy quarks. From the earliest days it was recognised that, together with $q\bar{q}$ annihilation processes, especially relevant at lower energies, an important role in the hadronic production of J/ψ has to be played by gluons. At the same time it was pointed out that,[17] since the J/ψ does not couple directly to two gluons, owing to its odd Charge Conjugation, the favoured $c\bar{c}$ states to be generated in hadronic processes should be the P-wave χ's ($C = +1$). The direct formation of J/ψ, requiring a three gluon vertex, would consequentially be less copious, and the experimentally observed ψ's would in fact be the product of the radiative decay $\chi \rightarrow \psi\gamma$. These qualitative considerations were put in a more rigorous form by Carlson and Suaya,[18] who adopted the concept of "χ dominance" and computed the cross section for charmonium production within the "colour singlet" hypothesis: the hard interaction among partons directly produces the $c\bar{c}$ pair in a colour singlet bound state. Given that the dynamics of the reaction are well defined, it is then possible to perform, within QCD, an unbiased calculation of the production cross section. Early experimental results[19] seemed to confirm this hypothesis, since the fraction of ψ's produced in conjucntion with a photon in the χ mass region was measured to be as high as 50 to 70%. Continuing the work of ref. 18, several authors expanded the colour singlet picture of heavy $q\bar{q}$ production: predictions were made for the relative rate of production of the three χ states[20] and for the shape of the angular distribution of the χ decay,[21] showing the different behaviour of the $q\bar{q}$ annihilation and gluon-gluon fusion diagrams. Following the evolution of QCD, more diagrams were included in the calculation,[22] and hard gluon radiation was included to predict the behaviour of charmonium production at high p_t.[23] Finally, attempts were also made to compute the direct production of ψ's, without the intervention of the cascade process.[24] Throughout the development of the model anyhow, the constant feature has been the prediction of the dominance of χ production over ψ: the situation is illustrated in fig.8, reproduced from ref. 23, showing the lower order diagrams giving raise to charmonium production. It appears clearly that there are several more ways to directly produce the P states rather than the $^3S_1$.

Concurrently, and in contrast, with these developments, a different model was being presented, where the lack of direct coupling of the ψ to two gluons was circumvented by invoking the emission of a soft gluon, to readjust the conservation of quantum numbers.

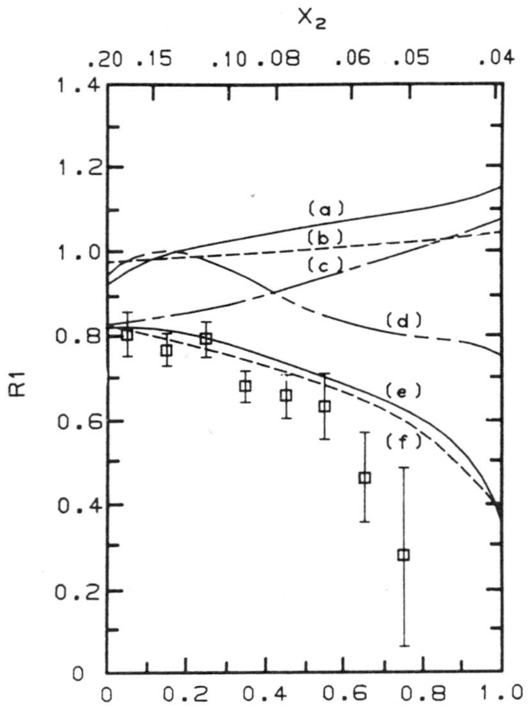

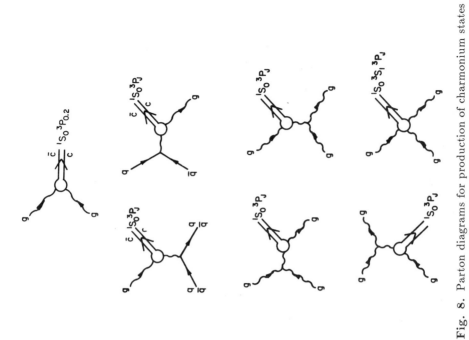

Fig. 7. Comparison of R1 ratio with various theoretical models: a, b, c EMC models, d shadowing (ref 15), e rescattering, (ref 16).

Fig. 8. Parton diagrams for production of charmonium states

This approach was first put forward by Fritzsch,[25] who postulated ψ production via a Drell-Yan-like process of $q\bar{q}$ annihilation into a virtual gluon, accompanied by soft gluon emission. This model erroneously predicted a very large cross section difference between p and $\bar{p}$, but agreement with experiment was attained by Gluck, Owens and Reya[26] by including the process of gluon-gluon fusion (plus soft gluon emission). The basic idea of the model is to compute the production of charmonium resonances by integrating the cross-section for a free $c\bar{c}$ pair over the mass interval $2m_c < m_{c\bar{c}} < 2m_D$, in a revival of the old duality concept. The quark pair is generally not produced in a colour singlet but in a colour octet state, and the final bound state is reached via the emission of a soft gluon, to provide the necessary colour "evaporation" or "bleaching". In a variation of the model, Afek, Leroy and Margolis[27] argued that a charmonium bound state can originate from a $c\bar{c}$ pair even if the pair is created above the $D\bar{D}$ mass threshold: the relative rate of bound versus associate production will be determined by the relative probability of gluon emission vs. light quark capture. Apart from the, somewhat ad hoc, requirement of soft gluon emission, one of the main weaknesses of the "semi-local duality" model is its inability to predict absolute cross sections for the individual charmonium states, since the formation of a particular bound state proceeds in an undefined way from the $c\bar{c}$ continuum. Different hypotheses have been put forward, ranging from an equipartition of the cross section among the 7 (or 8) charmonium states below $D\bar{D}$ threshold,[28] to the suggestion[27] that the three χ's should be produced in the ratio of their angular momentum states (i.e. 1 : 3 : 5). More recently, attempts have been made to treat gluon emission in a more rigorous way: high p_t behaviour was evaluated, in the context of semi-local duality, by computing the hard gluon radiation diagrams,[29] and an attempt was made to treat properly the contribution of soft gluon emission.[30] This last attempt nevertheless seemed to degrade the agreement of the theory with the data.

In spite of the rather fundamental differences between the various models, they all achieve reasonable agreement with the existing data, mainly because the experiments performed so far have been, almost exclusively, capable of bringing information on global ψ production only, without resolving the contributions of individual charmonium states. The crucial test would be to measure total and differential cross sections for the individually resolved χ states as well as for directly produced ψ's. The colour singlet approach makes rather definite predictions for the relative rates of production of the three χ states (e.g. 0 : 4 : 1 for $q\bar{q}$ annihilation or 3 : 0 : 4 for gluon gluon fusion, ref. 20), for the angular distributions of χ decay,[20,21] where again one can distinguish quark from gluon processes, as well as for x_F and p_t distributions. Such predictions, especially the relative cross sections and the angular distributions, are rather different for the case of colour evaporation models.

The reason why good data on the hadronic production of charmonium is not yet available is that one is dealing with a rather difficult experiment, needing to combine the conflicting requirements of high luminosity, large acceptance and, very critical to the success of the experiment, excellent mass resolution. The Cross Section X Branching Ratio for the detection of χ's in the most accessible channel $\chi \to \psi_{\to \mu^+\mu^-}\gamma$, is of a few nbarns at $\sqrt{s}$ around 20 GeV. When combined with the need to collect at least a few thousand events, with typical geometric acceptances of 20% or less, a sensitivity of at least 1 event/pbarn is required. This value is not prohibitive for a "beam dump" type of experiment, but is at the limit of the capability for an "open geometry" type of measurement. Moreover, in order to compare the data with the theoretical models, it is essential to produce total and differential cross sections for individually resolved charmonium states. The mass separation of only 40 MeV between the χ_1 and χ_2 requires a mass resolution of the order of 10 MeV (or $\sim$ .3% for the χ mass centered around 3530 MeV). The first χ state, with a mass of 3410 MeV, is more easily resolvable, but its branching ratio into $\psi\gamma$ is 20 to 30 times smaller than its two companions (.7 % vs. 25.8 and 14.8 % respectively). It is then not surprising if very limited data exists for the hadronic production of resolved or quasi-resolved χ states. The experimental situation is summarised in fig. 9, showing the results of the only two experiments that were capable of making a statement about individual χ's, CERN WA11[31] and Fermilab E-673.[32] WA11 was conceived as an early search for B states or other heavy particles decaying into J/ψ. Thanks to the large sample of ψ triggers collected, they were able to reconstruct

TABLE 3

Existing data on hadronically produced
and resolved χ states.

	WA 11	E 673
Beam, target	185 GeV π^-, Be	225 GeV π^-, Be
Number of events: χ_1	91	56
χ_2	66	28
Fraction of ψ from χ	.305 $\pm$ .05	.37 $\pm$ .10
χ_1/χ_2 production ratio	1.3 $\pm$ 0.4	0.9 $\pm$ 0.4

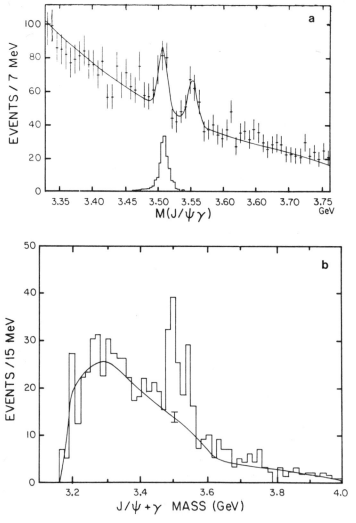

Fig. 9.　Existing data on hadroproduction of resolved χ states. a) CERN WA11, ref 31. b) FNAL E–673, ref 32.

the $\chi \rightarrow \psi\gamma$ decay by detecting the infrequent gamma conversions to e^+e^- in the target. Consequently, they could achieve the outstanding χ mass resolution of 7 MeV and produced a cleanly resolved, although statistics limited, χ_1, χ_2 spectrum (fig. 9a). E-673 was designed to detect and resolve the χ states as its primary goal: the experiment fell somewhat short of its expectations for yield and resolution, and produced the spectrum of fig 9b, where the two χ states are resolved only statistically: rather than claiming two peaks, the experimenters observe that the measured peak is broader than the computed spectrometer resolution, and separate the two components by means of a fit. Both experiments ran with π^-, at the incoming beam energies of 185 and 225 GeV/c respectively. E-673 also produced an upper limit for proton production. Table 3 reports the total event sample, the fraction of ψ's originating from χ decay, and the relative rate of χ_1 vs. χ_2 production. Within the errors, the results are consistent with each other, and the low measured value of ψ fraction from χ would appear to challenge the predictions of the χ dominance models. On the other side, the too large statistical errors on the relative rates of production prevent to draw any conclusion for or against any theoretical prediction.

It is in this scenario that E-705 finds its place. The Fermilab P-West spectrometer, first utilised to study dimuon physics, was upgraded with the sophisticated electro-magnetic calorimeter described elsewhere in these Proceedings.[4] Through an attempt to optimise the precision with which the photon's energy and position are measured, one aims at χ mass resolution ranging between 10 and 15 MeV/c2quite adequate for $\chi_1 - \chi_2$ separation. The dimuon trigger plays again an essential role in selecting the candidate events while running at a high interaction rate. Data taking was started in the late Summer 1987 and is due to end in January '88. The hope is to collect a resolvable χ sample exceeding by a large factor the existing world data, which will allow to make some definite progress in the understanding of an important aspect of strong interaction dynamics.

EXPERIMENT 771 AND THE PRODUCTION AND DECAY OF BEAUTY

The idea to trigger on the J/ψ and to search subsequently for new particles decaying into it is not a new one.[33] After a few attempts, it is now well understood that, at least for the case of Beauty search, a brute force approach is doomed to failure, due to the unmanageable combinatorial background hiding a very small signal. The difficulties encountered in the study of hadronic charm production, even though produced 1000 times more copiously than beauty, are a good indication of the problem at hand. Recent advances in the technology of Slicon Microstrip Detectors nevertheless seem to provide a new powerful handle towards the recognition of rarely produced new particles, if such particles do travel a reasonable distance before decay. The success of E-691 in charm photoproduction, together with positive preliminary indications from its charm hadroproduction successor, E-769,[34] show a promising path that could be followed to reveal the hadronic production of bare beauty. This challenge was undertaken by the E-771 collaboration,[3] who proposed to add a microstrip detector to the P-West spectrometer and to run the dimuon trigger with the highest energy beam available at Fermilab. The goal is to collect and reconstruct a large ψ sample, and to recognise ψ's coming from a secondary vertex as a sure indication of a beauty signal. An attempt would obviously follow to reconstruct the exclusive modes of $B \rightarrow \psi$ decay. The merits of running at the highest energy are made evident by the curve of fig. 10, showing the latest theoretical predictions[35] for B production by pions and protons: it is clear that the expected yield grows by a factor between 5 and 10 going from CERN to Fermilab energies. The agreement of the only existing data point[36] with the theoretical expectations, confirms the reliability of the calculation, and gives confidence to the B yield projections for E-771. At the same time, it is obvious that a more precise measurement of the total cross section, like E-771 intends to perform, is essential to resolve the outstanding uncertainties of the theory, as discussed by Berger. Apart from the B cross sections, the other essential parameters in the calculation of the projected E-771 B yield are the $B \rightarrow \psi$ branching ratio and the reconstruction efficiency. While the former has been measured to be 1.1 %,[37] the latter can only be estimated by Montecarlo. Fig. 11 shows the expected transversal and longitudinal resolutions of the microstrip detector proposed for E-771, and Fig. 12 the corresponding path

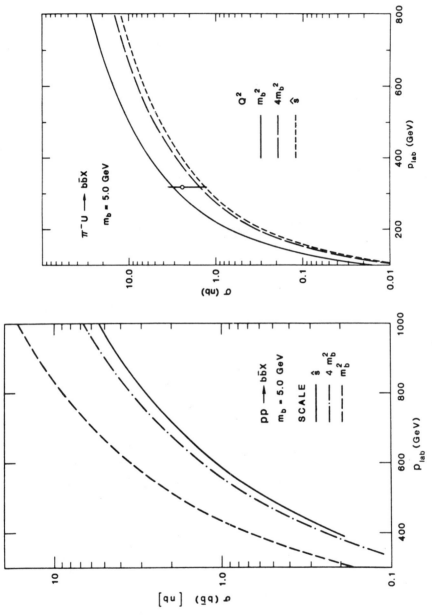

Fig. 10. Theoretical estimates for B hadronic production, from Ed Berger, these Proceedings. The data point is from CERN experiment WA78, ref 36.

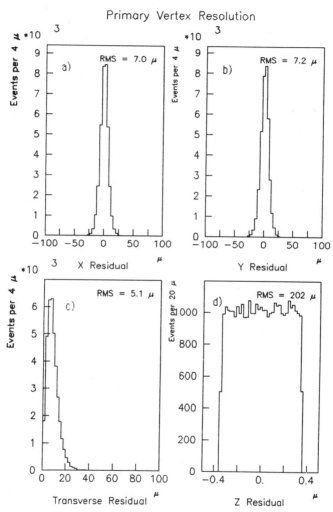

Fig. 11. Transversal and longitudinal vertex resolution expected from the E–771 Silicon Microstrip Tracking System

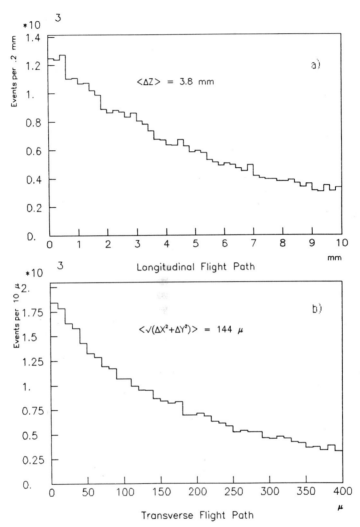

Fig. 12. Monte Carlo distributions of B mesons longitudinal and transverse flight path before decay.

lenghts of the produced B mesons before decay. The resolving power of the detector appears to be quite adequate to recognise secondary vertices. Combining all the factors, one can then estimate that, when running at 10^6 interactions/sec over the typical Fermilab period of $1 - 2 \times 10^3$ hours of data taking, it should be possible to collect a sample of several hundred $B \to \psi$ events, the main uncertainty coming from the knowledge of the cross section.

It is not necessary to stress the richness contained in the study of B physics. Apart from the already mentioned hadronic cross section, E-771's $B \to \psi$ sample will bring information on the B lifetimes and B hadronisation. Simultaneously, the dimuon trigger will also be sensitive to double semi-leptonic decays, allowing to study the properties of $B\bar{B}$ mixing. The running of the experiment will allow to study how higher and higher interaction rates can be achieved, and to investigate other effective ways of triggering the spectrometer, in addition to the detection of dimuons. The plans for the spectrometer are to collect the largest possible sample of reconstructable B's, an attempt to observe the effects of CP-violation in B decay being the final objective. The experiment is expected to run in the next Fermilab fixed target period, currently foreseen to start around Spring 1989.

The material presented here is the result of many years of work on the part of the E-537, E-705 and E-771 collaborators, whose contributions are here acknowledged. I would like to extend my thanks to the organisers of the St. Croix Meeting for their invitation, and to the Natural Science and Engineering Research Council of Canada and the Canadian Institute of Particle Physics for their financial support.

References

1. Fermilab Experiment 537, Athens-Fermilab-McGill-Michigan-Shandong collaboration
2. Fermilab Experiment 705, Arizona-Athens-Duke-Fermilab-Florence- Florida A&M-McGill-Northwestern-Shandong collaboration
3. Fermilab Experiment 771, Arizona-Athens-Berkeley-Duke-Fermilab-Lecce- McGill-Nanking-Northwestern-Pavia-Pennsylvania-Prairie View A& M-Shandong collaboration
4. D.E. Wagoner et al., these proceedings
5. E. Anassontzis et al., Nucl. Instr. and Methods A242, 215 (1986)
6. H.U. Bengsston and T. Sjostrand, Lund preprint LU–TP–87–3 (1987)
7. H. Areti et al., Nucl. Instr. and Methods 212, 135 (1983)
8. S. Conetti and K. Kuchela, IEEE Trans. Nucl. Sc., NS-32, 1318 (1985)
9. S. Conetti, M.Haire and K. Kuchela, IEEE Trans. Nucl. Sc., NS-32, 1326 (1985)
10. I. Gaines et al., Fermilab Conf. 87/21 (1987)
11. See also S. Katsanevas et al., Fermilab Pub 87/57-E, submitted to Phys. Rev. Lett.
12. E. Anassontzis et al., Phys Rev. Lett. 54, 2572 (1985)
13. See also F. Close, these proceedings
14. J. Badier et al., Z. Phys. C20, 101 (1983)
15. J. Qiu, Nucl. Phys. B291, 746 (1987)
16. B.Z. Kopeliovich and F. Niedermayer, JINR-E2-84-834, Dubna (1984). . See also A. Mueller, these proceedings.
17. S.D. Ellis, M.B. Einhorn and C. Quigg, Phys. Rev. Lett. 36, 1263 (1976)
18. C.E. Carlson and R. Suaya, Phys. Rev. D18, 760 (1978)
19. T.B.W. Kirk et al.,Phys. Rev. Lett. 42, 619 (1979) . C. Kourkoumelis et al., Phys. Lett. 81B, 405 (1979)
20. J.H. Kuhn, Phys. Lett. 89B, 385 (1980)
21. E.N. Argyres and C.S. Lam, Phys. Rev. D21, 143 (1980) B.L. Ioffe, Phys. Rev. Lett. 39, 1589 (1977)
22. R. Baier and R. Ruckl, Phys. Lett. 102B, 364 (1981) R. Baier and R. Ruckl, Nucl. Phys. B208, 381 (1982)
23. R. Baier and R. Ruckl, Zeit. Phys. C19, 251 (1983)
24. L. Clavelli, P.H. Cox, B. Harms and S.Jones, Phys Rev. D32, 612 (1985)
25. H. Fritsch, Phys. Lett. 67B, 217 (1976)
26. M. Gluck, J.F. Owens and E. Reya, Phys. Rev. D17, 2324 (1978)

27. Y. Afek, C. Leroy and B. Margolis, Phys. Rev. D22, 86 (1980)
28. M. Gluck and E. Reya, Phys. Lett. 79B, 453 (1978)
29. Z. Kunst, E. Pietarinen and E. Reya, Phys. Rev. D21, 733 (1980)
30. P. Chiappetta and P. Mery, Phys. Rev. D32, 2337 (1985)
31. Y. Lemoigne et al., Phys. Lett. 113B, 508 (1982)
32. S.R. Hahn et al., Phys. Rev. D30, 671 (1984)
33. CERN proposal WA11, SPSC 76–3 and D. Potter et al., Fermilab proposal P–471
34. J. Appel, these proceedings
35. E.L. Berger, these proceedings
36. G. Poulard, these proceedings
37. H. Albrecht et al., Phys. Lett. 162B, 395 (1985)

HEAVY-FLAVOUR PRODUCTION IN UA1

N. Ellis

(UA1 Collaboration at CERN)
University of Birmingham
United Kingdom

INTRODUCTION

Because of their high centre-of-mass energies, the proton–antiproton colliders at CERN and FNAL offer very large cross-sections for producing pairs of charm and bottom particles. At the CERN $p\bar{p}$ Collider ($\sqrt{s}$ = 630 GeV) the predicted cross-sections[1] are $\sigma(c\bar{c}) \approx 100$ μb and $\sigma(b\bar{b}) \approx 10$ μb. The integrated luminosity delivered to UA1 is ~ 1 pb^{-1}, so $\sim 10^8$ $c\bar{c}$ and $\sim 10^7$ $b\bar{b}$ pairs must have already been produced. The Antiproton Collector (ACOL) ring at CERN will increase the luminosity of the CERN Collider by about a factor of 20.

Whilst very large numbers of bottom and charm pairs have been produced at CERN in the UA1 detector, it is a difficult task to separate them from the enormous background of light quark and gluon jets, for which the cross-section is ~ 10 mb. The signature of 'prompt', high-p_T muons from semileptonic decays provides the required background rejection power. In UA1, an inclusive muon trigger selects events with candidate prompt muons, with a transverse momentum (measured relative to the direction of the beams) of $p_T^\mu > 3$ GeV/c in the pseudorapidity interval $|\eta| < 1.5$. In order to reduce the background to the muon sample, from decays in flight of pions and kaons, it is necessary to impose a harder cut on the muon transverse momentum, $p_T^\mu > 6$ GeV/c, when analysing single-muon events, although for dimuon events it is possible to work with muon transverse momenta down to 3 GeV/c.

By selecting $c\bar{c}$ and $b\bar{b}$ events containing a muon with $p_T^\mu > 6$ GeV/c, the background from light quark and gluon jets is reduced to an acceptable level. However, only a small fraction of the $c\bar{c}$ and $b\bar{b}$ events which are produced are selected. The heavy quark or antiquark must have a semimuonic decay (branching ratio $\sim 10\%$), and muons with $p_T^\mu > 6$ GeV/c are produced only by c-quarks or b-quarks of rather high p_T. Nevertheless, one is left with a large sample containing several thousand bottom events, the analysis of which is described in the first part of this paper.

There might be the temptation to perform a similar analysis using electrons rather than muons. However, the lepton from the semileptonic heavy-flavour decay is accompanied by hadronic fragmentation and decay products. In UA1 it is only possible to obtain adequate separation between electrons and the background from light quark and gluon jets when the electrons are isolated and at very high p_T: $p_T^e > 12$ GeV/c. It is therefore not possible to use the electron channel for studying charm and bottom production, although it can be used when searching for new, massive flavours of quarks.

The cross-section for producing top (t) at the CERN Collider is substantial, even for t-quark masses as large as $m_t = 50$ GeV/c^2, for which $\sigma(p\bar{p} \rightarrow W \rightarrow t\bar{b}) + \sigma(p\bar{p} \rightarrow t\bar{t}) \approx 1$ nb is calculated. This corresponds to $\sim 10^3$ events produced, although only a few per cent of these would be detected with semileptonic decays to electrons or muons in the UA1 experiment. Existing limits on the masses of new quarks from measurements at e^+e^- colliders are: $m_t > 26$ GeV/c^2 (from TRISTAN[2]) and $m_{b'} > 22.7$ GeV/c^2 (from PETRA[3]), where b' is a new charge $-\frac{1}{3}$ quark. Recent measurements of B^0–$\bar{B}^0$ mixing[4,5] have been interpreted[6] as an indication of a high t-quark mass, $m_t \gtrsim 50$ GeV/c^2. Measurements of the Standard Model parameters $\sin^2\theta_w$ and ϱ at the CERN Collider and in deep-inelastic scattering experiments can be interpreted to give an upper limit on the t-quark mass $m_t < 180$ GeV/c^2.

The sensitivity of experiments at proton–antiproton colliders to new, massive quarks depends on the production cross-section. If the t-quark mass is below the W mass, $t\bar{b}$ pairs will be produced in decays of W bosons. The cross-section for this process is reliably calculable since it can be normalized to measurements of $p\bar{p} \to W \to e\nu$, allowing for colour, phase-space, and radiative corrections. Direct production of $t\bar{t}$ pairs can be calculated using perturbative QCD, but with greater uncertainty than in the case of $W \to t\bar{b}$. A search for new quarks using events with muons or electrons and jets is described in the second part of this paper. Further details of this work can be found elsewhere.[7,8]

CHARM AND BOTTOM PRODUCTION

Results from Measurements of Dimuon Production

We have reported[9] on measurements of dimuon production in proton–antiproton interactions at $\sqrt{s} = 540$ GeV and 630 GeV. Events were selected if both muons were reconstructed with $p_T^\mu > 3$ GeV/c and with a dimuon mass $m_{\mu\mu} > 6$ GeV/c^2, and if the muons were accompanied by hadrons. For $b\bar{b}$ pairs with both quarks having $p_T > 5$ GeV/c and $|\eta| < 2.0$, the ISAJET Monte Carlo program[10] and a simulation of the UA1 detector were used to calculate the probability of producing and detecting dimuon events passing the above selection. The main process contributing to this event sample is flavour creation, mainly from gluon fusion. The higher-order processes — gluon splitting and flavour excitation — which are illustrated in Figs. 1b and 1c, are suppressed by the requirements on the dimuon mass and the muon transverse momenta, respectively: in gluon splitting, the b-quark and $\bar{b}$-quark tend to be collinear and cannot give rise to high-mass muon pairs. In flavour excitation, the 'spectator' b-quark normally has low p_T and cannot decay to give a high-p_T muon. The cross-section which is quoted,

$$\sigma(p\bar{p} \to b\bar{b}; \quad p_T^b > 5\,\text{GeV}/c; \quad |\eta| < 2.0) = 1.1 \pm 0.1 \pm 0.4\,\mu b,$$

is for the flavour-creation process only and is in reasonable agreement with the predicted cross-section of 1.7 μb. The small contributions to the dimuon sample from the flavour-excitation and gluon-splitting processes have been subtracted using the Monte Carlo calculation. Similarly, the small contribution from $c\bar{c}$ production has been subtracted. The large systematic error which is quoted is partly a result of the uncertainty in subtracting these background processes. Other uncertainties which are included are: the parametrization of the b-quark fragmentation function; the average branching ratio for bottom hadrons to muons; the subtraction of background from decays in flight of pions and kaons; and the luminosity measurement.

We have also reported the first evidence of B^0–$\bar{B}^0$ mixing[4] from the large fraction of like-sign muon pairs that are observed. We measured $\chi = 0.121 \pm 0.047$, where χ is the fraction of bottom hadrons which decay to 'wrong-sign' muons, i.e. $\chi = \text{BR}(b \to \bar{B}^0 \to B^0 \to \mu^+)/\text{BR}(b \to B \to \mu^\pm)$, where the denominator includes all bottom hadrons, including baryons. In the absence of B^0–$\bar{B}^0$ mixing, $\chi = 0$.

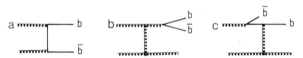

Fig. 1. Illustration of Feynman diagrams: a) gluon fusion, b) gluon splitting, c) flavour excitation.

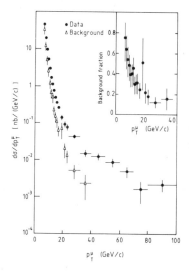

Fig. 2. Inclusive muon cross-section for $|\eta| < 1.5$ for raw data and background.

Inclusive Muon Data Sample

Events containing $p_T > 6$ GeV/c muon candidates were selected inclusively. This yielded $\sim$ 20,000 events, recorded at a centre-of-mass energy of 630 GeV for an integrated luminosity of 556 nb^{-1}. The pseudorapidity acceptance of the first-level muon trigger was limited to about ± 1.5 so as to maintain an acceptable rate. For $|\eta| < 1.5$, the geometrical acceptance times reconstruction efficiency for $p_T^\mu > 6$ GeV/c was about 33%.

Muons were identified in the central detector (CD) by a high-p_T track which, when extrapolated through more than 9 interaction lengths of lead and iron shielding, matched with a track in the muon chambers.

A number of sources of background to prompt muons were considered. Non-interacting hadrons and shower leakage can be neglected because of the very thick shielding between the interaction point and the muon chambers. Leakage through cracks in the shielding was eliminated by fiducial cuts. Cosmic-ray tracks which cross the CD, passing close to an interaction vertex, were identified and removed by software. Misassociation of high-p_T tracks in the CD with tracks in the muon chambers was rendered negligible by tight cuts in the matching, in position and angle, between the extrapolated CD track and the track in the muon chambers.

The only significant background to prompt muons comes from decays in flight of pions and kaons. This background to the inclusive muon sample was calculated using real events recorded with a minimum-bias trigger. Decays in flight of high-p_T particles in the events were simulated, assuming a charged-particle composition of 58% $\pi^\pm$, 23% $K^\pm$, and 19% p and $\bar{p}$, based on measurements from the CERN Collider.[11] All the selection criteria, including the trigger algorithm, were applied to the simulated muon events. The events were weighted according to the probability of the high-p_T pion or kaon decaying in flight before interacting in the shielding.

The inclusive muon cross-section is shown in Fig. 2 as a function of p_T^μ, both for the data and for the background. The acceptance was calculated assuming a flat pseudorapidity distribution for $|\eta| < 1.5$. Note that the background fraction falls rapidly with p_T, from over 70% at 6 GeV/c to about 35% for $p_T > 10$ GeV/c. The systematic error on the background calculation is estimated to be $^{+20}_{-30}$%, largely due to the uncertainty on the charged-particle fractions.

QCD-Based Monte Carlo Model

The Monte Carlo model[10] used in this analysis was the ISAJET program, version 5.23. This was tuned to e$^+$e$^-$ data for charm and bottom fragmentation and decays, and to UA1 data for the

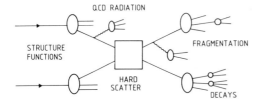

Fig. 3. Components of the ISAJET model.

underlying event. Technical modifications were made to improve the efficiency for generating heavy-flavour events with semileptonic decays. The generated Monte Carlo events were passed through a detailed simulation of the UA1 detector, and reconstructed using the same programs as for the real data.

Monte Carlo samples corresponding to an integrated luminosity of 10 pb^{-1} were produced for the following processes:

i) W and Z production with all decay modes simulated; the cross-sections were normalized to measurements[12] of W $\rightarrow \ell\nu$ and Z $\rightarrow \ell^+\ell^-$;
ii) the Drell–Yan mechanism, normalized to the measured cross-section;[9]
iii) high-p$_T$ Υ and J/ψ production, normalized to the measured cross-sections;[9,13]
iv) QCD production of charm and bottom particles. For this process, the cross-section was predicted by the ISAJET Monte Carlo calculation.

In each of the above cases, the generated events were processed only if they contained high-p$_T$ muons or electrons.

The ISAJET model for charm and bottom production is illustrated in Fig. 3. The momenta of the partons which enter the hard-scattering interaction are determined by the structure functions (EHLQ set 1).[14] The matrix element for the hard scattering is calculated[1] exactly to O(α_s^2), taking Q^2 = 2 stu/(s^2 + t^2 + u^2). The QCD radiation is included in the branching approximation.[15] This allows for *gluon splitting* (in the final state) into heavy-quark pairs, and *flavour excitation,* where a gluon splits into a heavy-quark pair in the initial state, as well as the lowest-order *flavour-creation* diagram, as illustrated in Fig. 1. Gluon radiation from the incoming and outgoing partons is also simulated. The outgoing partons are fragmented independently, using the Peterson et al. parametrization[16] with ϵ_c = 0.30 for charm and ϵ_b = 0.02 for bottom. The particles which are produced in the fragmentation are allowed to decay using measured branching ratios where possible; for bottom particles the measured average branching ratio of 12% is used. Semileptonic decays are simulated using V − A matrix elements; other decays are simulated using phase space only.

The underlying event due to the break-up of the spectator systems is simulated using a phenomenological model. It was found necessary to scale down the average charged-particle multiplicity in the spectator system so as to reproduce the level of activity in the data.

Comparison of Predictions and the Data

The predicted inclusive muon cross-section is compared with the data in Fig. 4 as a function of p$_T^\mu$, for $|\eta|$ < 1.5. The contribution from W $\rightarrow \mu\nu$ and Z $\rightarrow \mu^+\mu^-$ decays is shown separately and dominates at large p$_T$. The large error bars on the data points at low p$_T$ are due to the uncertainty on the background which is subtracted. Good agreement is observed between the prediction and the data over the entire range from 6 GeV/c to 100 GeV/c.

In order to study further the heavy-flavour content of the sample, a selection of events was made with 10 GeV/c < p$_T^\mu$ < 15 GeV/c and $|\eta|$ < 1.5. The decays W $\rightarrow \mu\nu$ and Z $\rightarrow \mu^+\mu^-$

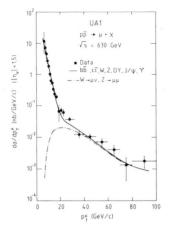
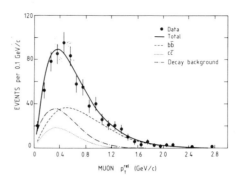

Fig. 4. Inclusive muon cross-section for $|\eta| <$ 1.5 after background subtraction.

Fig. 5. Distribution of p_T^{rel} for a subsample of $p_T^\mu > 10$ GeV/c muon events.

contribute very little to this sample, and the uncertainty on the background subtraction is not too large. The predicted composition of the data in this sample is 69% bottom decays, 24% charm decays, 5% Drell–Yan mechanism, and $J/\psi \to \mu^+\mu^-$ and $\Upsilon \to \mu^+\mu^-$ decays, and only 2% W $\to \mu\nu$ and Z $\to \mu^+\mu^-$ decays. The total predicted cross-section is 3.54 nb compared with the measured cross-section of 4.33 $\pm$ 0.20$^{+0.83}_{-0.56}$ nb. This agreement is surprisingly good considering the large theoretical uncertainties due to the choice of structure functions, to the Q^2 scale, and to unknown terms of higher order in α_s.

The predicted cross-section for producing high-p_T muons from bottom decays exceeds that from charm decays owing to the harder bottom fragmentation function. For muons with p_T^μ in excess of 10 GeV/c, only charm and bottom quarks with $p_T^Q \gg m_Q$ contribute, so there is little suppression due to the higher b-quark mass. The relative contribution of bottom and charm decays was measured for the subsample of events with muons of $p_T^\mu > 10$ GeV/c, containing at least one jet reconstructed in the calorimeters with $E_T > 10$ GeV, and with a well-defined charged-particle jet reconstructed near the muon in the CD.

The variable p_T^{rel} is defined as the momentum of the muon perpendicular to the direction of the accompanying charged-particle jet; it is sensitive to the mass of the object which decays to produce the muon and the jet. The distribution of the data in this variable is shown in Fig. 5, together with the predicted distribution for bottom and charm decays and for the background. The solid curve is the result of a fit to the data with a sum of the background (fixed) and the bottom and charm contributions, which were allowed to vary. The result of the fit is $b\bar{b}/(b\bar{b} + c\bar{c}) = 76 \pm 8 \pm 9$%, compared with the ISAJET prediction of 74% for this sample. Similar good agreement was found using an equivalent method for the dimuon data.[9]

It has been shown that the ISAJET model accounts well for the observed yield of high-p_T muons from bottom and charm decays, and that the relative contribution of bottom and charm is correctly described.

The use of the branching approximation in ISAJET makes possible the simulation of multijet events. A study has been made of how well this model describes the data.

Events were selected requiring a muon with 10 GeV/c $< p_T^\mu < 15$ GeV/c (to reduce the background from pion and kaon decays in flight, and W $\to \mu\nu$ and Z $\to \mu^+\mu^-$ decays), and at least one jet reconstructed in the calorimeters with $E_T > 12$ GeV and separated from the muon by at least one unit in pseudorapidity–azimuth space: $\Delta R = \sqrt{\Delta\eta^2 + \Delta\phi^2} > 1$. The number of jets ($N^{jets}$) in each event was then counted, including only those jets reconstructed in the calorimetry with

433

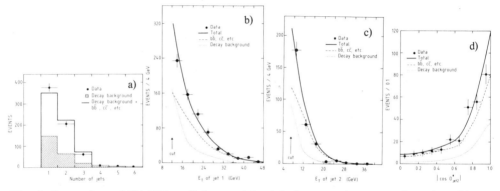

Fig. 6. Comparison of ISAJET predictions and the data for muon–jet events: a) number of jets, b) E_T of highest-E_T jet, c) E_T of second-highest-E_T jet, d) angular distribution of second-highest-E_T jet.

$E_T > 7$ GeV, validated by a charged track reconstructed in the CD with $p_T > 0.5$ GeV/c, and with $\Delta R > 1$ between the jet and the muon. Hadronic energy accompanying the muon, i.e. within the $\Delta R < 1$ cone, is considered separately.

The distribution of N^{jets} in the muon–jet events is shown in Fig. 6a, together with the predictions from the ISAJET model and the background calculation. Excellent agreement is observed. The E_T distributions of the highest-E_T jet (j_1) and of the second-highest-E_T jet (j_2) are plotted in Figs. 6b and 6c. Note that the decay background has E_T distributions softer than those of the data. The angular distribution of the second-highest-E_T jet is shown in Fig. 6d; $\theta^*_{j_2}$ is measured relative to the direction of the $\bar{p}$ beam in the centre of mass of the muon, the two jets, and the neutrino for which only the transverse components were used. The peak towards $|\cos\theta^*| = 1$, associated with initial-state radiation collinear with the beams, is well reproduced by the Monte Carlo.

The distribution of missing transverse energy E_T^ν measured in the muon–jet events is plotted in Fig. 7, together with the predictions from the ISAJET model and the background calculation. The distribution at low E_T^ν is determined mainly by the resolution of the detector, whilst the high-E_T^ν part of the distribution is due to neutrinos from bottom and charm semileptonic decays and from severe momentum mismeasurements caused by kinks in kaon decay tracks. The entire distribution is well reproduced.

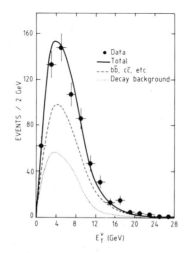

Fig. 7. Comparison of ISAJET prediction and the data for missing transverse energy in muon–jet events.

434

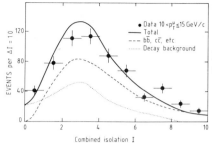

Fig. 8. Comparison of ISAJET prediction and the data for the isolation of the muons in muon–jet events.

The amount of activity near the muons in the muon–jet events can be studied using both the calorimeters and the CD. A combined isolation variable, $I = \sqrt{(\Sigma E_T/3)^2 + (\Sigma p_T/2)^2}$, is chosen because of the complementary geometrical acceptance of the two detectors; ΣE_T and Σp_T are the transverse energy and transverse momentum sums of calorimeter cells and charged tracks in a cone of radius $\Delta R = 0.7$ about the muon direction. The distribution of the data in this variable is shown in Fig. 8 together with predictions from the ISAJET model and the background calculation. The predictions account well for the data; this is also true if ΣE_T and Σp_T are plotted separately.

Conclusions on Charm and Bottom Production

The large rate of high-p_T^μ events in the UA1 experiment at the CERN $p\bar{p}$ Collider is well reproduced by the QCD-based ISAJET model of charm and bottom production with semileptonic decays. This follows earlier UA1 measurements in the dimuon channel.[4,9] Recent UA1 measurements[13] of high-p_T J/ψ production through $B \rightarrow J/\psi$ decays are well reproduced by the same Monte Carlo program.

The ISAJET model includes terms of orders higher than α_s^2 using the branching approximation. Predictions for a number of distributions which are sensitive to higher-order processes have been compared with the data and found to fit well.

SEARCH FOR NEW HEAVY QUARKS

Expected Features of Top-Quark Events

Top-quarks may be produced in $W \rightarrow t\bar{b}$ or $Z \rightarrow t\bar{t}$ decays if they are sufficiently light, or in direct QCD $t\bar{t}$ production. In either case, the t-quark is expected to decay semileptonically with a branching ratio of about 1/9 per lepton: $t \rightarrow \ell\nu b$. Owing to the high mass of the t-quark, the decay products will usually have high momenta and be well separated from each other. Because of this, the lepton is expected to be 'isolated', i.e. unaccompanied by hadrons. In the case of $W \rightarrow t\bar{b}(t \rightarrow \ell\nu b)$ decays, two jets are expected, one from the $\bar{b}$-quark and one from the b-quark. For $t\bar{t}$ production with $t \rightarrow \ell\nu b$, up to four jets are expected since the $\bar{t}$ will mostly decay to three lighter quarks which may be identified as separate jets. Thus a signature for t-quark decays is an isolated lepton with a high-p_T neutrino and at least two jets. The observation of six such events in data from the 1983 CERN Collider run, for which the integrated luminosity was 108 nb^{-1}, led to conjecture that top had been observed.[17] This paper describes a search for new heavy quarks in a larger sample, corresponding to about 700 nb^{-1}.

The ISAJET Monte Carlo program[10] was used to make predictions for top production. For $p\bar{p} \rightarrow W \rightarrow t\bar{b}$ production and decay, the cross-section was normalized to measurements[12] of $p\bar{p} \rightarrow W \rightarrow \ell\nu$. For direct QCD production, $p\bar{p} \rightarrow t\bar{t}$, the cross-section was normalized to a calculation made with the EUROJET Monte Carlo program,[18] which includes terms to $O(\alpha_s^3)$ at the tree level. The structure functions were EHLQ set 1,[14] and $Q^2 = m^2 + p_T^2$ and $\Lambda = 0.2$ GeV/c^2 were used. The sensitivity of the calculation of cross-sections to these parameters is described later.

To search for evidence of t-quark decays, events were selected with $p_T^\mu > 10$ GeV/c, at least one jet with $E_T > 12$ GeV, and $\Delta R > 1$, and by requiring the muon–neutrino transverse mass $m_T^{\mu\nu}$ to be less than 40 GeV/c^2 in order to remove $W \to \mu\nu$ decays. The distribution of these events in the isolation variable $I = \sqrt{(\Sigma E_T/3)^2 + (\Sigma p_T/2)^2}$ is shown in Fig. 9a. Also shown is the distribution expected from charm and bottom decays and from the other processes discussed earlier, plus the background from decays in flight of pions and kaons. For illustration, the predicted additional contribution from the t-quark is shown (shaded) for $m_t = 30$ GeV/c^2. The data are consistent with the prediction without top. The effects of additional cuts are shown in Figs. 9b and 9c, where the signal-to-background ratio predicted for $m_t = 30$ GeV/c^2 improves. In Fig. 9b the muon is required to have $p_T^\mu > 12$ GeV/c (instead of 10 GeV/c) and the jet is required to have $E_T^{j1} > 15$ GeV (instead of 12 GeV). In Fig. 9c, a second jet with $E_T^{j2} > 7$ GeV is required in addition. From the number of events observed with $I < 2$ in Fig. 9c, and taking the EUROJET normalization for the $t\bar{t}$ production cross-section, one obtains a limit on the mass of the t-quark: $m_t > 41$ GeV/c^2 at 95% CL.

A slightly improved limit can be obtained by performing a likelihood fit to the four event variables I, E_T^ν, E_T^{j1}, and $|\cos\theta_{j2}^*|$ which are plotted in Fig. 10 for the sample of events with $p_T^\mu > 12$ GeV/c, $m_T^{\mu\nu} < 40$ GeV/c^2, $I < 10$, $E_T^{j1} > 12$ GeV, and $E_T^{j2} > 7$ GeV; this yields the limits shown in Fig. 11. Systematic errors have been included for the energy scale, the selection efficiency, the integrated luminosity, the cross-section for $p\bar{p} \to W \to t\bar{b}$, and the background calculation. From the intersection of the predicted cross-section (EUROJET calculation for $t\bar{t}$ production) and the 95% CL contour, one obtains $m_t > 43$ GeV/c^2 from the muon channel only.

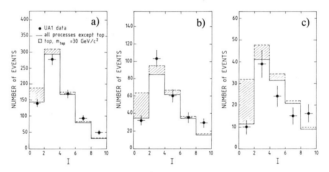

Fig. 9. Isolation distributions with progressively more restrictive selection criteria.

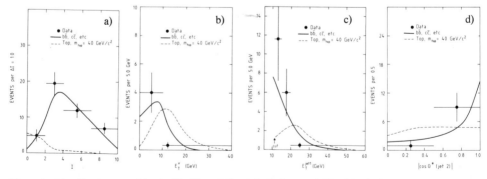

Fig. 10. Distributions used in the likelihood fit: a) isolation variable, b) missing transverse energy, c) E_T of highest-E_T jet, d) angular distribution of second-highest-E_T jet.

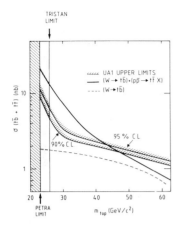

Fig. 11. Cross-section limits for t-quark production from the muon channel.

Search for the Top-Quark in the Electron Channel

The search for the t-quark in the electron channel closely parallels the one in the muon channel. The possibility to study non-isolated and semi-isolated muons was an important aspect of the preceding analysis, which is not possible here. However, semileptonic charm and bottom decays to muons and electrons should be similar, so the results from the muon channel can be taken over to the electron channel.

Electron candidates were selected by requiring a transverse energy cluster in the e.m. calorimeters with $E_T^e > 15$ GeV, matched in position and momentum by a charged track reconstructed in the CD with $p_T^e > 10$ GeV/c. The shower shape in profile and in depth was required to be compatible with an electron. A very strong isolation requirement was made on the electron:

$$\Sigma E_T < 0.1 E_T^e \text{ in } \Delta R < 0.7,$$

$$\Sigma p_T < 0.1 p_T^e \text{ in } \Delta R < 0.7,$$

$$\Sigma E_T < 1 \text{ GeV in } \Delta R < 0.4,$$

$$\Sigma p_T < 1 \text{ GeV/c in } \Delta R < 0.4.$$

This is considerably stronger than the isolation requirement made in the muon channel, and is required in order to control the background from light quark and gluon jets which fluctuate into a single, high-p_T charged particle and a number of π^0's. Photon conversions in the beam pipe or CD were removed by scanning.

The number of electron candidates selected was 205 for an integrated luminosity of 689 nb^{-1}. Of these, 119 events had $m_T^{e\nu} > 45$ GeV/c^2 and were removed as probable $W \rightarrow e\nu$ events, and a further 26 had a second candidate electron with $E_T^e > 6$ GeV and were removed as probable Drell–Yan or $\Upsilon \rightarrow e^+e^-$ decay events. Of the remaining 60 events, only 26 had a jet with $E_T^e > 12$ GeV outside a cone radius of $\Delta R = 1$ about the electron.

The background to the electron–jet sample from overlapping $\pi^\pm$ and π^0's was estimated to be $6.3 \pm 0.5 \pm 1.1$ events.[19] The background due to Dalitz pairs and to unrecognized converted photons, mainly from $\pi^0 \rightarrow \gamma\gamma$ decays, was estimated to be 1.8 ± 0.7 events.[19] The estimated contribution to the sample of events with an electron and at least one jet due to charm and bottom semileptonic decays and to the other processes discussed earlier, is 15.3 events.

Combining these sources of electron–jet events, the total number expected is $23.6 \pm 0.9 \pm 2.7$ compared with 26 observed, leaving little room for a contribution from t-quark decays. This leads

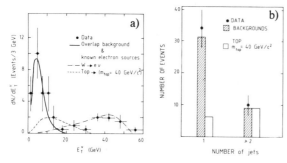

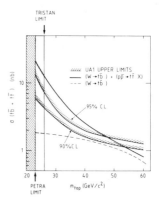

Fig. 12. Distributions used in likelihood fit: a) missing transverse energy, b) number of jets.

Fig. 13. Cross-section limits for t-quark production from the electron channel.

to a limit, based only on the number of events, of $m_t > 36$ GeV/c^2 at 95% CL, taking the EUROJET cross-section for $t\bar{t}$ production.

As for the muon channel, a better limit can be obtained using a likelihood fit to event variables. Fitting to the two variables E_T^ν and N^{jets} which are plotted in Fig. 12 yields the limits shown in Fig. 13. Systematic errors have been included for the energy scale, the selection efficiency, the integrated luminosity, the cross-section for $p\bar{p} \rightarrow W \rightarrow t\bar{b}$, and the background calculation. From the intersection of the predicted cross-section (EUROJET calculation for $t\bar{t}$ production) and the 95% CL contour, one obtains $m_t > 47$ GeV/c^2 at 95% CL from the electron channel only.

Mass Limits Combining Muon and Electron Channels

By combining the likelihood distributions from the muon and electron channels, the limits shown in Fig. 14 are obtained. From the intersection of the predicted cross-section (EUROJET calculation for $t\bar{t}$ production) and the 95% CL contour, one obtains $m_t > 56$ GeV/c^2 at 95% CL. Note that it is not yet possible to set a mass limit at 95% CL for $W \rightarrow t\bar{b}$ production only.

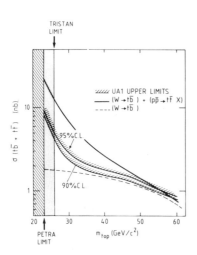

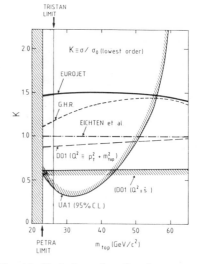

Fig. 14. Combined cross-section limits for t-quark production from the muon and electron channels.

Fig. 15. Variation of predicted cross-sections with choice of parameters. The UA1 cross-section limit is indicated.

438

Therefore the uncertainty in the calculation of the $t\bar{t}$ production cross-section must be estimated before giving final mass limits. The effects of different choices of structure function, of the Q^2 scale, and of including $O(\alpha_s^3)$ terms, are shown in Fig. 15, expressed as a K-factor relative to the $O(\alpha_s^2)$ calculation, with EHLQ set 1 structure functions and $Q^2 = m^2 + p_T^2$. The lowest cross-section which was obtained was with Duke and Owens (DO) set 1 structure functions,[20] taking $Q^2 = \hat{s}$, and with $O(\alpha_s^2)$ terms only. This yields a conservative limit of $m_t > 44$ GeV/c^2 at 95% CL. Note that the EUROJET calculation to $O(\alpha_s^3)$ which was used as a reference in this paper, and which corresponds to $K \simeq 1.5$, is not untypical, and that other choices of structure function give higher cross-sections.

Cross-Section Limits for b' Production

The production, via QCD, of a new, heavy, charged $-\frac{1}{3}$ quark b' is similar to t-quark production. However, it is not expected to be produced in W decays since its partner is presumably very heavy. Limits are therefore presented assuming strong production of b'$\bar{b}'$ pairs only. Taking the $O(\alpha_s^3)$ EUROJET prediction for b'$\bar{b}'$ production, one obtains $m_{b'} > 44$ GeV/c^2 at 95% CL. For the $O(\alpha_s^2)$ calculation with the most conservative choice of structure functions and Q^2 scale, $m_{b'} > 32$ GeV/c^2 at 95% CL is obtained.

Conclusions on the Search for New Heavy Quarks

No evidence for new, heavy quarks is found in the UA1 muon or electron data. Taking a representative estimate of the production cross-sections, using the EUROJET Monte Carlo program which includes $O(\alpha_s^3)$ tree-level terms and EHLQ set 1 structure functions, and taking $Q^2 = m^2 + p_T^2$, one obtains at 95% CL:

$$m_t > 56 \text{ GeV/c}^2,$$
$$m_{b'} > 44 \text{ GeV/c}^2.$$

Taking the lowest estimated production cross-sections, obtained using $O(\alpha_s^2)$ terms only, DO set 1 structure functions, and $Q^2 = \hat{s}$, one obtains at 95% CL:

$$m_t > 44 \text{ GeV/c}^2,$$
$$m_{b'} > 32 \text{ GeV/c}^2.$$

Future measurements at the CERN and FNAL proton–antiproton colliders should be sensitive to the process $p\bar{p} \to W \to t\bar{b}$ only, allowing mass limits to be set with much less theoretical uncertainty. The interesting possibility of a very heavy t-quark decaying into real W's is discussed in the paper presented by Eggert at this Workshop.

Acknowledgements

I would like to thank all my colleagues in UA1 who helped to make these results possible. Special thanks go to Michel Della Negra and Isabelle Wingerter. The financial support of NATO and the Royal Society, which allowed me to attend the Workshop, is gratefully acknowledged. Finally, I would like to express my appreciation to the Workshop organizers for their hospitality.

REFERENCES

1. B.L. Combridge, Nucl. Phys. **B151**:429 (1979).
2. H. Aihara (TOPAZ Collab.), 'Search for new heavy quarks and leptons', Univ. Tokyo preprint UT HE 87–10 (1987).
   K. Abe et al. (VENUS Collab.), 'Search for the top quark in e^+e^- annihilation at $\sqrt{s} = 50$ GeV: the first result from the VENUS detector at TRISTAN', KEK preprint P 87–79 (1987).
   H. Yoshida et al. (VENUS Collab.), 'Measurement of R and search for new heavy quarks in

e^+e^- annihilation at 50 and 52 GeV centre-of-mass energies', KEK preprint P 87–81 (1987).

H. Sagawa et al. (AMY Collab.), 'Measurements of R and a search for heavy quark production in e^+e^- annihilation at $\sqrt{s} = 50$ and 52 GeV', KEK preprint P 87–84 (1987).

3. H.J. Behrend et al. (CLEO Collab.), Phys. Lett. **144B**:297 (1984).

4. C. Albajar et al. (UA1 Collab.), Phys. Lett. **186B**:247 (1987).

5. H. Albrecht et al. (ARGUS Collab.), Phys. Lett. **192B**:245 (1987).

6. J. Ellis et al., Phys. Lett. **192B**:201 (1987);

   V. Barger et al., Phys. Lett. **194B**:312 (1987);

   I.I. Bigi and A.I. Sanda, Phys. Lett. **194B**:307 (1987).

7. C. Albajar et al. (UA1 Collab.), 'Study of heavy flavour production in events with a muon accompanied by jets at the CERN Proton–Antiproton Collider', preprint CERN–EP/87–189 (1987), to be published in Z. Phys. C.

8. C. Albajar et al. (UA1 Collab.), 'Search for new heavy quarks at the CERN Proton–Antiproton Collider', preprint CERN–EP/87–190 (1987), to be published in Z. Phys. C.

9. C. Albajar et al. (UA1 Collab.), Phys. Lett. **186B**:237 (1987).

10. F. Paige and S.D. Protopopescu, 'ISAJET Monte Carlo', Brookhaven report BNL 38034 (1986).

11. M. Banner et al. (UA2 Collab.), Phys. Lett. **122B**:322 (1983).

    G.J. Alner et al. (UA5 Collab.), Nucl. Phys. **B258**:505 (1985).

12. G. Arnison et al. (UA1 Collab.), Lett. Nuovo Cimento **44**:1 (1985).

    J.A. Appel et al. (UA2 Collab.), Z. Phys. **C30**:1 (1986).

13. C. Albajar et al. (UA1 Collab.), 'High-transverse-momentum J/ψ production at the CERN Proton–Antiproton Collider', preprint CERN–EP/87–175 (1987), to be published in Phys. Lett. B.

14. E. Eichten et al., Rev. Mod. Phys. **56**:579 (1984) and **58**:1065 (1986).

15. G.C. Fox and S. Wolfram, Nucl. Phys. **B168**:285 (1980).

    T. Sjöstrand, Phys. Lett. **157B**:321 (1985).

16. C. Peterson et al., Phys. Rev. **D27**:105 (1983).

17. G. Arnison et al. (UA1 Collab.), Phys. Lett. **147B**:493 (1984).

18. A. Ali et al., Nucl. Phys. **B292**:1 (1987).

19. UA1 Collaboration, Technical note UA1–TN 87–52 (1987), unpublished.

20. D.W. Duke and J.F. Owens, Phys. Rev. **D30**:49 (1984).

CHARM PRODUCTION FROM 400 AND 800 GEV/C
PROTON-PROTON COLLISIONS

A.T. Goshaw

For The NA27 and E743 Collaborations
Duke University, Durham, North Carolina 27706

I. INTRODUCTION

This is a report on charm particle production from pp
interactions at 400 and 800 GeV/c as measured in
experiments NA27 (CERN) and E743 (Fermilab). The 400
GeV/c data is final. It provides the most accurate charm
production cross section measurement to date. The 800
GeV/c data is preliminary. The combined data provide a
measurement, with small systematic error, of the energy
dependence of charm hadroproduction from 27 to 39 GeV.
These data are compared to the predictions of charm
particle production via the QCD fusion model with various
charm quark fragmentation schemes.

These experiments use the Lexan bubble chamber (LEBC) as
both target and high resolution vertex detector. LEBC
provides a volume of hydrogen 12 cm long by $7x5cm^2$. It
was exposed to a proton beam a few cm in height and a mm
in depth and triggered on all pp interactions producing
more than two charged particles. This trigger was
essentially unbiased for events containing charm
particles: a Monte Carlo estimate of this efficiency is
$(98^{+2}_{-3})\%$. The track images recorded on 50 mm film

consisted of 20 micron bubbles at a density of 80/cm. This allowed a reliable detection of tracks from a charm decay vertex with impact parameters down to 8 microns. A double scan of all the film resulted in the selection of charm decays surviving our cuts with an efficiency of over 90%.

The particles emerging from LEBC were momentum analyzed and identified by the European Hybrid Spectrometer (EHS) at CERN and by the Multi-Particle Spectrometer (MPS) at Fermilab. It is not appropriate in this brief report to discuss the detailed performance of these elegant spectrometers. However it should be noted that both had good $\pi^{\pm}/K^{\pm}/p^{\pm}$ separation via multiple Cherenkov counters, a transition radiation detector and an ISIS device (EHS only). In addition, the EHS has good π^{o}/γ and K^{o}/n detection capabilities via electromagnetic and neutral hadron calorimeters. For a more detailed description of these spectrometers see references [1] (EHS) and [2] (MPS).

II. SUMMARY OF CHARM DATA SAMPLE

Experiment NA27 had a sensitivity of 38.5 events/microbarn for the production of charm particles. A total of 324 events containing 557 charm decay candidates were selected by a double scan of the film for C1, C3, C5, V2, V4 and V6 decays. Experiment E743 has a sensitivity of about 15 events/microbarn for charm, but the data presented here come from a 9.0 events/microbarn subsample. The film in this experiment was scanned only for interactions producing C3, C5, V4 and V6 decays. This resulted in 165 events containing 210 charm decay candidates in the present subsample.

The actual number of events used from each experiment depends on the type of charm particle analysis being done. In this short report we will restrict ourselves to D meson production properties and an estimate of the total cross section for charm production.

III. D MESON AND TOTAL CHARM PRODUCTION CROSS SECTIONS

III.A. PP INTERACTIONS AT 400 GEV/C ($\sqrt{s}$ = 29 GeV)

The accuracy to which a charm production cross section can be measured is often limited by uncertainties in the exclusive and topological charm hadron decay branching ratios. This is a very severe problem for Λ_c and D_s production studies and is also a limitation for D meson production measurements if only a few decay channels are used. In order to avoid this problem we have measured the D meson production cross section by making use of C1, C3, C5, V2, V4 and V6 decays and thus our measurement is sensitive only to the branching ratio of D^0 decay to all neutrals (0.14 ± 0.04) [3]. In order to suppress Λ_c and D_s decays in the charged charm sample, we apply impact parameter and decay length cuts which tend to reject these particles because of their short lifetime and low Q values relative to a D^+. These cuts also help to clearly define the decay topologies. A summary of the cuts used and the 217 surviving $D/\bar{D}$ decays is given in Table 1. In addition, C1 and V2 decays were accepted only if one particle had a transverse momentum greater than 250 MeV/c in order to eliminate strange particle decays.

Table 1

Geometrical cuts applied to select the charm decays used for the D/D̄ total cross section calculations.

Decay channel	Decays surviving all cuts $N_i(D)$	Decay length (mm)		Transverse length (mm)	Minimum impact parameter lower limit (µm)	Maximum impact parameter range (µm)
		Lower limit	Upper limit			
C1	39	2	90	0.6		100–1500
C3	74	1		2.0	20	100–2000
C5	6	1		2.0	20	100–2000
V2	67	2	30	0.2	20	60– 500
V4	30	1		2.0	20	60–1500
V6	1	1		2.0	20	60–1500

The total cross section for D meson production was calculated as follows:

$$\sigma(D) = [\sum_{i=1}^{3} N_i(D) w_i(D)/\eta_i]/(S \cdot BR)$$

where $N_i(D)$ is the number of decays surviving the cuts (Table 1), w_i the weight which corrects for the loss of events due to the cuts, η_i the scanning efficiency, BR the branching ratio [$BR(D^+ \rightarrow C1 + C3 + C5) = 1.0$, $BR(D^0 \rightarrow V2 + V4 + V6) = 0.86 \pm 0.04$] and $S = (38.5 \pm 1.1)$ events/μb is the sensitivity of the experiment. The weights, w_i, were calculated from Monte Carlo simulations which determined the fraction of decays lost by our cuts. For more details see reference [4].

We have studied the stability of our calculated D cross sections to the cuts applied to the sample and to changes in the details of the Monte Carlo production/decay simulation. These uncertainties result in an estimated systematic error of 7% in the D^+/D^- and 9% in the $D^0/\bar{D}^0$ cross sections.

Using the above procedure, the calculated $D/\bar{D}$ inclusive cross sections (all X_F) for pp interactions at 400 GeV/c are:

$$\sigma(D^+/D^-) = (11.9 \pm 1.5) \; \mu b$$

$$\sigma(D^0/\bar{D}^0) = (18.3 \pm 2.5) \; \mu b$$

and $\sigma(D/\bar{D}) = (30.2 \pm 3.3) \; \mu b$

The errors quoted include the statistical plus systematic errors added in quadrature.

The $D/\bar{D}$ pair cross sections were measured using events with two charm decays surviving all our cuts. The resulting cross sections are:

$$\sigma(D^+D^-) \qquad = (2.5 \pm 0.7) \ \mu b$$

$$\sigma(D^o\bar{D}^o) \qquad = (5.9 \pm 1.4) \ \mu b$$

$$\sigma(D^+\bar{D}^o + D^-D^o) = (6.2 \pm 1.3) \ \mu b$$

$$\text{and} \quad \sigma(D\bar{D}) \qquad = (14.6 \pm 2.0) \ \mu b$$

Again, the errors include both estimated systematic plus statistical errors.

These measurements indicate that the $D/\bar{D}$ inclusive cross section is about twice the $D\bar{D}$ pair cross section. This means that the $\Lambda_c\bar{D}$ and $\bar{\Lambda}_cD$ cross sections cannot be large. A detailed analysis [5] allows us to establish the following limit (90% confidence level) on associated production: $\sigma(\Lambda_c\bar{D} + \bar{\Lambda}_cD) < 6.1 \ \mu b$.

The charm production cross sections quoted above are solid measurements made by the NA27 collaboration. I will next use these data to **estimate** the total charm production for pp interactions at 400 GeV/c in order to have a result that is directly comparable to QCD predictions (see Section V below).

The problem in obtaining an estimate of the pp -> $c\bar{c}X$ cross section from our pp -> $D/\bar{D}X$ measurement is that we have only an upper limit to the $(\Lambda_c\bar{D} + \bar{\Lambda}_cD)$ cross section and no measurement of the $\Lambda_c\bar{\Lambda}_c$ pair and $D_s/\bar{D}_s$ inclusive cross sections. The latter two are expected to be small. As a first order estimate, we use the predictions of the Lund scheme for charm quark fragmentation into charm hadrons in our pp production environment (see reference [6] and Section V). This prediction is that $\Lambda_c\bar{\Lambda}_c$, $\Lambda_c\bar{D}_s$ and $\bar{\Lambda}_cD_s$ production are negligable and $D_s/\bar{D}_s$ production is

8% of the total c/c̄ production cross section. Also, as measured from Ψ production [7], hidden charm production absorbs only about 1% of all cc̄ pairs. Therefore I **estimate**:

$$\sigma\ (pp\ ->\ c\bar{c}X)\ \sim\ \frac{\sigma(D/\bar{D})\ +\ \sigma(\Lambda_c\bar{D}\ +\ \bar{\Lambda}_cD)}{2(0.92)\ (0.99)}$$

$$=\ \frac{(30.2\ \pm\ 3.3)\ +\ (\ <\ 6.1)}{2(0.92\ \pm\ 0.04)\ (0.99)}\ \mu b$$

or $\sigma(pp\ ->\ c\bar{c}X)\ \sim\ (14\ to\ 23)\ \mu b$

at $\sqrt{s}\ =\ 27\ GeV$.

This cross section estimate is compared to fusion model predictions in Section V.

IIIB.   pp INTERACTIONS AT 800 GeV/C ($\sqrt{s}$ = 39 GeV)

The D/D̄ inclusive cross section was measured in experiment E743 using basically the same technique described in Section III.A. However, only C3, C5, V4 and V6 decays were used so the measurement is subject to a systematic error associated with these topological decay branching ratios. The data sample after cuts includes 49 C3 and 10 V4 decays for which we used the branching ratios [8] $BR(D^+\ ->\ C3)\ =\ (0.43\ \pm\ 0.10)$ and $BR(D^o\ ->\ V4)\ =\ (0.17\ \pm\ 0.04)$. This is 60% of our final data and represents a sensitivity of $(9.0\ \pm\ 1.0)$ events/μb. For more details of this calculation see reference [9]. The resulting D/D̄ inclusive cross sections (all X_F) for pp interactions at 800 GeV/c are:

$\sigma(D^+/D^-)\ =\ (34\ \pm\ 10)\ \mu b$

$\sigma(D^o/\bar{D}^o)\ =\ (28\ ^{+14}_{-11})\ \mu b$

and $\sigma(D/\bar{D}) = (62 \, ^{+17}_{-15}) \, \mu b$

The errors associated with the decay branching ratios have been added in quadrature with our experimental errors which are dominated by statistical uncertainties.

Again, I will **estimate** the total charm production cross section for pp $\to$ c$\bar{c}$X from the above measurements made by the E743 collaboration. The assumptions are those described in the discussion of the NA27 total cross section estimate. In addition, I assume that the ratio $\sigma(\Lambda_c\bar{D} + \bar{\Lambda}_c D)/\sigma(D/\bar{D})$ does not change from $\sqrt{s} = 27$ to 39 GeV and use the upper limit 6.1 $\mu b/(30.2 \pm 3.3) \, \mu b$ measured in NA27. Therefore, an estimate of the total charm production cross section is:

$$\sigma(pp \to c\bar{c}X) \sim \frac{\sigma(D/\bar{D})}{2(0.92)\,(0.99)} \quad [\; 1+ \frac{\sigma(\Lambda_c\bar{D} + \bar{\Lambda}_c D)}{\sigma(D/\bar{D})} \;]$$

$$= \frac{62 \, ^{+17}_{-15}}{2(0.92 \pm 0.04)(0.99)} \quad [\; 1+ \frac{<\; 6.1 \; \mu b}{(30.2 \pm \; 3.3) \; \mu b} \;]$$

or $\qquad \sigma(pp \to c\bar{c} \; x) \sim (29 \; to \; 55) \; \mu b$

at $\sqrt{s} = 39$ GeV

Finally, a measurement of the energy dependence of the cross section can be made. Using D/$\bar{D}$ meson production as a typical measurement of charm production, we find that:

$$\frac{\sigma(pp \to D/\bar{D} \; X) \; at \; \sqrt{s} = 39 \; GeV}{\sigma(pp \to D/\bar{D} \; X) \; at \; \sqrt{s} = 27 \; GeV}$$

$$= \frac{(62 \, ^{+17}_{-15} \;) \; \mu b}{(30.2 \pm 3.3)\mu b} \quad = \quad 2.05 \; ^{+0.60}_{-0.54}$$

This increase in the charm production cross section by a factor of about two between $\sqrt{s}$ = 27 and 39 GeV is consistent with QCD fusion model predictions. A detailed discussion of this comparison is made in Section V.

IV. D MESON DIFFERENTIAL PRODUCTION CROSS SECTIONS

The $D/\bar{D}$ differential production cross sections for 400 GeV/c pp collisions have been determined from 119 mesons with measured momenta and $X_F > 0$. The selection of these events makes use of kinematic fits to decay channels with particle identification information included to resolve fit ambiguities where possible. The details of the selection procedure have been described in reference 4. The data sample consists of 24 D^+, 27 D^-, 22 D^0 and 29 $\bar{D}^0$ decays plus 16 $D^0/\bar{D}^0$ and 1 D^+/D^- ambiguous decays. Each event has been assigned a weight which corrects for bubble chamber visibility and spectrometer acceptance efficiencies. These weights are slowly varying functions of X_F and P_T of the $D/\bar{D}$ mesons. Our experiment is sensitive to all D mesons produced with $X_F > 0$.

The measured $d\sigma/dX_F$ and $d\sigma/dP_T^2$ spectra for $D/\bar{D}$ mesons are presented in Figure 1. The solid curves in Figures 1a and 1b show the result of a fit of these distributions to the empirical form:

$$d^2\sigma/dX_F/dP_T^2 = A(1-X_F)^n e^{-aP_T^2}$$

This reproduces the measured differential cross sections well with n = (4.9 ± 0.5) and a = (1.0 ± 0.1) $(GeV/c)^{-2}$. A fit of the invariant cross section $(1/E)\ d^2\sigma/dX_F/dP_T^2$ to the same $(1-X_F)^m e^{-bP_T^2}$ form, yields m = (3.2 ± 0.6) and b

448

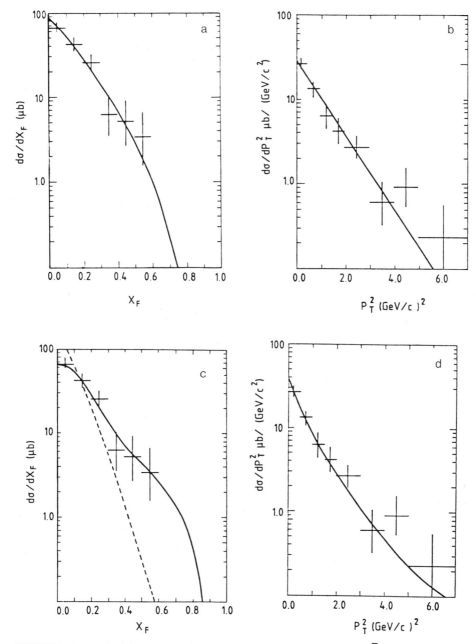

FIGURE 1. Differential cross sections for $D/\bar{D}$ production from pp collisions at 400 GeV/c: a) and b) show fits to the form $(1 - |X_F|)^n e^{-aP_T^2}$ with $n = 4.9 \pm 0.5$ and $a = (1.0 \pm 0.1) (GeV/c)^{-2}$; c) and d) show fusion model predictions normalized to the data (see the discussion in Section V for details)

$= (0.91 \pm 0.10) \ (GeV/c)^{-2}$. The average transverse momentum of the $D/\bar{D}$ mesons is $\langle P_T \rangle = (0.86 \pm 0.09) \ GeV/c$. The X_F distribution is very central with $\langle X_F \rangle = (0.15 \pm 0.02)$. We emphasize that our experiment has good acceptance for $D/\bar{D}$ mesons produced with large X_F and therefore the data show that there is little leading $D/\bar{D}$ production.

$$\frac{\sigma(D/\bar{D}) \ X_F > 0.5}{\sigma(D/\bar{D}) \ X_F > 0.0} = 0.02 \begin{array}{l} +0.02 \\ -0.01 \end{array}$$

Comparisons of these differential cross sections to QCD fusion model calculations are made in Section V.

All $D/\bar{D}$ mesons would have the same X_F and P_T spectra if there was no flavor dependence of the charm quark fragmentation into D mesons. Figure 2 shows the differential production spectra for various charm/charge states and the result of fits to $(1-X_F)^n \ e^{-aP_T^2}$. The D^- and D^o mesons have production spectra that agree with the average of the $D/\bar{D}$, but the D^+ has a harder spectrum in both X_F and P_T^2 while the $\bar{D}^o$ has a softer spectrum. These results are presented in more detail in Table 2 where we show the results of fitting the X_F distributions in both the invariant $(1/E) \ d\sigma/dX_F$ and non-invariant $d\sigma/dX_F$ forms. Both fits lead to the conclusion that the D^+ mesons have an appreciably flatter X_F distribution than the $\bar{D}^o$ mesons. This is in disagreement with the simplest model of leading meson production in which the $\bar{c}$ quarks would be boosted to large X_F when they are picked up by the valence u/d quarks in the proton. A possible explanation can be found if we assume the proton is composed of a quark and a di-quark. The di-quark (a 3-bar of color) can combine with a c quark to form a color singlet system. This could form either a

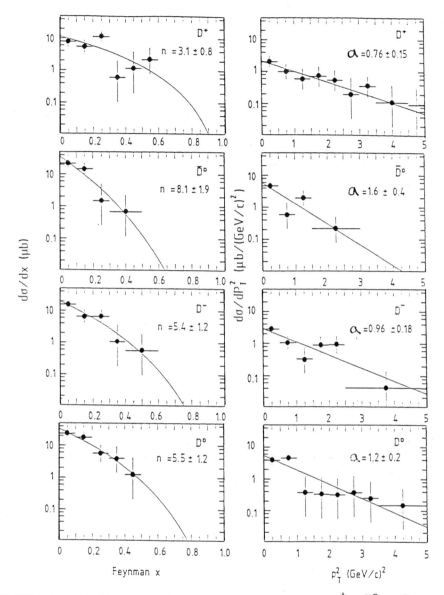

FIGURE 2. Differential cross sections for D^+, $\bar{D}^0$, D^- and D^0 meson production from pp collisions at 400 GeV/c. The solid curves are results of fits to the form $(1 - |X_F|)^n\, e^{-aP_T^2}$.

charm baryon or fragment into a D meson plus a baryon and thus produce rather forward charm mesons. Thus D's could be produced more leading than $\bar{D}$'s if the proton's valence di-quarks are appreciably faster than its valence quarks. For a more detailed discussion of this speculation see reference [10].

TABLE II

Fits to $\dfrac{d\sigma}{dx_F} \; \alpha \; (1-x_F)^n$ and to $\dfrac{1}{E}\dfrac{d\sigma}{dx_F} \; \alpha \; (1-x_F)^m$ for $x_F > 0$

	Number of Decays	n	m	$\langle p_T^2 \rangle$ $(GeV/c)^2$
All D	119	4.9±0.5	3.2±0.6	0.99±0.10
D^+	24	3.1±0.8	1.8±0.7	1.32±0.27
D^-	27	5.4±1.2	3.5±0.9	1.04±0.20
D^0	29	5.5±1.2	3.8±0.9	0.82±0.14
$\bar{D}^0$	22	8.1±1.9	6.2±1.4	0.62±0.14
D_{ambig}	17	3.9±1.1	2.9±0.9	0.93±0.30
$[D^+ + D^0]$	53	4.2±0.8	2.7±0.6	1.04±0.14
$[D^- + \bar{D}^0]$	49	6.6±1.1	4.6±0.8	0.84±0.12

V. COMPARISONS OF D/$\bar{D}$ PRODUCTION DATA TO THE QCD FUSION MODEL

The QCD fusion model can be used to calculate the D meson production spectrum. The calculation requires: 1) a description of the parton content of the proton, 2) calculation of the cross sections for the subprocesses gg -> $c\bar{c}$ and $q\bar{q}$ -> $c\bar{c}$ and 3) some approximation to the c -> D fragmentation process. Since the predictions of the model depend on the detailed treatment of the above components, we have allowed some variation of the input parameters. For a description of the parton content of the proton we use the structure functions determined by Duke and Owens (DO) [11] and Eichten et al (EHLQ) [12]. The intrinsic parton transverse momentum plus soft gluon radiation effects were included by giving the colliding partons a transverse momentum, k_T, with distribution $dN/dk_T^2 = e^{(-k_T^2/\langle k_T^2 \rangle)}/\langle k_T^2 \rangle$. The parameter $\langle k_t^2 \rangle$ was allowed to vary between 0 and $(0.8)^2$ $(GeV/c)^2$ [13]. The subprocess parton cross sections have been calculated by Combridge [14]. For the $c\bar{c}$ production threshold we use $(2m_c)^2$ where the charm quark mass is allowed to vary from 1.2 to 1.3 GeV/c^2. The strong interaction coupling constant is given by $\alpha_s = (12 \pi/25)/\log(Q^2/\Lambda^2)$ where Λ is taken to be 0.2 GeV and Q^2 is allowed to vary between $(2 m_c)^2$ and $\hat{s}$, the cm energy squared of the subprocess. Finally we investigated the sensitivity of the D meson differential production cross section to three different c -> D fragmentation processes: 1) a delta-function in which all the longitudinal momentum of the charm quark is transferred to the D meson, 2) an empirical fragmentation function determined by Peterson et al [15] and 3) a fragmentation which approximates the final state $c\bar{q}$ and $\bar{c}q$ formation via color string breaking as described by the Lund model [6]. The latter depends on adjustable string parameters but these were not changed from the values determined by the Lund group from other data.

Comparisons of some predictions of this QCD fusion model, normalized to our data are shown in Figures 1c and 1d. The calculated X_F and P_T^2 distributions are relatively insensitive to the choice of structure functions and charm quark mass. The curves shown in Figures 1c and 1d were obtained using the EHLQ set one structure functions and a charm quark mass of 1.25 GeV/c^2. The $d\sigma/dP_T^2$ distribution is sensitive to the effective parton intrinsic k_T and the QCD prediction with $\langle k_T^2 \rangle = 0$ does not fit the data. A value of $\langle k_T^2 \rangle = (0.8)^2$ $(GeV/c)^2$, similar to that determined from Drell-Yan measurements [13], was used in all our calculations.

The dashed curve in Figure 1c is the fusion model prediction with inputs fixed as described in the above paragraph and using a delta-function c -> D fragmentation. This illustrates the basic parton level prediction. If the isolated c -> D fragmentation function of Peterson et al [15] is added to the calculation, the predicted momentum spectrum is too soft and disagrees with both the measured $d\sigma/dX_F$ abd $d\sigma/dP_T^2$ spectra. The Lund c -> D fragmentation process is sensitive to the production environment of the $c\bar{c}$ pair since color singlet strings are formed between the charm quarks and the valence quarks and di-quarks left over from the proton after the hard scattering process. The predictions of this calculation [6] are given by the solid curves shown in Figures 1c and 1d. The fusion model plus Lund fragmentation clearly reproduces well the measured X_F and P_T^2 distributions of the total sample of $D/\bar{D}$ mesons. However, a closer examination of the predictions indicate some discrepancies between the model and our data. In particular, the fusion/Lund model does not predict our observation (Section IV) that the D mesons have a somewhat flatter X_F spectrum than the $\bar{D}$ mesons.

Finally, we compare the total charm production cross sections that were estimated in Section III to the predictions of the $0(\alpha_s^2)$ fusion model calculation. The difficulty with this comparison is that the fusion model prediction is uncertain for several reasons. First, the subprocess cross sections for gg -> $c\bar{c}$ and $q\bar{q}$ -> $c\bar{c}$ are very sensitive to the threshold determined by the charm quark mass [16]. For example, if m_c is varied between 1.2 and 1.4 GeV/c^2, the cross section for pp -> $c\bar{c}$ X varies by a factor of 2.2. Additional uncertainties arise from changes in the proton structure function and the choice of Q^2 in α_s and the evolution of the structure function. In the following calculation we fix the charm quark current mass at 1.25 GeV/c^2 and vary the proton structure function (D01 or EHLQ1) and the choice of Q^2 ($(2m_c)^2$ to $\hat{s}$). The result is:

$$\sigma(pp \rightarrow c\bar{c}X) \text{ at } \sqrt{s} = 27 \text{ GeV} = (8 \text{ to } 15) \ \mu b$$
$$\sigma(pp \rightarrow c\bar{c}X) \text{ at } \sqrt{s} = 39 \text{ GeV} = (15 \text{ to } 29) \ \mu b$$
$$\text{QCD fusion model with } m_c = 1.25 \text{ GeV/c}^2$$

Our data allows us to estimate the $c\bar{c}$ production cross sections to be (see Section III):

$$\sigma(pp \rightarrow c\bar{c}X) \text{ at } \sqrt{s} = 27 \text{ GeV} = (14 \text{ to } 23) \ \mu b$$
$$\sigma(pp \rightarrow c\bar{c}X) \text{ at } \sqrt{s} = 39 \text{ GeV} = (29 \text{ to } 55) \ \mu b$$
$$\text{from NA27 and E743 data}$$

The above suggests that the charm k-factor $\equiv [\sigma(pp \rightarrow c\bar{c}X)$ experiment$]/[\sigma(pp \rightarrow c\bar{c}X) \ 0(\alpha_s^2)$ QCD$]$ is:

$$k(\sqrt{s} = 27 \text{ GeV}) \sim \frac{18.5 \pm 4.5}{11.5 \pm 3.5} = 1.6 \pm 0.6$$

$$k(\sqrt{s} = 39 \text{ GeV}) \sim \frac{42 \pm 13}{22 \pm 7} = 1.9 \pm 0.8$$

That is, our data indicate that the charm k-factor is on the order of two if the charm quark current mass is 1.25 GeV/c^2. Note also that the $\sqrt{s}$ dependence of the pp -> $c\bar{c}$ X cross section is in agreement with the fusion model prediction.

It will be interesting to compare these results to recent calculations of higher order QCD contributions to $c\bar{c}$ production [17].

V. SUMMARY AND CONCLUSIONS

Experiments NA27 and E743 are providing reliable measurements of charm production from pp interactions at $\sqrt{s}$ = 27 and 39 GeV. The most accurate charm production cross section measurement made to date is $\sigma(pp \rightarrow D/\bar{D}X)$ = (30.2 ± 3.3) μb at $\sqrt{s}$ = 27 GeV. A preliminary result at $\sqrt{s}$ = 39 GeV is $\sigma(pp \rightarrow D/\bar{D}X)$ = $(62\ ^{+17}_{-15})$ μb.

The D/$\bar{D}$ meson production cross section increases by a factor of $2.05\ ^{+0.60}_{-0.54}$ from $\sqrt{s}$ = 27 to 39 GeV. This data plus some assumptions about Λ_c/D_s production allow us to estimate a charm k-factor of 1.8 ± 0.7 if the charm quark current mass is 1.25 GeV/c^2.

The measured differential cross section $(1/\sigma)\ d^2\sigma/dX_F dP_T^2$ for D/$\bar{D}$ production at $\sqrt{s}$ = 27 GeV is well represented by the empirical form $(1- |X_F|)^{(4.9\ \pm\ 0.5)} e^{-(1.0\ \pm\ 0.1)P_T^2}$. The production process is very central with $(98\ ^{+1}_{-2})$% of the D/$\bar{D}$ mesons having X_F < 0.5. Our data suggests that the D mesons have a somewhat harder production spectrum than the $\bar{D}$ mesons in disagreement with the naive expectation of a preference for $\bar{c}$ quark pickup by proton u/d quarks. The fusion model reproduces the overall D/$\bar{D}$ differential cross section shape if the Lund c -> D fragmentation scheme is used and an intrinsic k_T distribution with $\langle k_T^2 \rangle$ = (0.8 GeV/c)^2 is applied.

ACKNOWLEDGEMENTS

I was fortunate to have the opportunity to present these charm measurements at the St. Croix QCD Workshop. Obviously most of the credit for these results should go to the other members of the NA27 and E743 collaborations. I would like to add that it has been a pleasure both personally and professionally to work with this particular group of physicists.

REFERENCES

1) M. Aguilar-Benitez et. al., Nucl. Instr. Meth. A258 (1987) 26.

2) G. Canough, Ph.D. dissertation, Notre Dame University (1987), unpublished.

3) M. Aguilar-Benitez et. al., Phys. Lett. **B135** (1984) 237.

4) M. Aguilar-Benitez et al., Phys. Lett. **B189** (1987) 476.

5) M. Aguilar-Benitez et al., Charm Hadron Properties in 400 GeV/c pp Interactions, to be submitted to Z. Phys. C.

6) H.U. Bengtsson, CHARIS - The Lund Monte Carlo for charm production (private communication). H.U. Bengtsson and G. Ingelman, Comput. Phys. Commun. **34** (1985) 231.

7.) E.J. Siskind et al., Phys. Rev. D., **21** (1980) 628.

8) These branching ratios were extracted from SPEAR results. See M. Aguilar-Benitez et al., Phys. Lett. **B135** (1984) 237 for this analysis.

9) R. Ammar et. al., Phys. Lett. **B183** (1987) 110.

10) M. Aguilar-Benitez et al., Phys. Lett. **B201** (1988) 176.

11) D.W. Duke and J.F. Owens, Phys. Rev.D, **30** (1984) 49.

12) E. Eichten et. al., Rev. Mod. Phys. **56** (1984) 667.

13) B. Cox and P.K. Malhotra, Phys. Rev. D, **29** (1984) 63.

14) B.L. Combridge, Nucl. Phys. B, **151** (1979) 429.

15) C. Peterson et. al., Phys. Rev. D., **27** (1983) 105.

16) J.F. Cudell, F. Halzen and K. Hikasa, Phys. Lett. B,

17) K.R. Ellis et al., Proceedings of the Advanced Research Workshop on QCD Hard Hadronic Processes, St. Croix, Virgin Islands, October 8-13, 1987 (to be published).

PRODUCTION OF PARTICLES WITH HIGH TRANSVERSE MOMENTA

IN 800 GeV PROTON-NUCLEUS COLLISION, E605-FNAL

K. Miyake

Department of Physics, Kyoto University
Kitashirakawa, Sakyoku, Kyoto, Japan, 605

INTRODUCTION

E605 aims to study the production of single hadrons, pairs of hadrons and dileptons at large transverse momenta in 800 GeV proton-nucleus collisions. The design feature of the detector can be summarized as follows:
1. Measurement of particles with high p_T up to 10 GeV/c or higher.
2. High mass resolution of less than 0.2% for pairs.
3. Excellent identification of particle species of e/μ/π/K/p.
4. High luminosity and a large angular acceptance.
The physics subjects relevant to the experiment are
a. High p_T hadron production to study the hard scattering of constituents.
b. Atomic weight dependence of single and pair hadron productions.
c. Particle ratios for hadron production.
e. e^+e^- and $\mu^+\mu^-$ pair productions with the same beam. (e-μ universality)
d. Search for high mass dilepton resonances or new particles by measuring with an excellent mass resolution.

EXPERIMENTAL SET-UP

Fig. 1 shows our apparatus. In order to achieve the high luminonosity,

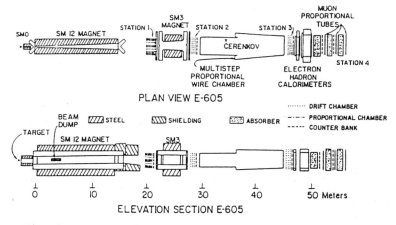

Fig. 1. Layout of the apparatus for the open geometry.

the fine mass resolution, and the excellent separation for e/μ/π/K/p, the aperture of the detector was required to be the "open geometry.

The design details are as follows:

i)   A large spectrometer magnet (SM12) having an aperture of 14.4 m long, 2.7 m wide, and 5 m high was placed immediately following the target.[1] SM12 provided the p_T kick of 9 GeV/c. The neutral particles produced at the target were well shielded by the beam dump placed inside the magnet, by the magnet york, and the lead teeth carefully placed on the inner wall of the magnet.   The beam dump were made of 2.74 m long copper and 7.32 m long concrete.

ii) Electrons and hadron were distinguished by the electromagnetic and hadron calorimeters of 19 radiation lengths and 9 absorption lengths, respectively.[2,3,4] The probability for both hadrons in the pair to mimic electrons was 1.4×10^{-6}. The hadron species π/K/p were identified by the ring imaging Cherenkov counter placed in front of the calorimeters.   Pure He gas enclosed in the large vessel of 15.2 m long was used as a Cherenkov radiator.[5,6,7,8,9,10] The Cherenkov photons of 8 eV were focused on two multi-step proportional chambers having the effective area of 40×80 cm^2 by the mirrors placed on the end wall of the vessel.   The multi-step proportional chambers were used to detect the radius of the Cherenkov ring. The resolution in measuring the radius of the Cherenkov ring was found to be 0.78 mm (standard deviation), while the radius for the particles with the light velocity was 68.78 mm. Since the direction of the particles were measured precisely by the drift chambers placed in front and of the helium vessel, so that the center of the Cherenkov ring could be precisely determined.   Therefore, the detection of only one photon was enough to determine the radius.   The average number of detected photons was about 2.5. The number of noise photons was 1.0 on an average. Thus, this counter was proved to be capable to identify pions and kaons up to 200 GeV/c, as is seen in Fig. 2.   The efficiencies for identification of particles was about 80%, 70%, and 50% for pions, kaons, and protons with p_T > 6 GeV/c.

iii) Trajectories of hadrons and electrons were recorded with a set of MWPC's in the station 1 and sets of drift chambers in the station 2 and 3.   The trajectories of muons were also recorded with three planes of proportional tubes placed between the absorbers of a 102 cm thick zinc wall and two 122 cm thick concrete walls behind the calorimeter.

iv) Our apparatus covered the kinetic region of 90° in the proton-nucleon center of momentum frame and x_T > 0.4 for single and $M/\sqrt{s} \sim 0.3$ for pair productions.

v) The second analyzing magnet (SM3) was placed between the stations 1 and 2 for an independent momentum measurement with moderate resolution so as to select the particles produced at the target from those produced at the beam dump or elsewhere.   Finally the momentum resolution of our apparatus was found to be $0.2 \sim 0.3$% depending on the target material.

vi) The beam dump, collimaters, and absorbers were carefully set up to reduce heavy backgrounds.

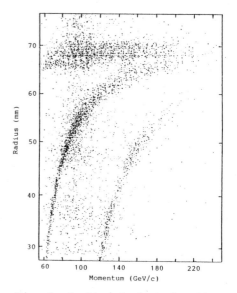

Fig. 2. Scattered plot of radius versus momentum for particles having one reconstructed photon.

460

DATA TAKING

The apparatus was tuned up in the spring of 1982 with the proton beam of 400 GeV. The beam intensity was reduced to less than 10^{10} protons per pulse due to heavy background rates. The data of single hadrons with p_T > 3.5 GeV/c and hadron pairs with M > 8 GeV/c were taken with Be, Cu, and W targets. After this run, the target was moved to upstream and a small magnet (SM0) was placed in front of the SM12 magnet so as to keep our acceptance centered near 90^0 in the proton-nucleon center of momentum frame. In the spring of 1984, TeV II started to accelerate protons successfully up to 800 GeV. The data for both hadron and lepton productions with LH_2, LD_2, Be, Cu, and W targets were taken in this period. The average beam intensity was about 2×10^{11} protons every 60 sec., and the total protons of 5×10^{15} were incident upon targets.

During the fall of 1984, lead absorbers of 48" thich were added at the exit of the SM12 magnet to reduce the electromagnetic and hadronic backgrounds, this is called as the "closed geometry". Drift tubes were also added at the exit of the SM12 magnet to maintain the momentum resolution of 0.3 %. In the spring of 1985, the data taking run for muon pairs started with a high beam intensity of 2×10^{12} protons per pulse at 800 GeV. By the late summer of 1985, about 1.5×10^{17} protons were incident upon a Be target and about 10,000 Υ events were collected. The summary of the data taking runs is give in Table 1.

Table I. Summary of runs

Target	LH2	LD2	Be	Cu	W	$\sqrt{s}$ (GeV)	Geometry
1982	–	–	0.2	0.3	0.3	27.4	open
1984	0.7	0.6	1.5	2.2	3.5	27.4	open
1984	0.8	–	12.9	–	2.2	38.8	open
1985	–	–	2800	–	–	38.8	closed

(Integrated luminosity / nucleons in 10^{39} cm^{-2}.)

RESULTS

The results of the hadron production measured in 1982 with the 400 GeV proton beam were already published,[4,11] [12] so that this talk will be concentrated on the results with the 800 GeV proton beams.

1. Single hadron production in proton-proton collision

Fig. 3 shows our results of the invariant cross section of single π^+ production in the proton-proton collision at 800 GeV together with the extraporation of previous measurements at Fermilab 400 GeV (CP)[13] and at ISR (CCRS).[14] The fits were made by using the formula:

$$\frac{d^3\sigma}{dp^3} = A \cdot p_T^{-N} \cdot f(x_T).$$

The fit shows a good agreement between our result and the extrapolations of the previous measurements.

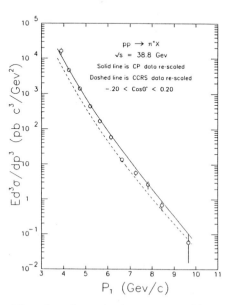

Fig. 3. Invariant cross section for π^+ production from hydrogen versus p_T.[15]

461

The same result was obtained
for single π^- production.[15]

2. Particle Ratios

The particle ratios of K/π
p/π for positive and negative
hadron productions from protons, Be, and W were measured
in the transverse momentum
range from 4 to 12 GeV/c.[15]
The results of the positively
charged particles for the proton target are shown in Fig.4,
together with the previous results at Fermilab 400 GeV (CP)
[13] and at ISR (ABCDHM)[16].

This figure shows a sort of
scaling for particle ratios as
a function of p_T. The
same behavior was observed

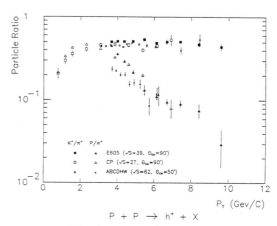

Fig. 4. K^+/π^+ and p/π^+ production ratios
from protons versus p_T.

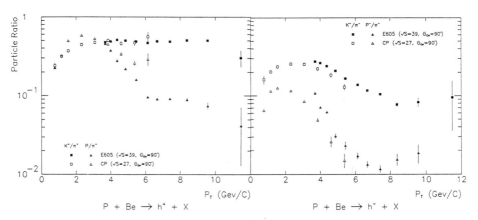

Fig. 5. Production ratios of K^+/π^+ and p/π^+ (left), and K^-/π^- and
$\bar{p}/\pi^-$ (right) versus p_T for the Be target.

for Be and W targets. The results for the Be target is shown in Fig. 5.
The ratio reaches to a constant in the region of $p_T > 4$ GeV/c for K^+/π^+
and $p_T > 6$ GeV/c for p/π$^+$. A simple explanation is that the dominant
valence u-quark in proton make the hard collision with high p_T and picks
up a $\bar{d}$-quark, a $\bar{s}$-quark or a (ud)-diquark during the fragmentation
process. However, our ratio of 0.48 ± 0.20 for K^+/π^+ is somewhat larger
than the prediction by the QCD-based models. The same behaviors were
observed for the negative hadron production as is seen in Fig. 5. The
ratios of K^-/π^- and $\bar{p}/\pi^-$ reach to the values of 0.08 and 0.01, respectively, at $p_T > 7$ GeV/c. The explanation of behavior of particle ratios for
negative hadrons is not so simple as that for the positive hadrons.
However, these ratios are expected to be very sensitive to the structure
functions at large x_T's.

3. Atomic weight dependence of single hadron productions

The atomic weight dependences were extracted from the measured yields
of single hadrons from the Be and W targets, because these two targets
have the same interaction lengths of 0.14. The α-parameters for both

positive and negative hadron production are plotted as a function of p_T in Fig. 6. As is seen in this figure the anomalous nuclear enhancement observed in the previous measurements seems to disappear at $p_T > 8$ GeV/c in both positive and negative hadron productions. This behavior is qualitatively expected from the prediction of the multi-scattering models,[17,18] because these models predict that the effects of the multi-scattering increase as p_T increases to around 2-4 GeV/c and then decrease at high p_T above 6 GeV/c and at high incident energies of above 600 GeV/c.

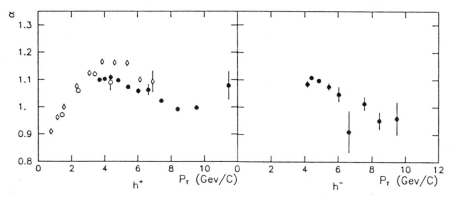

Fig. 6. Atomic weight dependence of positive (left) and negative (right) hadron productions at $x_F \simeq 0$. The production cross section is assumed to be proportional to A^α.
● E605(800 GeV/c); ◇ CP(Ref. 13); ○ CFS(Ref. 18)

4. Atomic weight dependence of hadron pair production

The atomic weight dependence for the hadron pair production was also extracted and are plotted as a function of $M_{h^+h^-}$ in Fig. 7, together with the previous results at Fermilab[11,19,20] and Serpukhov.[21] A sharp decrease at $p_T \gtrsim 6$ GeV/c is observed. There is a clear correlation between α_{pair} and p_{out} that α_{pair} increases as p_{out} increases. These features were also found in the present experiment at 400 GeV as was reported.[11]

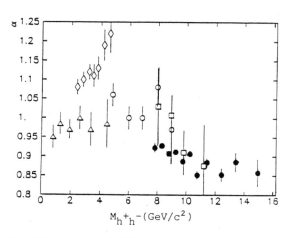

Fig. 7. Atomic weight dependence of hadron pair production versus $M_{h^+h^-}$.
$x_F \simeq 0.0$
$p_{out} < 1.20$

● E605(800 GeV/c)
□ E605(400 GeV/c Ref. 11)
○ CFS (Ref. 19)
◇ FNAL (Ref.20)
△ Serpukhov(Ref.21)

5. Dilepton production

The production of electron and muon pairs from the Be target were measured with the open geometry with the proton beam of 800 GeV/c. The rejection factor of pions by the electromagnetic calorimeter was 3×10^5, so

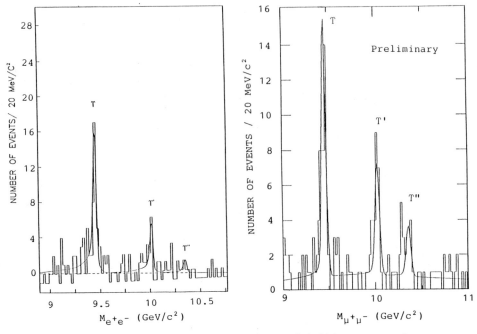

Fig. 8. Dielectron (left) and dimuon (right) mass spectra
measured with the open geometry.[22]

that the electron pairs were well
separated from the copious hadron
background.[22] Fig. 8 shows the mass
spectra of the electron and muon
pairs measured with the open geometry. The mass resolution for electron pairs was 0.17 %, so that the
Υ family is well separated. The data
of the muon pairs from the Cu target
with the "closed geometry" are being
processed, and the preliminary results on the raw mass spectrum is
shown in Fig. 9. The mass resolution of 0.3% was achieved. These results indicate that the Υ" peak is
clearly seen in proton-nucleon collision at the first time. In nucleon-nucleon collision Υ is produced
by the gluon-gluon fusion process
through the χ_b states, so that the
observation of Υ" indicates the existence of $\chi_{b}^{"}$ state below the open
B$\bar{\text{B}}$ threshold.[23]

The peaks corresponding the production of Υ, Υ', and Υ" are clearly observed, so that the ratios of
the cross sections among the Υ resonance production can be extracted
with a good precision. The results
are summarized in Table. 2 with the
prediction based on the QCD gluon

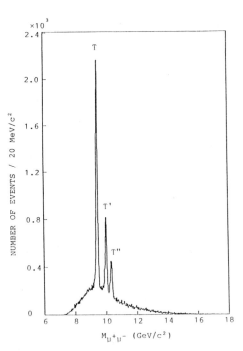

Fig. 9. Dimuon mass spectrum mea-
with the closed geometry.
(preliminary)

Table 2.   Ratios between the ϒ resonance production.

	e^+e^- [22] (open geometry, Be target)	$\mu^+\mu^-$ [**] (closed geometry, Cu target)	QCD [24]
$d\sigma/dy(\Upsilon'/\Upsilon)$	0.50 + 0.21	0.44 + 0.16	< 0.30
$d\sigma/dy(\Upsilon''/\Upsilon)$	0.08 + 0.09	0.10 + 0.05	< 0.15

* open geometry, Be target: ** closed geometry, Cu target (preliminary)

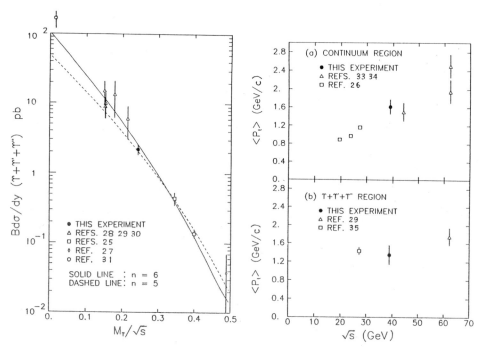

Fig. 10. Branching ratio times cross section for the ϒ resonances versus $M_T/\sqrt{s}$. The solid and dashed lines are the predictions.

Fig. 11. Mean transverse momentum of continuum versus the $\sqrt{s}$ for $M/\sqrt{s} = 0.22$ (a), and that for the ϒ resonances (b).

fusion model.[24]   Our results agree with the previous measurement at 400 GeV at Fermilab and are somewhat larger than the QCD prediction.

The production cross section are plotted as a function of $M_T/\sqrt{s}$ in Fig. 10, together with the previous results at different energies at Fermilab,[25,26] SPS,[27] ISR,[28,29,30] and UA1.[31]   The energy dependence of these production cross sections is well reproduced by the prediction of the QCD based calculation[32] with the gluon distribution function of

$$G(x) = 0.5(n+1)x^{-1}(1-x)^n, \quad \text{with } n = 6.$$

The mean transverse momentum $\langle p_T \rangle$ of the continuum and that of the ϒ resonances can be easily separated because of the high mass resolution. Fig. 11 shows $\langle p_T \rangle$ for both continuum and resonances as a function of $\sqrt{s}$, together with the previous results at different energies.[26,33,34,35]   This figure shows that $\langle p_T \rangle$ of continuum increases as $\sqrt{s}$ increases, while $\langle p_T \rangle$ of resonances is already large at the small $\sqrt{s}$, and shows no significant increase with $\sqrt{s}$.

465

SUMMARY

As was stated, our data analysis are in progress. However, the following tentative conclusion can be drawn at this moment.

For the hadron production;

1. The invariant cross section for π^+ and π^- production from hydrogen are in agreement with earlier data when they are extrapolated to our energy.
2. The single particle ratios seem to saturate at high p_T. $K^+/\pi^+ = 0.48$ at $p_T > 4$ GeV/c, and p/π^+, K^-/π^- and $\bar{p}/\pi^-$ are also constant at $p_T > 7$ GeV/c.
3. An anomalous nuclear enhancement on A-dependence of single hadron production seems to decrease for both positive and negative hadrons at $p_T > 4$ GeV/c and α reaches 1 at $p_T > 7$ GeV/c.
4. Atomic weight dependence, α of hadron pair production decreases at $M \geq 8$ GeV/c.
5. There is a correlation between α_{pair} and p_{out}.

For the dilepton production;

1. The hadronic production of dileptons has been measured with a high mass resolution of $0.2 \sim 0.3$ % by the 800 GeV proton beam.
2. The three Υ resonances were clearly seen for the first time in the proton-nucleon collision. Therefore, χ_b'' is proved to be below the open $B\bar{B}$ threshold.
3. There is no narrow peak observed in the dilepton mass spectrum from 11 to 16 GeV/c^2.
4. The production ratios of three Υ resonances are consistent with the previous results at 400 GeV. However, the ratio for Υ'/Υ is somewhat larger than the prediction based in the gluon fusion model.
5. The mean transverse momentum of continuum and that of the Υ resonances have different energy dependences reflecting the different production processes.

ACKNOWLEDGEMENTS

This was supported in part by the U.S. Department of Energy, the National Science Foundation, the Commissariat a l'Energies Atomique, and the Japan Ministry of Education, Science and Culture (Monbusho).

E605 Collaborators ; M. Adams[8], C. Brown[4], G. Charpak[2], S. Childress[4], J. Crittendon[3], D.Finley[4], H.Glass[8], R.Gray[9], Y.Hemmi[7], Y.Hsing[3], J.Hubbard[1], K. Imai[7], A. Itoh[4], D. Jaffe[8], A.Jonckheere[4], H. Jöstlein[4], D. Kaplan[5], J. Kirz[8], L. Lederman[4], K.Luk[9], R.McCarthy[8], A.Maki[6], Ph.Mangeot[1], G.Moreno[4], K.Miyake[7], T.Nakamura[7], R.Orava[4], A.Peisert[2], R.Plaag[9], J.Rutherfoord[9], J. Rothberg[9], F. Sauli[2], S.Smith[3], K.Sugano[4], N.Sasao[7], Y.Sakai[6], B. Straub[9], K. Ueno[4], N. Tamura[7], T. Yoshida[7], K. Young[9]

E605 Institutes

[1]CEN Saclay, Gif-sur-Yvette, France.
[2]CERN, Geneva, Switzerland.
[3]Columbia University, New York, USA.
[4]Fermilab, Batavia, Illinois. USA.
[5]Florida State Universty, Tallahassee, Florida, USA.
[6]KEK, Ibaraki-ken, Japan.
[7]Kyoto University., Kyoto, Japan.
[8]State University of New York at Stony Brook, New York, USA.
[9]University of Washington, Washington, USA.

REFERENCES

1. W. Fast et al., IEEE Trans. Magnetics **MAG-7,** 1903 (1981)
2. H. Glass et al., IEEE Trans. Nucl. Sci. **NS-28,** 514 (1981)
3. Y. Sakai et al., IEEE Trans. Nucl. Sci. **NS-28,** 528 (1981)
4. Y. Sakai, Memoirs of the Faculty of Sci., Kyoto Univ. **36,** 401 (1985)
5. G. Coutrakon et al., IEEE Trans. Nucl. Sci. **NS-29,** 323 (1982)
6. R. Bouclier et al., Nucl. Instr. and Meth. **205,** 403 (1983)
7. Ph. Mageot et al., Nucl. Instr. and Meth. **216,** 79 (1983)
8. H. Glass et al., IEEE Trans. Nucl. Sci. **NS-30,** 1983)
9. M. Adams et al., Nucl. Instr. and Meth. **217,** 237 (1983)
10. H. Glass et al., IEEE Trans. Nucl. Sci. **NS-32,** 692 (1985)
11. Y. B. Hsing et al., Phy. Rev. Lett. **55,** 457 (1985)
12. A. Crittendon et al., Phys. Rev. **D34,** 2584 (1986)
13. D. Antreasyan et al., (CP collaboration), Phys. Rev. **D19,** 764 (1979)
14. F. W. Büsser et al., (CCRS collaboration), Phys. Lett., **B106,** 1 (1976)
15. D. Jaffe, Ph. D. Thesis, State University of New York at Stony Brook, 1987
16. A. Breakstone et al., (ABCDHW collaboration), Phys. Lett. **135B,** 510 (1984) and **147B,** 237 (1984)
17. M. Lev and B. Peterson, Z. Phys. **C12,** 155 (1986)
18. T. Ochiai et al., Prog. Theor. Phys. **75,** 288 (1986)
19. H. Jostlein et al., (CFS collaboration), Phys. Rev., **D20,** 53 (1979)
20. D. Finley et al., Phys. Rev. Lett, **42,** 1031 (1979)
21. V. Abramov et al, JETP Lett. **38,** 352 (1983) and Z. Phys. **C24,** 204 (1984)
22. Y. Yoshda, Ph. D. Thesis, Kyoto University, 1987, and Memoirs of the Faculty of Sci., Kyoto Univ. **38,** 93 (1988)
23. J. L. Resner, Proceedings of the 1985 International Symposium on Lepton and Photon Interactions at High Energies, p. 448, Aug. 19-24, 1985, Kyoto.; S. Cooper. Proceedings of the XXIII International Conference on High Energy Physics, p. 67, July 16-23, 1986, Berkely
24. R. Baier and R. Ruckl, Z. Phys, **C1,** 251, (1983)
25. K. Ueno et al., Phys. Rev. Lett., **42,** (1979)
26. J. Yoh et al., Phys. Rev. Lett., **41,** 684 (1978)
27. J. Badier et al., Phys. Lett. **86B,** 98 (1979)
28. A. Anglis et al., Phys. Lett. **87B,** 398 (1979)
29. C. Kourkoumelis et al., Phys. Lett., **91B,** (1980)
30. L. Camiller, Proceedings of the 1978 International Symposium on Lepton Photon Interactions at High Energies. p.232.Fermilab, 1979
31. C. Albajar et al., Phys. Lett., **186B,** 237 (1987)
32. V. Barger et al., Z. Phys., **C6,** 169 (1980)
33. D. Antreasyan et al., Phys. Rev. Lett., **47,** 12 1981)
34. A. Anglis et al., Phys. Lett, **147B,** 472 (1984)
35. D. Kaplan, Ph. D. Thesis, State University of New York at Stony Brook (1979)

EXPERIMENTAL STUDY OF B$\bar{\text{B}}$ HADROPRODUCTION

IN THE WA75 AND WA78 EXPERIMENTS

G. Poulard

for the WA75 and WA78 collaborations
CERN, Geneva, Switzerland

1.   Introduction

The two collaborations WA75 and WA78 are studying the hadroproduction of *beauty* particles in fixed target experiments at CERN SPS.

For the first time an associated production of beauty particles B^- and $\bar{B}_0$ was observed in an emulsion target by the WA75 collaboration [1]. The status of the experiment is reviewed.

A signal of $B\bar{B}$ associated production in π^- U interactions at 320 GeV was observed in WA78 experiment looking at events with three muons in the final state [2]. Further results including analysis of two like − sign muons are presented. The experimental energy and p_t distributions of muons as well as the cross section are in good agreement with QCD predictions. The effect of $B^0\bar{B}^0$ mixing is discussed.

2.   WA75

Experiment WA75 was designed to observe directly the decay of *beauty* particles and to measure their lifetimes. The experiment used a combination of electronic and emulsion techniques in order to get a high selectivity (previous limits on production cross section were of the order of a few nb [3]) and a high sensitivity to short lifetimes (of the order of 10^{-13} to 10^{-14}sec).

a.   Set − up

The set − up comprised a Beam Hodoscope and a Vertex Detector in both front and back of the emulsion target , a muon filter and a magnetic muon spectrometer.

The Beam Hodoscope gave the position of the incoming beam particles with a precision of the order of 15 μm in the direction perpendicular to the beam.

The purpose of the Vertex Detector was twofold : first, together with the Beam Hodoscope, to give the position of the interaction in the emulsion target ; secondly to provide the topology of the events, hence permitting a match with the tracks found in the emulsion, and establishing the temporal correlation with the primary interaction of tracks from neutral particles decaying in emulsion.

The muon filter was either a passive one, consisting of a two meters iron block with a central cone in tungsten, or a active one composed of a hadronic calorimeter (designed for the WA78 experiment [4]) and followed by a short dump in iron and tungsten.

The magnetic spectrometer provided the energy E_μ as well as the p_t of the muon.

The emulsion was exposed in the form of stacks composed of pellicles oriented either in directions perpendicular or parallel to the beam. The stacks were mounted on a moveable stage, the position of which was measured with a precision of 5 μm. The stage moved during the burst, at a speed depending on the beam intensity, in order to obtain a uniform irradiation of the emulsion.

The standard trigger was defined as a coincidence between an interaction trigger and a muon trigger.

80 litres of emulsion were exposed at a CERN SPS 350 GeV/c π^- beam. The number of events recorded was 1.5×10^6 corresponding to a total number of interactions of 2.6×10^8.

b.   Analysis

After reconstruction of muons in the spectrometer the energy E_μ and the transverse momemtum p_t were corrected for energy loss in the dump and multiple scattering effects. A first cut was applied by requesting $p_t > 1.$ GeV/c (> 0.6 GeV/c for a subsample).

Then, tracks were reconstructed in the Vertex Detector and a good matching was requested for the muon track with a Vertex Detector track. A track was said to be associated when the angular difference with the closest track was less than 4 standard deviations and this occured for 75 % of the events. The 25 % of events with no associated muon could be divided into 5% with the muon out of the acceptance of the Vertex Detector and 20 % due to inefficiency or π/K decay.

Finally, events with muons coming from the primary interaction due mostly to π decay with an undetectable kink angle were rejected. This was the case for about 90 % of the events.

The remaining 10 % events were fully analysed. The position of the vertex in the emulsion was predicted with typical errors of 50 μm in transverse directions and 500 μm along the beam. All minimum ionising tracks from the primary vertex were followed in the forward direction for three millimeters to search for decays of charged particles. A cylinder of radius 150 μm and at least 3 mm long was scanned for neutral topologies. Tracks measured in emulsion were extrapolated to the Vertex Detector and compared with Vertex Detector tracks. For events with tracks non associated, a complementary scanning effort was performed in order to find additional secondary decay vertices or interactions.

Out of 11335 selected events, 9267 were found in emulsion. The total number of single, respectively (multi) decay events was 306 (226) with 187 (198) with a muon coming from a decay and we identified 13 events with 3 or 4 decay vertices and a muon coming from one of the secondary vertices.

Out of these 13 events we found :
  −   1 beauty event ($B^- \overline{B}{}^0$) [1]
  −   2 associated pairs of charm particles [5]

c.   Beauty Event

The beauty event has the following signatures :
  −   2 negative muons with respectively $P_{\mu 1} = 27.7$ GeV/c , $P_{t_{\mu 1}} = 1.9$ GeV/c and
      $P_{\mu 2} = 11.5$ GeV/c , $P_{t_{\mu 2}} = 0.45$ GeV/c
  −   4 decay vertices compatible with the decay cascade :
      $B^- \to \mu^- + D^0$; $D^0 \to V4$
      $\overline{B}{}^0 \to D^- + H^+$; $D^- \to \mu^-$

A sketch of the events is given in Fig. 1. The estimated lifetimes of the two $B's$, calculated by the impact parameter method or assuming plausible decay schemes, are found to be :
$$t_{B^-} = (0.8 \pm 0.1) \ 10^{-13} \text{ sec}$$
$$t_{\overline{B}{}^0} = (5^{+2}_{-1}) \ 10^{-13} \text{ sec}$$

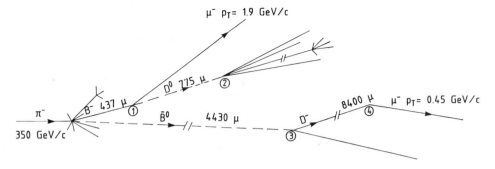

Figure 1.   Sketch of the $B^- \bar{B}^0$ event.

Finally the weighted mean flight time of the B^- and $\bar{B}^0$ is :

$$t_B = (3^{+2}_{-1}) \, 10^{-13} \, \text{sec}$$

3.   WA78

The study of hadroproduction of heavy quarks is considered to be an important tool to test QCD theory predictions. Indeed the comparison between theory and experimental results on charm [6] shows a poor agreement while, due to the increase of the quark mass, beauty hadroproduction is expected to be well described in terms of the leading order QCD perturbation theory [7] .

a.   Experimental method

In the WA78 experiment the search for $B\bar{B}$ pairs was done looking for multi−muon final states of the kind $\mu^+ \mu^+$, $\mu^- \mu^-$, $3\,\mu$ or $4\,\mu$ resulting from the semileptonic decay chains :

$\pi^- U \to B\bar{B} + X$

$\qquad B \to \mu^- D + X$; $D \to \mu^- + X$

$\qquad \bar{B} \to \mu^+ D + X$; $D \to \mu^+ + X$

Such events are characterised by large transverse momenta of the muons (p_t) coming from the B decays as well as by a large "missing energy" associated to energetic escaping neutrinos.

The WA78 experimental set up [4] comprised a dump calorimeter followed by a magnetic spectrometer to provide the muon energy (E_μ). The missing energy E_{miss} was determined by comparing the beam energy E_{beam} with the total energy of outgoing muons (ΣE_μ) and the hadronic energy measured in the dump calorimeter E_{cal} : $E_{miss} = E_{beam} - E_{cal} - \Sigma E_\mu$. The calorimeter, consisting of a uranium−scintillator sandwich, was constructed with flexible mechanics so that it was easily expandable in order to vary its mean density (ρ) . This facility was used in order to measure directly the non prompt muon background using the $1/\rho$ extrapolation method.

The standard acquisition trigger was achieved in two levels. The first level trigger was obtained with a system of hodoscope counters requiring at least two muons in the spectrometer. The second level trigger used a hardware processor in order to allow the online rejection of those events having a calorimeter energy exceeding a threshold value usually set at 280 GeV. With this trigger, 2.2 x 10^7 events were recorded on tape corresponding to 5.5 x 10^{11} effective interactions in the dump.

After the off−line muon reconstruction requiring at least two muons with $E_\mu > 15$ GeV and $80 < E_{cal} < 260$ GeV the following numbers of events were obtained :

$$N (\mu^+ \mu^-) = 3.5 \text{ x } 10^6$$
$$N (\mu^+ \mu^+) = 8193$$
$$N (\mu^- \mu^-) = 15539$$
$$N (> 3\mu) = 5589$$

471

b. Trimuon Sample

The sample of 3μ events was fully scanned using a graphical representation of each event. As a result 3700 clean trimuon events survived this selection. Further cuts as explained in [2] reduced the final sample to 1466 events.

The background in the trimuon sample is mainly due to the association of a direct $\mu^+ \mu^-$ pair with a third "non$-$prompt" muon coming from a π or K decay.

Direct $\mu^+ \mu^-$ pairs originate from two major processes : Drell$-$Yan pair production (a) and $D\bar{D}$ production and their subsequent semileptonic decay (b).

It has to be noted that, while no associated neutrino is present in process a), process b) is accompanied by two neutrinos. Therefore the latter process becomes increasingly important at higher missing energies. This fact is shown in Fig.2 where the $\mu^+ \mu^-$ experimental E_{miss} distribution is compared with a Monte Carlo simulation obtained as the superimposition of the two processes previously described.

For the estimate of the trimuon background, a sample of pseudo$-$events was generated by combining the experimental $\mu^+ \mu^-$ events with a third μ extracted from the non prompt distributions obtained with the $1/\rho$ extrapolation method. For this purpose, data were taken in special runs, with the calorimeter expanded, in order to obtain single muon yields at different densities as shown in Fig. 3.

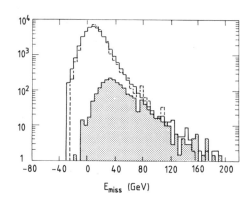

Figure 2

E_{miss} distribution of $\mu^+\mu^-$.
The dashed histogram represents a Monte Carlo calculation obtained combining Drell$-$Yan and $D\bar{D}$ processes. The shaded area refers to $D\bar{D}$ contribution. A $D\bar{D}$ cross section of $30\mu b\ A^{0.80}$ has been assumed.

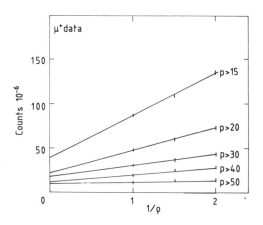

Figure 3.   WA78 density extrapolation.
Separation of prompt and non$-$prompt yields as a function of momentum using the $1/\rho$ extrapolation method. Measurements were performed with a single$-$muon trigger at dump densities $\rho = 1$, $2/3$, $1/2$ ($\rho = 1$ was the standard conFiguration, corresponding to ~ 10.6 g/cm^3).

472

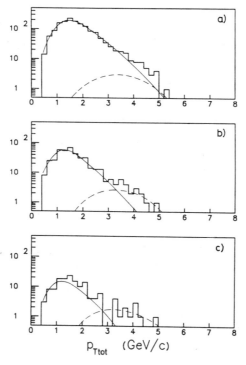

Figure 4. $p_{t_{tot}}$ for 3 μ events

a) without a cut in E_{miss} ; b) E_{miss} > 30 GeV ; c) E_{miss} > 50 GeV ;

The solid line represents the background estimate. The dashed line is the Monte Carlo prediction for $B\bar{B}$ produced according to expression (2) with α = 2.5 as explained in [2]. The Monte Carlo events are normalised to the experimental data at E_{miss} > 50 GeV and $p_{t_{tot}}$ > 3.2 GeV/c.

As has been pointed out already, the muons coming from the B decays are expected to have a larger p_t than the background muons and to be accompanied by energetic neutrinos, giving rise to high missing energy. It was found that a strong enhancement of the signal to noise ratio can be achieved using a cut on the variable $p_{t_{tot}} = p_{t1} + p_{t2} + p_{t3}$ combined with a cut on missing energy.

In Fig. 4, $p_{t_{tot}}$ distributions of the background pseudo — events are compared with experimental data for different cuts on missing energy. The relative normalisation of the two samples was done equalizing the areas of the two histograms of Fig. 4a (no missing energy cut) where the contribution of the $B\bar{B}$ signal is expected to be negligible.

As we can see in this Figure, without a missing energy cut the background estimate represents well the data over the whole $p_{t_{tot}}$ range, while for E_{miss} > 50 GeV an excess of events with respect to the background is present at large $p_{t_{tot}}$ (>3 GeV/c) where the effect of $B\bar{B}$ decay is expected.

An excess of events is also present for $p_{t_{tot}}$ < 2.5 GeV/c and E_{miss} > 50 GeV. This effect can be explained as being due to the production of four charmed particles, a process already observed in emulsion [5] as mentionned in the first part of this talk.

For $p_{t_{tot}}$ > 3.2 GeV/c and E_{miss} > 50 GeV we obtain a signal of 13 events against $1.4^{+1.0}_{-0.7}$ estimated background events. The estimated error in the background estimate takes into account :
- the π/K ratio uncertainty
- the uncertainty of non — prompt p_t distribution
- the possible four — charms contribution.

c. Dimuon sample

The $\mu^+\mu^+$ and $\mu^-\mu^-$ samples have been analysed in a similar way: 8193 $\mu^+\mu^+$ and 15539 $\mu^-\mu^-$ were respectively reduced to 2685 and 6847 events by applying appropriate cuts on the hit multiplicity in the counter hodoscopes, rejecting essentially punch−through events and upstream interactions. The remaining samples were scanned looking at the event display, this selection resulted in 2157 $\mu^+\mu^+$ and 3744 $\mu^-\mu^-$ clean events.

The background in the two muons like−sign sample is mainly due to the contribution of a single μ coming from D decay, or, from Drell−Yan $\mu^+\mu^-$ pair with one muon not detected in the spectrometer and a second "non prompt" muon.

A sample of background pseudo events was generated by combining the experimental single muon events taken in special runs with a second muon extracted from the "non prompt" muon distribution obtained with $1/\rho$ extrapolation method.

The like−sign muon events arising from $B\bar{B}$ decay are expected to have in average larger values of muon energies and transverse momenta with respect to background. Moreover also the missing energy due to the escaping neutrinos is expected to be higher for the $B\bar{B}$ events.

To enhance the signal−to−noise ratio in the sample of $B\bar{B}$ events combined cuts were applied to the sum of transverse momenta $p_{t_{tot}} = p_{t1} + p_{t2}$, E_{miss} , and $E_{lept} = \Sigma E_\mu + E_{miss}$. Requiring $p_{t_{tot}} > 2.7$ GeV/c , $E_{miss} > 20$ GeV , and $E_{lept} > 100$ GeV , we are left with 35 $\mu^+\mu^+$ and 49 $\mu^-\mu^-$; the corresponding contamination estimated from the background pseudo−events sample turns out to be 6 $\mu^+\mu^+$ and 14 $\mu^-\mu^-$.

The contribution of other processes giving rise to two like sign muons like $\psi\psi$, $\psi D\bar{D}$ or $D\bar{D}D\bar{D}$ production has been estimated to be negligible after the quoted cuts.

After background subtraction 29 $\mu^+\mu^+$ and 35 $\mu^-\mu^-$ events remain.

d. Production mechanism and cross−section

The distributions of our like−sign dimuon and three muon events are compared with the QCD predictions for $B\bar{B}$ production in π U interactions given in [11] .

The x_F and p_t behaviour of the b quark given by QCD are representable with the expression (1) where the shift of x_F distribution towards positive x_F values reflects the difference in quark and gluon x distributions for pions and nucleons.

$$\frac{d^2\sigma}{dx_F dp_t^2} \propto exp(-\frac{(x_F - 0.05)^2}{A_\pi^2}) \; exp(\frac{-p_t^2}{B}) \qquad (1)$$

$[\, A_\pi \approx 0.30 \; ; B \approx 6.9 \; (GeV/c)^2 \,]$

Correlations between the two particles as predicted by QCD [12] have been taken into account. Smearing due to the b quark fragmentation into physical particles and the B and D decay bringing to the leptons in the final states have been folded with eq. (1).

We have included a 20 % probability of a B_0 ($\bar{B}_0$) decaying as $\bar{B}_0$ (B_0) , as observed by the UA1 [13] and ARGUS [14] collaborations and we have assumed that charged and neutral B mesons are produced in equal numbers[1] .

Introducing the mixing we obtain a better agreement in the like−sign sample between Monte Carlo predictions and experimental data particularly in the distribution of $p_{t_{min}} [p_{t_{min}} = min(p_{t_i})]$ the variable most sensitive to these effects (Fig. 5).

---

[1] In our case the neutral B state is an unknown mixture of B_d^0 and B_s^0

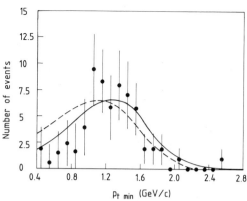

Figure 5

$p_{t_{min}}$ for like − sign dimuons after background. subtraction. The lines represent the QCD predictions with (solid line) and without (dotted line) mixing effects.

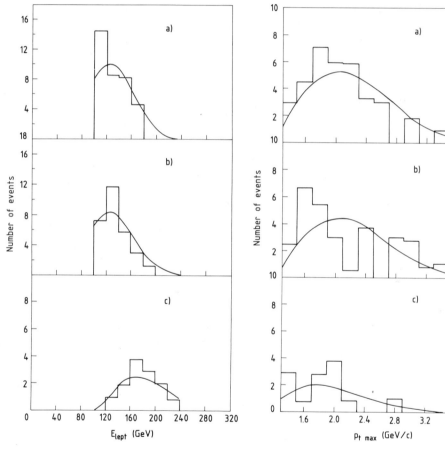

Figure 6. E_{lept} after background subtraction. *Figure 7.* $p_{t_{max}}$ after background subtraction.
a) $\mu^-\mu^-$; b) $\mu^+\mu^+$; c) 3μ. a) $\mu^-\mu^-$; b) $\mu^+\mu^+$; c) 3μ.
Continuous lines represent the Continuous lines represent the
QCD predictions. QCD predictions.

The experimental distributions of the total leptonic energy E_{lept} is shown in Fig. 6, the $p_{t_{max}}[p_{t_{max}} = max(p_{ti})]$ distribution is given in Fig. 7, for the 3 μ , $\mu^+\mu^+$ and $\mu^-\mu^-$ events, compared with the QCD predictions. We obtain good agreement between the data and predictions in all channels for all the variables of interest.

To compute the total cross section, the acceptance has been evaluated using the QCD prediction for $B\bar{B}$ production including the correlation between the two particles. The absolute normalization has been determined using a sample of reconstructed ψ events.

The following values for the branching ratio have been used :

$$BR\ (B \rightarrow \mu + X) = 11.0\ \%$$
$$BR\ (D \rightarrow \mu + X) = 12.6\ \%$$
$$BR\ (B \rightarrow D \quad) = 100\ \%$$

Under these assumptions the total cross section computed separately for 3 μ , $\mu^+\mu^+$ and $\mu^-\mu^-$ for a linear dependence of the atomic number is found to be :

$$\sigma\ (\mu^+\mu^+\ ) = \quad 2.55 \pm 0.6 \quad nb \quad per\ nucleon$$
$$\sigma\ (\mu^-\mu^-\ ) = \quad 2.1 \pm 0.5 \quad nb \quad per\ nucleon$$
$$\sigma\ (\ 3\mu\ ) \quad = \quad 1.6 \pm 0.5 \quad nb \quad per\ nucleon$$

where the errors are the statistical ones.

The present value of the cross section σ (3 μ) is smaller than the value given in [2] due to the different production mechanism used to compute the acceptance. In particular the expression (1) brings to larger acceptance of the apparatus because of the asymmetric x_F distribution. The systematic error is mainly due to the uncertainty in the absolute normalization and in the acceptance of the apparatus. These effects factorise in the three channels, the other systematic errors for each channels have been estimated to be smaller than the corresponding statistical errors; therefore combining the three values we obtain :

$$\sigma\ = 2.0 \pm 0.3 \pm 0.9\ nb\ per\ nucleon$$

the first error being statistical and the second one systematic. The uncertainties from B and D semileptonic branching ratios have been not taken into account.

4.   Conclusions

For the first time the WA75 collaboration has observed an associated production of *beauty* particles B^- and $\bar{B}_0$. The weighted mean flight time of the B^- and $\bar{B}^0$ was found to be :

$$t_B \quad = \quad (\ 3^{+2}_{-1}\ )\ 10^{-13}\ sec$$

The total cross section measured by the WA78 collaboration in π^- U interactions at 320 GeV $\sigma\ = 2.0 \pm 0.3 \pm 0.9$ nb per nucleon is in good agreement with the QCD prediction which provides a value of σ between 1 and 3 nb per nucleon.

Similar conclusions are obtained by the UA1 collaboration at $\sqrt{s} = 630$ GeV.

One can conclude that within experimental and theory uncertainties the QCD predictions describe the main features of the $B\bar{B}$ hadroproduction over a very wide interval of c.m. energy.

References

1.   J.P. Albanese et al.: Phys. Lett. 158B(1985) 186.

2.   M.G. Catanesi et al.: Phys. Lett. 187B(1987)431.

3.   Diamant Berger et al.; Phys. Rev. Lett. 44 (1980) 507.

      J. Badier et al.; Phys. Lett. 158B (1985) 85.

      A. Ereditato et al.; Phys. Lett. 157B (1985) 463.

4.   M.G. Catanesi et al.: Nucl. Instr. & Meth. A253(1987)222 .

5.   S. Aoki et al.: Phys. Lett. 187B(1987) 185.

6.   For a recent review on heavy flavour hadroproduction see :
S.P.K. Tavernier, Inter University Institute for High Energies, VUB − ULB Brussels
Preprint sub. to Report on Progress in Physics (1986) .

7.   R.K. Ellis: Proc. 21st Rencontres de Moriond, Strong interaction and Gauge Theories
(Editions Frontieres Gif − sur − Yvette 1986 p. 339)
J.C. Collins et al.: Nucl. Phys. B263(1986)37
R.K. Ellis and C. Quigg: FNAL Report FN − 445(1987)
E.L. Berger ANL − HEP − CP − 87 − 53 (June 1987), presented at the 22nd Rencontre de
Moriond, Hadrons, Quarks and Gluons, Les Arcs, France, 1987.

8.   M. Basile et al.: Lett. Nuovo Cimento 31(1981)97; Nuovo Cimento 68A(1982)289 .

9.   UA1 coll.; C. Albajar et al.: Phys. Lett. 186B(1987)237 .

10.   P.Bordalo et al.: presented at the EPS International Conference on High − Energy Physics
Uppsala 1987.

11.   E.L. Berger, ANL − HEP − PR − 90, presented at the Topical Seminar on Heavy Flavours,
San Miniato, Italy, 1987.

12.   E.L. Berger: private communication .

13.   UA1 coll.; C. Albajar et al.: Phys. Lett. 186B(1987)247 .

14.   ARGUS coll.; H. Albrecht et al.: Phys. Lett. 192B(1987)245 .

HEAVY QUARK PRODUCTION IN HADRON COLLISIONS

A THEORETICAL OVERVIEW

Davison E. Soper

Institute of Theoretical Science, University of Oregon

Eugene, OR 97403 USA

I discuss the production of heavy quarks (or, more generally, other heavy colored particles.) By "heavy," I mean that the quark mass M_Q is large compared to the 1 GeV scale of the strong interactions. When the quark mass is heavy, QCD perturbation theory is applicable to the calculation of the production cross section [1]. (I will discuss the issues involved in QCD calculability more fully later in this talk.) Presumably, the bottom quark is heavy enough for this purpose, but the charm quark is to be treated with extreme caution.

Modes of Heavy Quark Production in Hadron Collisions

There are three essentially different ways in which heavy quarks can be produced in high energy hadron collisions. These different production mechanisms correspond, on one hand, to different calculational methods, and on the other hand, to different event structures. The mechanisms are illustrated below.

This figure illustrates the mechanism that makes the dominant contribution to heavy quark production, namely, parton-parton fusion. This mechanism produces heavy quark pairs with the two heavy quarks having opposite transverse momenta, P_T, that are normally on the order of the heavy quark mass, M_Q.

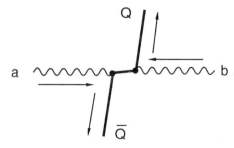

Parton-parton fusion

The parton-parton fusion mechanism can also produce heavy quarks $P_T \gg M_Q$, but with a much smaller cross section, comparable to the cross sections for the mechanisms discussed next.

Another mechanism is the scattering of a heavy quark that already exists as a constituent of one of the incoming hadrons. Here one heavy quark appears with very large transverse momentum, $P_T \gg M_Q$. The other has transverse momentum of order M_Q, and thus appears to be much more closely associated with the beam jet. Heavy quarks

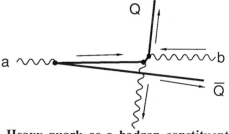

Heavy quark as a hadron constituent

as constituents are also important for other high virtuality processes, such as Higgs production.

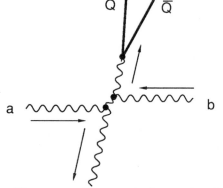

The final mechanism is the production of a heavy quark pair as part of the decay products in a light parton jet, where the P_T of the jet is much larger than the heavy quark mass. One of the lowest order diagrams is shown in the figure.

Heavy quark as part of a parton jet

Why Are Most Heavy Quarks Produced with $P_T \sim M_Q$?

When $P_T \sim M_Q$, there is only one large momentum scale involved in the calculation of the production cross section, namely M_Q. I stated in the previous section that $P_T \sim M_Q$ is the dominant momentum region in which heavy quarks are produced. Let us see why this is so. Suppose, to the contrary, that $M_Q \ll P_T \ll \sqrt{s}$. The standard factorization formula for the cross section is

$$\frac{d\sigma}{dP_T^2 dy} = \int \frac{dx_A}{x_A} x_A f_{a/A}(x_A) \int \frac{dx_B}{x_B} x_B f_{b/B}(x_B) \frac{d\hat{\sigma}}{dP_T^2 dy} \ .$$

(For an accurate calculation of the cross section when $M_Q \ll P_T$, one would normally factor the M_Q dependence into parton distribution and decay functions involving the heavy quarks, but for this analysis, I include all heavy quark subdiagrams as part of $\hat{\sigma}$ calculated to a high order.) Dimensional analysis and Lorentz invariance allows us to extract a factor $1/P_T^4$ and write the parton level cross section in the form,

$$\frac{d\hat{\sigma}}{dP_T^2 dy} = \frac{1}{P_T^4} H(\frac{x_A x_B s}{P_T^2}, \frac{x_A}{x_B} e^{-y}) \ .$$

480

Define scaled variables $\xi_A = x_A/(P_T/\sqrt{s})$, $\xi_B = x_B/(P_T/\sqrt{s})$. Then

$$\frac{d\sigma}{dP_T^2 dy} = \frac{1}{P_T^4} \int \frac{d\xi_A}{\xi_A}\, x_A f_{a/A}(x_A) \int \frac{d\xi_B}{\xi_B}\, x_B f_{b/B}(x_B)\, H(\xi_A\xi_B, \frac{\xi_A}{\xi_B}e^{-y}),$$

with

$$x_A = (P_T/\sqrt{s})\xi_A, \quad x_A = (P_T/\sqrt{s})\xi_A.$$

We conclude from this formula that if the distribution functions obey $xf(x) \sim const.$, then the dependence of the cross section on P_T is

$$\frac{d\sigma}{dP_T^2 dy} \propto \frac{1}{P_T^4} \quad \text{for} \quad P_T \gg M_Q.$$

Thus, most of the heavy quarks are produced with the smallest P_T possible: $P_T \sim M_Q$.

Characteristics of Heavy Quark Production with $P_T \sim M_Q$

(1) P_T distribution. A rough power counting estimate is:

$$\frac{d\sigma}{dP_T^2 dy} \propto \frac{1}{(P_T^2 + M_Q^2)^2}.$$

This reflects the $1/P_T^4$ for $P_T \gg M_Q$ obtained above, together with a cutoff when $P_T \sim M_Q$. In the figure, I show the distribution as predicted by lowest order perturbation theory [2,3], in terms of the quantity

$$\sigma(P_T^{min}) = \int_{P_T^{min}}^{\infty} dP_T^2 \int_{-2}^{2} dy\, \frac{d\sigma}{dP_T^2 dy}.$$

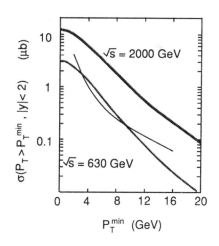

This quantity is plotted against P_T^{min} for b quark production at $\sqrt{s} = 630$ GeV and 2 TeV. The light line illustrates a pure $1/P_T^2$ dependence. Note the cutoff when $P_T^{min} \sim M_Q$. When $P_T \gg M_Q$, the curves are a bit steeper than the $1/P_T^2$ expected from the dimensional argument because really $xf(x)$ is not constant, but decreases with increasing x.

481

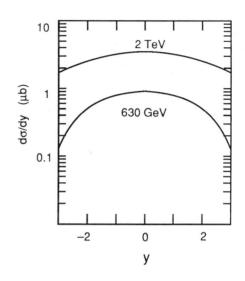

(2) y distribution. The lowest order cross section is fairly flat over the mid range of rapidity, but decreases rapidly for large rapidities since there aren't many partons at large momentum fraction. This is illustrated for b quark production at $\sqrt{s} = 630$ GeV and 2 TeV in this figure from [2,3].

Total Production Cross Section

To see roughly what to expect, we apply the same sort of dimensional reasoning used earlier. On dimensional grounds, the parton level total cross section scales like

$$\hat{\sigma} = \frac{1}{M_Q^2} H(\frac{x_A x_B s}{M_Q^2}) \ .$$

Define scaled variables ξ so that $x_A = (M_Q/\sqrt{s})\xi_A$, $x_B = (M_Q/\sqrt{s})\xi_B$. Then

$$\sigma = \frac{1}{M_Q^2} \int \frac{d\xi_A}{\xi_A} x_A f_{a/A}(x_A) \int \frac{d\xi_B}{\xi_B} x_B f_{b/B}(x_B) H(\xi_A \xi_B) \ .$$

This would give $\sigma \propto 1/M_Q^2$, independent of s, if $xf(x)$ were independent of x. Since $xf(x)$ grows with decreasing x, we expect that

(1) σ increases with s at fixed M_Q^2

(2) $M_Q^2\sigma$ decreases with increasing M_Q^2 at fixed s.

This is illustrated in the graph at the right, which I have adapted from Berger [3].

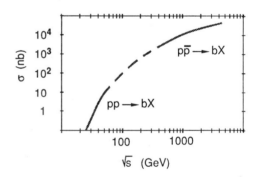

Comparison with Experiment

The UA1 group has reported a result for b production with $|y| < 2$ and $P_T > 5$ GeV at the CERN collider [4]. In the graph to the right, I show this result along with the lowest order QCD prediction [2,3] for $|y| < 2$ and P_T greater than a variable P_T^{min}. There is good agreement, but I should caution the reader that the relation between the experimental datum and the actual measurements of muons from the b-decays is rather indirect. In addition, the theory curves in both this figure and the next should be regarded as having factor-of-two errors associated with them.

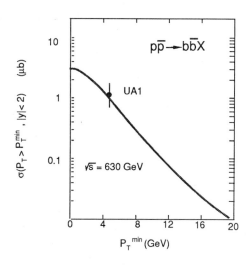

The WA78 group has reported a measurement [5] of the cross section for b production in $\pi^- U$ collisions at $\sqrt{s}_{\pi N} = 25$ GeV, based on the detection of electrons and muons in the final state. In order to get a total cross section, one must extrapolate from a limited acceptance. When WA78 assumes a cross section shape as given by lowest order QCD, they obtain a cross section per nucleon $\sigma = 2.4 \pm 0.7 \pm 0.8$ nb. This is illustrated in the figure at the right, in which the result is compared to the lowest order QCD prediction by Berger [3].

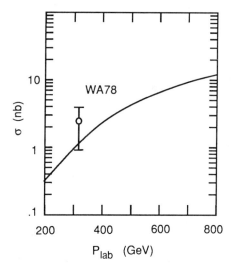

Higher Order Corrections

Within the framework of the usual factorization formula,

$$\frac{d\sigma}{dP_T^2 dy} = \int dx_A \; f_{a/A}(x_A) \int dx_B \; f_{b/B}(x_B) \; \frac{d\hat{\sigma}}{dP_T^2 dy} \;,$$

there are higher order corrections to $\hat{\sigma}$, such as

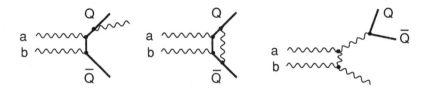

There is some reason to think that these corrections may be large, but only a full calculation including the virtual graphs can decide. The corrections have not yet been calculated, but two groups are working on them. See the talk of R.K. Ellis in these proceedings.

Flavor Excitation

Some papers in the earlier literature have claimed big cross section contributions for $M_Q \gg 1$ GeV from physical mechanisms suggested by the picture at the right, in which the exchanged gluon is very soft.

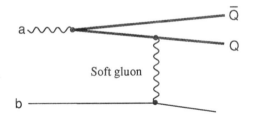

Should one add this to the perturbative mechanism discussed earlier? An argument in [1] shows that this is not, in fact, a new mechanism.

In the figure, we see the same Feynman diagram drawn in such a way as to suggest that the low transverse momentum gluon at the bottom should be regarded as a constituent of hadron B. The subdiagram at

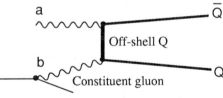

the top is the hard scattering, since the virtual heavy quark is forced to be far off-shell by the kinematics. Thus this diagram is a contribution to the ordinary gluon-gluon fusion process, rather than something new and non-perturbative that is to be added to the usual cross section.

Diffractive Production

Can one produce a heavy quark with $M_Q \gg 1$ GeV while one of the hadrons is left intact? If so, should one add a "diffractive" term to the cross section? The figure below shows the creation of a heavy quark pair by diffraction, visualized as the exchange of a pomeron, $\mathbb{P}$. The proton at the bottom loses a samll fraction of its energy (less than 10%, say) and gains 300 MeV or so of transverse momentum, but is otherwise not affected. It is still kinematically possibl: to produce the heavy quark pair because 10% of the proton energy is still very large.

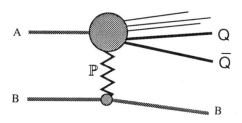

Quantum chromodynamics suggests that this can indeed happen, as first proposed by Ingleman and Schlein [6]. The mechanism is ordinary gluon-gluon fusion (and also quark-antiquark fusion). One simply has to find one of the gluons as a part of the exchanged pomeron. The process is illustrated at the right. There are several points to be noted:

• This is included as part of the usual parton fusion hard scattering process.
• A similar process can also produce high P_T jets.
• By measuring the cross section, one can learn about the distribution of gluons in the pomeron.
• A detailed analysis can be found in ref. [7]. Somewhat different perspectives can be found in [8] and [9].

An estimate from [7] for the diffractive component of bottom quark production at $\sqrt{s}$ = 2 TeV is shown at the right. This estimate is based on a guess for the distribution of gluons in the pomeron. Future experimental results will enable us to *measure* this intriguing quantity.

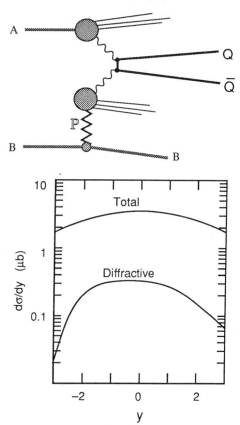

Higher Twist Effects

If M_Q is not large enough, contributions to the cross section suppressed by a power of 1 GeV/M_Q become important. Such effects may be expected to be important for charm production. Let us look at two possible higher twist contributions.

(1) Intrinsic Charm We imagine that the heavy quark exists as part of the hadron wave function [10], as illustrated in the figure at the right. This mechanism may give a significant contribution to charm production at large momentum fraction. It is, however, suppressed by $(1\ GeV/M_Q)^2$, as argued in [11]. The size of this effect is not known theoretically.

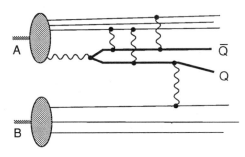

(2) Coalescence Enhancement This effect, illustrated below, was investigated in ref. [12] in a model with scalar partons and a non-relativistic Coulomb potential for the exchanged gluon. Suppose that the spectator quark is detected and has velocity $v \ll 1$ in the rest frame of the heavy quark. Then the final state interaction modifies the Born cross section by a factor

$$1 + c\,\frac{\pi\alpha_s}{v} + \dots\,,$$

where c is an appropriate color factor (positive for an attractive channel, negative for a repulsive channel.)

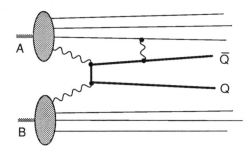

Suppose now that the spectator quark is *not* detected. Then one finds that the Born cross section is not affected in the large M_Q limit. That is, the probability to create the heavy quark is not changed, although the light quark spectators are strongly pulled in toward the heavy quark, or pushed away if they have the wrong color. There is a remaining higher twist effect, suppressed by one power of $(1\ GeV/M_Q)$. The size, sign, and distribution of this effect is a function of the hadron wave functions. With reasonable assumptions, we expect a fairly large positive effect for $y_Q \sim y_{max} - 1$.

Heavy Quarks as Hadron Constituents

The distribution function $f_{Q/A}(x,\mu)$ is relevant for heavy quarks Q that participate in a hard interaction with momentum scale much larger than M_Q. Here x is the momentum fraction and the renormalization scale μ is chosen to be something on the order of the large momentum scale. An example is the production of a heavy quark with $P_T \gg M_Q$, as illustrated in the figure. Other examples include heavy Higgs production at the SSC. See refs. [13,14] for calculational methods.

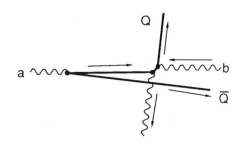

What, then, is the distribution $f_{Q/A}(x,\mu)$? We first note that $f_{Q/A}(x,\mu)$ obeys the Altarelli-Parisi equation

$$\frac{d}{d\ln(\mu)} f_{a/A}(x,\mu) = \frac{\alpha_s}{\pi} \sum_b \int_x^1 \frac{d\xi}{\xi} P_{a/b}(\tfrac{x}{\xi}) f_{b/A}(\xi,\mu) \ ,$$

which can be written in diagrammatic form as

$$\frac{d}{d\ln\mu} \quad \underset{A \qquad a}{\bigcirc\!\!\!\!-\!\!\!=\!\!\!\!-^x} \quad = \quad \underset{A \qquad b \qquad a}{\bigcirc\!\!\!\!-\!\!\!^\xi\!\!\!-\bigcirc\!\!\!\!-\!\!=\!\!\!-^x}$$

The term in the kernel of this equation that describes heavy quark creation is

$$P_{Q/g}(z) = \frac{1}{2} [z^2 + (1- z)^2] \ .$$

The Altarelli-Parisi equation gives $f_{Q/A}(x,\mu)$ for $\mu \gg M_Q$ if we know it for $\mu \approx M_Q$. For $\mu \approx M_Q$, calculation gives [13](using the simple $\overline{MS}$ definition for distribution functions)

$$f_{Q/A}(x,\mu) = \frac{\alpha_s}{\pi} \int_x^1 \frac{d\xi}{\xi} P_{Q/g}(\tfrac{x}{\xi}) \ln(\tfrac{\mu}{M_Q}) f_{g/A}(\xi,\mu) \ .$$

Thus the needed boundary condition is

$$f_{Q/A}(x,M_Q) = 0 + O(\alpha_s^2)$$

The resulting distribution of top quarks in the proton is illustrated at the right as a function of μ for fixed x = 0.01.

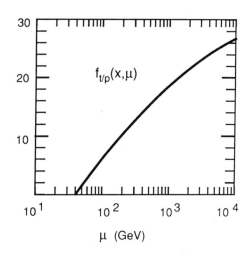

$f_{t/p}(x,\mu)$

μ (GeV)

Heavy Quarks as Fragments in Jets

The final heavy quark production mechanism arises from heavy quarks appearing as fragments of jets. The heavy quark, and also an antiquark, are inside a jet having $P_T \gg M_Q$. This involves the decay function for gluons (or light quarks) $d_{Q'/g}(z,\mu)$ evaluated at a renormalization scale μ that is on the order of P_T. In general, one must use this function as part of a calculation that includes the precise definition of what is meant experimentally by a jet. However, at the lowest level of approximation, the decay function can be interpreted as the probability to find a heavy quark inside a jet and carrying a fraction z of the jet P_T.

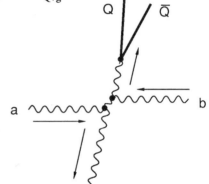

In the notation $d_{Q'/g}(z,\mu)$, the Q' denotes a heavy quark in the final state. There is some ambiguity here since only mesons and baryons really appear in the final state. To a good approximation, a $Q\bar{q}$ meson carrying momentum p^μ is equivalent to a Q carrying momentum p^μ. Roughly, a $Q\bar{Q}$ meson carrying momentum p^μ is equivalent to a Q carrying momentum $p^\mu/2$, since the Q carries about half of the meson momentum. There are, however, problems associated with this approximation that are not yet well understood.

What is the distribution $d_{Q'/g}(z,\mu)$? First of all, $d_{Q'/g}(z,\mu)$ obeys the Altarelli-Parisi equation

$$\frac{d}{d\ell n(\mu)} d_{Q'/a}(z,\mu) = \frac{\alpha_s}{\pi} \sum_b \int_x^1 \frac{d\xi}{\xi} P_{b/a}(\frac{z}{\xi}) f_{Q'/b}(\xi,\mu)$$

$$\frac{d}{d \ln \mu} \frac{1}{a} \underset{Q'}{\bigcirc}\!\!=\!\!z = \frac{1}{a} \underset{b}{\bigcirc}\!\!\overset{\frac{z}{\xi}}{=}\!\! \underset{Q'}{\bigcirc}\!\!=\!\!z$$

with, for instance,

$$P_{Q/g}(y) = \frac{1}{2} [y^2 + (1-y)^2] .$$

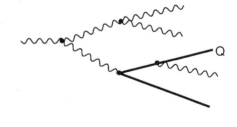

The Altarelli-Parisi equation describes jet evolution processes such as those illustrated on the right.

The evolution equation gives $d_{Q'/a}(z,\mu)$ for $\mu \gg M_Q$ if we know it for $\mu \approx M_Q$. Now we need to know what to do for virtualities of order M_Q. For $\mu \approx M_Q$, calculation based on the figure at the right gives (using the simple $\overline{\text{MS}}$ definition for distribution functions)

$$d_{Q'/g}(z,\mu) = \frac{\alpha_s}{\pi}\,P_{Q/g}(z)\,\ell n(\frac{\mu}{M_Q}) + O(\alpha_s^2)$$

$$d_{Q'/Q}(z,\mu) = \delta(1-z) + O(\alpha_s) \ .$$

Note the absence of large logarithms when $\mu \approx M_Q$. This happens because M_Q provides an infrared cutoff. We conclude that the needed boundary condition for the Altarelli-Parisi equation is

$$d_{Q'/g}(z,M_Q) = 0 + O(\alpha_s^2) \qquad\qquad d_{Q'/Q}(z,M_Q) = \delta(1-z) + O(\alpha_s) \ .$$

Problem at higher orders. There is an unresolved problem at higher orders in perturbation theory. Graphs like this have a threshold singularity at $k^2 = 4M_Q^2$:

$$\mathcal{A} \sim \mathcal{A}_{Born} \times \{\,1 - C_F\frac{\pi\alpha_s}{v} + \dots\,\} = \mathcal{A}_{Born} \times \frac{-C_F 2\pi\alpha_s/v}{1-\exp(C_F 2\pi\alpha_s/v)} \quad \text{with } v = \frac{\sqrt{k^2 - 4M_Q^2}}{\sqrt{k^2}}$$

The $1/v$ singularities build up disastrously. This is because the heavy quarks have a long time to repel each other when v is small. If one of the heavy quarks were to emit a soft gluon first, then the $Q\bar{Q}$ could be in a color singlet state and the heavy quarks would attract each other ($\alpha_s \to -\alpha_s$ in the formula). I offer the conjecture that this problem goes away when we integrate over k^2 and all of the momentum fractions. A singularity at $z = 1/2$ will remain, so we will have to average over a range of z, that is, view $d_{Q'/g}(z,\mu)$ as a distribution instead of an ordinary function.

Multiplicity of Heavy Quarks in a Gluon Jet. Mueller and Nason [15] have analyzed the total number of heavy quarks in a gluon jet, that is, the quantity

$$N(\mu) = \int_0^1 dz\, d_{Q'/g}(z,\mu) \ .$$

Let us examine their result and its relation to the discussion given above. First of all, we shall neglect the possibility that a heavy quark is created in the jet, emits a gluon, and that gluon or one of its daughters gives birth to another heavy quark, which is the one detected. This amounts to neglecting the $P_{g/Q}(z/\xi)$ term in the evolution equation. With this approximation, the evolution equation gives

$$N(\mu) = \frac{1}{3} \int_{M_Q}^{\mu} \frac{d\bar{\mu}}{\bar{\mu}} \frac{\alpha_s(\bar{\mu})}{\pi} \frac{G(\mu)}{G(\bar{\mu})} \qquad \text{where} \qquad \frac{d}{d\ell n(\mu)} G(\mu) = \frac{\alpha_s(\mu)}{\pi} \int_0^1 dz \; P_{g/g}(z) \times G(\mu).$$

However, the integral over z is divergent. Following Mueller and Nason, we use instead

$$\frac{d}{d\ell n(\mu)} G(\mu) = \left\{ \sqrt{\frac{3\alpha_s(\mu)}{2\pi}} + c \; \frac{\alpha_s(\mu)}{\pi} \right\} \times G(\mu),$$

where c is a certain constant. This gives the result of Mueller and Nason except that they have a threshold factor $(1 + 2M_Q^2/\bar{\mu}^2)(1 - 4M_Q^2/\bar{\mu}^2)^{1/2} \; \theta(\bar{\mu}^2 > 4M_Q^2)$ in the integral for $N(\mu)$. This threshold factor results from the exact definition used and does not affect the result at the leading approximation.

Mueller and Nason go on to show that higher twist effects are not important for this problem, so that the result should, surprisingly, be reliable even for charm quarks.

Numerical Results. In the figure on the next page, I show results for the distribution of charm quarks (not including antiquarks) in a gluon jet with a transverse momentum of 30 GeV, where the charm quark mass is taken to be 1.2 GeV. The box on the left shows the total number of charm quarks according to the formula of Mueller and Nason [15]. I also show an experimental result for jets of roughly this P_T from the UA1 group [16]. In the right hand box, I show the quantity

$$\int_x^1 dz \; d_{Q'/g}(z, 30 \text{ GeV})$$

as a function of the momentum fraction x. I have calculated this function as described above, except that I have neglected the $P_{g/Q}(z/\xi)$ term in the Altarelli-Parisi kernel. We see that, indeed, this quantity diverges if we try to take x to 0. If we wanted a result for the decay function that would be valid for very small x, we would have to properly account for soft gluon interference. However, the result in its simple form should be quite reliable for x > 0.1. Notice that x = 0.1 corresponds to $k_T = 3$ GeV. For charm quarks with this little transverse momentum the simple picture in which the charm quark is definitely associated with the jet begins to break down, and a more sophisticated calculation involving the whole collision process becomes necessary.

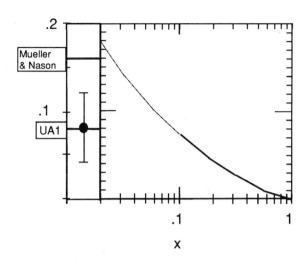

x

Acknowledgement: I would like to thank Ruiben Meng for help with the jet decay calculation, and I would like to thank G. Sterman, J.C. Collins, E. Berger, S. Brodsky and J. Gunion for helpful conversations.

References

[1] J.C. Collins and G.Sterman and D.E. Soper, Nucl. Phys. B263 (1986) 37

[2] E.L. Berger and J.C. Collins and D.E. Soper, Phys. Rev. D35 (1987) 2272

[3] E.L. Berger, Argonne Preprint ANL-HEP-PR-87-90, to be published in Nucl. Phys. B, Proceedings Supplements: Proc. of the Topical Seminar on Heavy Flavors, San Miniato Italy, May, 1987

[4] UA1 Collaboration, C. Albajar *et al.*, Phys. Lett B186 (1987) 237

[5] WA78 Collaboration, M.B. Catanesi *et al.*, CERN report CERN-EP/86-177

[6] G. Ingleman and K.-H. Schlein, Phys. Lett. 152B (1985) 297

[7] E.L. Berger, J.C. Collins, D.E. Soper, and G. Sterman, Nucl. Phys. B286 (1987) 704

[8] H. Fritsch and K.H. Streng, Phys. Lett. 1689 (1985) 391

[9] A. Donnachie and P.V. Landshoff, Manchester preprint M/C TH 87/05

[10] S.J. Brodsky, C. Peterson and N. Sakai, Phys. Rev D23 (1981) 2745; S. J. Brodsky, P. Hoyer, C. Peterson and N. Sakai, Phys. Lett 93B (1980) 451

[11] S.J. Brodsky, J.C. Collins, S.D. Ellis, J.F. Gunion and A.H. Mueller, in *Proceedings of the Workshop on the Design and Untilization of the SSC*, 1984, edited by R. Donaldson and J.G. Morfin (Fermilab, Batavia, IL, 1985), p. 227

[12] S.J. Brodsky, J.F. Gunion and D.E. Soper, Phys. Rev. D36 (1987) 2710

[13] R.M. Barnett, H.E. Haber and D.E. Soper, in V. Barger, T. Gottschalk and F. Halzen, eds., Proc. Workshop on Physics Simulations at High Energy, Madison, 1986 (World Scientific, Singapore, 1987); Oregon preprint OITS 365 ("Ultra-Heavy Particle Production from Heavy Partons at Hadron Colliders")

[14] J.C. Collins and W.-K. Tung, Nucl. Phys. B278 (1986) 934; F.J. Olness and W.-K. Tung, IIT preprint IIT-TH/87-17

[15] A.H. Mueller and P. Nason, Phys. Lett. 157B (1985) 226; Nucl. Phys. B266 (1986) 265

[16] UA1 Collaboration, K. Eggert, private communication

CHARM AND BEAUTY DECAYS VIA HADRONIC PRODUCTION

IN A HYBRID EMULSION SPECTROMETER (FERMILAB E653[a])

Noel R. Stanton

Physics Department
Ohio State University
Columbus, Ohio 43210

The goals of E653 are the measurement of lifetimes (especially those of charged and neutral beauty particles and of charmed baryons); study of production dynamics (total cross-section for charm and beauty, Feynman x and transverse momentum distributions, pair correlations, and multiple charm production); search for new states of charm (excited states, exotics); and search for unusual decay modes such as D_S to τ (double-kink topology) which require a high-resolution visual technique. Because this visual technique can see nearly all decay vertices, the additional possibility of surprises is always present.

E653 had its skakedown run with 800 GeV protons in 1985, and has just completed a very successful second run with 600 GeV pions. Run 1 resulted in 42 million interactions after fiducial and quality cuts, and Run 2 about 200 million. The estimated yields of pairs found in the emulsion are 1500-2000 charm and a few beauty from Run 1, and 5000-10000 charm and about 20 beauty pairs (assuming 10 nb/pair[1]) from Run 2.

This experiment studies heavy quark decays selected topologically, rather than through selected decay modes. The decays are observed in emulsion, but are selected and located by reconstructing them in a spectrometer with a silicon microstrip vertex detector. Seeing the decays in emulsion has several strong advantages. Decay tracks are unambiguously associated with the correct vertex, even in many-vertex events such as beauty decays. Because the background of false decays is at the 1% level, a few examples of a rare decay are statistically significant. The direction of the heavy-quark parent is well-measured, so that transverse momentum balance is a strong constraint in assigning neutral pions and kaons to decay vertices. A large fraction of 0-C fits should be usable (as was the case in E531[2]) for lifetime and dynamics studies.

The strategy of the experiment is indicated in Fig. 1. In this scheme the "trigger decay" is semimuonic, while the other decay or decays in the event are unbiased as to decay mode. The events recorded on tape have an interaction in the target region and a muon candidate with a range of 5 to 6 GeV of iron. Offline, all recorded events in the spectrometer are reconstructed, with the silicon microstrip detectors providing precision tracking. Events to be

[a]The E653 collaboration: Aichi University of Education, University of California at Davis, Carnegie–Mellon University, Chonnam National University, Fermilab, University of Gifu, Gyeongsang National University, Jeonbug National University, Kobe University, Korea University, Nagoya University, Ohio State University, Okayama University, University of Oklahoma, Osaka City University, Osaka Prefecture Science Education Institute, Sookmyong Women's University, Toho University, Wan Kwang University, Yokohama National University.

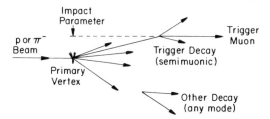

Fig. 1.   Event selection strategy.

searched for in the emulsion are those in which the muon candidate has appreciable transverse momentum p_T and a significant impact parameter b_μ. The criteria are specified in Table 1, along with the fraction and number of the Run 1 triggers selected. One of the four selection types demands a secondary vertex containing the muon and at least one hadron.

Table 1. Criteria for Selection of Events to be Scanned in Emulsion

Event Type	Selection Criteria	Fraction Selected	Total Number
3	Secondary vertex containing trigger muon $p_\mu \geq 8$ GeV/c, $p_{T\mu} \geq 0.20$ GeV/c, $b_\mu \geq 0.05$ mm $p_{had} \geq 5.0$ GeV/c, $b_{had} \geq 0.05$ mm Decay length ≥ 2.0 mm	29%	17,500
2	High p_T muon (mainly for beauty) $p_\mu \geq 8$ GeV/c, $p_{T\mu} \geq 0.80$ GeV/c, $b_\mu \geq 0.10$ mm	6%	3,600
1	Single muon, no secondary vertex $p_\mu \geq 8$ GeV/c, $p_{T\mu} \geq 0.25$ GeV/c, $b_\mu \geq 0.10$ mm	42%	25,500
4	Secondary vertex outside emulsion K^0, Λ^0 for calibration (not scanned)	(23%)	(14,000)

In the emulsion, it is first established whether the muon candidate is indeed absent from the tracks coming from the primary vertex. If this muon candidate does not come from the primary it is followed back from the downstream end of the emulsion to locate the semimuonic decay candidate, which is visually inspected to eliminate obvious secondary interactions. Other tracks which are seen in the spectrometer but not at the primary vertex in the emulsion are also followed back to search for other decay candidates.

A plan view of the spectrometer is shown in Fig. 2. Its strong points are precision tracking in a high-multiplicity environment, reconstruction of electromagnetic showers, and identification of muons, which are momentum-analyzed both before and after the iron absorber to greatly reduce incorrect matches between the downstream muon candidate and the corresponding track upstream of the iron. The heart of the experiment, almost invisible on the scale of this drawing, is the precision front end consisting of the emulsion target and silicon microstrip vertex detector. Multisampling drift chambers[3] downstream of the analyzing magnet have position and track-pair resolutions which are well-matched to the vertex detector. The liquid argon calorimeter can resolve electromagnetic showers less than a centimeter apart using the small pads in its central region. An excellent time-of-flight system can distinguish protons and pions up to 7 GeV/c.

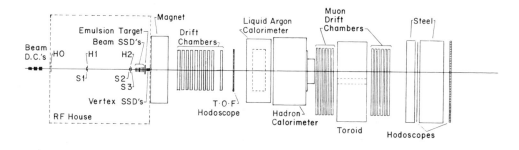

Fig. 2.   Plan view of the spectrometer.

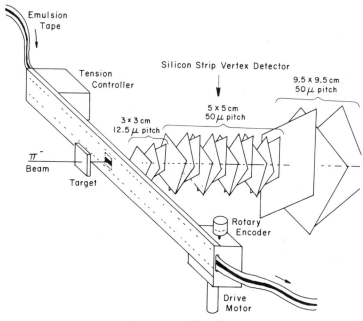

Fig. 3.   Schematic drawing of vertex detector.

The precision front end of the spectrometer as configured for the 1987 run of the experiment is shown schematically in Fig. 3. The vertex detector consists of 18 planes of silicon microstrip detectors divided equally among three stereo views. The three most upstream detectors, which have 12.5 micron pitch, and the three most downstream ones, which have 50 micron pitch and are 9.5 cm × 9.5 cm, are additions for the 1987 run. They replace six of the eighteen 5 cm × 5 cm, 50 micron pitch detectors used for the first run.

Microstrip detector information for two events in which charm pairs have been found is shown in Figs. 4 and 5. The event in Fig. 4 has relatively low multiplicity, with 11 tracks traversing the magnet, compared to an average of 17 tracks for all events. The event in Fig. 5 is more challenging, having 23 such reconstructed tracks. In these figures the hits are marked by bars proportional in length to the pulse height in the struck microstrips. This pulse height information is used to find the hit positions from charge sharing in the outer regions of the detector, in which only every nth strip is read out (n=2,3,5 at progressively larger distances

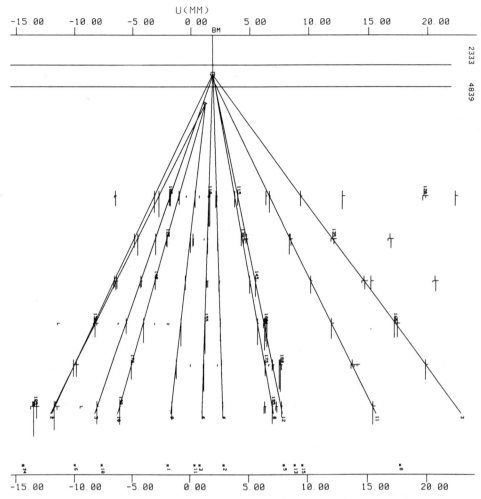

Fig. 4. Reconstruction of a charm pair event in one view of the microstrip vertex detector. The main emulsion stack is marked by a pair of horizontal parallel lines. Hits are marked by bars with length proportional to the pulse height in the struck microstrips. The lines through the hits are tracks reconstructed by the program. A trident reconstructed outside the emulsion stack is marked with a small triangle while a neutral two-prong decay in the emulsion is barely resolved by the vertex detector.

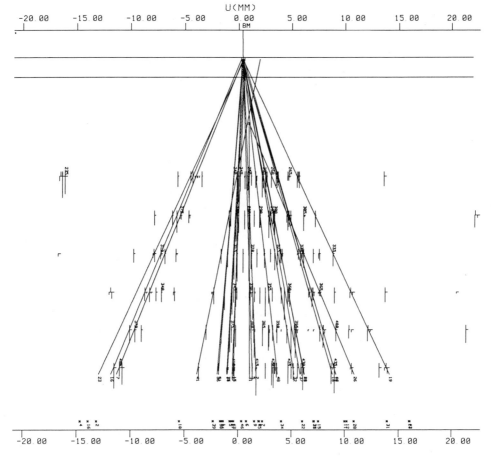

Fig. 5. Reconstruction of a charm pair event with high primary multiplicity. A trident downstream of the emulsion is marked by a small triangle. The second charm decay occurs in the emulsion, close to the primary vertex.

from the beam). In the central ±4 mm, in which every strip is read out, the pulse height measurement both improves the resolution and tags situations in which more than one particle registered in the same strip. In this central region of each detector the position resolution is ± 8.5 microns.

The multisampling drift chambers downstream of the magnet must also deal with high multiplicities and track densities. The fact that many tracks cross each other due to the bend in the magnet complicates the pattern recognition. Fig. 6 shows a typical event reconstructed in one view of these chambers. They are able to resolve tracks less than a millimeter apart, and the "minivectors" reconstructed from the five hits in each chamber are seen to greatly simplify the pattern recognition.

An emulsion target module, depicted as a simple box in Fig. 3, is shown in more detail in Fig. 7. This is a module of the "vertical" type, in which the tracks are incident perpendicular to the emulsion sheets. (Modules of the "horizontal" type, with tracks largely in the plane of the sheets, were also exposed). The main emulsion block consists of twenty sheets, each 25 cm by 25 cm and 700 microns thick, occupying 14 mm along the beam direction. Each such sheet is composed of two 300 micron layers of emulsion bonded to a thin plastic base. To extend the depth of the fiducial region for decays without adding a great deal of mass, there are in

497

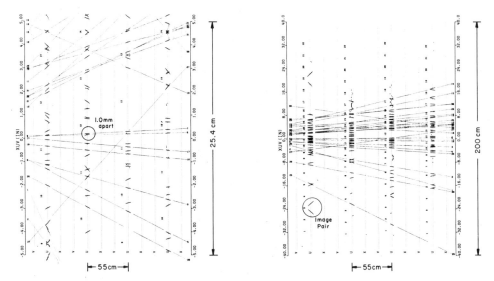

Fig. 6.   Reconstruction of a typical event in one of three stereo views of the multi-sampling drift chambers downstream of the magnet. The picture at the left shows the full width of the chambers (approximately 2 meters), while the one at the right shows an enlargement of the central 25 cm. The short heavy lines are "minivectors" fit to the five position measurements in each chamber; the dotted lines are reconstructed tracks connecting the minivectors.

addition five "analysis sheets" spaced apart by honeycomb paper to occupy an additional 8 mm. The main emulsion block and the low-mass analysis plates are clearly seen in Fig. 8, which shows distributions in the z coordinate (along the beam direction) of interactions found by the vertex detector. Maintaining a constant track density in the emulsion requires the target to be kept in constant motion during the beam spill by a precision mechanical device whose speed is proportional to the beam intensity.

Several techniques have been developed to increase the speed and efficiency of finding decay events in the emulsion. The most downstream analysis sheet, which has a thicker plastic base and thinner emulsion coatings than the others, is being successfully used as a vernier device similar in concept to the "changeable sheet"[4] used by members of this collaboration in Fermilab Experiment 531. Spectrometer tracks are extrapolated into this sheet with an accuracy of a few tens of microns; candidate tracks in this emulsion sheet are then re-extraplolated into the main emulsion block. The improvement in impact parameter measurement resulting from this technique is illustrated in Fig. 9. The improvement is greatest (up to a factor of 5) for low momentum tracks due to the larger multiple scattering in the silicon.

The Nagoya group is now using a completely automatic trackfinder to search for tracks in this vernier sheet. This device couples a video camera to electronics which senses the depth in the emulsion at which exposed grains are in best focus. This grain information at sixteen values of depth (which corresponds to the beam direction in vertical emulsion) is stored in sixteen 512 × 512 frame memories. Tracks are reconstructed by software from this matrix of grain positions. This trackfinder is now operating 24 hours a day without human intervention. Preliminary measurements place its efficiency at approximately 90% per track at typical exposure densities. To exploit this automatic scanning capability more efficiently for Run 2 data, the "emulsion tape" shown in Fig. 3 was installed for that run. This moving tape, which has emulsion on both sides of a plastic base similar to the vernier sheet, has a lower track density because it moves through the beam at a more rapid rate.

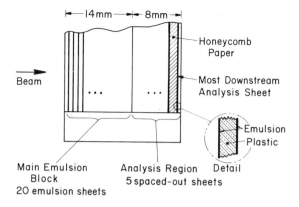

Fig. 7. Emulsion target module of the "vertical" type.

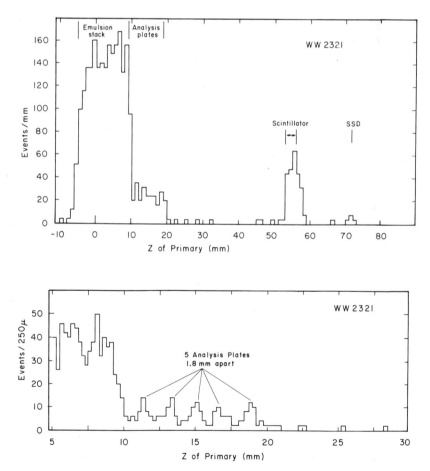

Fig. 8. Distribution along z (beam) direction of interactions, showing the main emulsion block and the low-density analysis plates. The lower portion of the figure is expanded in z by a factor 3.

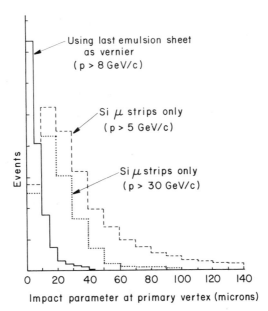

Fig. 9. Impact parameter of tracks at the primary vertex, as extrapolated from the most downstream emulsion sheet (solid histogram), and from the microstrip detectors (dashed and dotted curves).

At this writing, all the data from Run 1 has passed through the spectrometer track reconstruction program, and the emulsion physicists have begun production searching for charm and beauty. Twenty-eight charm events have so far been found in the 2% of the data scanned which has been scanned systematically. (A comparable number of charm events have also been found during the learning process.) Retooling of the spectrometer code for Run 2 is in progress, and the reconstruction pass for this data will begin early in 1988.

REFERENCES

1. E. L. Berger, Dynamics of Bottom Quark Production in Hadron Collisions, ANL-HEP-PR-87-90 (1987).

2. Alain Gauthier, "Charmed-Particle Lifetime Measurements and Limits for Neutrino Oscillations and the Existence of the Tau Neutrino," Ph.D. Thesis, The Ohio State University (1987), and references contained therein.

3. S. F. Krivatch, N. W. Reay, R. A. Sidwell and N. R. Stanton, Design and Performance of Large Multisampling Drift Chambers for Fermilab Experiment 653, in "Proceedings of the XXIII International Conference on High Energy Physics", Stuart C. Loken ed., World Scientific Co., Singapore (1987).

4. N. Usida et al., Experimental Details on Lifetime Measurements of Neutrino-Produced Particles in a Tagged Emulsion Spectrometer, Nucl. Instr. and Meth. 224:50 (1984).

HEAVY FLAVOR PRODUCTION

Edmond L. Berger

High Energy Physics Division
Argonne National Laboratory
Argonne, IL 60439

1. INTRODUCTION

Discovery and exploration of the properties of new particles including heavy quarks is a principal goal of high energy investigations being carried out at both fixed target and collider energies. The specification of reliable cross sections for heavy quarks, including their production spectra in longitudinal and transverse momentum, and comparisons with data test the quantum chromodynamic (QCD) mechanisms by which all heavy objects are expected to be produced. Strategies in the search for new flavors such as top are predicated on best estimates of cross sections and of momentum distributions in phase space not only of the new flavor but, perhaps more importantly, of lighter flavors which contribute deceptive backgrounds. Those considering hadronic experiments to establish flavor-antiflavor mixing and possible CP violation require a detailed understanding of expected production spectra and momentum correlations.

Addressed to these and related questions, presentations in the session on heavy flavor production included reports on

- hadroproduction of charm[1,2,3] and bottom[4] in fixed target experiments at CERN and Fermilab, including plans for new experiments[2,3,5-7] ,

- experimental investigations of the photoproduction[5,7] of charm, and

- charm and bottom production at collider energies, as well as lower limits on the mass of the top quark.[8]

Theoretical issues bearing directly on heavy flavor production were treated by D. Soper[9] and R. K. Ellis[10] as well as by J. D. Bjorken[11] and A. H. Mueller[12] in the workshop's opening and closing lectures.

My purposes in this paper are to pull together topics discussed during the session by showing direct comparisons where possible of data and theoretical results, and to summarize specific open questions deserving of further detailed theoretical and experimental investigation.

A significant result reported at the workshop was the completion[13] of a calculation of the heavy flavor production cross section through order α_s^3 in QCD perturbation theory. Here α_s^3 is the running coupling strength in QCD. In this paper I will present comparisons of the $O(\alpha_s^3)$ cross sections with data on hadroproduction of charm and bottom. As I will show, the $O(\alpha_s^3)$ contributions are larger in many cases of interest than the $O(\alpha_s^2)$ terms. Not yet available are $O(\alpha_s^3)$ distributions in rapidity and transverse momentum. These are eagerly awaited inasmuch as the $O(\alpha_s^3)$ $Q\bar{Q}$ jet contributions provide different event topologies[14-16] and may be very much larger than the $O(\alpha_s^2)$ contributions when $p_{T,Q} \gg m_Q$. These distributions are important, especially at collider energies, for a proper estimation of the bottom quark background to a possible top quark signal.

In Section 2, I provide a brief summary of the results of the $O(\alpha_s^3)$ computation. Comparisons with data on charm production and on bottom production are presented in Secs. 3 and 4. Comments on top quark production are made in Sec. 5, and conclusions are summarized in Sec. 6.

2. TOTAL CROSS SECTIONS

In hadron hadron interactions, the lowest order parton-parton subprocesses leading to production of a pair of heavy quarks are

$$q\bar{q} \rightarrow Q\bar{Q} \tag{1a}$$

and

$$gg \rightarrow Q\bar{Q}. \tag{1b}$$

The square of the invariant matrix element for these two-to-two subprocesses is proportional to α_s^2, where $\alpha_s = g^2/4\pi$ and g is the coupling strength in QCD. In QCD, α_s is a logarithmic function of the renormalization/evolution scale Q^2, which is only determined to be of order the mass m_Q of the heavy quark. Phenomenological applications of the lowest order subprocesses have been reviewed elsewhere[14-17], and I will not repeat that discussion here.

Additional subprocesses enter in the next order in the QCD perturbation expansion. These are:

$$
\begin{array}{lll}
q\bar{q} \rightarrow Q\bar{Q} & \alpha_s^2 \text{ and } \alpha_s^3 & (2a) \\
gg \rightarrow Q\bar{Q} & \alpha_s^2 \text{ and } \alpha_s^3 & (2b) \\
q\bar{q} \rightarrow Q\bar{Q}g & \alpha_s^3 & (2c) \\
gg \rightarrow Q\bar{Q}g & \alpha_s^3 & (2d) \\
gq \rightarrow Q\bar{Q}q & \alpha_s^3 & (2e) \\
g\bar{q} \rightarrow Q\bar{Q}\bar{q} & \alpha_s^3 & (2f)
\end{array}
$$

I list the two-to-two subprocesses in Eqs. (2a) and (2b) as a reminder that both virtual and real gluon emission amplitudes are present in the full calculation at order α_s^3.

The total cross section for $ab \rightarrow Q\bar{Q}X$, the inclusive production of a pair of heavy quarks, is

$$\sigma_{ab}(s) = \sum_{ij} \int dx_1 \int dx_2 f_i^a(x_1, Q^2) f_j^b(x_2, Q^2) \hat{\sigma}_{ij}(\hat{s}, Q^2). \tag{3}$$

502

In this equation, $f_i^a(x_1, Q^2)$ represents the density of partons of type i in incident hadron a, and Q^2 is the evolution scale; $\hat{s} = x_1 x_2 s$ is the square of the energy in the parton-parton collision. I use the same evolution scale in the parton-parton cross section $\hat{\sigma}_{ij}$ and in the parton density functions. As expressed in Ref. 13, $\hat{\sigma}_{ij}$ is written as

$$\hat{\sigma}_{ij}(\hat{s}, Q^2) = \frac{\alpha_s^2(Q^2)}{m_Q^2} F_{ij}\left(\rho, \frac{Q^2}{m_Q^2}\right), \tag{4}$$

where $\rho = 4m_Q^2/\hat{s}$, and m_Q is the mass of the heavy quark. The dimensionless functions F_{ij} are written in the form

$$F_{ij}\left(\rho, \frac{Q^2}{m_Q^2}\right) = F_{ij}^{(0)}(\rho) + 4\pi\alpha_s(Q^2)\left[F_{ij}^{(1)}(\rho) + \bar{F}_{ij}^{(1)}(\rho)\ln\frac{Q^2}{m_Q^2}\right] + O(\alpha_s^2). \tag{5}$$

Explicit expressions for the set of functions $F_{ij}^{(0)}$, $F_{ij}^{(1)}$, and $\bar{F}_{ij}^{(1)}$ may be found in Ref. 13. Two choices for $F_{ij}^{(1)}$ are given in Ref. 13; I adopt the "physical" choice appropriate to the case in which quark and antiquark densities are defined in terms of the structure function $F_2(x, Q^2)$ measured in deep inelastic lepton scattering.

The lowest order QCD contributions are represented by $F_{ij}^{(0)}(\rho)$. Note that $F_{qg}^{(0)}(\rho) = 0$ because there is no QCD subprocess in order α_s^2 in which an incident quark and gluon interact to produce a $Q\bar{Q}$ system. The qg contribution enters first at order α_s^3.

The contributions of order α_s^3 are represented by the functions $F_{ij}^{(1)}$ and $\bar{F}_{ij}^{(1)}$ in Eq. (5). These contributions are particularly significant in two regions of phase space: (i) the threshold region $\sqrt{\hat{s}} \sim 2m_Q$, and (ii) the large $\hat{s}$ region $\sqrt{\hat{s}} \gg 2m_Q$. The threshold region is important because there are large logarithmic terms proportional to $\ln^2(\sqrt{1-\rho})$ and $\ln(\sqrt{1-\rho})$. The large $\hat{s}$ region is important because the $\hat{s}$ dependence of the order α_s^3 two-to-three subprocess cross section $\hat{\sigma}_{2\to3}(\hat{s})$ differs dramatically from that of the order α_s^2 two-to-two subprocess cross section $\hat{\sigma}_{2\to2}(\hat{s})$. The higher order 2 to 3 subprocesses are mediated by gluon exchange in the crossed ($\hat{t}$) channel, implying that $\hat{\sigma}_{2\to3}$ approaches a constant as $\hat{s}$ increases, whereas the lowest order 2 to 2 subprocess is controlled by heavy quark exchange in the crossed channel, with $\hat{\sigma}_{2\to2} \to 1/\hat{s}$.

While the large contributions from the $O(\alpha_s^3)$ 2 to 3 subprocesses for $\sqrt{s} \gg 2m_Q$ may appear alarming, they do not *per se* cast doubts on the reliability or applicability of perturbation theory for heavy quark production. These large contributions at order α_s^3 arise from new subprocesses, with different $\hat{t}$ channel exchanges, not from simple "corrections" to the lowest order subprocesses.

For the reasons just described, the order α_s^3 contributions will be larger for bottom production than for top production at collider energies, and even greater for charm production.

Several sources of uncertainty beset attempts to make definite predictions. These include choice of the heavy quark mass; choice of parton densities (particularly the gluon density); and specification of the evolution scale Q^2. The last is an intrinsic theoretical uncertainty. In this report I will show results for different choices of Q^2, $Q^2 = m_Q^2$ and $Q^2 = 4m_Q^2$. When the cross section is computed to order α_s^3, changes in Q in the vicinity of m_Q result in "errors" of order α_s^4. As I will show, these differences are not always small.

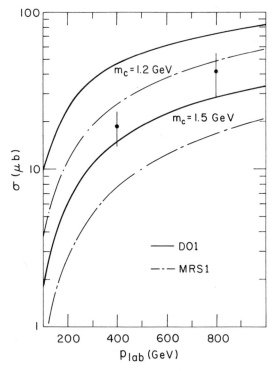

Fig. 1. Calculated cross sections for $pp \rightarrow c\bar{c}X$ as a function of the laboratory momentum of an incident proton for two choices of the charm quark mass m_c and two different sets of parton densities. I obtained these results from the full QCD expression[13] through order α_s^3. Duke-Owens set 1 and Martin-Roberts-Stirling set 1 parton densities were used, and the evolution scale Q^2 was chosen as $Q^2 = 4m_c^2$ in the parton densities, in Eq. (5), and in the evaluation of $\alpha_s(Q^2)$. The data at $p_{\text{lab}} = 400$ and 800 GeV are from the LEBC-EHS[18] and LEBC-MPS[19] experiments analyzed as described by A. Goshaw[1].

3. HADROPRODUCTION OF CHARM

In Figs. 1–3 I present calculations of cross sections for charm quark production in hadronic collisions at fixed target energies. I will comment below on results for the ISR energy of $\sqrt{s} = 63$ GeV.

The first point to be made is that the QCD contributions in order $O(\alpha_s^3)$ are large. This is illustrated in Fig. 3 where I show the ratio K of the full cross section computed through order α_s^3 with the result obtained in lowest order, order α_s^2;

$$K = \sigma\left(O(\alpha_s^2) + O(\alpha_s^3)\right) / \sigma\left(O(\alpha_s^2)\right). \qquad (6)$$

The curves in Fig. 3 suggest that values of K are typically 3 for charm at fixed target energies.

It should be emphasized that K is not a *constant*. Indeed K may be a strong function of the choice of evolution scale which I have arbitrarily chosen to be $Q^2 = 4m_c^2$. It has been argued that the best choice of Q^2 is obtained through an optimization procedure. I know of no general rule which guarantees that there exists such an optimal value in all cases of physical interest. Moreover, for charm it is difficult to explore the variation of K with evolution scale inasmuch as m_c itself is so small. For example, I would not decrease Q^2 to the "physical" choice $Q^2 = m_c^2$ since there are doubts as to whether any set of parton densities is meaningful at such small values of Q^2. Therefore I intend Fig. 3 to be illustrative; it should not be misunderstood as a demonstration that $K \simeq 3$ for charm at fixed target energies independently of considerations having to do with choices of Q^2 and of parton densities.

It has been known for some years that the lowest order calculations in QCD provide cross sections which are significantly below experimental measurements. The large increase provided by the $O(\alpha_s^3)$ contributions helps to remedy this discrepancy. At comparison is presented in Fig. 1 of calculations through order $O(\alpha_s^3)$ with data from $pp \rightarrow c\bar{c}X$. Before discussing the comparison I should first comment on the limited set of data selected in Fig. 1.

The data in Fig. 1 are from the LEBC-EHS[18] and LEBC-MPS[19] experiments. In these experiments the $D/\bar{D}$ inclusive and $D\bar{D}$ pair cross sections are measured, and a limit is placed on the associated production of $\Lambda_c\bar{D}$ and $\bar{\Lambda}_cD$. Beginning with these measurements, the Lund scheme for charm quark fragmentation to estimate the $\Lambda_c\bar{\Lambda}_c$ pair and $D_s/\bar{D}_s$ inclusive cross sections, and measurements of J/ψ production, Goshaw[1] obtained estimates of the cross sections for $pp \rightarrow c\bar{c}X$. His values are $\sigma(pp \rightarrow c\bar{c}X) = 14$ to 23 μb at $p_{lab} = 400$ GeV/c and 29 to 55 μb at $p_{lab} = 800$GeV/c. Obviously $D/\bar{D}$ and Λ_c cross sections have been determined in many other experiments. However, I choose not to show them in Fig. 1 for two reasons. First, many of those measurements were made with nuclear targets, and the precise nuclear A dependence is not known in each case. Second, I cannot work backwards in most cases to obtain the $c\bar{c}$ cross sections of interest.

A glance at Fig. 1 shows the considerable sensitivity of predictions to the choices of the charm quark mass m_c and of the parton densities. For a given set of parton densities, the results in Fig. 1 show that a decrease of the mass from $m_c = 1.5$ GeV to $m_c = 1.2$ GeV results in an increase in cross section by about a factor of three. For a given m_c, there is about a factor of two increase in predicted cross section when the

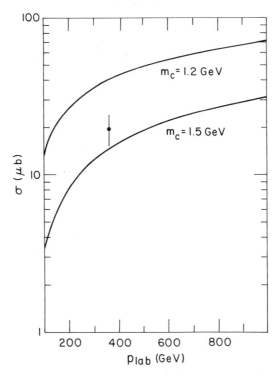

Fig. 2. Calculated cross sections for $\pi^- p \to c\bar{c}X$ as a function of the laboratory momentum of the incident π^- for two choices of the charm quark mass m_c. I obtained these results from the full QCD expression[13] through order α_s^3. The Duke-Owens set 1 parton densities were used for the proton and the Owens set for the pion. The evolution scale Q^2 was chosen as $Q^2 = 4m_c^2$ everywhere. The datum at $p_{lab} = 360$ GeV is from the LEBC-EHS experiment[23], reinterpreted as described in the text.

Duke-Owens[20] set 1 (DO 1) parton densities are used instead of the Martin-Roberts-Stirling[21] set 1 (MRS 1). Both sets have relatively "soft" gluon distributions, but they have different values of Λ. The value of Λ also enters in the evaluation of $\alpha_s(Q^2)$. Defining the ratio

$$r_\alpha \equiv \alpha_s(Q^2, \Lambda_{\text{DO 1}} = 200 \text{ MeV}) \Big/ \alpha_s(Q^2, \Lambda_{\text{MRS 1}} = 107 \text{ MeV}) , \qquad (7)$$

I find $r_\alpha^2 = 1.57$ if $Q^2 = 4m_c^2$ and $m_c = 1.2$ GeV; and $r_\alpha^2 = 1.52$ if $Q^2 = 4m_c^2$ and $m_c = 1.5$ GeV. Since the $c\bar{c}$ cross section is proportional to α_s^2, a substantial fraction of the difference of predicted yields may be attributed to the different values of Λ in the DO 1 and MRS 1 parametrizations.

The gluon distributions in both the Duke-Owens (DO) set 1 and the Martin-Roberts-Stirling (MRS) set 1 are parametrized at the starting value Q_0^2 such that $xg(x, Q_0^2) = $ constant as $x \to 0$. In the MRS set 3, by contrast, a very different behavior is assumed in which $xg(x, Q_0^2) \to 1/\sqrt{x}$ as $x \to 0$. Since the values of Q^2 chosen in Fig. 1 are not far above $Q^2 = Q_0^2$, it may be imagined that the shape and magnitude of the charm cross section would differ significantly if the MRS set 3 is used. This turns out not to be the case. For each of the mass values shown in Fig. 1, results obtained with the MRS set 3 lie roughly midway between the Duke Owens and MRS 1 curves.

Calculated charm cross sections at the ISR energies of $\sqrt{s} = 63$ GeV are listed in Table 1 for $Q^2 = 4m_c^2$.

Table 1. $\sigma(pp \to c\bar{c}X; \sqrt{s} = 63 \text{ GeV})$

Parton Densities	$m_c = 1.2$ GeV	$m_c = 1.5$ GeV
MRS 1 (soft glue, $\Lambda = 107$ MeV)	96 μb	39 μb
MRS 3 ($x^{-1/2}$ glue, $\Lambda = 178$ MeV)	130 μb	49 μb
DO 1 ($\Lambda = 200$ MeV)	122 μb	56 μb

Using the DO 1 ($m_c = 1.5$ GeV) and MRS 1 ($m_c = 1.2$ GeV) curves in Fig. 1 to bracket uncertainties from below and above, I estimate that $\sigma(pp \to c\bar{c}X; \sqrt{s} = 63 \text{ GeV})$ lies most likely in the range of 55 to 100 μb. Even with $O(\alpha_s^3)$ contributions included, it is difficult to accommodate a charm cross section greater than $\sim$130 μb at $\sqrt{s} = 63$ GeV.

Calculations for $\pi^- p \to c\bar{c}X$ are presented in Fig. 2. I used the Owens[22] parton densities for the pion, with $\Lambda = 200$ MeV. For consistency, nucleon parton densities having $\Lambda = 200$ MeV must also be used. The one datum in Fig. 2 is derived from the measurements[23] of the CERN-NA27 LEBC/EHS collaboration with a hydrogen target at 360 GeV/c. They report production cross sections of neutral and charged D mesons at $x_F > 0$ of $\sigma(D^0/\bar{D}^0) = 10.1 \pm 2.2\mu$b and $\sigma(D^\pm) = 5.7 \pm 1.5\mu$b, as well as an estimate of $\sigma(\Lambda_c/\bar{\Lambda}_c) = 4^{+5}_{-3}\mu$b for $x_F > 0$. Treating these numbers in a fashion similar to that described by Goshaw for pp interactions, and using $O(\alpha_s^2)$ calculations of the x_F distribution to estimate contributions from $x_F \leq 0$, I derive $\sigma(\pi^- p \to c\bar{c}X) \sim 15$ to 24μb for all x_F.

The results in Figs. 1 and 2 demonstrate that defensible QCD calculations reproduce the magnitude of the measured total charm cross section at fixed target energies. However, because of the sensitivity to the choice of the parton densities, we cannot use the results to "pin down" the charm quark mass appropriate in perturbative calculations to better than $1.2 < m_c < 1.5$ GeV. It does appear possible, however, to discard a mass as large as $m_c = 1.8$ GeV.

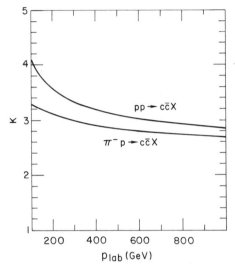

Fig. 3. Ratio K of the full cross section through order α_s^3 to that through order α_s^3 for charm production as a function of laboratory momentum. Here $m_c = 1.5$ GeV and $Q^2 = 4m_c^2$, the Duke-Owens set 1 parton densities were used for the nucleon and the Owens set for the pion.

The agreement between theory and data in Figs. 1 and 2 is an indication that charm production is on the way towards being "understood" in terms of perturbative QCD. There are several open issues, however. First, discomfort with the large size of the K factor shown in Fig. 3 should be addressed by a study of contributions at order α_s^4. Because of remarks I made earlier about new processes entering at order α_s^3, it is possible but not guaranteed that contributions in order α_s^4 will be small. Second, it is known that the experimental longitudinal momentum distribution of charm *particles*, $d\sigma/dx_F$, is not reproduced by perturbative calculations of charm *quark* production.[1] Final state interactions are required between the charm quark and either antiquark or "diquark" spectator subsystems. It has been argued[11,24] that these effects fall as a single power of the heavy quark mass, $\propto 1/m_Q$, but whether and how these final state effects affect the total cross section is not clear to me. Third, the nuclear dependence of charm production is not yet understood. Data[25-27] indicate that $\sigma \propto A^\alpha$, with $\alpha \simeq 0.75$ for $x_F \gtrsim 0.1$, whereas $\alpha \simeq 1$ would be expected in the naive parton model. The Fermilab E-537 collaboration[3] has provided detailed information on the x_F and

p_T variations of the A dependence observed in J/ψ production by π and $\bar{p}$ beams. Again, shadowing is observed in the sense that $\alpha < 1$ except for $p_T \gtrsim 1$ GeV. Can a quantitative relationship be established between the "shadowing" phenomenon observed in J/ψ and in charm production and the onset of shadowing now confirmed for $x \lesssim 0.2$ in deep inelastic lepton scattering?[29-28] Interpretation of the A dependence is not only of interest on its own merits, but it is also essential for establishing that the mechanisms for hadroproduction and photoproduction of the J/ψ and χ charmonium states are well understood. Unless the A dependence is explained quantitatively, it may not be possible to assert that the cross sections are determined by parton subprocesses with gluons in the initial state, even in measurements made with proton targets. This is important because data on J/ψ and χ production are potentially valuable sources of precise information on the gluon density distribution. Fourth, it is believed that higher-twist effects[30] are important in deep-inelastic lepton scattering for values of $Q^2 \simeq 4m_c^2$, accounting in the low Q^2 range for perhaps half of the observed Q^2 dependence. This suggests that additive higher twist contributions[24] should be present also in heavy flavor production, with a strength of order $(1 \text{ GeV}^2/m_Q^2)$ relative to the leading twist effects included in the perturbative calculations discussed above.

4. HADROPRODUCTION OF BOTTOM

In Figs. 4–7 I provide calculations of $b\bar{b}$ pair cross sections as a function of energy in $\bar{p}p$, $\pi^- p$, and pp interactions. As in the case of charm discussed in Sec. 3, the contributions in order α_s^3 are significant. This point is illustrated in Fig. 4 for $\bar{p}p \to b\bar{b}X$ at collider energies and in Fig. 7 for $\pi^- p \to b\bar{b}X$ and $pp \to b\bar{b}X$ at fixed target energies. In Fig. 7 the value of K in pp interactions is seen to be larger than that in $\pi^- p$ interactions. This is related to the more important role of gluon initiated subprocesses in pp interactions at fixed target energies.

In Fig. 5, I present results for the total $b\bar{b}$ pair cross section in $\pi^- p$ interactions at fixed target energies for a particular choice of bottom quark mass $m_b = 5$ GeV. The one datum on Fig. 5 is the measurement of the CERN WA78 collaboration discussed by G. Poulard.[4] Originally the WA78 group had published[31] a value of $\sigma(\pi^- N \to b\bar{b}X$; $p_{\text{lab}} = 320$ GeV$) = 4.5 \pm 1.4 \pm 1.4$ nb per nucleon. A subsequent reevaluation was made of their acceptance and efficiency based on a model which incorporates production properties in transverse and longitudinal momenta consistent with those predicted[14,32] in perturbative QCD. These improvements result in a reduction of the cross section[4] to $\sigma(\pi^- N \to b\bar{b}X$; $p_{\text{lab}} = 320$ GeV$) = 2.0 \pm 0.3 \pm 0.9$ nb per nucleon.

Another group[33] has reported observation of a signal consistent with bottom production and quotes a "model dependent $b\bar{b}$ production cross section" $\sigma(\pi^- N \to b\bar{b}X$; $p_{\text{lab}} = 286$ GeV$) = 14^{+7}_{-6}$ nb per nucleon, considerably larger than that of the CERN WA78 collaboration. The model adopted by the NA10 group to simulate $B\bar{B}$ production does not seem correct to me. In particular, their assumed form of the production spectrum in p_T yields b's with smaller values of transverse momentum than those expected[14-16] in QCD. Thus, the NA10 group has probably underestimated their efficiency and acceptance. Correspondingly, I imagine that the NA10 cross section will decrease by at least a factor of two after a reanalysis is done with a production model more consistent with theoretical expectations.

The calculations shown in Fig. 5 are appropriate for $\pi^- p \to b\bar{b}X$ whereas the one

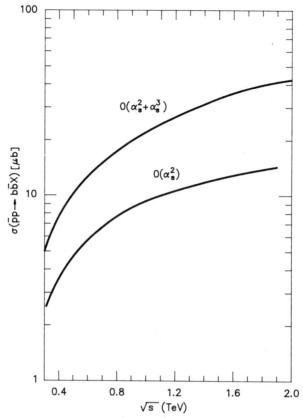

Fig. 4. Calculations of bottom quark production in proton-antiproton collisions as a function of $\sqrt{s}$. I show both the full answer through order α_s^3 and the lowest order $O(\alpha_s^2)$ result for $m_b = 5$ GeV and evolution scale $Q^2 = m_b^2$. The Duke-Owens set 1 parton densities were used.

datum is derived from interactions on a uranium target. The experimenters assumed a linear dependence, $\sigma_A \propto A\sigma_N$ to extract σ_N, the cross section per average nucleon. This procedure is probably fine for bottom quark production in spite of the fact that it is known not to work for charm. Nevertheless, if they are to be compared directly with the data, the theoretical results in Fig. 5 should be modified for the fact that the target is a mixture of neutrons and protons. This effect was studied in Ref. 16 where I showed that the cross section per average nucleon is *smaller* than the cross section for production from proton targets; the factors are 0.68, 0.80, and 0.87 at $p_{\text{lab}} = 200$, 400, and 600 GeV/c.

The choice of bottom quark mass is one uncertainty associated with the theoretical prediction. I examined this sensitivity before[14-16] for cross sections at order α_s^2. Similar sensitivity should be true also at order α_s^3. As reported previously[14-16], over the range $400 < p_{\text{lab}} < 1000$ GeV/c, the expected cross section $\sigma(pp \to b\bar{b}X)$ is decreased by about a factor of two when the b quark mass is increased from $m_b = 5$ GeV to 5.4 GeV and increased by about a factor of two if the b quark mass is decreased from 5 GeV to 4.6 GeV.

For $\pi^- p \to b\bar{b}X$ there is limited opportunity to examine sensitivity to the choice of parton densities. In the Owens set[22] for the pion, $\Lambda = 200$ MeV. For consistency, nucleon parton densities having $\Lambda = 200$ MeV should also be used. As stated earlier[14], calculations of $\sigma(\pi^- N \to b\bar{b}X)$ with the Eichten et al. (EHLQ) set 1 parton densities[34] differ by less than 5% from those done with the D0 1 set for 300 GeV/c $< p_{lab} <$ 1000 GeV.

Of greatest interest for cross sections computed through order α_s^3 is sensitivity to the choice of evolution scale Q^2. For $\pi^- p \to b\bar{b}X$, I present a comparison in Table 2 of results for the choices $Q^2 = m_b^2$ and $Q^2 = 4m_b^2$. Comparison with the datum in Fig. 5 favors the choice $Q^2 = 4m_b^2$ if $m_b = 5$ GeV.

No data is available on bottom quark production in pp collisions at fixed target energies, but the greater range of sets of parton densities allows broader exploration of the sensitivity of predictions to various parameters. I commented above on sensitivity to the choice of m_b. In Fig. 6 I show curves obtained with different choices of evolution scale. A tabulation of results is provided in Table 3. There is roughly a factor of 1.7 increase in expected cross sections when D0 1 instead of MRS 1 parton densities are used at $Q^2 = 4m_b^2$, and a further factor of two increase when Q^2 is dropped from $4m_b^2$ to m_b^2 in the D0 1 set. Evaluating the ratio r_α defined in Eq. (7) at $Q^2 = 4m_b^2$ with $m_b = 5$ GeV, I find $r_\alpha^2 = 1.35$. Thus, half of the increase in cross section at fixed $Q^2 = 4m_b^2$ is attributable to the different values of Λ in the D0 1 and MRS 1 parametrizations.

The range of values shown in Table 3 may be a useful guide to those estimating rates for future experiments. Calculated bottom quark cross sections at the ISR energy of $\sqrt{s} = 63$ GeV are listed in Table 4 for $m_b = 5$ GeV.

Table 2. Cross sections in nanobarns through order α_s^3 for $\pi^- p \to b\bar{b}X$; $m_b = 5$ GeV.

Evolution Scale	300 GeV/c	400 GeV/c	500 GeV/c	600 GeV/c
$Q^2 = m_b^2$	6	12	20	30
$Q^2 = 4m_b^2$	3.7	7.5	12.3	18.1

Table 3. Cross sections in nanobarns through order α_s^3 for $pp \to b\bar{b}X$; $m_b = 5$ GeV.

Parton Density; Evolution Scale	400 GeV/c	500 GeV/c	600 GeV/c	700 GeV/c	800 GeV/c
MRS 1; $Q^2 = 4m_b^2$	0.4	1.1	2.1	3.5	5.2
MRS 3; $Q^2 = 4m_b^2$	0.9	2.1	3.8	6.0	8.7
DO 1; $Q^2 = 4m_b^2$	0.7	1.8	3.6	6.1	9.4
DO 1; $Q^2 = m_b^2$	1.6	4.1	8.0	13	20

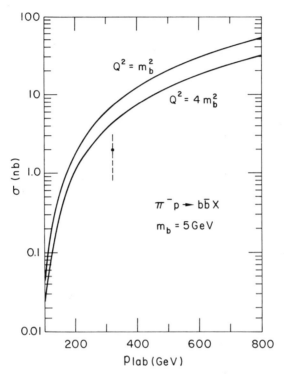

Fig. 5. Curves showing the full cross section through order α_s^3 for bottom quark production in $\pi^- p$ interactions as a function of laboratory momentum.
Here $m_b = 5$ GeV, and two choices are made for
the evolution scale, $Q^2 = m_b^2$ and $Q^2 = 4m_b^2$.
The one datum is the result reported by the
WA78[31] collaboration, as discussed by Poulard[4].
The solid vertical bar indicates statistical uncertainty, and the dashed line denotes the systematic uncertainty.

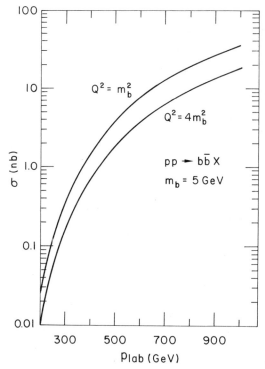

Fig. 6. Calculations of bottom quark production in proton-proton interactions through order α_s^3. Here $m_b = 5$ GeV, and two choices are made for the evolution scale, $Q^2 = m_b^2$ and $Q^2 = 4m_b^2$.

Table 4. $\sigma(pp \to b\bar{b}X;\ \sqrt{s} = 63$ GeV$)$.

Parton Densities	$Q^2 = 4m_b^2$	$Q^2 = m_b^2$
MRS 1	55 nb	85 nb
MRS 3 $(x^{-1/2}$ glue$)$	72 nb	118 nb
DO 1	100 nb	172 nb

Turning to collider energies, I show calculations of $\sigma(\bar{p}p \to b\bar{b}X)$ vs. $\sqrt{s}$ in Fig. 4. Presented are the cross sections integrated over all phase space in lowest order, $O(\alpha_s^2)$, and through order α_s^3, both for $Q^2 = m_b^2$. Results for different choices of parton densities are provided in Table 5. Included is a calculation at $\sqrt{s} = 300$ GeV appropriate perhaps for the RHIC facility at Brookhaven. For bottom production at $\sqrt{s} \gtrsim 300$ GeV, the pp and $\bar{p}p$ cross sections are nearly equal.

Table 5. Cross sections in microbarns through order α_s^3 for $\bar{p}p \to b\bar{b}X$; $m_b = 5$ GeV.

Parton Density; Evolution Scale	$\sqrt{s} =$ 300 GeV	$\sqrt{s} =$ 630 GeV	$\sqrt{s} =$ 1000 GeV	$\sqrt{s} =$ 1800 GeV
MRS 1; $Q^2 = 4m_b^2$	3.0	9.5	17.0	
MRS 3; $Q^2 = 4m_b^2$	3.5	12.5	25.1	
DO 1; $Q^2 = 4m_b^2$	4.4	12.8	22.6	43.7

One observation is evident regarding the numbers in Table 5. At $\sqrt{s} = 630$ GeV, there is not a large spread in predictions associated with the choice of different parton densities—the lowest and highest predictions shown in Table 5 are 9.5 μb and 12.8μb. This is a small effect when compared with the large ($\sim \times 3$) increase in predicted cross section in going from $O(\alpha_s^2)$ to $O(\alpha_s^3)$, as shown in Fig. 5. (No predictions for the MRS sets of parton densities are shown at $\sqrt{s} = 1.8$ TeV because integrations over all phase space for b production at $\sqrt{s} > 1$ TeV require values of $x < 10^{-4}$, which is below the range of applicability of the MRS and EHLQ parton densities.)

The only measurement of bottom production at $\bar{p}p$ collider energies is that reported by the UA1 collaboration[35,8] based on an analysis of dimuon production. The $b\bar{b}$ pair cross section is provided in a limited region of phase space:

$$\sigma(p\bar{p} \to b\bar{b}X; \ p_{T,b} > 5 \text{ GeV}, \ |\eta_b| < 2.0) = 1.1 \pm 0.1 \pm 0.4 \ \mu\text{b}. \tag{8}$$

The data appear to have been analyzed in such a way that this cross section excludes[8] at least some of the $Q\bar{Q}$ plus jet topologies associated with the $O(\alpha_s^3)$ contributions in QCD. As discussed elsewhere[36], the cross section in Eq. (8) agrees with $O(\alpha_s^2)$ predictions, but it appears to fall at the lower edge of the range of expectations.[14,15]

5. TOP QUARK

For a fixed value of $\sqrt{s}$, the contributions through $O(\alpha_s^3)$ in perturbative QCD result in smaller increases in predicted yields as the quark mass is increased. For example, at $\sqrt{s} = 630$ GeV, *typical K* factors[13] are in the range $1.2 \lesssim K \lesssim 1.7$ for a top quark of mass $m_t = 40$ GeV and $1.1 \lesssim K \lesssim 1.3$ for $m_t = 80$ GeV. At $\sqrt{s} = 1.8$ TeV, the numbers[13] are $1.3 \lesssim K \lesssim 1.8$ at $m_t = 40$ GeV and $1.2 \lesssim K \lesssim 1.7$ at $m_t = 80$ GeV.

Based on a detailed analysis of events in which muons are observed at large transverse momentum in coincidence with hadronic jets, the UA1 collaboration[8] derived a lower limit of 44 GeV at 95% confidence level for the mass of the top quark. There are at least two reasons that numbers quoted above cannot be used in a straight-forward way to derive a lower bound on the mass of the top quark which is more stringent. First, contributions in order $O(\alpha_s^3)$ were already included in an approximate way in the UA1 analysis through the UA1 simulation of $O(\alpha_s^3)$ "gluon splitting" and "flavor excitation" contributions. Second, and more important, a better determination of the bound from the data requires that the correct $O(\alpha_s^3)$ distributions in rapidity and transverse momentum be used in the simulation programs both for top production and for the backgrounds from bottom and charm production.

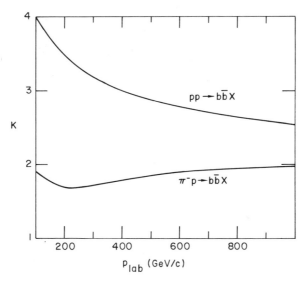

Fig. 7. Ratio K of the full cross section through order α_s^3 to that through order α_s^2 for bottom quark production. Here $m_b = 5$ GeV and $Q^2 = m_b^2$.

6. CONCLUSIONS AND DISCUSSION

Heavy flavor production has been advanced to a new level of precision in perturbative QCD now that full next-to-leading order calculations have been completed. The contributions in order α_s^3 are large in many cases of practical interest. These calculations offer the potential of better agreement with data on charm quark production, but they also raise questions about convergence of the expansion of the cross section in perturbative QCD. No doubt these questions will be addressed over the coming months. The new level of theoretical achievement is matched, if not exceeded, by the remarkable developments in experimental techniques, as reported in detail during this session.[1-7,37] The convergence of these developments promises substantial progress in the coming years in our quantitative understanding of the basic mechanisms responsible for heavy flavor production.

Applicability of the perturbative results requires that the mass of the heavy flavor be "large". Whether charm, with $1.2 < m_c < 1.5$ GeV, satisfies this restriction is not clear. Issues include a quantitative understanding of the nuclear A dependence of charm production and quantitative estimates of the role of (additive) higher twist terms, proportional to 1 GeV$/m_Q$ and expected to be substantial for charm production. The size of the $O(\alpha_s^3)$ contributions introduces a new condition of applicability of the perturbative results. The contributions are particularly significant when the integration over phase space is dominated by values of the parton-parton subenergy *either* close to threshold, $\hat{s} \sim 4m_Q^2$, or very large, $\hat{s} \gg 4m_Q^2$. Correspondingly, predictions will be most stable in restricted intervals of $2m_Q/\sqrt{s}$. According to these criteria, predictions for top quark production at the current collider energies of $\sqrt{s} = 630$ GeV and $\sqrt{s} = 1.8$ TeV would seem particularly reliable, as would those for bottom quark production at Fermilab fixed target energies. It is desirable to develop an understanding and/or techniques which will lead to confidence in calculations extended into the regions $\sqrt{s} \gg 2m_Q$ and/or $\sqrt{s} \simeq 2m_Q$.

If the reservations discussed above are set aside, the calculations through order α_s^3 may be compared with data on charm production at fixed target energies, as was done in Sec. 3. As shown, good agreement is obtained for values of m_c in the range $1.2 < m_c < 1.5$ GeV. Uncertainties include the choice of $\Lambda_{\rm QCD}$, the magnitude and shape of the gluon density $g(x, Q^2)$ in the relevant interval of x, and the choice of the evolution scale Q^2 in both α_s and the parton densities. Only much more precise data over a broad range of energies will permit tighter bounds on the "parameter set" (m_c, Q^2, Λ, and $g(x, Q^2)$).

The importance of the gluon density distribution $g(x, Q^2)$ was also emphasized in the session on prompt photon production. Predictions depend most sensitively on assumptions made about the magnitude of $g(x, Q^2)$ in a limited range of x probed in a particular experiment. This range is centered near $x = 2p_T/\sqrt{s}$ for prompt photon production and near $x \simeq 2m_Q/\sqrt{s}$ for heavy flavor production. In spite of this, results of fits are often reported in terms of global properties of an assumed parametrization for the gluon density, such as the values of α, β, and c is a fit such as $xg(x) = x^{-\alpha}(1 + cx)(1 - x)^\beta$ (which are, in turn, related by a normalization constraint on the integral $\int dx\, xg(x)$). This practice is unfortunate since rarely are experiments sensitive either to the very small x behavior represented by α or to the very large x behavior controlled by β. It would be preferable if results of fits to data could be stated in terms of *values* of the function $g(x, Q^2)$ in the specified interval of x probed by the experiments in question. Since next-to-leading order calculations have been done both for prompt photon production and for heavy flavor production it will be instructive to compare the gluon densities extracted from the separate sets of data.

"Diffractive production" of heavy flavors has been studied both theoretically[9,38] and experimentally.[39] By "diffractive", I mean the presence of a large gap in rapidity ($\Delta y \gtrsim 2$) between a quasi-elastically scattered p or $\bar{p}$ and the rest of the final state particles: $\bar{p}p \rightarrow \bar{p} + $ gap $ + X$. In diffractive hard scattering, one is interested in the production of a $Q\bar{Q}$ subsystem or jets with large p_T in the system X: $\bar{p}p \rightarrow \bar{p} + $ gap $ + Q\bar{Q}X$. Such events can be analyzed in terms of the hypothesized parton substructure of the Pomeron exchange. Defining a gluon density distribution of the Pomeron, one is interested in measuring at least three properties: the behavior of $g_P(x)$ for $x \rightarrow 1$; its behavior as $x \rightarrow 0$; and its normalization, $\int xg_P(x)dx \simeq 1$? Precise data on the rapidity distribution of either heavy flavors or jets will be required to address these three questions.

Measurements of correlations in rapidity between a produced Q and $\bar{Q}$ will provide valuable additional checks on production dynamics.[32] They may be particularly valuable in the case of charm production for separating perturbative effects from those associated with final state interactions.

ACKNOWLEDGMENTS

I have benefitted from valuable discussions with R. K. Ellis. Research done at Argonne National Laboratory is supported by the Department of Energy, Division of High Energy Physics, Contract W-31-109-ENG-38. Part of this work was done at the Institute for Theoretical Physics, Santa Barbara, California, where research is supported in part by the National Science Foundation under Grant No. PHY82-17853, supplemented by funds from the National Aeronautics and Space Administration.

REFERENCES

1. A. T. Goshaw, "Charm Production from 400 and 800 GeV/c Proton-Proton Collisions", these proceedings.

2. N. R. Stanton, "Charm and Beauty Decays Via Hadronic Production in a Hybrid Emulsion Spectrometer (Fermilab E653)", these proceedings.

3. S. Conetti, "Dimuon Experiments at the Fermilab High Intensity Laboratory", these proceedings.

4. G. Poulard, "Experimental Study of $B\bar{B}$ Hadroproduction in the WA75 and WA78 Experiments", these proceedings.

5. J. A. Appel, "QCD: Photo/Hadroproduction of Heavy Flavors; Fermilab E691, E769 and Beyond", these proceedings.

6. K. Miyake, "Production of Particles with High Transverse Momenta in 800 GeV/c Proton-Nucleus Collisions, E605-FNAL", these proceedings.

7. D. Treille, "Charm Photoproduction and Lifetimes from the NA 14/2 Experiment", these proceedings.

8. N. Ellis, "Heavy-Flavor Production in UA1", these proceedings.

9. D. E. Soper, "Heavy Quark Production in Hadron Collisions", these proceedings.

10. R. K. Ellis, "Heavy Quark Production in QCD", these proceedings.

11. J. D. Bjorken, "QCD: Hard Collisions are Easy and Soft Collisions are Hard", these proceedings.

12. A. H. Mueller, "Hard Processes in QCD", these proceedings.

13. P. Nason, S. Dawson, and R. K. Ellis, Fermilab report FERMILAB-Pub-87/222-T (Dec. '87), to be published in Nucl. Phys. B.

14. E.L. Berger, Nucl. Phys. B (Proc. Suppl.)1B: 425 (1988).

15. E. L. Berger, "Benchmark Cross Sections for Bottom Quark Production", Argonne report ANL-HEP-CP-87-121, Proceedings of the Fermilab Workshop on High Sensitivity Beauty Physics, Nov., 1987.

16. E. L. Berger, in *Hadrons, Quarks, and Gluons*, Proceedings of the XXII Rencontre de Moriond, Les Arcs, France, edited by J. Tran Thanh Van (Editions Frontières, France, 1987) pp. 3–40.

17. R. K. Ellis and C. Quigg, Fermilab report FN-445/2013.000 (1987).

18. M. Aguílar-Benitez *et al.*, Phys. Lett. 135B: 237 (1984), 189B: 476 (1987), and Z. Phys. C. to be published.

19. R. Ammar *et al.*, Phys. Lett. 183B: 110 (1987).

20. D. W. Duke and J. F. Owens, Phys. Rev. D30: 49 (1984).

21. A. D. Martin, R. G. Roberts, and W. J. Stirling, Phys. Rev. D37: 1161 (1988).

22. J. F. Owens, Phys. Rev. D30: 943 (1984).

23. M. Aguilar-Benitez *et al.*, Z. Phys. C31: 491 (1986).

24. S. J. Brodsky, J. F. Gunion, and D. E. Soper, Phys. Rev. D36: 2710 (1987).

25. H. Cobbaert *et al.*, CERN report CERN-EP/88-34 (March, 1988), submitted to Phys. Lett. B.

26. H. Cobbaert *et al.*, Phys. Lett. B191: 456 (1987).

27. M. E. Duffy *et al.*, Phys. Rev. Lett. 55: 1816 (1985).

28. J. Ashman *et al.*, Phys. Lett. B202: 603 (1988).

29. E. L. Berger and J.Qiu, Phys. Lett. B206: 141 (1988).

30. L. F. Abbott, E. L. Berger, R. Blankenbecler, and G. L. Kane, Phys. Lett. 88B: 157 (1979); L. F. Abbott, W. B. Atwood, and R. M. Barnett, Phys. Rev. D22: 582 (1980).

31. M. G. Catanesi *et al.*, Phys. Lett. 187B: 431 (1987).

32. E. L. Berger, Phys. Rev. D37: 1810 (1988).

33. P. Bordalo *et al.*, CERN report CERN-EP/88-39 (March, 1988), submitted to Z. Phys. C.

34. E. Eichten, I. Hinchliffe, K. Lane, and C. Quigg, Rev. Mod. Phys. 56: 579 (1984).

35. C. Albajar *et al.*, Phys. Lett. 186B: 247 (1987).

36. E. L. Berger, J. C. Collins, and D. E. Soper, Phys. Rev. D35: 2272 (1987).

37. See also P. H. Garbincius, Fermilab report FERMILAB-Conf-88/39, Proceedings of the Workshop on High Sensitivity Beauty Physics at Fermilab, Nov., 1987.

38. E. L. Berger, J. C. Collins, D. Soper, and G. Sterman Nucl. Phys. B286: 704 (1987).

39. K. Eggert, UA1 Collaboration, private communications.

HARD PROCESSES IN QCD[*]

A.H. Mueller

Physics Department
Columbia University
New York, New York 10027

INTRODUCTION

I shall try and restrict the topics which are covered in this summary to those topics on which there has been some significant discussion at this Workshop. Thus, this is in no sense a review of hard processes in QCD.

Much has happened in the last few years as concerns hard processes. Data on charm production at fixed target energies has improved considerably. We now have good values for the total charm production cross section and the theoretical debate is now turning to whether or not QCD should be applicable to this production. In Sec.2, this is discussed assuming QCD is a good approximation for processes as hard as charm production, at least for the total charm cross section, even though it is difficult to make precise predictions at fixed target energies because of a strong dependence on the value chosen for the charm mass. Indeed, the cross section measurements give a very interesting determination of that mass.

Progress in direct photon production has been enormous and much more is yet to come. The theoretical analysis includes a complete higher order calculation. Probably, nowhere in QCD phenomenology is the situation in such good shape. This has given us the luxury to focus on questions of what are the best ways to determine the running coupling and factorization scales. "Optimized" perturbation theory identifies the higher order correction as being small, of size α/π. Questions relating to direct photons are discussed in Sec.3.

In Sec.4, a few brief comments are given on μ-pair production. Overall, the situation has been rather satisfactory for quite some time. However, there does appear to be some difficulty with the angular dependence of the cross section.

Production of high $p_\perp$ jets, up to 150 GeV or so, tests QCD at very high energies. The 3-jet to 2-jet cross section is now well measured, but determinations of Λ are difficult because of an unknown K-factor in the three-jet cross section. Deviations of the angular dependence of the two jet cross section from the Rutherford formula test the running of α

---

[*]Summary talk given at NATO Research Workshop on QCD Hard Hadronic Processes, Oct. 8-13, 1987 in St. Croix.

but do not yet give a good Λ-determination. Production of jets when combined with a knowledge of the quark distribution in the proton allow a determination of the gluon distribution of the proton. It will be very interesting to see what $xG(x,p_\perp^2)$ looks like at very small x-values. Four jet events have been analyzed both at the ISR and at the CERN Collider. Double quark scatterings appear to have been seen at the ISR while the dominant mechanism for four jets at the collider seems to be a single hard scattering, at least in the $p_\perp$ range examined. Sec.4 concludes with a brief discussion of an inclusive approach to multi-jet phenomena.

In Sec.5, some topics relating to A-dependence are covered. The physical pictures underlying shadowing in deeply inelastic scattering are reviewed. The conclusion that A^{eff}/A depends only weakly on Q^2, for small x, is emphasized. The common claim that the A-dependence of J/ψ production determines the J/ψ cross section on nucleons, $\sigma(J/\psi,N)$ is disputed. Rather, it is claimed, the J/ψ is only formed after the $C\bar{C}$ pair leave the nucleus, and that while in the nucleus the $C\bar{C}$ pair are very compact, in transverse dimensions, and so interact much more weakly than would a physical J/ψ. Finally, the A-dependence of jet production is discussed briefly.

I have not attempted to cover all the topics discussed at this workshop. Many of these topics have been included in Bjorken's talk. In other cases, I am simply not expert enough and the topic may have been reviewed in detail in one of the theoretical overviews by a true expert, as for example in the case of sum rules relating to the spin content of the proton.

1. Heavy Quark Production

There is rapid progress being made, both theoretically and experimentally, in understanding heavy quark production. In particular, data from NA 27 and E743[1-3] at CERN and Fermilab now give solid values on $D\bar{D}$ production rates and good limits on $D\bar{\Lambda}_c + \bar{D}\Lambda_c$ production thus allowing reasonable estimates of charm production cross sections at fixed target energies. The estimate given by Goshaw[2] is $\sigma(pp \to C\bar{C} + x) = (21 \pm 5)\mu b$ at $\sqrt{s}$ = 27 GeV. This is a very important finding as it firmly establishes charm production cross sections in the microbarn rather than in the millibarn range as had been suggested earlier at the ISR. As one increases the center of mass energy from $\sqrt{s} \approx 27$ to $\sqrt{s} \approx 60$ one may expect a gentle rise in σ, perhaps by a factor of about 3, but still far below the millibarn region. Data from E 743 suggests such a slow rise.

Now what can one say on theoretical grounds? The dominant contribution to heavy quark production in QCD[4,5] is given by

$$\sigma = \int dx_1 \, dx_2 \, G(x_1,M^2) \, G(x_2,M^2) \, \sigma_{gg \to Q\bar{Q}} \tag{1}$$

with $\sigma_{gg \to Q\bar{Q}}$ the order α^2 cross section for gluons having momentum fractions x_1 and x_2 to produce a $Q\bar{Q}$ pair. In (1) M is the mass of the heavy quark and $G(x,M^2)$ is the gluon distribution, for the hadronic of interest, at momentum fraction x and virtuality M^2. Thus, the dominant cross section is of size $\alpha^2(M)/M^2$. Corrections to (1) are of two distinct types. (i) There are higher order perturbative corrections to (1) of which the leading correction is of size $\alpha^3(M)/M^2$. These leading corrections have now been calculated by Dawson, Ellis, and Nason.[6,7] A preliminary conclusion of this calculation is that the "K-factor" may be something like a factor of 2.[8] This is reassuring as it had been feared that the process $gg \to g\,g^* \to g + Q + \bar{Q}$ might give a very big contribution because of the large 90° elastic gg cross section. (ii) There are also higher twist effects giving corrections of higher powers in Λ/M. The leading twist term, given in (1), exhibits factorization explicitly. If there is factorization at the next-to-leading level in twist then the power corrections should be of size $(\Lambda/M)^2$ compared to the leading twist terms. Brodsky, Gunion, and Soper[9,10] have carried out a calculation in a nonrelativistic approximation and find a term of size Λ/M. It is very important to do this calculation more completely to see if factorization is valid. My guess would be that the first power corrections are of size $(\Lambda/M)^2$ in which case it is very natural to expect such corrections to be small, even when M is the charm quark mass.

Thus, it is worthwhile trying to confront (1) with the experimental data. (Of course, it is even better to include the leading QCD corrections and we may expect a detailed discussion to appear soon.) However, one quickly realizes that QCD cannot yet be used to make a real prediction. The problem is that the cross section given by (1) is exceedingly dependent on the value of the charm quark mass used in the QCD calculation,[11] at least at fixed target and ISR energies. As one varies the charm quark mass between 1.2 GeV and 1.8 GeV, the total cross section decreases by almost an order of magnitude.[11] This rapid variation is due in large part to the strong x-dependence of the Gluon structure functions in the region in question. This severe mass dependence of the cross section goes away at collider energies.

If one takes the charm production cross section to be about 20 µb at 400 GeV laboratory momentum, then (1) is about a factor of 1.5 to 2.0 low for a charm quark mass $M_c = 1.2$ GeV. Since the higher order corrections are positive this is all very reasonable and it may well be true that perturbative QCD is reliable in this circumstance. Let us take this point of view, that perturbative QCD is reliable for charm production. Then what we learn by comparing (1) to the experimental cross section is that the charm quark mass is about 1.2 GeV.

Now what does it mean to say $M_c = 1.2$ GeV? Is this a bare mass or a constituent mass? Though the concept of constituent mass is not unambiguous it always it always involves a nonperturbative condensate.[12-15] In the present process the heavy quarks can be viewed as off-shell by the amount proportional to M and so the nonperturbative parameters $\langle \alpha F^2 \rangle$ and $\langle m \bar{\psi}\psi \rangle$, for the gluon and light quarks, must occur in the ratios $\langle \alpha F^2 \rangle / M^4$, $\langle m \bar{\psi}\psi \rangle / M^4$ which correspond to very small corrections. (For example, if the heavy quark propagator is written as $iS^{-1}(p) = A(p^2)\,\gamma \cdot p - B(p^2)$ and the running mass, $M(\mu)$, is defined by

$$M(\mu) = \frac{B(-\mu^2)}{A(-\mu^2)} , \tag{2}$$

call $M_0(\mu)$ the perturbative contribution to M and $M_1(\mu)$ the part involving condensates. Then it is easy to see, in this Euclidean definition, that

$$M_1(\mu) = M_0(\mu)\left[1 + c_1 \frac{\langle \alpha F^2 \rangle}{M_0^4} + c_2\, \alpha(M_0) \frac{\langle \tilde{\psi}\psi \rangle}{M_0^4}\right]. \tag{3}$$

Of course, when p^2 is in the Minkowski region and $B(p^2)/A(p^2)$ is near M_0 nonperturbative condensate corrections can be large. This means that in charmonium calculations, where $\Lambda/E_{binding}$ is not small the bare and "effective" constituent masses may be very different.)

Of course, the bare mass is also prescription dependent. In the present application, one is defining the mass, M, by the pole position in the heavy quark propagator. Thus, the value $M_c \approx 1.2$ GeV seems a little lower than the mass obtained using QCD sum rules. The sum rule mass[15] corresponding to the mass definition used here is $M_c = (1.46 \pm 0.05)$ GeV.

I take this discrepancy to be real, subject to a better determination of the mass from charm production using the higher order corrections of Ref.6, and interesting. Thus, although it is impossible to make a unique prediction of charm production using only parameters determined in high energy reactions, it is possible to determine a charm quark mass which can be directly compared to sum rule determinations of the same mass. We are also able to see that potential models may require a significantly different charm mass because such calculations involve heavy quark propagation not very far from mass-shell.

In addition to measuring charm cross sections NA27 and E743 also determine the spectrum of D's in $\pi^- P$ and PP collisions at CERN and Fermilab.[2] The issue here is the x_F dependence of the observed D in a parametrization $d\sigma/dx_F \propto (1-x_F)^n$. In $\pi^- P$ reactions large x D's are predominately D^0 and D^-'s as expected if a $C(\bar{C})$ picks up a $\bar{u}(d)$ valence quark from the π^-. However, in PP reactions the D^- and $\bar{D}^0$ spectrum is found to be softer than the D^+, D^0 spectrum, the opposite from what one would expect for single quark pickup. If this result holds up it is clearly very interesting. However, the spectrum analysis is based on very few events and I think the result is not yet entirely convincing.

Brodsky, Gunion and Soper[9,10] are involved in an interesting program of building a "QCD inspired" model which takes into account nonperturbative final states interactions in an approximate way. A $Q\bar{Q}$ heavy quark pair certainly has strong interactions with other virtual quarks and gluons which are present at the time of production. For very heavy quarks these final state interactions do not modify total cross sections, but they can distort the final states. However, the amount of distortion of the heavy hadron spectrum is limited because the momentum transfers involved can only be comparable to those of soft hadron physics, on the order of 400 MeV or so, before power suppressions become important. Brodsky et al. find the possibility of strong enhancements when a heavy quark has a small relative velocity compared to a light quark and when the light and heavy quarks are in an attractive channel.

2. Direct Photons

During the course of the workshop it has become quite clear that direct photon production has come of age.[16,17] We have seen a wide variety of data presented covering the $p_\perp \approx 3-12\,\text{GeV}$ range from fixed target and ISR experiments up to the $p_\perp \approx 50\,\text{GeV}$ range from the CERN Collider. It also now seems clear that a measurement of low mass high $p_\perp$ $\mu^+\mu^-$ pairs at collider energies should furnish a very good way of getting precise data on high $p_\perp$ direct photons.[13] Thus, we may expect data on direct photons to become very good over the next few years in a wide range of $p_\perp$ and $x_\perp$. This should allow an excellent determination of gluon distributions for both π's and protons, and we may expect direct photon analysis to be a major testing ground of QCD.

Indeed, already the complete higher order corrections for direct photon production have been done and used in fits to the large sample of existing data.[17,19] Presently, there is excellent agreement between existing data and the theoretical calculations performed by Aurenche et al. and the success of these fits using Duke-Owens set 1 distributions is strong evidence for a soft gluon distribution. The reader can find detailed discussions of this phenomenology in the many reports on individual experiments as well as in the overviews by Baier[20] and Owens[21] in these proceedings. I would like to focus attention here on some of the subtle, and not yet understood, issues raise by the work of Aurence et al.,[17,19] especially in regard to the fixing of the various scales appearing in the calculation.

To illustrate the issues involved consider the (schematic) picture of a higher order correction to direct photon production shown in Fig.1. The direct photon is assumed to have a fixed transverse momentum, $p_\perp$, and we suppose that this momentum is large enough for perturbative techniques to apply. The gluon labelled by g is assumed to be hard, having a transverse momentum on the order of $p_\perp$, and we suppose that this gluon is always associated with the "hard part," when factorization is done. The gluon labelled by k is also assumed to have a momentum on the order of $p_\perp$ and, after factorization, it partially appears in the "hard part" and partially as a higher order correction to the structure function. We may write the cross section as

$$E \frac{d\sigma}{d^3p} = \frac{1}{p_\perp^4} \int dx_L \, dx_R \, q(x_L,M) \, H \, (x_L,x_R, \, p_\perp^2/M^2, \, p_\perp^2/\mu^2, \, \alpha(\mu)) \, \bar{q}(x_R,M) \quad (3)$$

where q and $\bar{q}$ are the quark and antiquark distributions, with a term where $q \leftrightarrow \bar{q}$ understood, and where we neglect the Compton process for simplicity of discussion. μ is the scale at which we choose to define the coupling in the hard process while M reflects our choice in separating what we call the "hard part" of the process from structure function evolution. Roughly speaking, $k_\perp < M$ regions are included in q and $\bar{q}$ while $k_\perp > M$ regions are included in the hard part, H, of the factorization expression. In order to make the factorization useful M and μ should be chosen to be of the same order of magnitude as $p_\perp$, but they are otherwise arbitrary. We suppose that a particular renormalization scheme, say $\overline{MS}$, has been chosen in defining q and $\bar{q}$. The $b \ln p_\perp^2/\mu^2$ dependence in H reflects the fact that the running coupling depends on $p_\perp$, with $\alpha(\mu)$ defined as $\alpha(\mu) \sim \frac{1}{b \ln \mu^2/\Lambda^2} + \ldots$, while the $\ln p_\perp^2/M^2$ in H reflects the fact the designation of collinear logarithms to lie below the scale M is arbitrary, and that if M is chosen too small such logarithms become important in H.

Now if the coupling did not run there would be no M dependence in (3) even if the expansion in H, in terms of $\alpha(\mu)$, were terminated at a finite order. However, because of the running of α this is not the case. The problem is the following: the contribution of the line k in Fig.1 occurs with a factor $\alpha(\mu)$ when it appears in H while it appears with a running coupling, in an anomalous dimension term, when it is put in q. That is, in taking the derivative, with respect to M, in (3) one obtains a factor

$$M^2 \, \frac{d}{dM^2} \, (\alpha(\mu) \, \ln \frac{p_\perp^2}{M^2} \, + \int^{M^2} \frac{d\lambda^2}{\lambda^2} \, \alpha(\lambda)), \quad (4)$$

the first term coming from H and the second term coming from q(or $\bar{q}$). If α were constant the expression in (4) would be zero, however, because of the running of the coupling one obtains $\alpha(M) - \alpha(\mu)$ which is not zero, but which is of higher order in α if M and μ are of the same order. Thus, in an evaluation of $d\sigma/d^3p$, at the level if next-to-leading logarithms, the dependence on M only occurs at the next-to-next-to-leading terms, terms which have not been systematically included in (3).

As far as the μ dependence of $d\sigma/d^3p$, it is all contained in H. Suppose we write H as an asymptotic expansion in $\alpha(\mu)$ through order α^2. Then one finds[17]

$$H(\ln p_\perp^2 / M^2, \ln p_\perp^2/\mu^2, \alpha(\mu))$$

$$= H^{(1)}(\alpha(\mu) - b\alpha^2(\mu)\ln p_\perp^2/\mu^2) + H^{(2)} (\ln p_\perp^2/M^2)\alpha^2(\mu), \quad (5)$$

523

where we have suppressed dependence on x_L and x_R in (5). Using

$$\mu^2 \frac{\partial}{\partial\mu^2} \; \alpha(\mu) = - \, b \; \alpha^2(1 + c\alpha \,) \tag{6}$$

with

$$b = \frac{33-2N_f}{12\pi} \, , \; c = \frac{153 - 19N_f}{(66-4N_f) \; \pi} \, , \tag{7}$$

one sees immediately that μ dependence in H occurs only at order α^3, a level which is not systematically included in our discussion.

Thus, if higher order corrections have been done through next-to-leading level in the factorized expressions, (3), M and μ dependence occurs only at the next-to-next-to leading order. In general, this is the best we can do without invoking new principles or rules. If α is truly small the ambiguity due to the arbitrariness of M and μ choices should also be small. However, in practice, the dependence on M and μ is not necessarily small, since intermediate values of α occur. Thus, one must choose values for M and μ in order to get definite predictions. Except for the requirement that M and μ be of order $p_\perp$ there is no fundamental requirement limiting that choice. Some time ago, Stevenson[22] and Politzer[23] suggested that a good choice of M and μ is one where predictions are stable with respect to small variations about the particular values chosen. Aurenche et al.[17] follow this procedure, called optimized perturbation theory, and demand that the $d\sigma/d^3p$ of (3), calculated through next-to-leading corrections, obey

$$\mu^2 \frac{\partial}{\partial\mu^2} \; \frac{d\sigma}{d^3p} \; = M^2 \frac{\partial}{\partial M^2} \; \frac{d\sigma}{d^3p} \; = 0. \tag{8}$$

These two equations should then be used to calculate μ and M. Let us see how this works in our simplified factorization given in (3).

Begin by requiring μ-independence. Since μ-dependence only occurs in H we require

$$\mu^2 \; \frac{\partial}{\partial\mu^2} \; H = 0. \tag{9}$$

Using (6) and (5) we find

$$0 = H^{(1)}(-bc\alpha^3 + 2b^2\alpha^3 \; \ell n \; p_\perp^2 /\mu^2) - 2b\alpha^3 \; H^{(2)} \tag{10}$$

or

$$H^{(2)} = H^{(1)}(- \frac{c}{2} + b \; \ell n \; p_\perp^2 /\mu^2) \tag{11}$$

which equation is to be used to determine μ as a function of M. (We do not discuss here the role of the x-variables. For a detailed description of what is exactly done see Ref.17.) We suppose Λ to have been determined from other processes and the discussion is always in the framework of a given renormalization scheme, say $\overline{MS}$.

Substituting (11) into (5) already gives the interesting result

$$H = H^{(1)}(\alpha - \frac{1}{2} \; c\alpha^2) \tag{12}$$

with $\alpha = \alpha(\mu)$ satisfying (11). Thus, in the end the higher order calculation of the hard part, $H^{(2)}$, is used only to determine μ. The ratio of the higher order to lowest order terms in H is $- \, 1/2 \; c\alpha$, a small number. Finally, using (12) in (3) the value of M is determined numerically to satisfy $M^2 \; d/dM^2 \; d\sigma/d^3p = 0$.

Now that μ and M are determined (12) can be used in (5) to confront theory with experiment. General conclusions reached by Aurenche et al.[17] are: (i) Rather good fits to all available data are obtained. (ii) The region of stability in M is rather large about the optimized result so long as μ is determined by (11). (iii) K-factors are small suggesting that optimized perturbation theory is a very efficient way to do the asymptotic expansion in α. (iv) A rather soft gluon distribution, like Duke-Owens set 1, is favored in the fits.[21]

I find the work of Aurenche et al., extremely impressive. It brings QCD phenomenology to a new level of completeness and sophistication. It also raises a number of interesting questions. (i) The optimization given by (10), for example, is not a fundamental equation. Though the exact QCD answer should be independent of μ an approximate μ-independence, in a perturbative expansion, does not guarantee correctness. That is, in order that higher perturbative corrections be small it is <u>necessary</u> but <u>not sufficient</u> that a calculation at a given order be stable in μ and M. (ii) If optimized perturbation theory works better, in a variety of processes, than fixing μ and M at physical scales we have learned something very important about the asymptotic series in QCD, beyond terms which have been calculated specifically. For that reason it is important to try and compare, for many processes, optimized perturbation theory with the usual more naive procedure of fixing scales at some physical, but somewhat ad hoc, value.

Finally, over the last 5 years or so, Contogouris and his collaborators[24,25] have been carrying out a program which could be very interesting. The procedure is to make a soft and collinear gluon approximation in the higher order calculation and then to extract the π^2 factor which arises in this approximation. It appears that such a procedure usually comes rather close to exact calculations so long as the $x_\perp$ values of the process are not too small. Of course, if a complete higher order calculation exists it should be used. However, if it is true that the dominant contributions to higher order corrections are really given by real and virtual gluons which are soft and collinear, even if such is the case only in a limited kinematic region, the possibility of doing <u>approximate</u> higher order calculations, beyond orders where exact calculations can be done, is quite exciting. The key point is to try and understand why the soft gluon approximation should be reliable.

3. μ-Pair Production

μ-pair production continues to be an excellent testing ground for QCD. It is not my purpose here to try and review the complete situation but simply to very briefly comment on the features discussed at this workshop. Overall, QCD has been remarkably successful in describing the normalization and $p_\perp$ shape of massive vector production in a wide mass range extending from masses of about 4 GeV for μ-pairs to W and Z production at the CERN collider.[26,27] In general, this description has taken the form of predictions, the theoretical results being somewhat ahead of the experimental confirmations.

At this workshop four recent results have been discussed which I would like to briefly review here. (i) NA10 extracts a value of Λ by fitting scaling violations as a function of τ.[28] They find $\Lambda = 115 {}^{+145}_{-94}$. This is a nice indication of the level of precision of modern μ-pair data. (ii) A clear difference in the $p_\perp^2$-distributions of μ-pairs has been seen in tungsten and deuterium. NA10 finds[28,29] $<p_\perp^2>_W - <p_\perp^2>_D = 0.15 \pm 0.02$ (stat) ± 0.03 (syst) GeV.^{2} This is clear indication that the antiquark, from the incident π^-, undergoes significant scattering in a nucleus before

it annihilates to produce the μ-pair. Unfortunately, it is not yet possible to extract quark-nucleon scattering cross sections as even the simplest
models involve both an average quark-nucleon scattering cross section and
an average momentum transfer per scattering in the nucleus.[30,31] More
detailed A dependence of $<p_\perp^2>_A - <p_\perp^2>_D$ would be very useful in estimating
the quark nucleon cross section. (iii) At very large values of x it appears that the higher twist term, discussed by Berger and Brodsky,[32] has
been definitively seen.[33] Fitting f_π according to

$$f_\pi(x) = (1-x)^2 + \frac{2}{9}\frac{<k_\perp^2>}{M^2} \tag{13}$$

near x=1, for fixed M^2, E615 finds $<|k_\perp|> \stackrel{\sim}{\sim} 1/2$ GeV. (iv) Finally, a
problem with spin dependence in μ-pair production has emerged.[33,34] The
angular dependence of the cross section can be written as

$$\frac{d\sigma}{d\Omega} \alpha\ 1 + \lambda\cos^2\theta + \mu\sin 2\theta\cos\phi + \frac{1}{2}\nu\sin^2\theta\cos 2\phi \tag{14}$$

with θ and φ the polar and azimuthal angles of a muon in the dimuon rest
system. The parton model predicts $\lambda=1, \mu=\nu=0$. QCD corrections give the
result $\lambda=1-2\nu$. Experimentally, $\lambda-1$ and μ seem to be small for a range of
$p_\perp$, a range over which ν grows with $p_\perp$, thus violating the above relation.[35-37] I am far from being expert on the angular dependence of μ-pair
production, however, the experimental result is clearly very interesting
and calls for some better theoretical understanding of the details of μ-
pair production.

4. Jets

Over the past decade or so jets produced in e^+e^- annihilation and in
hadronic collisions have furnished a very fruitful area for confronting
QCD with experiment. There is in general very good qualitative agreement
between theory and experiment at collider energies for jets ranging between
$p_\perp \approx 5$ GeV and $p_\perp \approx 150$ GeV, however, this agreement is not better than a
factor of 2 in normalization, the limit of experimental accuracy.[38]

Λ-values and running coupling effects

For example, attempts to determine Λ in hadronic collisions are not
yet successful. Experimentally, the 3-jet to 2-jet cross section
ratio[39-41]

$$R = \sigma_3/\sigma_3$$

is well measured experimentally. However, one expects relatively large
next-to-leading corrections in both σ_3 and σ_2. The K-factor for σ_2 has
been calculated, in principle,[42] but the K-factor for σ_3 must await significant improvement in calculational technique. This is an example of a
process where some approximate determination of K-factors, perhaps by calculating only soft, collinear gluon effects, could be interesting and
important.

The running of α_{QCD} has been seen from the angular dependence of two
jet events.[38,43] Such an analysis provides a very rough determination of
Λ. This deviation from the "Rutherford" formula is a clear evidence for
scaling violations but one needs a much better experimental determination
of the angular dependence before one can claim more than a qualitative
agreement with QCD.

Gluon distribution determination

UA(1) has determined the gluon distribution in a proton by analyzing
jet data.[38] Thus, if one writes the two jet cross section as

$$\frac{d\sigma}{dy_1 dy_2 d^2 p_-} = \frac{1}{\hat{s}} F(x_1, p_\perp^2) \, F(x_2, p_\perp^2) \, \hat{\sigma}_{gg}(\hat{s}, \hat{t}), \tag{15}$$

where $\hat{\sigma}$ is the Born term for gluon-gluon scattering in the EHLQ normalization, and with[44]

$$F(x, Q^2) = xG(x, Q^2) + \frac{4}{9} [xQ(x, Q^2) + x\bar{Q}(x, Q^2)], \tag{16}$$

then, experimentally, one determines $F(x, p_\perp^2)$. Taking $Q + \bar{Q}$ from deeply inelastic experiments then gives xG. Using this procedure UA(1) determines xG and finds a rather rapidly increasing function of x.

This method should allow one to obtain $xG(x, Q^2)$ at quite small values of x. For example, at the Tevatron with $x_1 \approx x_2$ and $p_\perp \approx 10$ one already reaches x values of about 0.01. If $x_1 x_2$ is fixed and x_1/x_2 is varied it should be possible to go to x_1 values of about 0.001. At such small x values $Q + Q$ is not known but as a first estimate of xG it should be reasonable to suppose that xG is significantly bigger than $x(Q + \bar{Q})$. It is extremely interesting to see how xG behaves at very small values of x. In particular, if $xG(x, p_\perp^2) \approx x^{-\lambda}$ then $d\sigma/dy_1 dy_2 d^2 p_\perp$ will not change much as x_1/x_2 is varied for fixed $x_1 x_2$, at least when both x_1 and x_2 are reasonably small. Collins[45] has argued that $\lambda \approx 1/2$ and one might imagine that this behavior has already set in when $x \approx 0.01$.

Double parton scattering

Both the AFS collaboration[46] at the ISR and the UA(2) collaboration[39] at the CERN collider have studied the question of double parton scatterings by looking at 4-jet events. 4-jet events can occur either as part of a single hard scattering (one impact parameter), as schematically illustrated in Fig.2a for gluon jets, or from two hard scatterings at distinct impact parameters,[47-49] as schematically illustrated in Fig.2b. These two different mechanisms of 4-jet production can be distinguished kinematically. In the case of two separate hard scatterings, whose cross section we shall denote by $\sigma_{2hs}^{(4)}$, there should be a transverse momentum balance between pairs of jets while this is not expected to be the case for a single hard scattering producing 4 jets.

Define $\sigma_{eff} \equiv \pi R_{eff}^2$ by the relation

$$\sigma_{2hs}^{(4)} = \frac{[\sigma^{(2)}]^2}{\sigma_{eff}} \tag{17}$$

where $\sigma^{(2)}$ is the two jet cross section for $p_\perp$ in a comparable momentum range. In the UA(2) analysis the jets are predominately gluon jets while in the AFS analysis the jets are presumably mostly quark jets. Then σ_{eff} is a measure of the two parton correlations in a proton, with small σ_{eff} indicating a strong two parton correlation. One expects R_{eff} to be approximately a hadronic radius. UA(2) reports $R_{eff} > 0.6$ fm while AFS finds $R_{eff} \approx 0.3$ fm. These different values of R_{eff} are not in contradiction. After all, the R_{eff} determined by UA(2) corresponds to gluonic correlations at small x values while the R_{eff} determined by AFS presumably refers to valence quark correlations at intermediate values of x. In addition, the transverse momentum regions of the jets considered by AFS and UA(2) are very different. These are very interesting analyses and are just beginning to furnish us with the parton-parton correlations of the proton.

Multiple "minijet" production

At the energies of the CERN collider and above the production of jets

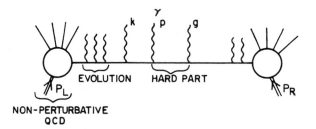

Figure 1

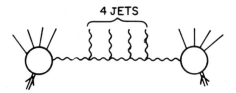

Figure 2(a)

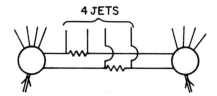

Figure 2(b)

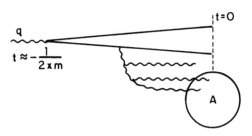

Figure 3

having $p_\perp/\sqrt{s} \ll 1$, minijets,[38,53] should have many of the characteristics of soft hadron production. The structure of the possible observables should be rather rich and, in addition, most of the features of such jet production should be understandable in terms of perturbative QCD, though for some of the most interesting aspects one touches a new, nonperturbative, domain of QCD.

Let me discuss one such measurement, a two jet inclusive measurement, which can be used to extract the "perturbative Pomeron." Let $\sigma(y_1,y_2,M)$ be the two jet inclusive cross section defined by

$$\sigma(y_1,y_2,M) = \int d^2p_{1\perp}d^2p_{2\perp} \frac{d\sigma}{dy_1 dy_2 d^2p_{1\perp}d^2p_{2\perp}}$$

$$\cdot \quad \oplus (p_{1\perp}^2-M^2) \, \oplus \, (p_{2\perp}^2-M^2) \tag{18}$$

Then, using factorization

$$\sigma(y_1,y_2,M) = F(x_1,M^2) \, F(x_2,M^2) \, \sigma_{gg}(\hat{s},M^2) \tag{19}$$

where (15) and (19) have identical information if $\sigma_{gg}(\hat{s},M^2)$, with $y=|y_2-y_1|=\ell n \frac{\hat{s}}{4M^2} = \ell n \frac{x_1 x_2 s}{4M^2}$, is evaluated in Born approximation. However, our interest here is in a region where $\frac{\hat{s}}{4M^2} \gg 1$ in which case higher corrections to σ_{gg} may be important because such corrections come in powers of $\alpha(M^2)y$. In Ref.50, all terms of the type $(\alpha(M^2)y)^n$ were summed using the techniques of Lipatov et al. The result is

$$\sigma_{gg} \sim [(\frac{2}{M})^2\pi] \frac{e^{\frac{12\alpha}{\pi} y \ell n 2}}{\sqrt{\frac{21}{2} \alpha y \, \zeta(3)}} , \tag{20}$$

with σ_{gg} having the interpretation of a gluon-gluon total cross section at impact parameter about $1/M$. When $M \approx 10$ GeV, $12\alpha/\pi \, \ell n 2 \approx 1/2$ and one has an effective power growth of σ_{gg} like $(\hat{s})^{1/2}$, similar to that suggested by Collins[45] for the small-x growth of structure functions.

The growth indicated in (20) is the growth due to the perturbative QCD Pomeron.[51,52] However, the first factor, $(2/M)^2\pi$, is the geometric cross section for the gluon-gluon scattering being discussed. Thus, when the second factor in (20) is of order 1 there must be new, nonperturbative, corrections which slow the rapid growth of σ_{gg}. This "saturation" region[53] corresponds to the production of a high density system of (predominately) gluons of transverse size $\sim 1/M$ which are able to interact strongly because the number of overlapping gluons times α is of order 1. (This is the type of system that one expects to produce in a heavy ion collider in the central region,[54] though in an ion collider the volume of the system will be much larger and the transverse size of the parton is expected to be about $1/$GeV.)

Theoretically, I feel this is quite an exciting domain of QCD to study. Now, how is one to measure, say σ_{gg}, experimentally? The 100% safe way, the way suggested in Ref.50, is in a ramping run where the x_1 and x_2 in (19) are held fixed as s, the center of mass energy^2 of the collision, is varied. In that case, $F(x_1,M^2)$ and $F(x_2,M^2)$ are fixed and the energy dependence of $\sigma_{gg}(x_1 x_2 s,M^2)$ is directly measured. However, one may not have to be so conservative in extracting σ_{gg} from the factorized form (19). If $F(x,M^2)$ is determined through measurement and QCD evolution, then

$\sigma_{gg}(\hat{s}, M^2)$ can be extracted using (19). (Remember, that when $\hat{s}/M^2$ is not large one may use the Born approximation for σ_{gg} to determine F.) The energy dependence of $\sigma_{gg}(x_1 x_2 s, M^2)$ is then obtained by varying x_1 and x_2. The only danger with this latter procedure is that at very small values of x the whole factorization procedure may break down because of single logarithmic terms. Kwiecinski,[55] by explicit calculation, claims that the usual QCD procedure remains valid in evaluating $F(x, M^2)$ up to very small values of x so that (19) may work up to such very small values, at least if M is not too small. As a general comment I think it may be quite useful to begin seriously treating the characteristics of multijet production in a more inclusive way than we theorists have been accustomed to do in the past.

5. A-Dependence

There are a number of experiments now reporting new results on A-dependence in a variety of reactions. In my short review here I would, briefly, like to go back over the physical picture of shadowing in the parton model, a topic already discussed by Bjorken,[56,57] and in QCD. The bottom line is that shadowing persists to large values of Q^2 so long as x is small. Then I would like to discuss J/ψ A-dependence as well as that of jet production.

Shadowing[58] – the physical pictures

Let me discuss shadowing, as it appears in a deeply inelastic scattering of a nucleus, in two different frames.[59] Call frame (i) the frame where the nucleus is at rest, the laboratory frame, and frame (ii) the Bjorken frame where the nucleus has a large longitudinal momentum with the virtual photon having predominately a transverse momentum. It is in frame (i) where $A^{eff}/A < 1$ looks like a shadowing reaction, and it is in frame (ii) where the same physical effect appears as a depletion of the quark sea content of the nucleus. Bjorken[56] has already given a discussion of the issues involved, from the point of view of the parton model, viewed from frame (i).

In frame (i) the picture is as shown in Fig.3. The deeply inelastic process is characterized by variables Q^2 and x, with x very small. In this frame the lifetime of the virtual photon is $\tau \approx 1/2mx$ since the photon has a four momentum $q \approx (Q^2/2mx - mx, o, o, Q^2/2mx)$. m is the mass of the proton. We suppose $1/2mx \gtrsim 2R$, with R the radius of the nucleus. Then in order for a reaction to take place the virtual photon must be in a hadronic state before it reaches the nucleus. Let t=o be the time when this hadronic system passes over the center of the nucleus. The fluctuation of the photon into a quark-antiquark pair then occurs at a time $t \approx -1/2mx$. Bjorken has discussed the interaction of this quark-antiquark component of the photon with the nucleus and this discussion can be extended to include the more general configurations, with gluons and extra quark-antiquark pairs of QCD. As one moves from left to right along the QCD evolution, a backward evolution, shown in Fig.3, one moves from greater to lesser virtuality and from smaller Bjorken x-values to larger x-values. (In this frame the longitudinal momentum of a quark or gluon is _inversely_ proportional to its Bjorken x.) If the virtual photon has evolved into a hadronic component having spatial extent $\Delta x_\perp$ when it reaches the hadron it will interact with a cross section proportional to $\Delta x_\perp^2$. Now the probability of such a component of the photon is $1/\Delta x_\perp^2 Q^2$, so that the total probability of the virtual photon to interact with the nucleus is proportional to $1/Q^2$ as expected. However, if $\Delta x_\perp^2 << 1/fm^2$, the hadronic component of the photon interacts weakly with the nucleus and we expect an A-dependence going as A^1, while $\Delta x_\perp^2 \sim 1$ fm^2 components interact strongly and should vary like

$A^{2/3}$. Indeed, since $\Delta x_\perp^2$ can be anywhere in the range $1/Q^2 < \Delta x_\perp^2 \lesssim 1$ fm^2, and since evolution takes place as the hadronic system moves across the nucleus, the A-dependence is in general very complicated. However, what is clear is that A^{eff}/A does not decrease like $1/Q^2$ and hence is expected to be less than 1 for very large Q^2. In this frame $A^{eff}/A < 1$ truly takes on the appearance of a shadowing phenomena.

However, it is in the infinite momentum frame, frame (ii), that one can use the usual partonic description, and where one relates deeply inelastic scattering at small x to the sea quark distribution of the nucleus. In this frame a nucleon in the nucleus has momentum $p \approx (p+m^2/2p,o,o,p)$ while the photon has momentum $q \approx (q^2/2px, q,o)$ and where $Q^2 \approx q^2$ and p is assumed to be very large. As illustrated in Fig.4, the photon strikes a sea quark with momentum $k_o + k_z = x\,p$ with $1/fm^2 < k_\perp^2 < Q^2$ the allowed range of $k_\perp$. Now, in this frame, the longitudinal extent of the valence quarks in the nucleus is about $(\Delta z)_v \approx 2R\,m/p$ while the longitudinal spread of the struck quark is $(\Delta z)_k \approx 1/xp$. Thus, when $x \lesssim 1/2Rm$ all sea quarks having such x-values and lying within the transverse tube $\Delta x_\perp \lesssim 1/k_\perp$ spatially overlap independently of which nucleon one might try to attribute the different sea quarks. That is, the spatial density of sea quarks grows like $A^{1/3}$. When $dN/d^2x_\perp \cdot 1/k_\perp^2 \gtrsim 1$, with N the sea quark number, we may expect the sea quark number to saturate,[53,58] through annihilation, thus leading to total sea quark distributions growing less rapidly than A^1. In this frame, also, the situation is quite a bit more subtle than the above description might indicate. If the struck sea quark has a large $k_\perp^2$ then $1/k_\perp^2 \; dN/d^2x_-$ will certainly be less than one. However, if at any point in the QCD evolution to this large $k_\perp^2$ one goes through a regime $1/k_\perp'^2 \; dN'/d^2x_\perp \gtrsim 1$, for a parent of the quark k, similar saturation effects will be important.

We may try to better visualize this in Fig.5[59] where two different paths of evolution are illustrated. Q_o^2 is the initial point where the quark and gluon distributions must be given and we suppose Q_o^2 to be on the border between the perturbative and nonperturbative regimes. Then at Q_o^2 we expect strong shadowing for small x-values. Evolution along path I will give an A^1-dependence because the evolution goes to very small transverse parton sizes when $x \gg 1/2Rm$. Through this complete region of evolution $1/k_\perp^2 \; dN/d^2x \ll 1$ while the large x region of the initial distribution should not exhibit shadowing. Evolution along path II will exhibit a dependence closer to $A^{2/3}$ because the initial distribution will exhibit shadowing and a correct evolution will include annihilation in the low Q^2-regime as one moves away from the initial distribution to the observed point, P.

The physical ideas here have been known for some time and the recent small-x data[60] from EMC seems to qualitatively support the picture presented above.

Modified evolution equation

In the region where parton densities become large, but not too large, it is possible to find the modification[53,61] of the usual QCD evolution, as exemplified by the Altarelli-Parisi equation. Call

$$g(x,Q^2) = \frac{xG^A(x,Q^2)}{A}, \qquad (21)$$

the gluon number distribution per nucleon of the nucleus. The nonlinear aspects of evolution are dominated by gluonic interactions for which one has

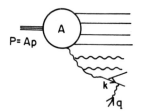

$P = Ap$

Figure 4

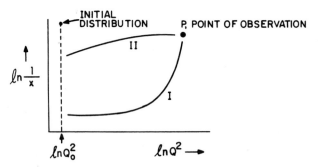

INITIAL
DISTRIBUTION

P, POINT OF OBSERVATION

II

$\ln \frac{1}{x}$

I

$\ln Q_o^2$

$\ln Q^2 \longrightarrow$

Figure 5

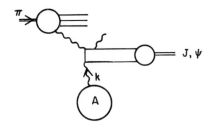

π

J, ψ

k

A

Figure 6

$$Q^2 \frac{\partial}{\partial Q^2} g(x,Q^2) = \frac{\alpha C_A}{\pi} \int_x^1 \frac{dx'}{x'} g(x',Q^2)$$

$$- \frac{9}{8} \alpha^2 \pi \left(\frac{2}{Q}\right)^2 \int_x^1 \frac{dx'}{x'} g^{(2)}(x',Q^2) \qquad (22)$$

with $g^{(2)}$ the two gluon correlation in the nucleus. For large A

$$g^{(2)}(x,Q^2) = \frac{A}{\frac{8}{9}\pi R^2} [g(x,Q^2)]^2 \qquad (23)$$

When (23) is used in (22) the gluon distribution exhibits shadowing in its evolution, due to the negative nonlinear term, which ultimately limits the possible growth of g. The first and second terms on the right hand side of (22) become comparable at $Q^2 \approx 1$ GeV^2 for a large nucleus[54] indicating a strongly interacting dense gluonic system. These are the hardest gluons that have a chance to thermalize when freed in a high energy heavy ion collision. For $Q^2 >> 1\,$GeV^2 the usual Altarelli-Parisi equation is quite accurate even for a large nucleus, at least in all x-ranges of practical interest, however, A^{eff}/A remains less than 1 because the shadowing effects on low virtuality partons are always felt when one integrates (22) from some initial distribution.

If one were to fix x and let $Q^2 \to \infty$ shadowing would go away slowly[58] because this limit emphasizes paths of type I as shown in Fig.5. However, when $Q^2 \to \infty$ and $x \to o$ shadowing remains because one is always sensitive to the shadowing present in the initial distribution. Qiu[62] has made a physical parametrization of initial distributions of quarks and gluons and evolved these distributions to large Q^2 using a generalization of (22) which includes quark interactions. His conclusions are much as I have stated above. It should be emphasized, however, that it is not really possible to make precise predictions of shadowing effects because we don't yet know the initial distributions, at Q_o^2, from which to start the evolution.

J/ψ production in nuclei

There has been data on J/ψ production in nucleon targets, both by incident photons and by incident hadrons, for some time.[63-66] The conclusion reached in analyzing this data is that $\sigma(J/\psi,N) \approx 1-2$mb. As we shall indicate below it is not necessary to make any assumption on vector meson dominance to reach the above cross section. Questions of J/ψ production in nuclei have become especially interesting recently because of the suggestion[67] that lack of J/ψ production in ion-ion collisions might be a good signal for plasma formation. An essential ingredient in this conclusion is the assumption that the J/ψ interacts very weakly with normal nuclear matter, corresponding to $\sigma(J/\psi,N) \approx 1-2$mb, and thus if a plasma were not formed J/ψ production should go as $A_1 \cdot A_2 \cdot \sigma_{NN \to J/\psi + x}$ for collisions involving ions with atomic numbers A_1 and A_2 respectively.

Let me begin by describing briefly how the J/ψ cross section with nucleus is extracted from incoherent J/ψ production data. One uses the formula[68]

$$A^{eff} = \int_o^R 2\pi\rho d\rho n \int_{-\sqrt{R^2-\rho^2}}^{\sqrt{R^2-\rho^2}} dz\ e^{-\sigma[\sqrt{R^2-\rho^2} -z]n} \qquad (24)$$

corresponding to J/ψ being created at impact parameter ρ and longitudinal coordinate z with respect to the center of the nucleus. n is the nuclear density, $n = A/\frac{4}{3}\pi R^3$ for a large nucleus. Using (24) the result $\sigma \approx 1-2$mb results from measuring A^{eff}.

Two essential assumptions have gone into (24). (i) The $C\bar{C}$ pair is created locally in z. (ii) The J/ψ becomes a physical particle immediately after the creation of the $C\bar{C}$ pair. (i) is a reasonable assumption so long as the momentum of the resulting J/ψ is less than 100 GeV or so. However, (ii) is simply not correct. For expliciteness, let me assume that the J/ψ is being photoproduced and that the J/ψ carries most of the momentum of the photon. Then it is reasonable to suppose that at some point the photon creates a $C\bar{C}$ pair which after some time, to be determined, suffers a rather hard collision allowing the $C\bar{C}$ pair ultimately to become a J/ψ. Now, the lifetime of the $C\bar{C}$ pair, before it must suffer a hard collision, is

$$\tau_{Prod.} \simeq \frac{1}{M} \cdot \frac{p_\gamma}{M} \tag{25}$$

where M is the mass of the $C\bar{C}$ pair which we shall take to be $M_{J/\psi}$. For p_γ = 100 GeV, M = 3 GeV, $\tau_{Prod.} \approx$ 2fm a moderately short distance. For p_γ < 100 GeV the production of the $C\bar{C}$ is local with respect to the granularity of a nucleus.

Now, how long does it take the $C\bar{C}$ pair, after suffering the hard collision, to become a normal sized J/ψ. We require this time, τ_{Form}, to obey

$$v_\perp \tau_{Form} = r_{J/\psi} . \tag{26}$$

Now we estimate[69] $r_{J/\psi} \approx$ 1/2 fm and $v_\perp \approx 0 \cdot 5 GeV/p_\gamma$, where $v_\perp$ is the transverse velocity of a charmed quark in the J/ψ, and so obtain

$$\tau_{Form} = 1 \text{ fm} \cdot p_\gamma \tag{27}$$

with p_γ in GeV. Thus, even for $p_\gamma \approx$ 10 GeV this formation time is such that what passes through the nucleus is not a normal J/ψ and hence, the σ extracted using (24) presumably has little to do with J/ψ scattering on a nucleon. The situation here is very similar to the transparency effects predicted for newly born π's and nucleons in nuclei.[59] Thus, if the radius of the J/ψ is actually about 1/2 fm I see nothing wrong with supposing that the actual cross section is something like

$$\sigma(J/\psi, N) \approx \pi r^2 \approx 8\text{mb} \tag{28}$$

or so. In any case, I do not believe that present experiments have actually extracted a physical J/ψ cross section.

What then causes $A^{eff}/A \neq 1$. I am not really sure. In the low energy SLAC experiment[66] I could imagine that the $C\bar{C}$ system had separated enough before leaving the nucleus to produce cross sections of a mb or so. However, the J/ψ's coming from E691 and the large x_F J/ψ's coming from E537 are quite relativisitc, momenta of 50-150 GeV, so the transverse separation of the C and $\bar{C}$ should be extremely small during passage over the nucleus.

The mechanism for J/ψ production is not really known. However, suppose a simple graph like that shown in Fig.6 were responsible for J/ψ production in a π A reaction. The x-values of the gluon, k, seem too large for any appreciable shadowing of the input gluon distribution, at least for the energies covered by E537. So the A-dependence seems not to be in the initial parton distributions.

One final possibility. Suppose, say in J/ψ photoproduction off a nucleus A, a J/ψ is produced in the forward direction with momentum p. The J/ψ becomes physical at a time given by (27) in the laboratory system. At this time there are also lower momentum pions which are becoming physical and which spatially overlap the J/ψ. (Without a good quark model of the pion it is difficult to say exactly what this pion momentum should be.) Thus, there could be a J/ψ breakup due to interaction with such pionic systems long after the C$\bar{C}$ system has left the nucleus. (This is not taken into account in (24) which is not really a correct field theoretic description of J/ψ interaction with nuclear matter.) It would be interesting to try and do a rough model calculation along these lines.

High $p_\perp$ jets produced in nuclei

It has been known for some time that high $p_\perp$ inclusive hadron production off nuclei grows faster than A^1 as one increases the atomic number of the nucleus.[71] A similar, and even stronger, growth with A has been seen in the $d\sigma/dE_T$ dependence on A.[72] For inclusive high $p_\perp$ hadron production this effect is likely due to the high $p_\perp$ being reached by two medium hard scatterings rather than a single hard scattering for some of the events. However, no reliable calculation has yet been done for the effect. The A-dependence of $d\sigma/dE_T$ is complicated by the competition between the energy coming from a hard collision and that coming from fluctuations in the transverse momentum coming from soft hadrons.

Recent experiments at Fermilab now indicate that at least our qualitative understanding of these hard scattering in nuclei is correct. E557 sees an A^1 behavior in $d\sigma/dE_T$ for high planarity events. This is very reassuring. The high planarity discriminates against a lot of E_T coming from soft hadrons and thus is a good jet trigger. E605 has also seen an A^1 behavior at very high $p_\perp$ for single particle measurements. Now that the experimental situation is clear in such a broad range of $p_\perp$ and E_T, perhaps it would be a good time to try to improve our threoretical understanding of jet production in nuclei and bring it up to the level demanded by this new data.

References

1. R. Ammar et al., Phys. Lett. 183B (1987) 110.
2. A. Goshow, thes proceedings.
3. M. Aguilar-Benitez et al., CERN/EP 87-45 (1987).
4. For a review see E.L. Berger, proc. XXII Rencontre de Moriond (1987).
5. J.C. Collins, D.E. Soper and G. Sterman, Nucl. Phys. B263 (1987) 37.
6. P. Nason, S. Dawson and R.K. Ellis, Fermilab-Pub-87/222 T(1987).
7. R. K. Ellis, these proceedings.
8. The authors of Ref.6 do not plot cross sections for charm production. The factor of 2 quoted in the text is my attempt to scale masses and energies down to charm production at fixed target energies and so this number should be viewed with some suspicion.
9. S. J. Brodsky, J.F. Gunion and D.E. Soper, SLAC-Pub-4193 (1987).
10. D.E. Soper, these proceedings.
11. R.K. Ellis, proc. XXI Rencontre de Moriond (1986).
12. H. Georgi and D. Politzer, Phys. Rev.D14 (1976) 1829.
13. M.A. Shifman, A.I. Vainshtein and V.I. Zakharov, Nucl. Phys. B147 (1979) 385.
14. D.J. Broadhurst and S.C. Generalis, OUT-4102-12 (1984).
15. S. Narison, CERN-TH-4746/87.
16. See the many contributions to this workshop.
17. P. Aurenche, R. Baier, M. Fontannaz and D. Schiff, LPTHE Orsay 87/30 (1987).
18. K. Eggert, these proceedings.
19. P. Aurenche, R. Baier, A. Douiri, M. Fontannaz and D. Schiff, Nucl. Phys. B286 (1987) 553.
20. R. Baier, these proceedings.

21. J.F. Owens, these proceedings.
22. P.M. Stevenson, Phys. Rev. D23 (1981) 2916.
23. H.D. Politzer, Nucl. Phys. B194 (1982) 493.
24. A.P. Contogouris and H. Tanaka, Phys. Rev. D33 (1986) 1265.
25. A.P. Contogouris, N. Mebarki and H. Tanaka, McGill preprint (1987).
26. G. Altarelli, R.K. Ellis, M. Greco and G. Martinelli, Nucl. Phys. B246 (1984) 12.
27. G. Altarelli, R.K. Ellis and G. Martinelli, Phys. Lett. 151B (1985) 457.
28. K. Freundenreich, these proceedings.
29. NA10 collaboration, CERN preprint, CERN-EP/87-68.
30. G.T. Bodwin, S.J. Brodsky and G.P. Lepage, Phys. Rev. Lett. 47 (1981) 1799.
31. C. Michael and G. Wilk, Z. Phys. C10 (1981) 169.
32. E.L. Berger and S.J. Brodsky, Phys. Rev. Lett. 42 (1979) 940.
33. K. McDonald, these proceedings.
34. NA10 Collaboration, CERN preprint, CERN EP/87-199.
35. J.C. Collins, Phys. Rev. Lett. 42 (1979) 291.
36. P. Chiappetta and M. LeBellac, Z. Phys. C32 (1986) 521.
37. C.S. Lam and W.K. Tung, Phys. Lett. 80B (1979) 228.
38. W. Scott, these proceedings.
39. K. Meier, these proceedings.
40. G. Arnison et al., Phys. Lett. 177B (1986) 244.
41. J. Appel et al., Z. Phys. C30 (1986) 341.
42. R.K. Ellis and J. Sexton, Nucl. Phys. B269 (1986) 445.
43. C.J. Maxwell, these proceedings.
44. B.I. Cambridge and C.J. Maxwell, Nucl. Phys. B239 (1984) 429.
45. J.C. Collins in proc. SSC Workshop, UCLA (1986).
46. H. Boggild, these proceedings.
47. P.V. Landshoff and J.C. Polkinghorne, Phys. Rev.D18 (1978) 3344.
48. B. Humbert, Phys. Lett. 131B (1983) 461.
49. N. Paver and D. Treleani, Z. Phys. C28 (1985) 187.
50. A.H. Mueller and H. Navelet, Nucl. Phys. B282 (1987) 727.
51. E.A. Kuraev, L.N. Lipatov and V.S. Fadin, Sov.Phys. JETP 45 (1977) 199.
52. Ya. Ya. Balitsky and L.N. Lipatov, Sov.J. Nucl. Phys. 28 (1978) 822.
53. L.V. Gribov, E.M. Levin and M.G. Ryskin, Phys. Reports 100 (1983) 1.
54. J.P. Blaizot and A.H. Mueller, Nucl. Phys. B289 (1987) 847.
55. J. Kwiecinski, Z. Physics. C29 (1985) 561.
56. J.D. Bjorken, these proceedings.
57. J.D. Bjorken, in the proceedings of the Summer Institute on Theoretical Particle Physics, Hamburg (1975).
58. L. Frankfurt and M. Strickman, Phys. Reports (to be published).
59. A. H. Mueller in proceedings of the XVII Recontre de Moriond (1982).
60. F. Close, these proceedings.
61. A.H. Mueller and J. Qiu, Nucl. Phys. B268 (1986) 427.
62. J. Qiu, Nucl. Phys. B291 (1987) 746.
63. U. Camerini et al., Phys. Rev.35 (1975) 483.
64. B. Knapp et al., Phys. Rev. Lett. 34 (1975) 1040.
65. R. Anderson et al., Phys. Rev.Lett. 38 (1977) 263.
66. J. Branson et al., Phys. Rev. Lett. 38 (1977) 1334.
67. T. Matsui and H. Satz, Phys. Lett. B178 (1986) 416.
68. T. Banner, R. Spital, D. Yennie and F. Pipkin, Rev. Mod. Phys. 50 (1978) 383.
69. The precise values of $r_{J/\psi}$ and $v_\perp$ are not crucial for this agument, only that $r_{J/\psi} \gg 1/M_{J/\psi}$.
70. S. Conetti, these proceedings.
71. J. Cronin et al., Phys. Rev. D11 (1975) 3105.
72. C. Bromberg, et al., Phys. Rev. Lett. 42 (1979) 1209.
73. A. Zieminski, these proceedings.

PARTICIPANTS

Professor Bo Andersson
Lund University
Department of Theoretical Physics
Solvegatan 14A, S-223 62
Lund, Sweden

Dr. Jeffrey A. Appel, Computing
Fermi National Accelerator
  Laboratory
P.O. Box 500, M.S. #120
Batavia, Illinois 60510

Professor Patrick Aurenche
LAPP, Lab. d'Annecy-le-Vieux de
  Phys. des Particules
Chemin de Bellevue-BP 909, F-74019
Annecy-le-Vieux, France

Professor Rolf Baier
University of Bielefeld
Fak. fur Physik, Postfach 8640
D-4800 Bielefeld 1, W. Germany

Professor Edmond L. Berger
High ENergy Physics, Bldg. 362
Argonne National Laboratory
Argonne, Illinois 60439

Professor James D. Bjorken
Fermi National Accelerator Lab.
P.O. Box 500, M.S. #106
Batavia, Illinois 60510

Professor Hans Boggild
Niels Bohr Institute
Blegdamsvej 17
DK-2100 Copenhagen 0
Denmark

Dr. William Carithers
Lawrence Berkeley Laboratory
Bldg. 50A-2129, 1 Cyclotron Road
Berkeley, California 94720

Dr. Allan Clark
Fermi National Accelerator Lab.
Computing Department
P.O. Box 500, M.S. #120
Batavia, Illinois 60510

Professor Frank Close
(Rutherford Appleton Lab)
University of Tennessee
Department of Physics
Knoxville, Tennessee 37996

Professor John Collins
Department of Physics
Illinois Institute of Technology
Chicago, Illinois 60616

Dr. Sergio Conetti
McGill University
High Energy Physics Group
3600 University
Montreal, PQ H3A 2T8
Canada

Professor Andreas P. Contogouris
McGill University
High Energy Physics Group
3600 University
Montreal, PQ H3A 2T8
Canada

Professor Rod Cool
The Rockefeller University
Experimental Physics Department
1230 York Avenue
New York, New York 10021

Dr. Flavio Costantini
INFN, Sez. di Pisa
Via Livornese, 582/A
S. Piero a Grado, I-56010
Pisa, Italy

Dr. Bradley Cox (Fermilab/Univ. of Virginia)
University of Virginia
Physics Department
Charlottesville, Virginia 22901

Dr. P. Timothy Cox
CERN, EP-Division
Bldg. 28, 1-028
CH-1211, Geneve 23
Switzerland

Dr. Luigi DiLella
CERN, EP-Division
CH-1211, Geneve 23
Switzerland

Professor Karsten Eggert
CERN, EP-Division
CH-1211, Geneve 23
Switzerland

Dr. Keith Ellis
Fermilab, P.O. Box 500
Theoretical Physics Dept.
Mail Station #106
Batavia, Illinois 60510

Professor Nick Ellis
University of Birmingham
Physics Dept., P.O. Box 363
Birmingham, B15 2TT
England

Professor Stephen D. Ellis
Theoretical Physics
University of Washington
Seattle, Washington 98195

Dr. George Fanourakis
University of Rochester
c/o Fermilab, E-706
M.S. #221, P.O. Box 500
Batavia, Illinois 60510

Professor Thomas Ferbel
University of Rochester
Physics Department
Rochester, New York 14627

Dr. Melissa Franklin
Harvard University
Department of Physics
Jefferson Lab. 364
Cambridge, Massachusetts 02138

Professor Klaus Freudenreich
Eidg. Technische Hochschule
Laboratorium Fur Kernphysik
Honggerberg, CH-8093
Zurich, Switzerland

Dr. Angela Galtieri
Lawrence Berkeley Laboratory
1 Cyclotron Blvd.
Berkeley, California 94720

Dr. Arthur Garfinkel
Purdue University
Physics Department
West Lafayette, Indiana 47907

Professor Alfred Goshaw
Duke University
Department of Physics, Science Drive
Durham, North Carolina 27706

Professor Michel Martin
University of Geneva
Dept. de Phys. Nucleare et
  Corpusalaire 24
Quai E. Ansermet, CH-1211 Geneve 4
Switzerland

Dr. Chris Maxwell
University of Durham
Department of Physics, South Road
Durham City, DH1 3LE
United Kingdom

Professor Kirk McDonald
Princeton University
Jadwin Hall, P.O. Box 708
Princeton, New Jersey 08544

Dr. Karlheinz Meier
CERN, EP-Division
CH-1211, Geneve 23
Switzerland

Professor Kozo Miyake
Kyoto University, Faculty of Science
Department of Physics
Sakyoku, Kyoto 606 Japan

Professor Benoit Mours
LAPP, Lab d'Annecy-le-Vieux de
  Phys. des Particules
Chemin de Bellevue-BP909
F-74019, Annecy-le-Vieux
CEDEX, France

Professor Alfred Mueller
Columbia University
Physics Department
538 W. 120th Street
New York, New York 10027

Professor Jeff Owens
Florida State University
Theoretical Physics Department
Tallahassee, Florida 32306

Professor Ewald Paul
Phys. Inst. der Univ. Bonn
Nussallee 12, D-5300 Bonn 1
West Germany

Dr. Pio Pistilli
Univ. of Rome I, Ist. di Fisica
"G. Marconi", INFN, Roma
Piazzale A. Moro 2
I-00185 Pisa, Italy

Professor Bernard Pope
Michigan State University
Physics Department
East Lansing, Michigan 48824

Dr. Gilbert Poulard
CERN, EP Division, Bldg. 17, R-021
CH-1211, Geneve 23
Switzerland

Dr. Klaus Pretzl
Max-Planck Institute
Werner Heisenberg Inst. fur Physik
Fohringer Ring 6, D-8000
Munchen-40, West Germany

Dr. Leo Resvanis
University of Athens
Physics Laboratory
104 Solonos
Athens, Greece

Dr. Ramon Rodrigues
University of Puerto Rico
Department of Physics
Mayaguez, Puerto Rico

Professor Dominique Schiff
Orsay, Lab de Phys. Theor. et
    Particules Elementaires
Univ. de Paris-Sud
Batiment 211, F-91405
Orsay, France

Professor William Scott
University of Liverpool
Department of Physics
Oliver Lodge Lab., P.O. Box 147
Liverpool, L69 3BX, England

Professor Peter Seyboth
Max Planck Institute
Werner Heisenberg Inst. fur Physik
Fohringer Ring 6, D-8000
Munchen-40, West Germany

Professor Davison E. Soper
University of Oregon
Inst. of Theoretical Science
Eugene, Oregon 97403

Dr. Noel Stanton
Ohio State University
Physics Department
Columbus, Ohio 43210

Dr. Michael Tannenbaum
Brookhaven National Laboratory
Bldg. 510A, 20 Pennsylvania St.
Upton, New York 11973

Professor Julia Thompson
University of Pittsburgh
Physics Department
Pittsburgh, Pennsylvania 15260

Dr. Daniel Treille
CERN, EP-Division
Bldg. 22, 1-011
CH-1211, Geneve 23
Switzerland

Dr. Guy VonDardel
CERN, EP-Division
CH-1211, Geneve 23
Switzerland

Dr. David Wagoner
Prairie View A & M University
High Energy Physics, Box 355
Prairie View, Texas 77446-0355

Dr. William Willis
CERN, EP-Division
CH-1211, Geneve 23
Switzerland

Professor Andrzej Zieminski
Indiana University, Physics Department
Swain Hall, West 117
Bloomington, Indiana 47405

E615 (continued)
  muon pair production, 57–65
    angular distribution, 62–63
    continuum analysis, 58–65
    resonance analysis, 65
    structure functions, 59–60, 62
    transverse momentum, 62
  parton distribution in pions, 72–74
E629, 172
E653, charm and beauty decays via
      hadron production, 493–500
E672, 391
  jets
    A-dependent events, 387–389
    apparatus, 380, 385–386
    beam jet fragmentation, 383–384
    event rates, 388
    physics goals, 383, 385
  psi production, 258–259
E673, 421–422
E691, heavy flavor production,
      399–408
E705, 91, 205, 220, 427
  charmonium production, 419–423
  dimuon trigger performance, 414
  direct photons, 303–313
    apparatus, 304–312
    event sensitivity, 312–313
    goals, 304
  FNAL P-West spectrometer, 412–413
E706, 91, 205
  direct photons, 291–301
    calorimeters, 294–301
    Cherenkov Counter, 291–294
    data acquisition and analysis,
      301
    proportional wire chamber, 294
    silicon strip detector, 294–295
  jets, 385–386
E711, 288
E743, 520, 522 (see also Charm
      production, proton–proton
      collisions)
E762, 301
E769
  charm hadroproduction, 406–407
  heavy flavor production, 399–408
E771, 414, 423–427
E772, 257
ECL-CAMAC, 415
EHLQ, 281
Electromagnetic calorimeter
  E705, 306–310
  E706, 297–299
Electron channels, 438
Electrons
  heavy quark decays, 265
  virtual photon decay, ISR, 166–
      169
Electroweak Lagrangian, 1, 67
Electroweak processes, 76–77
Ellis-Jaffe sum rule, 6, 259–260

Ellis-Sexton calculation, 353
EMC effect, 14, 75–76, 254, 257
      (see also specific experi-
      ments)
  muon-nucleon DIS, 5–6
  NA10 deuterium data, 50–51
  parton distribution
    d/u ratios, 281–283
    lambda measurement, 281
  polarised proton spin distribution,
      259
Energy-Energy Correlation (EEC), 395
EUROJET, 438

Factorization hypothesis, 52
Factorization mass scale, 82
Fastest Apparent Convergence (FAC),
      213
Fermi motion corrections, 77–79
Final states, 11
  A-dependent effects, 18–19
  direct photons from proton-
      antiproton collisions, 132
  photon hard scattering, 200
  three-body, 288
Fixed angle two-body scattering
      processes, 19
Fock-space description, 19
  hadron, 9–10
  two-gluon, 23
Fragmentation functions, jet analysis,
      317–318
Fritiof model of hadronic interactions,
      See String fragmentation,
      Fritiof model
FS ambiguity, 82

Global transverse energy trigger, 401
Gluon bremsstrahlung, and multijet
      events, 317, 319
Gluon distribution, 91, 206
  determination of, 283–286, 526–527
  optimization and ambiguity in,
      212–213
Gluon distribution function, 73–74,
      85
  charm hadroproduction, 507–508
  R806, R807, and R808, 182–189
  shadowing, 533
Gluon field, 18
Gluon fragmentation, 202, 316
Gluon-gluon subprocesses, 22, 65, 82,
      257–258
  heavy quark production, 271–273
  UA1, 430, 432
Gluon jets, heavy quarks in, 489–490
Gluon-photon fusion, 257–258, 275
Gluon-quark contributions, heavy
      quark production, 271–273
Gluons
  Bethe-Heitler photon-gluon fusion, 15

Quarks spins, 259
Quasi-elastic fixed angle
        scattering, 19

R108, 205
R110, 92, 205, 220
R806, 92, 205, 220
   prompt photon cross sections,
        86, 88
   theory vs. experiment, 89
R806, R807, R808, 181-190
   vs. CDHS, 188-189
   vs. Duke-Owens, 189-190
   event reconstruction efficiency,
        186
   gluon distribution function,
        182-189
   subprocesses, 184
   systematic errors, 186-188
   transverse momenta, 183, 185
Rapidity gap, 22
Reggeon diagram, 23, 25
Renormalization mass scale, 82
Resonance analysis, Berger-
        Brodsky effect, 65
Ring Imaging Cerenkov Counter
        (RICC), 150, 158
Rutherford scattering, Fritiof
        model, 234-238

Scale, Principle of Minimal
        Sensitivity optimization,
        211-212
Scale uncertainty, 215-216, 218
Scaling behavior, direct photon
        cross sections, 175-176
Scaling violations, 72
   dimuon pair production, NA10,
        44-48
   in lepton production, 18-19, 43
Sea parameters, 73-75
Secondary partons, 18
SGE approximation, 48, 52
Shadowing, 14-16, 254, 530-531
   charm hadroproduction, 508
   gluons, 257
   Mueller-Qiu theory, 258
   at small x, 18
Single-diffraction experiment, 23
Single photon cross section, 287
Single photon spectra, 205
Singlets, color, 23
Singularities, two- and three-body
        final states, 288
SLAC, 281-282
S-matrix formulation, perturbative
        QCD, 2
Soft gluon approach, 72-73, 82,
        213-214
Soft gluon exchange, heavy quark
        production, 485

Soft gluonic bremsstrahlung, Fritiof
        model, 232-234
Space-time domains
   Drell-Yan process, 6-7
   perturbative, 9-10
Spectator hadrons, 333
Spectrometers
   axial field, 315-326
   direct photon measurement, WA70, 114
   FNAL P-West, 411-413
   tagged photon, 399-401
Spin asymmetry, 5-6, 18-19
Spin-dependent structure function, 8
Spin distributions, polarised
        protons, 259-262
Stable photino, 334
Standard Model, 394
   extensions of, 76-77
   W and Z bosons, 334-335
        accuracy in UA2, 340-341
        deviations from, 338-339
        neutrino limits, 337-338
Strict perturbative criterion, 2-4
String fragmentation
   Bose-Einstein correlations, 247-249
   Fritiof model, 227-241
        experimental data, 238-241
        mechanisms, 227-231
        Rutherford scattering, larger
                transverse momentum, 234-238
        soft gluonic bremsstrahlung, 232-
                234
   Lund model, 242-247
String models, 11, 350
Strong interactions, 1
Structure functions, 8-10, 91-92
   anomlous photon, 148
   atomic number and, 14
   cross section at zero rapidity and,
        221
   Duke-Owens, *See* Duke-Owens structure
        functions
   muon pair production
        E615, 58-60, 62
        NA10, 52
   Pomeron, 20
   prompt photon-jet correlations, 20
   second-order QCD, 85
   terminology, 68
   UA1 direct photons, 140, 144
Sum rules, 6, 73, 259-262
SU(N), 266, 350
Supersymmetric particles, 77

Tagged photon spectrometer, 399-401
Tamm-Dancoff Fock-space theory, 9
Tau, 79
   K factor and, 74
   semileptonic decay, 9-10
Tevatron, 392, 397
   apparatus, 33-35